WITHDRAWN

HARNESSING
3ds MAX™ 8

HARNESSING

3ds MAX™ 8

AARON ROSS
MICHELE BOUSQUET

THOMSON

DELMAR LEARNING

Autodesk·

Australia • Canada • Mexico • Singapore • Spain • United Kingdom • United States

Harnessing 3ds MAX™ 8
Aaron Ross and Michele Bousquet

Vice President, Technology and Trades SBU:
David Garza

Director of Learning Solutions:
Sandy Clark

Acquisitions Editor:
James DeVoe

Product Manager:
John Fisher

Marketing Director:
Deborah S. Yarnell

Channel Manager:
William Lawrensen

Marketing Coordinator:
Mark Pierro

Production Manager:
Larry Main

Production Editor:
Jennifer Hanley

Technology Project Specialist:
Linda Verde

Editorial Assistant:
Tom Best

Library of Congress Cataloging-in-Publication Data:

ISBN: 1-4180-4813-5

NOTICE TO THE READER

Publisher does not warrant or guarantee any of the products described herein or perform any independent analysis in connection with any of the product information contained herein. Publisher does not assume, and expressly disclaims, any obligation to obtain and include information other than that provided to it by the manufacturer.

The reader is expressly warned to consider and adopt all safety precautions that might be indicated by the activities herein and to avoid all potential hazards. By following the instructions contained herein, the reader willingly assumes all risks in connection with such instructions.

The publisher makes no representation or warranties of any kind, including but not limited to, the warranties of fitness for particular purpose or merchantability, nor are any such representations implied with respect to the material set forth herein, and the publisher takes no responsibility with respect to such material. The publisher shall not be liable for any special, consequential, or exemplary damages resulting, in whole or part, from the readers' use of, or reliance upon, this material.

CONTENTS

ABOUT THIS BOOK

Students of 3D computer graphics need a combination of *explanation, demonstration,* and *hands-on* work, all in roughly equal measure. Some books on the market consist only of tutorials, while others are heavy on explanation with few hands-on exercises. While these texts have value to the intermediate or advanced user, they aren't much use to the beginner. You can click all the buttons the tutorial tells you to and still not know why you did it or how to reproduce the same effect with variations for your own project.

The sequence of topics contained in this book has been worked out over years of instruction to make the learning process easy and gradual. The tutorials have been tested by hundreds of students. We know from experience in the classroom that these exercises result in the all-important light bulb going on over the head.

Computer graphics are all about *creating*. Nearly every student comes to class with a vision for a CG project. Clicking buttons alone doesn't aid the creation process. Understanding what you're doing, why you're doing it, and how you can do it again later to make your own vision come alive– that's what learning 3D graphics should be, and what you'll find in these pages.

In this book, we have endeavored to provide everything a student needs to go from baby steps to great leaps of imagination. Our goal is to help each artist gain the skills to create the things he or she has visualized. This book is primarily designed to be used in a classroom environment, but anyone who wants to learn on their own can also use it to get from zero to sixty quite quickly. All you'll need besides this book is practice and imagination.

HOW TO USE THIS BOOK

This book is divided into chapters progressing from simple to advanced topics. The ideal way to learn 3ds Max with this book is to start at the beginning and work your way to the end, rather than jumping around.

Chapter 1, *3D Basics,* gives a general explanation of the elements you are likely to encounter in any 3D program. Instruction and practice with 3ds Max starts with an exercise, **3ds Max Quick Start**, at the end of Chapter 1.

Each chapter starts with *Objectives* and ends with a *Summary* and *Review Questions* pertaining to that chapter. The body of the book contains explanations of concepts and techniques. These explanations are supplemented by hands-on *Exercises*.

In many cases, these exercises build upon one another, so that as you proceed through the book, the scenes become progressively more complete. When a scene is to be used later on in the course, you are instructed to save the file with a particular file name. If you are not instructed to save the file, you may save it if you like, but it's not necessary to do so.

Many exercises start with an instruction to load a file that you created earlier. If you haven't created the earlier file, you can find a workable version of the file on the disc that comes with this book.

Some exercises allow you much freedom to experiment with your own creativity. For this reason, the images shown to illustrate a tutorial might not match your screen exactly.

ALERTS

You will often see graphic icons in the margins of this book. Each of these indicates an *alert*. An alert is a piece of text that expands upon concepts discussed in the main body of the book. The meaning of these alert graphics is explained in the following.

A **DISC** icon in the margin indicates that you'll be asked to load a file from the disc that accompanies this book.

A **TIP** icon appears next to a hint or tip that provides interesting or helpful information.

A **WARNING** icon cautions you about pitfalls in 3ds Max, or common errors related to the current topic.

A **NOTE** icon indicates the answer to a frequently asked question, or a more in-depth explanation of the terms and concepts at hand.

STYLE CONVENTIONS

The following type style conventions are used in this book.

Boldface Type

Bold type indicates permanent interface elements within 3ds Max, such as buttons and spinners. To make the book easier to read, we don't use bold type every time an interface element is mentioned. Instead, we use bold whenever a new interface element is introduced, or when we feel it is appropriate to remind you of it. Usually, after the first mention of an interface element, we simply capitalize it in regular type.

Examples:

Viewports	Activate the **Top** viewport.
Buttons	Click **Select and Move**, and move the object.
Dialogs	Open the **Render Scene** dialog.
Parameter spinners	Locate the **Height** spinner.
Panel rollouts	Expand the **Basic Parameters** rollout.

Italic Type

Italics are used whenever a new vocabulary term is introduced. Generally speaking, these terms are universal to the field of 3D graphics and are not specific to the 3ds Max interface. Terminology is defined whenever it is used for the first time. Additionally, most new terms are defined in the Glossary at the end of the book.

Chapter titles also appear in italics. Occasionally, we use italics for emphasis.

Sans Serif Type

To distinguish the main text of the book from the exercises, we use different fonts. The main body text is set in the Adobe Caslon typeface. Exercises and alerts are set in the **Gill Sans** typeface. Gill Sans is a *sans serif* font, because it lacks ornaments on the letters, called *serifs*.

Monospaced Type

Data that you type into the computer is indicated by a special typeface, `Letter Gothic Bold`. This font was chosen for its similarity to old-style computer printouts. All of its letters are the same width, so it is called a *monospaced* typeface.

Examples:

Filenames	Save the scene as `Table.max` in your folder.
Folder paths	Load the file from the `3dsMax8\maps\Backgrounds` folder.
Object names	Select the object named `Teapot01`.
Numerical values	Change the cylinder's **Height** to **50** units.

Hotkeys

Many commands in 3ds Max have keyboard shortcuts, called **Hotkeys**. In this book, Hotkeys are indicated by boldface type enclosed in angled brackets.

Examples:

3ds Max Hotkeys   Press the **<G>** key to deactivate the Grid display.
Hold down the **<CTRL>** key and press **<Z>** to Undo.

generic keystrokes   Type a value of **50** and press **<ENTER>**.

Menus and Dialogs

There are many menus and dialogs within 3ds Max. Menus have submenus, and dialogs often have many tabs and sections within them. We use a right-facing arrow (a "greater-than" symbol) to indicate steps in a menu or dialog structure.

Examples:

Menus   Choose **File > Save** from the **Main Menu**.

Dialogs   From the Main Menu, open the **Customize > Preferences** dialog, and locate the **Gizmos > Scale Gizmo** section.

USING THE DISC

The disc that comes with this book contains many files to help you with the exercises, and to help you continue to develop your skills with 3ds Max.

The disc is organized by chapter. Each chapter in the book has an accompanying folder on the disc. Within each chapter folder are numerous subfolders, one for each exercise. These folders contain scene files needed to start the exercise, examples of finished scenes, and examples of finished images and animations.

Occasionally, a disc icon will appear in the body text of the book, but usually it will appear within an exercise.

SCENE FILES

When a disc icon appears at the beginning of an exercise, this means it uses a *scene file* from the disc. A scene file contains 3D objects and has the filename extension **.MAX**. You must load the scene file from the disc to begin the exercise.

In some cases, an exercise builds upon an earlier exercise. If you haven't completed the previous exercise, you can still proceed. For your convenience, the necessary scene file from the previous exercise is provided in the current exercise folder.

In addition, each exercise folder contains an example scene file that shows you what the exercise should look like when it is finished. You will find the finished scene file in a subfolder named **Completed**.

IMAGES AND ANIMATION

Still images and animation movies that are results of exercises can be found in some of the exercise folders. You are invited to view these files so you can see the results of the finished exercise. These images and movies are also found in the subfolder named `Completed`.

MAPS

Some exercises in this book rely on image files that are applied as maps onto objects in the scene. They are necessary for the exercise to function correctly. Most of the required map files should already be located in the `\3dsMax8\maps` folder on your hard drive. 3ds Max should always be able to find the necessary maps. If a custom map is used in an exercise, you are instructed to copy it from the disc to your hard drive. For your convenience, copies of all maps used are included in each exercise folder.

If you see a **Missing External Files** dialog when loading a scene from the disc, it means that 3ds Max could not find the required image file. In this case, you need to configure the bitmap paths within 3ds Max. See the topic *Bitmap Path*s in Chapter 4, *Materials*, for a full explanation of bitmap paths and how to configure them.

INSTRUCTIONAL .AVI MOVIES

The disc that comes with this book also contains numerous instructional `.AVI` movies. These are essentially lectures presented by Aaron Ross, including moving screen captures of the 3ds Max interface. They are designed to present hard-to-explain topics, showing actual procedures in 3ds Max. Watching these movies is almost like taking a college-level course in 3D graphics. They are found within each chapter folder on the disc.

To watch the movies, the Techsmith video compressor/decompressor (codec) must be installed on your computer. The disc includes software to install the codec.

ACKNOWLEDGMENTS

Thanks to the crew at Autodesk Press: James Devoe, John Fisher, Tom Best, and Jennifer Hanley.

Thanks to Technical Editor Jon McFarland for his thoroughness in checking our work for technical accuracy. Special thanks to Copy Editor and friend Stephanie Provines.

Artwork for the color section was contributed by the following:

Matt Highison
Game Art and Design student at the Art Institute of California - San Francisco

Planet Nine Studios
David Colleen, CEO

Joseph Wehland
Animation student at the Art Institute of California - San Francisco

The cover art is by Joseph Wehland, animation student at the Art Institute of California - San Francisco.

Aaron Ross gives extra special thanks to the following friends and relations who have helped and encouraged him over the years: Mom and Boardy, Anna, and Shaz.

REVIEWERS

During the development of this book, reviewers were asked to take a look and give their comments and suggestions. A big thank you goes out to the reviewers for their time and attention in helping to make this book as useful as possible.

Alexi Balian
Rick Hansen Secondary School, Mississauga, Ontario, Canada

Pat Gombarcik
Southern Nevada VoTech Center, Las Vegas, NV

Eric Joseph
Pittsburgh Technical Institute, Oakdale, PA

E.RESOURCE FOR INSTRUCTORS

This book was designed to be used as either a course text or a learning tool for the student working on his/her own. A disc has been developed to assist teachers in planning and implementing their instructional programs for the most efficient use of time and other resources.

The *Harnessing 3ds Max 8 e.resource*™ disc contains tools and instructional materials that enrich the classroom experience and shorten instructors' preparation time. The elements of the *e.resource* link directly to the textbook to provide a unified instructional system. The *e.resource* allows you to spend your time teaching, not preparing to teach. It is available to instructors only.

The *e.resource* includes the following materials.

- **Lesson Plans** for each part of the course help you present the material in the text. Areas in which students often need more practice are highlighted, along with additional exercises they can be given to help them keep up with the class. These lesson plans can easily fit in with course outline you have already developed.

- **Answers to Review Questions** enable you to grade and evaluate end-of-chapter quizzes.

- **Supplemental Instructional .AVI Movies** explore concepts and techniques more deeply. Special attention is given to common problems and complex procedures. These additional **.AVI** movies are designed to augment those on the disc that comes with this textbook.

HOW TO ORDER

To order the *Harnessing 3ds Max 8 e.resource*™ disc, contact Autodesk Press at (800) 347-7707. Use the following ISBN number when ordering:

ISBN 1-4180-4814-3

3DS MAX SUPPORT

Support for 3ds Max is available from the user forum at **www.autodesk.com**. This service is free to all users of 3ds Max.

On the forum, users post messages regarding 3ds Max. Autodesk support personnel and other users read the messages regularly and respond to questions. Often, a question is answered within a few hours. Questions range from the use of specific functions to troubleshooting and hardware issues.

The forum is also an excellent place to browse for questions and answers from other users to improve your knowledge of 3ds Max.

Before posting a question on the forum, be sure to check this book or the 3ds Max User Reference to see if you can find the answer to your question.

ABOUT THE AUTHORS

AARON ROSS

Aaron Ross is a college instructor and artist living in San Francisco, California. He holds an MFA in Film/Video from California Institute of the Arts, and a BFA from the School of the Art Institute of Chicago. He has worked professionally in numerous fields, including video production, print production, sound design for computer gaming, and web design.

Aaron's passion is abstract art. His experimental videos have screened at festivals and museums around the world, including SIGGRAPH, the Inter-Society for Electronic Art, and Prix Ars Electronica. His personal art gallery can be found at **www.dr-yo.com**. He also has a site devoted to computer graphic education: **www.cglearn.com**.

MICHELE BOUSQUET

Michele Bousquet is an animator, writer and instructor based in southern New Hampshire. She got into animation in 1990 with the first release of 3D Studio for DOS. She has taught 3ds Max in the United States and Australia since 1995. Michele is the author of several books on 3ds Max, and contributed to the 3ds Max User Reference.

Michele invites you to visit her website created especially for 3ds Max users, **www.mbousquet.com**.

READER FEEDBACK

If you have questions, problems, or comments regarding this book, we encourage you to contact:

The CADD Team
c/o Autodesk Press
Thomson Learning - CPG
25 Thomson Place
Boston, MA 02210

Or visit our Web site at **www.autodeskpress.com**.

Chapter 1
3D Basics

OBJECTIVES

In this chapter, you will learn:

- The main phases of a 3D project
- Key 3D concepts
- Common terms used in 3D software
- The basics of 2D bitmap images

Working with 3D software requires you to think in new ways. Although you've lived in a three-dimensional world all your life, you might not be accustomed to looking at it in a 3D manner. With a little practice, this type of thinking will come naturally to you. This chapter will tell you what you need to know to start working in this new world. Some of these concepts may be familiar to you, while others will be new.

WORKING IN 3D

3D graphics involves many complex processes that can take a lot of time. With any big project, whether making a film or building a house, it's helpful to break the complexity down into more manageable chunks. This is absolutely necessary in the case of 3D graphics. In the following section, we will look at the primary phases of a 3D graphic project and how they fit together to allow artists and engineers to create their best work.

PRE-PRODUCTION

Production is simply the act of producing something. The world of computer graphics (CG) uses the term *production* similar to the way it is used in the motion picture arts. In filmmaking, production is the actual shooting of live action footage. In CG, it's the creation of new material that is intended to be used in the final product, such as a film, game, simulation, or illustration.

Pre-production is the planning and design phase that must occur before production begins. In a student or personal art project, this mostly means research and design. For example, if the production features a character or creature, that character must be designed on paper before it can be created in the computer. If the piece tells a story, then that narrative must be written before the production can begin.

In a professional environment, where a great deal of money may be at stake, it's even more critically important that pre-production be given the necessary consideration. Funding must be procured, people must be hired, schedules and budgets must be created in detail. But whether the project is a blockbuster film or a hobbyist's modification to a computer game, the success of the project rests in large measure on pre-production. Proper planning and design results in a production that runs smoothly despite unforeseen circumstances. In general, the more time is spent planning, the less time (and money) is wasted during production.

Reference Materials

To help with the design of characters, objects, and environments, artists seek out *reference materials* in libraries, bookstores, on the Internet, in motion pictures, and in real life. Creating a representation of something requires a full understanding of what that thing is. So, if one is designing a character such as an elephant, then a trip to the zoo is in order. Observing, drawing, and taking photographs and movies of the animals will give the artist something to work from. Without adequate study of reference material, there's little hope of creating a convincing representation. It doesn't matter how good the artist's skills on the computer are if she doesn't have a firm idea of what she's doing.

SCENES

A computer graphic *scene* is like a scene in a film: a collection of elements that combine to create a visual illusion. Of course, in computer graphics, the objects are not real, but merely mathematical representations of reality. CG objects are often referred to as *virtual*, to distinguish them from the real world. In a CG scene, we commonly find many virtual objects, some of which may be moving. The scene also defines how those objects will look, based on their surface properties, virtual lighting, camera positions, etc.

Scenes are sometimes called *projects*, although that can be confusing. In general terms, a project usually refers to a *production* such as a movie or video game. Most computer graphic projects use many 3D scenes throughout the process of creating the final production.

3ds Max stores almost all of the information for a given scene in a single file. The 3ds Max scene file format ends in a **.MAX** extension. The **.MAX** file is a convenient package that encapsulates nearly all of the data needed to construct the scene. However, 2D images are not stored within the scene file, but are saved as separate files.

Other 3D programs store various parts of the scene in separate files, in a structure of folders that is predefined by the application. This is an added complication that 3ds Max users need not worry about.

MODELING

The objects in a 3D graphic program are geometric representations of real objects, generally called *models*. Models are often simply referred to as *geometry*. The art of making 3D objects in the computer is called *modeling*.

Beginning 3D modelers often don't know where to begin. There are many techniques for modeling objects. Although there is usually more than one solution to a modeling challenge, certain techniques lend themselves more readily to certain types of objects. In this book you will be exposed to all of the major modeling techniques in 3ds Max and will gain the experience necessary to apply these procedures quickly and efficiently.

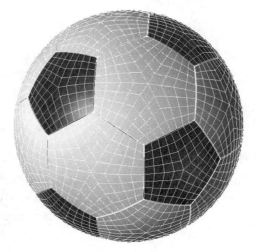

Figure 1-1: 3D model of a soccer ball

 **TIP:** Modeling begins with analysis and imagination. Before you set out to model an object, look closely at a real-world example, and study images in books or on the Internet. Imagine what the object would look like. Draw sketches and diagrams of the object to familiarize yourself with its structure. Are its edges sharp or rounded? What are its separate pieces? These questions will help you figure out how to go about modeling the object.

MATERIALS AND MAPS

Geometric models define the shapes, contours, and volumes of objects, but raw models do not have any surface properties. Visual qualities such as color, shininess, and bumpiness are added through the use of *materials*. Materials are the "paint" or "wallpaper" applied to objects in a scene. Other properties that are determined by materials include pattern, transparency, and reflection.

Materials can be made in a variety of ways. A common method is to take a scanned or drawn image and paste it on an object. An image applied to an object in this way is called a *map*.

Figure 1-2: Teapots with various materials

Materials and mapping are a very important part of creating an effective illusion. Fine details that could not possibly be created using geometry modeled in 3D can be easily "faked" with materials. The creation and editing of maps, in particular, is such an important part of 3D CG that it is a career specialization. Some artists create maps for 3D objects as their primary source of income.

LIGHT AND CAMERA

Lighting goes hand-in-hand with materials to create a visual tone or mood. Whether or not the goal is to reproduce photographic realism, lighting is essential to a computer graphic production. Even the most beautifully constructed model can look uninteresting if it is not lit well.

Although the tools for lighting within CG programs work differently than the way light does in the real world, we can and must still apply the principles of artistic lighting developed over the centuries. The visual arts of painting, theater, photography, and motion pictures rely in part on lighting to convey emotion. A horror film is lit with dark, contrasty lighting, while a comedy is lit brightly and colorfully.

In the case of a production that has noninteractive elements, such as a film or a magazine illustration, then the point of view of the audience is established through the use of a camera. Virtual cameras work pretty much the same way that cameras work in the real world. A camera frames the scene, defining exactly what audiences can see and what they can't. The camera operator is essentially placing and moving the eyes of the audience.

Just as with lighting, camera manipulation is a very powerful tool to create an emotional response in the audience. An extreme closeup on a character's face can pack a much greater emotional impact than a shot from far away. A wide-angle shot of a landscape can take in the beauty of the vista in a way that a shot of a single feature of that landscape cannot.

Effective use of cameras is also a critical stage in a 3D CG production. Quite often, the perspective of the camera defines much of the other work to be done on the project. If the camera can't see it, then the audience can't either, so anything that's outside of the camera's field of view should not be built in the first place.

ANIMATION

If the end result of the production is a moving image or an interactive experience such as a game, then something in the scene must be *animated*. The word "animate" literally means "to give life." Even if there are no characters or other moving geometric objects in the scene, animation may take the form of moving lights or cameras.

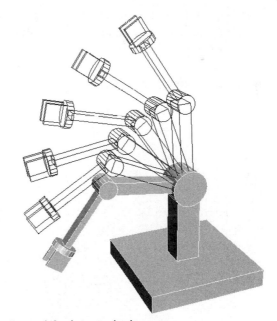

The most common method of computer animation is called *keyframing*. This is a term borrowed from traditional hand-drawn animation. Creating a hand-drawn animation involves drawing thousands of individual frames, so it often requires the skills of many people. Some of the artists on an animation crew are responsible for drawing the most important frames, showing

Figure 1-3: Animated robot arm

extreme or representative poses of a character, for example. These artists are known as *lead* or *key animators*, and the images they draw are *keyframes*.

After the keyframes are drawn, junior animators begin the task of drawing the many images in-between the keyframes. This process is called *in-betweening*, or simply *tweening*. In computer animation, the artist merely defines the keyframes, and the tweening is done more or less automatically by the computer.

RENDERING

In traditional art, *rendering* means drawing a representation of something. This term has been around for centuries.

In the world of 3D, rendering is done by the computer. The artists, of course, have built the scene, applied materials and set up the lights and camera. When the scene is done, the computer takes all the scene information and creates a finished picture from it.

If the production results in a still or moving image, then the rendering is done by a 3D graphics application such as 3ds Max. If it's an animated scene, each frame of the movie is stored on disk as a separate file, and each file is given a number. This is called an *image sequence*.

The 3D rendering program takes its time with each frame, working long hours to get the best possible image quality. Depending on the complexity of your scene and your render settings, a single image can render in a second, a minute, an hour, or even days. It is easier than you may think to throw more at your computer than it can handle. In this case, the image will *never* render.

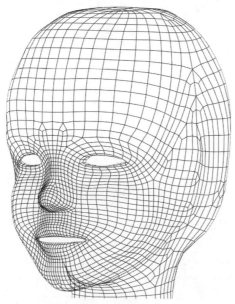
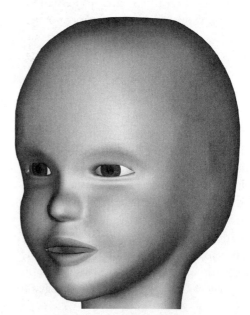

Figure 1-4: A model before and after rendering

For an interactive experience such as a game or a visual simulation, the rendering is done in *real time*, at many frames per second. Action games such as Quake or Unreal have very advanced game engines that animate and render many objects in real time. The goal is to achieve adequate image quality at the highest possible frame rate.

Currently, an acceptable render time for a feature film is about a half hour per frame. Compare that with real-time games that render at 30 or more frames per second. The film frame holds about seven times more information than a game screen, and the quality of the film rendering is far superior.

It is reasonable to assume that non-real-time rendering will always have higher quality than real-time rendering. However, there will come a day when human eyes won't be able to tell the difference in quality, and so someday we can look forward to photorealistic moving images rendering in real time on personal computers.

2D IMAGES

In order to work effectively in 3D, it's necessary to first have a handle on how 2D images work on the computer. In the following section, we'll look at the technical details of computer graphic images.

BITMAPS

A *bitmap* is a file that contains an image. The file is defined by a grid or mosaic of colored dots or *pixels*. The word pixel is an abbreviation of *picture element*. A picture element is the smallest piece of an image, a single cell in the grid or matrix. Many very small pixels, packed closely together, give the viewer the illusion of a single, unbroken image.

A bitmap is a flat 2D image. For 3D work, bitmaps are very useful for materials and backgrounds. Bitmaps can be created in a number of ways, such as scanning a photograph or drawing a picture on the computer with a paint program. A rendered image is also a bitmap.

Figure 1-5: Zooming in on a bitmap reveals the mosaic of pixels.

Not all computer images are bitmaps. Some images are created from mathematical data representing points, curves, and flat colors. These are called *vector graphics*, and some examples include Macromedia Flash movies, Adobe Illustrator images, PostScript files, and most typographic fonts.

Bitmap File Formats

Bitmaps can have many formats. A file's format is identified by its filename extension. Common bitmap types are **.BMP** (Windows Bitmap), **.TGA** (Truevision Targa), **.TIF** (Tagged Image File Format), and **.JPG** (Joint Photographic Experts Group).

Most bitmap formats are about the same. Your choice of file format will be most likely guided by the program you're using or the industry you're working in. The video production workhorse is the Targa file, whereas graphic design programs prefer **.TIF**.

WARNING: One thing to be certain of is whether or not the bitmap uses *lossy compression*. This is a scheme for intelligently discarding information to reduce file size. JPG uses lossy compression. In 3D graphics production, one should never render directly to a lossy format, because the lost information can never be recovered. Sometimes JPG still images of 3D graphics are needed; for example, to place on a web page. In this case, render to an uncompressed Targa and then convert to JPG using image manipulation software such as Adobe Photoshop.

Resolution

Every bitmap has a *resolution*. This is the number of pixels across and down the image. A bitmap graphic cannot be divided into smaller spatial chunks than individual pixels, so it is said that we cannot *resolve* anything smaller than a pixel. A common bitmap size is 640 x 480. Higher resolution images, such as 1024 x 768, are potentially sharper and crisper because they hold more information.

Pixels per Inch

When an image is scanned or printed, each pixel must have some size; it must take up some area on the page. This is measured *pixels per inch*, or ppi. An image with a low ppi setting will print out larger than an image that has the same resolution, but a higher ppi. A setting of 72 ppi results in bigger pixels than 600 ppi, so the image printed at 600 ppi would look crisper, but would print much smaller.

Color Channels

Most bitmap images are divided into three *channels*, one for each primary color. Primary colors are those that can create all other colors. You're familiar with the primary colors of pigment: red, yellow, and blue. But in 3D graphic software, the primary colors are red, green, and blue. Computer monitors actually project light to your eyes. Colors projected with light behave differently than those created with paint or ink.

With pigments, mixing red, yellow, and blue together will yield black. Computer color works in the opposite way. When red, green, and blue light are projected together, the result is white. This is called *additive color*, because when the primaries are added together, we get white. It is also sometimes referred to as *RGB color*.

The *bit depth* of an image determines how much information is stored. In a standard bitmap file such as a `.TGA`, each channel has a depth of 8 bits. An 8-bit number yields 256 possible numerical values, from zero to 255. Since an RGB image has three channels, and each channel holds eight bits, the combined bit depth is 24 bits (3 x 8 = 24).

As the number of bits rises, the number of possible values expressed by those bits rises exponentially, as shown in the following table.

8 bits = 2^8 =   256

16 bits = 2^{16} =   65,536

24 bits = 2^{24} =   16,777,216

So a 24-bit image is capable of expressing over 16 million different colors. Sometimes 24-bit images are called *true color*, because 16 million variations in color are about the maximum that human beings can perceive. Some specialized systems use higher bit depths than 24, especially in the film world, but most computer displays can only handle 24 bits of color.

Computer programs that deal with color have *color pickers*, where colors can be mixed interactively to get the shade you want. These include places to enter number values for red, green, and blue, and usually include a colored area where a color can be picked from the screen, or a slider where the red, green, or blue values can be interactively increased and reduced.

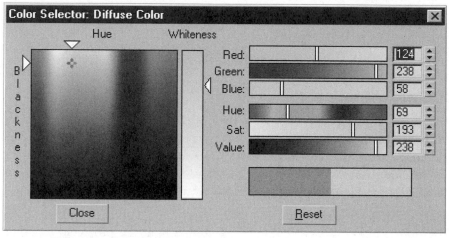

*Figure 1-6: The 3ds Max **Color Selector***

However, some artists find the method of mixing RGB colors to be confusing and non-intuitive. Therefore, software designers have included a method of picking colors that is easier to use. This method was originally developed for use with video equipment. The user defines three properties of a color. These are *hue*, *saturation*, and *luminance*, or HSL for short. On a computer, each of these color properties can also range from 0 to 255.

Hue is the tint of a color. Hues are colors of the spectrum, or rainbow, starting with red and going through yellow, green, cyan, blue, magenta and back to red.

Saturation is the relative purity of a hue. A color that is a single, pure hue, like a vibrant forest green, is highly *saturated*. If we reduce the saturation, we are actually mixing in all other hues to the color, and it becomes washed out and less intense. A pastel color like mint green is said to be *desaturated*.

Luminance is the brightness of a color. A luminance of zero is pure black, no matter what the hue is. Luminance is also sometimes called *value*, which is a term borrowed from drawing and painting. Therefore, you may often see the abbreviation HSV instead of HSL. They mean the same thing.

TIP: Saturation and value can be adjusted together in the 3ds Max Color Selector, using the **Whiteness** slider. As you add white, the luminance goes up while the saturation goes down. Sometimes this type of color picker is called HSW, for hue, saturation, and whiteness.

Many programs that deal with color allow you to use both RGB and HSL methods together in setting a color. For example, you can use the red, green, and blue values to get the general color you want, then use the luminance parameter to make the color a little darker or lighter.

Alpha Channel

A bitmap sometimes has an *alpha channel*. An alpha channel is transparency information built into a bitmap. This information can be used when layering images. The alpha channel acts as a mask or matte, making some parts of the image opaque and others transparent. An alpha channel holds 8 bits of data, so there are 256 possible levels of transparency. This range of gray values allows for partial transparency, which is particularly important for smoothness at the edges of objects. Usually, black pixels in an alpha channel are fully transparent, and white pixels are fully opaque.

Figure 1-7: A rendered image and its alpha channel

The only RGB images containing alpha channels are those with 32 bits per pixel. The first 24 bits contain the color information, while the transparency data is stored in the last 8 bits. **.TGA** and **.TIF** files can have up to 32 bits, but not all **.TGA** and **.TIF** files have alpha channels. Note that **.JPG**, **.PCX** and many other file formats can't have alpha channels.

Image Editing

When making materials and backgrounds, you will often find that you need to scan or draw an image. An *image editor* or *paint program* is the best way to do this. With a paint program, you can draw an image from scratch, change an existing image, or combine images in new and interesting ways.

Scanned images often need to be touched up to remove unwanted elements. An image can also be much improved by altering its brightness, contrast, or use of colors. Image editing programs can convert file formats, resize images, and layer images over one another. They can even perform special effects, such as mimicking real world materials like paint and canvas.

The industry standard paint program is Adobe Photoshop. Photoshop has a wide range of features and will accomplish almost any task you choose to perform on a bitmap. It's almost impossible to be a computer graphics artist without at least knowing the basics of Photoshop.

Compositing

In noninteractive productions such as film and video, it is common for multiple moving images to be combined in a process called *compositing*. Perhaps a live actor needs to be superimposed over a computer-generated background. In this case the actor is filmed or videotaped on a special stage that is all blue or all green. Using a compositing program, an artist replaces the green or blue stage set with the CG background. This particular type of compositing is called a *chroma key*.

Even if all scene elements are computer-generated, each element is often rendered out separately, and saved with its own alpha channel. The scene elements are then layered in a compositing program. This results in greater freedom to make changes, such as color correction. For example, if a CG character is rendered separately from its background, then it is a simple matter to change the background brightness and contrast, without affecting the character. Or, it may be necessary to re-render the character if the director or client requires changes. In this case, the re-rendering will be much faster, since only part of the frame needs to be calculated.

There are many compositing programs available, each with its own features and intended uses. Autodesk and its subsidiary, **Discreet**, are at the cutting edge of award-winning compositing technologies. Autodesk offers a program called **combustion**, which runs on Windows and Macintosh. Another common compositing application is Adobe After Effects.

EXERCISE 1.1: 3ds Max Quick Start

By now, you're probably itching to get started with 3ds Max. In the next chapter, you'll learn more about how 3ds Max works. Here, you'll get started by learning a few commands.

1. Load 3ds Max. To do this, click the **Start** button at the lower left of the Windows screen. Click **Programs**, choose **Autodesk**, and choose **3ds Max 8**.

   As 3ds Max loads, you see a *splash screen* which shows you some of the default keyboard shortcuts. After a few moments, the 3ds Max user interface appears.

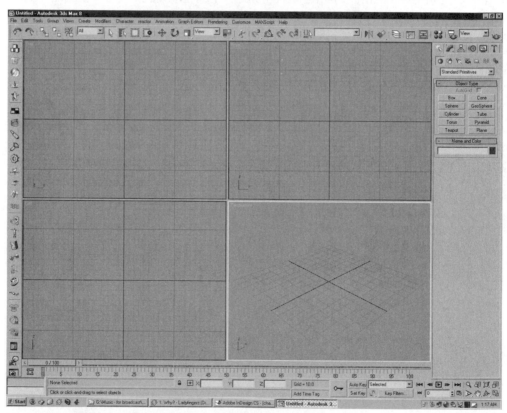

Figure 1-8: 3ds Max user interface

The majority of the screen consists of four viewports where you can build a 3D scene. As you click in each viewport, the viewport's border highlights to indicate that it is active. In this exercise, you'll be using the **Perspective** viewport, which is in the lower-right quadrant. Click the Perspective viewport to activate it.

2. Look in the lower-right corner of the 3ds Max interface. You'll see a collection of icons that are known as the **Viewport Controls**. As you place your cursor over a button in 3ds Max, you'll see a **ToolTip** which tells you the name of the icon. The button at the extreme lower right is called **Maximize Viewport Toggle**. Press this button to make the Perspective viewport fill the screen.

Figure 1-9: **Maximize Viewport Toggle** *button in the Viewport Controls*

3. Look to the upper right of the interface. Under **Standard Primitives > Object Type** you will see a series of buttons labeled with the names of objects. Click the button labeled **Sphere**. Then click and drag in the viewport to create a sphere. Your click point defines the sphere's center location, and the drag distance defines the sphere's radius. Release the mouse button when the sphere is the size you desire.

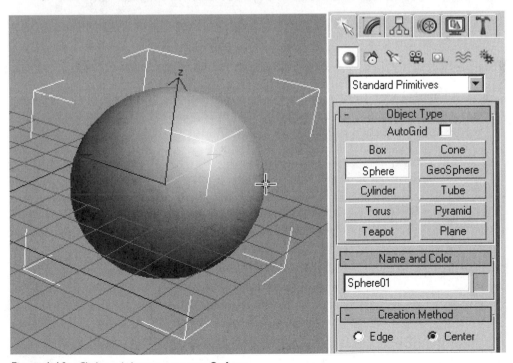

Figure 1-10: Click and drag to create a **Sphere**

4. As long as the **Sphere** button is highlighted, you can continue creating spheres. Instead, click the right mouse button to exit the Sphere creation tool. Then activate the Box creation tool by clicking the button labeled **Box**.

    Because a sphere has only one measurement (its radius), you only have to drag once. For a box, you must define length, width, and height. To create a box, click and drag to create a rectangle that defines the box's length and width. Release the mouse button, and then move the mouse to define the height of the box. Finally, click a second time to complete the operation.

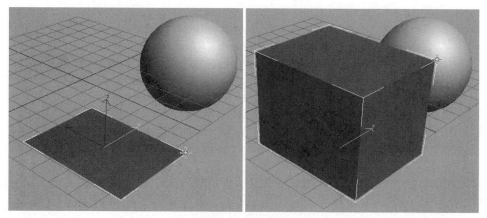

*Figure 1-11: Click and drag to define **Box** length and width, release to define height*

5. In the **Viewport Controls** you will find more buttons enabling you to view the scene in the Perspective view from different angles. The Viewport Controls are designed to act like the controls of a real-world camera.

 Zoom moves the viewpoint closer or farther away from the scene. Click and drag in the viewport to zoom in and out.

 Pan moves the viewpoint left/right and up/down. Click and drag in the viewport to pan around the scene.

 Arc Rotate orbits the viewpoint around the scene. Click and drag inside the big circle in the viewport.

 Zoom Extents automatically zooms the active viewport to so that all objects in the scene are visible.

These and other Viewport Controls are explained more thoroughly in Chapter 2. For now, just experiment with the controls to get a feeling for how they operate.

6. Near the top of the screen is a row of many icons. This is called the **Main Toolbar**. Look in the Main Toolbar for the **Select and Move** tool.

Figure 1-12: **Select and Move** *tool in the Main Toolbar*

Activate the Move tool. Click on an object to select it. Your cursor changes to resemble the Move icon when it is over a selected object. You should see three perpendicular axes with arrowheads. This is called the **Move Gizmo**, and it allows you to control the direction of object movement.

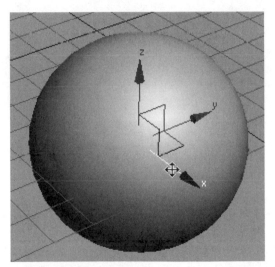

Figure 1-13: **Move Gizmo** *with the X axis highlighted*

As you move your mouse cursor over an axis of the Move Gizmo, the axis is highlighted in yellow. Click and drag to move the object in that direction.

(If you don't see the Move Gizmo as pictured here, it may be hidden. Press the **<X>** key on the computer keyboard to show the Gizmo.)

7. Use the object creation tools, viewport controls, and the Move tool to create a few objects and position them in the scene.

    You have just created your first 3D scene. There is no need to save this scene because you'll make plenty more in the chapters that follow. In the next chapter, you'll learn more about creating objects and how to save a scene.

SUMMARY

Pre-production is the planning and design phase of a production, and it is necessary for the smooth functioning of the project. *Reference materials* such as drawings and motion pictures are invaluable in the design process.

The creation of objects in a 3D program is called *modeling*. A modeled 3D object is called a *model*, or simply *geometry*. *Materials* are applied to objects to give them colors and patterns, or to make them appear bumpy or shiny.

Lighting and *camera* techniques add emotion and direction to a scene.

An artist creates *animation* to make objects move over time. The animator defines *keyframes* for the objects, and the computer fills in the in-between frames.

All of the models, animations, materials, lights, and cameras combined together are referred to as a *scene*. After the scene is finished, it is *rendered* to show the final result. For an interactive game, the rendering is done in real time by the game engine. For a noninteractive production such as film, video, or print, the 3D program renders the images and stores them as files on disk.

A *bitmap* is a picture stored on the computer, whether a picture drawn with a paint program, a scanned image, or a rendered image. A bitmap can be stored in any one of many formats. An *image editor* or paint program can be used to create or alter the colors, brightness, or contrast of a bitmap. Bitmaps have three main characteristics: *resolution* (number of pixels), *pixels per inch* (size of pixels on a page), and *bit depth* (color accuracy).

Colors are determined using red, green, and blue color *primaries*. Computer artists can pick colors using the RGB system, or with more intuitive color properties such as *hue* (the color of the rainbow), *saturation* (the purity of the color), and *luminance* (brightness or value).

A bitmap may also have transparency information embedded in it, in the form of an *alpha channel*. A *compositing* program can combine image sequences, often using alpha channels to layer the bitmaps into a single composition.

REVIEW QUESTIONS

1. What is *pre-production*?

2. Why would you use *reference materials?* Where can you get them?

3. What is a *scene?*

4. What is a *model?*

5. What is the name for surface properties applied to a model?

6. Why is lighting important?

7. How does the placement of a camera influence the audience?

8. Why isn't it necessary for a computer animator to define the positions of objects on each and every frame?

9. What is *rendering?*

10. What are the three *primary colors* used when working with computers?

11. What is a *pixel?*

12. What is an *image sequence?*

13. How many *color channels* are in a 24-bit image?

14. What is an *alpha channel?*

15. What *bit depth* does a bitmap with an alpha channel generally have?

16. What is an *image editor?*

17. What is *compositing?*

Chapter 2
Introduction to 3ds Max

OBJECTIVES

Welcome to the world of 3ds Max! In this chapter, you'll learn about:

- Setting up your computer to run 3ds Max
- The 3ds Max interface
- Basic modeling techniques
- Applying a material to an object
- Animating objects
- Rendering an animation

This chapter goes over the major functions of 3ds Max in a general way. In later chapters you'll learn about each feature in more depth, and you'll get a chance to try out the more advanced features.

SYSTEM CONFIGURATION

To optimize 3ds Max and avoid any potential problems, it is very important to set up your computer and the Windows operating system for best performance. When you start working on important projects, you don't want any unpleasant surprises.

Like all 3D graphics programs, 3ds Max is very complex. In fact, 3D graphics are the toughest job you are likely to give your computer. So, it's necessary that you have a fast computer. Learning 3D computer graphics is not a task to be taken lightly, and you must be well-equipped for the many hours of work ahead of you.

PROCESSOR

For best results, your processor clock rate should be at least 1 GHz. This will allow the many complicated calculations in 3ds Max to happen faster, giving you improved performance. The choice of processor brand is much less important than the clock rate. Also, you will experience a substantial performance boost if you invest in a computer with dual processors.

MEMORY

For serious graphic work, you will need at least 1 GB (gigabyte) of RAM. Increasing the amount of RAM in your computer is the most cost-effective upgrade you can make.

WARNING: If your computer doesn't have enough RAM to process a scene, it will start using *virtual memory*. Data that won't fit into available physical memory is saved to the hard disk in a hidden file called the *swap file* or *page file*. If this happens, your computer will suddenly become much, much slower, and might even crash. Needless to say, this is something to avoid. Make sure you have more than enough memory in your computer.

Try not to have any other programs running while you have 3ds Max open. Those other programs reserve memory for themselves, and 3ds Max has to take what's left over. Check the Windows Taskbar to see if there are other programs running, and close them before launching 3ds Max. Likewise, it's a very bad idea to have more than one instance of 3ds Max running simultaneously.

Some programs run in the background, and/or appear as icons in the System Tray at the right side of the Windows Taskbar. Examples are instant messaging programs and "agents" for media players such as RealPlayer. Disable these programs, and you will have fewer system crashes and fewer distractions keeping you away from your important work.

3DS MAX GRAPHICS DRIVERS

3ds Max renders images in real time in its viewports. To do this, 3ds Max can use one of several different technologies called *graphics drivers*. You must choose the best driver for your system to get the best performance.

Graphics drivers come in two main flavors: hardware and software. If you have a graphics card with 3D acceleration (nearly all of them do today), then you should take advantage of it by choosing a hardware driver. When you use a hardware driver, the graphics card does the work of rendering images onscreen, which reduces the amount of work the main processor must perform. If you use software rendering, then the overall performance of 3ds Max will probably be worse, because the hard work is being done by the system processor instead of the graphics card.

When you start 3ds Max for the very first time, you are presented with a choice of graphics drivers. To change your graphics driver settings later, you can access this same setting from the Main Menu. Select **Customize > Preferences**, click the **Viewports** tab, and click the **Choose Driver** button. After you make any changes, you must close and restart 3ds Max.

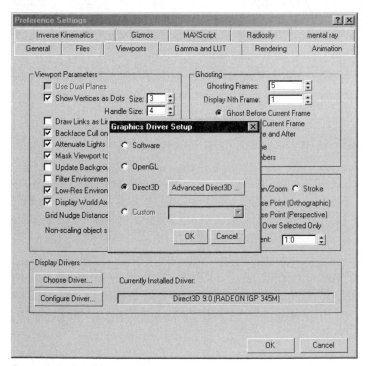

Figure 2-1: 3ds Max graphics driver setup

In most cases, Direct3D will be the best choice, because 3ds Max is optimized to take advantage of the acceleration available on consumer graphics cards for computer gaming. However, if you have an expensive graphics card specifically designed for 3D production, OpenGL may be faster. Take a half hour and experiment with the different settings, and see what works best with your system. Create a lot of objects and **Arc Rotate** in the Perspective view to see if the system bogs down. Open the **Customize > Preferences > Viewports > Configure Driver** window to choose different options for the driver. A little bit of experimentation with your 3ds Max graphics driver preferences may pay off with greatly improved performance.

Finally, there may be settings specific to your graphics card that can optimize performance. Consult your hardware documentation; there may even be a preset designed specifically for 3ds Max. These settings are typically accessed through the Windows Control Panel **Display Properties > Settings > Advanced**.

3DS MAX LAYOUT

Although you may be eager to start on a complex scene, it's important to become familiar with the fundamentals of this program first. Once you understand the basics, it will be much easier for you to grasp the more advanced uses of 3ds Max. In fact, the ease of your work later on depends largely on your mastery of the basics.

To prepare to use 3ds Max, you will need to familiarize yourself with the layout of the screen. This section will help you navigate around 3ds Max and find the components you'll need for creating scenes.

DEFAULT USER INTERFACE

3ds Max has a highly customizable *user interface* (UI) that allows artists to set up the workspace to suit their individual needs. However, this is a mixed blessing. For new users it is difficult to learn the program if the user interface is not set to the factory default. This is particularly important if you are using a computer that has multiple users, such as at a school.

This book uses the factory default 3ds Max interface throughout. If your 3ds Max screen does not resemble the illustrations in this book, you need to restore the interface to the factory default. From the Main Menu at the top of the screen, select **Customize > Load Custom UI Scheme**. In the pop-up dialog box that opens, select `DefaultUI.ui` from the file list and click **Open**.

WARNING: Until you are very familiar with 3ds Max, it is not advisable to customize the interface. To prevent accidental changes to the interface, lock it to the current state. This prevents you from inadvertently dragging an interface element away from its default position. Select **Customize > Lock UI Layout** from the Main Menu at the top of the screen. For more information about customization, consult the appendix and the 3ds Max product documentation.

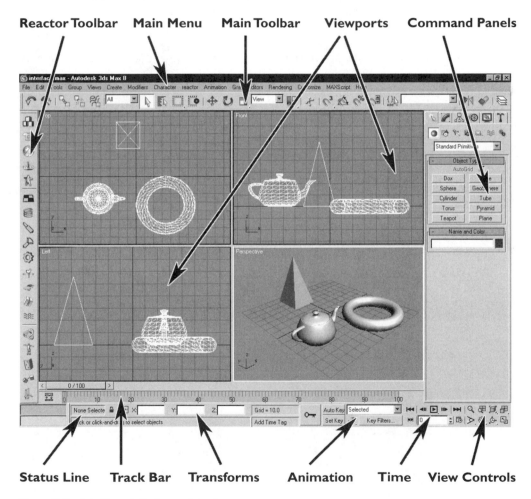

Figure 2-2: 3ds Max default user interface

MENUS

Like all Windows programs, 3ds Max has a menu across the top of the application window. To distinguish it from menus found elsewhere in the program, the menu at the top of the 3ds Max program window is called the **Main Menu**. Main Menu items are used for many purposes, such as loading and saving files, launching panels, and issuing commands. For example, **File > Reset** erases the current scene and resets all displays and values in 3ds Max to their defaults.

The Main Menu also allows you to access the electronic help system for 3ds Max. Choose **Help > User Reference** (or press the **<F1>** key) to read the help file. The help system has a convenient search engine that makes it extremely easy to locate the documentation for a specific feature.

DIALOGS

When you select certain items from the Main Menu, a pop-up window appears, requesting further information from you. This is called a *dialog box*, or simply a *dialog*. An example is **Customize > Units Setup**. This dialog lets you assign units of measurement for your scene. You can work in millimeters, feet and inches, miles, or whatever is appropriate for the scale of the objects in the scene.

Some dialogs are *modal*, which means that when the dialog is displayed, you can't click other areas of the screen and continue working. To resume your work, you must first close the modal dialog. Other dialogs are *modeless*, and allow you to interact with objects and other interface elements while the dialog is open.

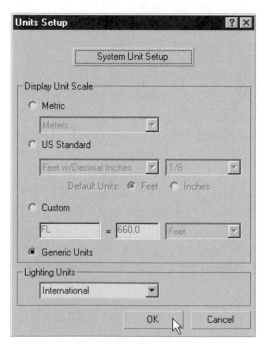

Figure 2-3: **Units Setup** *dialog*

FLOATERS

A *floater* is a modeless dialog that can remain on the screen while you continue your work. Floaters were developed as an alternative to certain frequently used modal dialogs and other commands. A floater does not have a **Cancel** button. Clicking the **X** at the top right of the floater closes the box but doesn't cancel the choices made on the floater.

An example of a floater can be found in the Main Menu item **Tools > Display Floater**. This floater consolidates object display commands found elsewhere in 3ds Max. Its big advantage is that it can stay on the screen while you perform other tasks. Since you don't have to hunt through the interface to find the display commands, your workflow is easier and faster.

Figure 2-4: ***Display Floater***

MAIN TOOLBAR

Below the menu is a row of many icons called the **Main Toolbar.** There are other toolbars in 3ds Max, but the Main Toolbar is special because it contains buttons that aren't found anywhere else.

Figure 2-5: Part of the ***Main Toolbar***

In any given project, you will use many of the functions on the Main Toolbar, some more than others. Here are some of the more commonly used buttons. You will learn more about each of these functions later in this book.

Locate each of the following buttons on the **Main Toolbar**. Depending on the resolution of your display, you might have to scroll the **Main Toolbar** to the right to see the buttons at the rightmost end. To do this, move the cursor to an area of the **Main Toolbar** between buttons, or above or below a button, until your cursor turns into a little hand. Click and drag to the left to view the buttons at the rightmost end of the **Main Toolbar**.

 Undo cancels the most recent action performed. You can also right-click this button to undo a series of commands. The keyboard shortcut for Undo is **<CTRL> + <Z>**. Hold down the **<CTRL>** key and press **<Z>** to undo the most recent action. Be warned, however: some commands are not undoable.

 Redo reverses the last Undo and reperforms the most recent action. **<CTRL> + <Y>** is the keyboard shortcut for Redo.

 Select Object is used to select one or more objects.

 Select and Move is used to move an object in a viewport. If the object isn't already selected, you must click the object before moving it. Move the cursor over an object, and the selection cursor appears. Click to select the object, then click and drag the **Move Gizmo** to reposition the object. Using gizmos is discussed later in this chapter.

 Select and Rotate is used to rotate an object. It works similarly to the Select and Move tool, except that the **Rotate Gizmo** appears when an object is selected. The Rotate Gizmo acts like a trackball.

 Select and Scale is used to resize an object. It works similarly to the Select and Move tool, except that the **Scale Gizmo** appears.

 Material Editor launches the Material Editor, which is used to "paint" objects with colors or patterns.

 Render Scene accesses a dialog that can be used to produce a rendering of the scene.

 TIP: You can easily find the name of any button in 3ds Max. When you park your cursor over a button for a moment, the name of the button appears in a tiny pop-up window called a **ToolTip**. Once you know the name of the button, you can go to **Help > User Reference** and use the **Search** feature to learn about the button.

In addition, there are other toolbars available in 3ds Max. You can hide or unhide them by right-clicking an empty space in any toolbar. A pop-up menu appears, from which you can choose which toolbars to view. If you wish, you may *dock* a toolbar to the top, bottom, left, or right of the screen. Simply drag the toolbar to the desired location. If Customize > Lock UI Layout is enabled, you won't be able to dock or undock any toolbars. For more information on interface customization, see the appendix at the end of this book.

Flyouts

Some buttons in 3ds Max have more than one function. If you see a little triangle in the lower-right corner of a button, it means more than one button is available. Click and hold the button, and you will see a *flyout* appear. If you continue holding the mouse down, you can select one of the other buttons on the flyout. As you do so, look to the bottom of the screen. There you will see a text field called the **Prompt Line**, which shows you the name of the currently selected button.

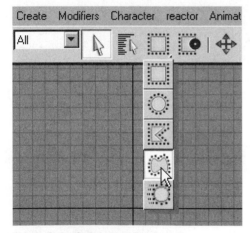

Figure 2-6: **Selection Region** *flyout*

When you choose a button from a flyout, that button remains active and visible until you click an drag the flyout to select a different button.

Snap Controls

Near the right side of the Main Toolbar are a series of buttons that look like magnets. These are the **Snap** controls.

Figure 2-7: **Snap** *controls*

Snaps allow you to precisely control actions in 3ds Max. For example, when **3D Snap** is enabled, moving an object is more exact because your cursor will "snap" to points on the Home Grid, or to other items on screen if you so choose from the **Customize > Grid and Snap Settings** dialog.

When **Angle Snap** is turned on, objects rotated in the viewports will only turn in set increments. The default is 5 degrees. This feature is handy when you need to rotate an object by exactly 45, 90, or 180 degrees.

EDITORS

Some functions of 3ds Max require additional interfaces that are so large and comprehensive that they are almost like programs within a program. In 3ds Max, you are always in the main user interface, but you might have a big editor window open, perhaps filling the screen. With some editors, you can minimize or maximize the window using the standard Windows buttons at the top right of the editor window.

NOTE: In the past, most 3D graphics application interfaces were *modular*, meaning they had separate program screens for doing different tasks. For example, modeling was done in a different program or screen interface than animation or material creation. The first release of the application, then called 3D Studio MAX, broke new ground, introducing numerous specialized modules in windows instead of totally isolated interfaces. This progressive interface style has now been adopted in the other high-end 3D software packages.

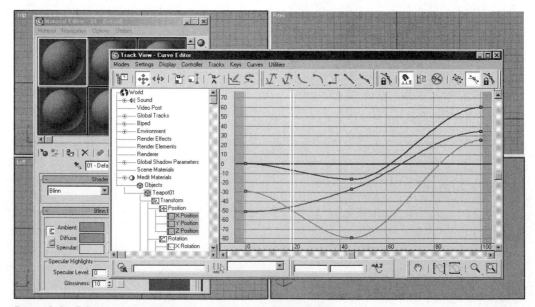

*Figure 2-8: Editor panels: **Material Editor** and **Curve Editor***

An editor panel in 3ds Max is basically a large modeless dialog with many powerful features. Some examples include:

- **Material Editor** for designing surface materials
- **Edit UVWs** for placing maps on 3D objects
- **Dope Sheet** for managing animation keyframes
- **Curve Editor** for editing animation curve data
- **Schematic View** for viewing and manipulating object/scene relationships
- **Environment and Effects** for creating special effects
- **Render Scene** for choosing render output options
- **RAM Player** for viewing rendered image sequences

VIEWPORTS

The main portion of the 3ds Max window consists of one to four *viewports* where geometric objects, lights, cameras, and other scene elements are seen and controlled. Each viewport has a *grid* for reference, a red-green-blue *axis tripod* in the lower-left corner to indicate spatial orientation, and a *viewport label* at the upper left.

A solid yellow line appears around the active viewport. Only one viewport can be active at a time. To activate a viewport, you can either left-click or right-click in it. Right-clicking is usually better, since left-clicking can alter the current selection of objects.

Perspective View

A **Perspective** viewport depicts the scene as it would appear to a real-world camera or to a human eye. Each viewport has grid lines to aid in object construction, location, and orientation. The grid lines in a perspective image converge as they go back. As you may have learned in a drawing class, parallel lines in a perspective drawing appear to meet in the distance. This is called a *vanishing point*. Of course, the lines don't really meet; they just look like they do.

Figure 2-9: Parallel lines appear to converge in a perspective image

Perspective views are perfect for representing naturalistic scenes, because that's how our eyes see the world. Real-world cameras also represent the world in perspective. Here's how it works.

NOTE: Objects in a scene, whether a real-world scene or a virtual 3D scene, exist in three dimensions. However, our eyes can actually only see two dimensions. Depth perception, or the ability to perceive distance, results from the brain synthesizing two separate 2D views from different viewpoints: our two eyes. This is called *stereoscopy*, and it's a fascinating topic, but it's beyond the scope of this book. The important thing to realize is that each eye can really only see in two dimensions.

Converting a 3D scene into a 2D image is called *projection*. This process happens every time you open your eyes, or take a picture with a camera, or render a 3D graphic. In your eye, an image of the outside world is focused through a lens and projected onto the inside of your eyeball, to a light-sensitive area called the *retina*. The same thing happens with a camera, except that the image is projected onto a piece of film or a light-sensitive computer chip.

Of course, CG software doesn't have film or any other physical properties. Instead, the software tries to mimic the process of capturing images from the real world. Images of 3D objects are projected onto a 2D surface known as the *picture plane*. In your eye, the picture plane is the retina. The picture plane of a camera is the film or light-sensitive chip. In computer graphics, the picture plane is the viewport or the rendered image.

A perspective view requires a single point of view, often simply called the *viewpoint*. As the viewpoint moves, the view changes. The viewpoint of a 3D CG perspective projection is often represented by an icon resembling a camera. In a 3ds Max Perspective View, you can't see the camera you are manipulating, but as you'll learn later, you can easily create a **Camera** object that you can see and manipulate from within another viewport.

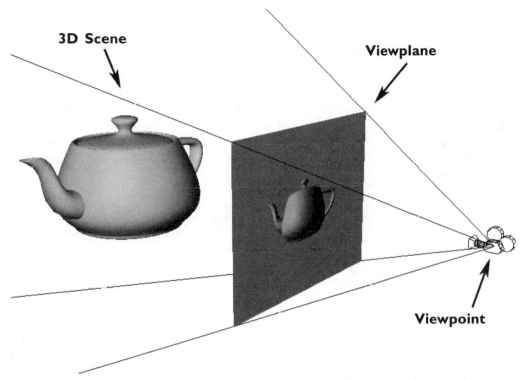

Figure 2-10: Diagram of a perspective projection

There is a problem with perspective projections. Because of the single viewpoint, often a perspective view can create distortions, leading to visual confusion. For example, as an object moves farther away from the viewpoint, it appears to become smaller. (This is actually the same thing as the illusion of converging parallel lines.) Also, the exact position of an object is sometimes difficult to determine due to misleading visual cues in a perspective projection.

Axonometric and Orthographic Views

Computer graphics work should be accurate and precise, with no confusion as to scale or distance. To solve the problem of perspective distortion, 3D programs use another type of projection called *axonometric*. In this type of projection, there is no distortion of size or location, because objects do not appear smaller as they recede into the distance. In an axonometric or straight-line projection, parallel lines never converge, but remain parallel as they go off into infinity.

The key to the difference between perspective and axonometric projections lies in the single point of view found in a perspective projection. With a single point of view such as a camera, all of the rays of light coming from objects in a scene must converge at the viewpoint. These rays are called the *lines of projection*, and in a perspective view they are not at 90 degrees (perpendicular) to the picture plane. Instead, the lines of projection strike the picture plane at varying angles. As an object moves away from the viewpoint, the angles of the lines of projection become shallower, and the projection of the object takes up less area on the picture plane. The end result is that the object looks smaller.

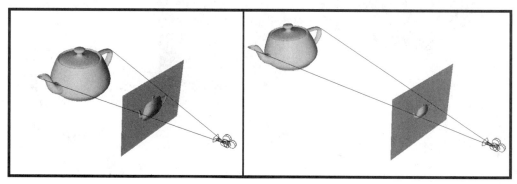

Figure 2-11: Lines of projection in a perspective view have variable angles

To see an animation that illustrates this effect more clearly, view the file **Perspective Projection.avi** from the disc that came with this book.

In an axonometric view, the lines of projection are always perpendicular to the picture plane. The lines of projection do not converge at a single location, so an axonometric projection doesn't have a single point of view. Instead, the entire picture plane is also the *viewplane*.

Because the lines of projection are always parallel to one another, axonometric views do not distort object size and position like perspective views do. As an object moves away from the viewplane, it does not appear to shrink. This makes it much easier to establish exactly where something is and what size it is.

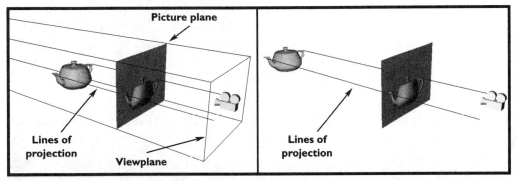

Figure 2-12: *Lines of projection in an axonometric view are parallel to one another*

There is an animation on the disc that illustrates the fact that objects don't appear to shrink as they move away from an axonometric viewplane. View the file **Axonometric Projection.avi**.

To see how perspective views can give misleading visual cues, and why you need axonometric projections, view the following files from the disc:

Viewports.avi

Perspective Scale Distortion.avi

Perspective Position Distortion.avi

The **Top**, **Left**, and **Front** viewports seen by default in 3ds Max are special cases of an axonometric projection and are called *orthographic* views. The only difference between an axonometric view and an orthographic view is that orthographic views are always aligned with the world coordinate system. In 3ds Max, the Top view is locked to the XY plane, the Left view is locked to the YZ plane, and the Front view is locked to the ZX plane.

If you use the **Arc Rotate** tool in an orthographic view, it is instantly converted into a **User** view. As you Arc Rotate around the scene in a User view, notice that the grid lines always remain parallel to one another.

Viewport Display Options

If you right-click the viewport label at the top left of a viewport, you can access a pop-up menu of viewport display options. At the top of the menu are the shading options, which control how geometry is rendered in viewports. The most commonly used options are **Smooth+Highlights**, **Wireframe**, and **Edged Faces**. By default, Perspective views are rendered with smooth shading and highlights, to make the objects look solid. Orthographic views are rendered as wireframes by default, so you can only see the *polygonal edges* of objects. This allows you to better understand the geometric structure of the objects. You'll learn about polygons and edges in Chapter 3, *Modeling*.

In general, wireframe renderings will yield faster onscreen interaction than shaded views, but that might not be true with certain types of 3D accelerators. As always, performance is dependent on many factors. To put it another way, "your mileage may vary."

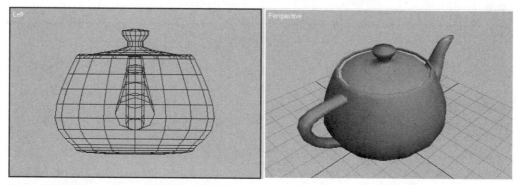

*Figure 2-13: Perspective in **Smooth+Highlights**; orthographic views in **Wireframe***

TIP: Edged Faces is a display option which combines shaded and wireframe modes to give you the best of both worlds. You can also change display modes with Hotkeys. **<F3>** switches back and forth between Smooth+Highlights and Wireframe modes, and **<F4>** turns Edged Faces on or off.

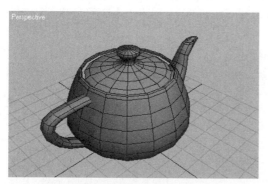

*Figure 2-14: **Edged Faces** display mode*

Viewport Navigation Controls

Several buttons at the bottom right of the screen offer many options for changing the views in viewports. In the course of any project, you'll have to change your view of the objects many times to complete the scene. 3ds Max provides ample tools to achieve this.

With any of the operations described here, your view of the objects in the scene changes, but the objects themselves don't move or change. Some of these tools are only available from flyouts.

 Zoom moves the point of view forward or back. Click this button, then click and drag upward in any viewport to move forward, or click and drag downward to move back. Unlike a zoom lens in a real camera, the Zoom tool moves the viewpoint, and does not change the angle of view. Shortcut: **<CTRL> + <ALT> + middle mouse button**, or **mouse wheel**.

 Zoom Extents moves the current point of view so all objects in the scene can be seen.

 Zoom Extents All moves the point of view in all viewports to the extents of the entire scene. Keyboard shortcut: **<Z>**.

 Region Zoom moves the viewpoint to enclose a selected area. This button is available by default in orthographic viewports such as **Top**, **Front**, and **Left**. Click this button, then click and drag in a viewport to draw a rectangular area. The point of view moves to frame the selected area in the viewport.

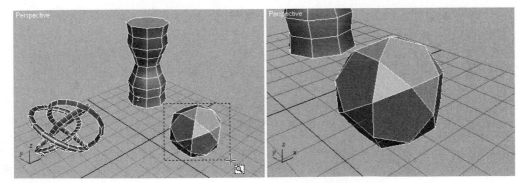

Figure 2-15: **Region Zoom** *zooms to rectangular bounding area*

 Field-of-View is available only when the **Perspective** view is active. This button acts like the zoom lens of a real camera. You will learn more about Field-of-View in the *Lights and Cameras* chapter. In a Perspective view, you should generally use the Zoom button instead.

Pan moves the point of view left, right, up or down. Click this button, then click and drag in a viewport to move it in the direction of the drag. Shortcut: **middle mouse button**.

 Maximize Viewport Toggle toggles the view between a full-screen view of one viewport and the multiple-viewport layout. Shortcut: **<ALT> + <W>**.

 Arc Rotate orbits the point of view around the scene. Click this button, and a circle appears in the viewport. Click and drag inside the circle to orbit the point view in two axes, like a trackball. Click and drag directly on one of the square handles at the top, bottom, left, or right of the circle to orbit the point of view in one axis only. If you click and drag outside the circle, you tilt the point of view, which is usually not what you want. Keyboard shortcut: **<ALT> + middle mouse button**.

 **WARNING:** Using **Arc Rotate** on an orthographic view will convert the viewport to a **User** view. The view is no longer lined up with the world coordinate axes, but is rotated at an angle relative to the grid. User views take some getting used to. For this reason, it is recommended that you use **Arc Rotate** only with the Perspective view when you're first learning 3ds Max.

Some of the viewport controls also have flyouts to further customize the view.

 To use **Region Zoom** in a Perspective view, you must select it from the Field-of-View flyout. Region Zoom is generally more useful than the Field-of View button, anyway.

 Zoom Extents Selected is a button available on a flyout from the Zoom Extents button. Clicking this button zooms the currently active viewport to the extents of selected objects. This cube appearing on this button is white, in comparison to the gray cube on the Zoom Extents button.

 Zoom Extents All Selected is available on a flyout from the Zoom Extents All button. This button moves the point of view in all viewports to the extents of selected objects. This button also features a white cube.

 Arc Rotate Selected is available on a flyout from the Arc Rotate button. This button works the same as Arc Rotate, except it orbits the point of view around selected objects rather than around the entire scene. You can also choose **Arc Rotate Sub-object** from the flyout. This is very helpful when modeling scenes, because you can orbit the point of view around part of an object.

Axis Tripod

Each viewport has an *axis tripod* that sits at the lower left of the viewport. The axis tripod consists of three lines originating from one point. The lines point in three perpendicular directions labeled X, Y, and Z.

In the orthographic views, one of the axes goes straight into or out of the screen. You can't see this axis, but you can see its letter label. For example, in the **Front** viewport, the X and Z axes are visible, but the Y axis is not because it points straight out toward you.

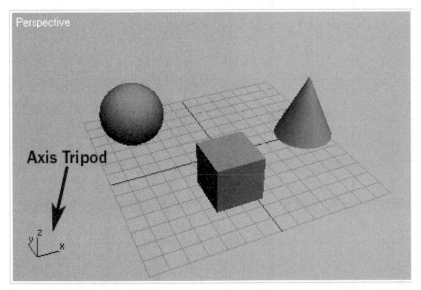

Figure 2-16: **Axis Tripod**

The axis tripod displays your orientation relative to the world, giving you a firm visual reference for orienting yourself in viewports. It is particularly helpful in an angled view, such as the Perspective or User views. Compare the orientations of the axis tripods in different viewports so you always know in which direction you're looking.

COORDINATES

The location of any object can be defined with a *coordinate system*. This is a set of numbers that can be used to set a location in relationship to another location. One example of a coordinate system is the longitude/latitude or *polar* coordinate system, which is used to specify locations on earth. A navigator on a ship or airplane uses this system to specify where the craft is and where it's headed.

All computer graphic programs use the *Cartesian coordinate system*. You are most likely familiar with this coordinate system from your geometry studies.

A flat surface, such as a piece of paper, is an example of a geometric *plane*. The plane is two-dimensional, or 2D. The word *dimension* simply means "measure." A 2D surface can be measured in two exclusive, perpendicular directions, commonly referred to as length and width. However, unlike a piece of paper, a geometric plane has no thickness at all. A geometric plane doesn't exist in the real world, but it is an abstract concept that is necessary for the science of geometry to exist.

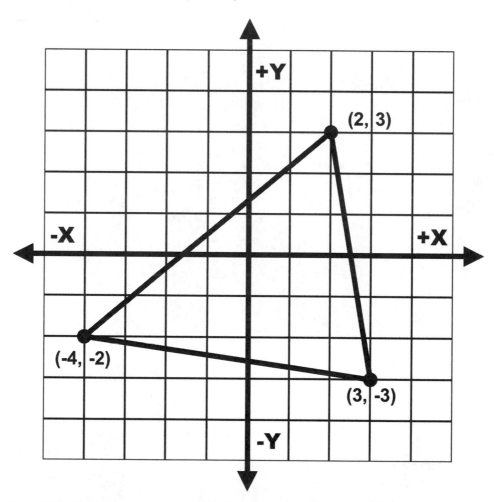

Figure 2-17: Cartesian coordinate system in two dimensions

In the Cartesian system, the 2D plane is divided into quadrants by two perpendicular lines. Each line represents one of the dimensions of measurement, and is called an *axis*. The plural for axis is *axes*. The axes are labeled X and Y. The X axis is horizontal, running left to right, and the Y axis is vertical, running bottom to top.

Any point on the plane can be located by measuring its distance from the X and Y axes. This gives each point two numbers, which can then be used in algebraic equations. These numbers can be either positive or negative, depending on which side of the axis the point is located. The numbers are separated by commas, and listed in the order (X,Y). The location where the axes cross is the reference point for the system. It is called the *origin*, and it has a value of (0,0).

In a 3D Cartesian coordinate system, a third axis is added. This third axis is labeled Z, and it is perpendicular to the plane defined by the X and Y axes. The orientation of the three axes varies among different applications. In 3ds Max, the XY plane represents the ground, and the Z axis points up toward the sky. Measurements along the Z axis define the height of an object. This is called the "Z-up" convention. Another common convention is "Y-up," in which the ground plane is defined by the X and Z axes, and the Y axis points upward.

Figure 2-18: Cartesian coordinate system in three dimensions

Using the three dimensions or axes, you can specify any point in space. To locate a point, measure its distance from the X, Y, and Z axes. This yields three numbers, expressed in the order (X,Y,Z). You can also use two or more points to define an object in three dimensions. For example, you can define a three-dimensional box by specifying two points at opposite corners of the box.

NOTE: The famous French philosopher and mathematician René Descartes described the Cartesian coordinate system in 1637. With this discovery, Descartes joined the two disciplines of algebra and geometry into a unified system, allowing equations to be applied to geometric objects. This was a great breakthrough that enabled many later developments, including computer graphics.

In a 3D coordinate system, any two axes can combine to form a plane. So, there are three intersecting planes in a 3D system: XY, YZ, and ZX. When looking down on the XY plane from above, X points to the right, while Y points to the top of the screen.

In any 3D graphics program, there is a Cartesian coordinate system that serves as the master reference point for all objects. The locations of objects are always expressed relative to this master system, which is called the *world coordinate system*, or simply *world space*.

To see an animation that illustrates the Cartesian coordinate system, double-click the file `world_coords.avi` on the accompanying disc.

QUAD MENUS

Right-clicking any object in the viewports launches a *context-sensitive* **Quad Menu**. Context-sensitive means that the menu may be different depending on the situation. The Quad Menus are customizable shortcuts to commonly used commands. Holding down the **<SHIFT>**, **<CTRL>**, and/or **<ALT>** keys while right-clicking an object will bring up different Quad Menus, specialized for different tasks such as modeling or animation.

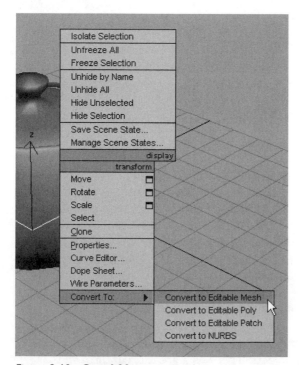

Figure 2-19: **Quad Menu**

STATUS BAR AREA

At the bottom of the 3ds Max interface are a number of toolsets that can be collectively referred to as the Status Bar area. We will discuss them in order from left to right. Later in the book we will thoroughly cover the usage of these tools; for now we will simply identify them and mention a few words about their purpose.

MAXScript Mini-Listener

In the far lower-left corner of the 3ds Max interface are two empty text fields. This is the **MAXScript Mini-Listener,** where you can type text commands in 3ds Max's scripting language, MAXScript. With MAXScript, programmers can make 3ds Max do all sorts of amazing things.

This book does not deal with MAXScript, but just to demonstrate the concept, click in the white area at the bottom of the Mini-Listener, and type the following command exactly as follows.

`teapot radius:50`

When you hit the **<ENTER>** key on your keyboard, a Teapot with a radius of 50 units is created at the origin. The Mini-Listener echoes your command with more precision, telling you that a Teapot has been created at world space coordinates **(0,0,0)**.

Status Line and Prompt Line

To the right of the Mini-Listener are two text readouts that give you information about the current action or command. If your desktop is set to a resolution below 1280 pixels wide, you might not be able to see these two text fields. There is a movable bar separating these two fields from the Mini-Listener. Simply move the bar to the left to minimize the Mini-Listener.

The top text field is the **Status Line**. It tells you what objects you currently have selected. The bottom field is more interesting; it's called the **Prompt Line**. It tells you what you need to do next to complete an operation. If you don't know what to do, the first place to look is the Prompt Line. It doesn't lead you through every single step of creating 3D graphics, but it gives you valuable feedback, especially when you are just learning 3ds Max.

Transform Type-In

When an object is selected with the **Move, Rotate,** or **Scale** tools on the Main Toolbar, numbers can be seen in the three fields labeled X, Y, and Z in the Status Bar at the bottom of the screen. This is called the **Transform Type-In** area, because position, rotation, and scale are collectively known as *transformations* or *transforms* in computer graphics. You can enter values in these fields to control the location, orientation, or size of selected objects. Transforms will be discussed more thoroughly later in this chapter.

Animation Controls

This section of the Status Bar area deals with animation. Using these tools, you can create animation keyframes manually or automatically. You can also choose which objects you want to animate and what types of keyframes you wish to create, such as position or rotation keys.

Figure 2-20: Animation Controls

Time Controls

The **Time Control** buttons are similar to those on a desktop media player, or even a CD player in your home audio system. They allow you to play, pause, step, or skip through your animation in the viewports. The current frame is displayed in this area.

You can also find the **Time Configuration** 🗔 button here. This button launches a dialog in which you can control the frame rate, display speed, and number of frames in an animation.

Figure 2-21: Time Controls

Track Bar

Below the viewports is an area known as the **Track Bar**. It is labeled with numbers that represent frames of an animation. The Track Bar shows a visual representation of the current time in an animation, and also displays keyframes for the selected objects.

Above the Track Bar is the **Time Slider**, which lets you quickly "scrub" or "shuttle" through time to view animation in the viewports or locate a specific point in time. The Time Slider shows you the current frame and the number of total frames in the current animation segment.

COMMAND PANELS

Down the right side of the screen is the area for **Command Panels**. Tools within the Command Panels are used for many different purposes, including creating and modifying objects, controlling animation, and miscellaneous utilities.

At the top of this area are six tabs. When the cursor is placed over a tab for a moment, the name of the tab appears as a ToolTip.

Figure 2-22: ***Command Panel*** *tabs*

To choose a Command Panel, simply click on one of the panel tabs. Some panels have several subpanels within them; selecting an icon or text button at the top of the panel sends it into a different mode.

Create constructs objects such as Geometry, Lights, Cameras, and Helpers.

Modify edits geometry or parameters of objects after creation. Model and animate the shapes of objects.

Hierarchy determines how objects that are linked to one another will behave when animated. You will learn about object links in Chapter 6, *Keyframe Animation*.

Motion controls animated objects by editing keyframes and **Controllers**. Animation Controllers are discussed in Chapter 10, *Advanced Animation*.

Display sets options for object display. Hide or unhide objects, freeze objects, and so on.

Utilities offers various tools, such as the **Asset Browser** and the **Color Clipboard**.

The items within each panel are organized into *rollouts*, which can be minimized to make room for other items. Click on the Rollout title to expand or contract the rollout.

Panels can be scrolled vertically to reveal the various rollouts. There is a tiny black line at the right of the Command Panel; this is a *scrollbar*. Click and drag it to scroll up and down. You can also click and drag in an empty area of a Command Panel to scroll the Command Panel. When your cursor turns into a hand, the **Pan** tool is active.

Finally, the entire Command Panel area can be expanded horizontally, revealing more than just one column of rollouts. Click and drag the left edge of the Command Panel to resize it.

The Command Panel and toolbars can be detached from their default *docked* positions and placed anywhere on the screen. If they are not docked, user interface elements are said to be *floating*. This is especially useful if you have two or more computer monitors. Until you are very comfortable working in 3ds Max, you should lock the UI in its default state by selecting **Customize > Lock UI Layout** from the Main Menu, as mentioned previously.

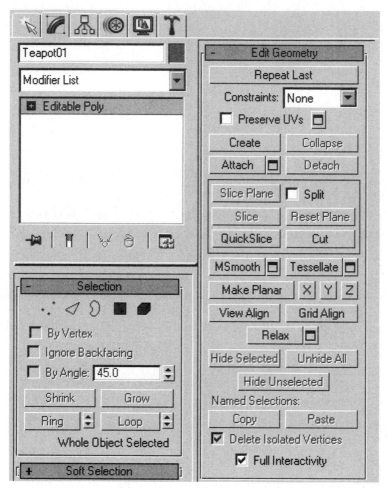

*Figure 2-23: Expanded **Command Panel** and **Rollouts***

Create Panel

By default, the **Create** tab is selected. Below the **Create** panel are buttons for creating different types of objects. The **Geometry** button is selected by default. This is the option you will use for now.

Parameters

When you click a button on the Create panel such as Sphere or Box, several entry areas and values appear on the panel. A number value on the Command Panel, such as **Radius**, is called a *parameter*. Parameters are used to set an object's dimensions and other characteristics. You can change a parameter by typing directly into its entry box and pressing the <**ENTER**> key.

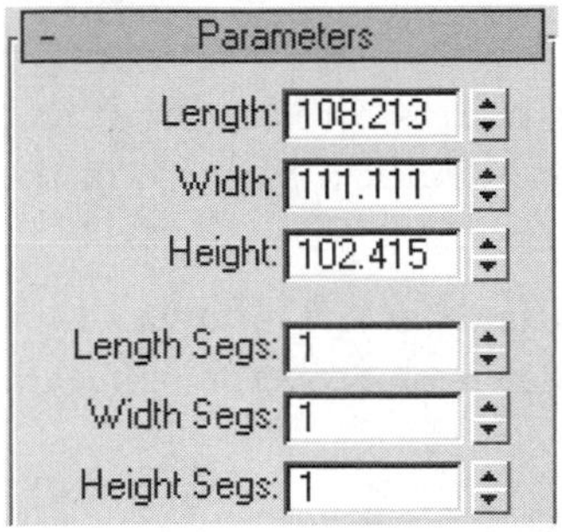

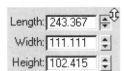

Figure 2-24: ***Parameters***

Spinners

You can also change a parameter by clicking on the small up and down arrows next to the entry box. These arrows are called *spinners*.

You can make a parameter change quickly by clicking a spinner arrow and dragging upward or downward. Parameters and spinners appear not only on Command Panels, but in other areas of 3ds Max as well.

Figure 2-25: ***Spinners***

Units

When an object is created, the parameters on the panel are used to set the object's dimensions. These dimensions are expressed in **Units**. A Unit can represent one inch, six meters, three miles, or any unit of measurement you wish to use.

Units are important when a scene needs to be made to architectural or engineering specifications, or when the scene will be shared with other users working on the same project. If you want 3ds Max to represent objects using a specific unit of measurement, you can open **Customize > Units Setup** from the Main Menu and define your **Display Unit Scale**. Simply select what measurement system you wish to use.

For now, do not alter the settings in the **System Unit Setup** dialog found within the Units Setup dialog. Changing the System Unit settings could cause serious problems with your scenes.

Modify Panel

After an object has been created, you can change its parameters by selecting it and going to the **Modify** panel. The parameters found here are identical to those in the Create panel.

At the top of the Modify panel is an area called the **Modifier Stack**. This area allows you to add functions and effects, called **Modifiers**, to geometric objects. You will learn more about modifers later in this chapter.

MODELING BASICS

There are many ways to make objects in 3ds Max. All object creation methods are available from the **Create** panel. Under the **Create** tab are seven buttons. The first button is **Geometry**. When **Geometry** is on, the label **Standard Primitives** appears below it by default.

PULLDOWN MENU

The **Standard Primitives** option appears on a *pulldown menu*. To see more options on the pulldown menu, click **Standard Primitives** or click the down arrow to the right of **Standard Primitives**. More options appear.

These are different types of geometry (3D objects) that can be created with 3ds Max. Each of these types of objects will be covered in more detail later in this book. In this chapter, you will create Standard Primitives.

*Figure 2-26: Geometry pulldown in the **Create** panel*

STANDARD PRIMITIVES

A *primitive* is a basic building block for a 3D scene. To create a primitive, you can click a text button on the Create panel. Immediately after clicking the button, click and drag in a viewport to begin creating the primitive. Different primitives require different numbers of mouse clicks and drags.

Immediately after you create an object, you can change its parameters (Radius, Length, Height, and so on) on the Create panel. If you deselect the object, you can still change its parameters by going to the Modify panel, as you will learn later in this chapter.

SEGMENTS AND SIDES

Each primitive has a certain number of **Segments**, or divisions along or around the object. Many objects also have **Sides**, which are the divisions running at right angles to the segments.

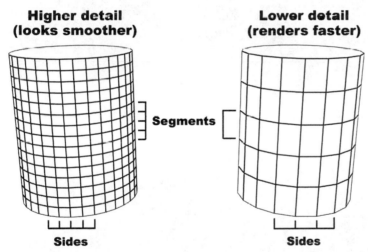

Figure 2-27: *Segments and Sides*

By default, the number of segments or sides is set to a low number such as zero or **16**. You can increase this number to raise the amount of detail on the object. For example, increasing the **Segments** value for a sphere will make it appear more round.

The number of Segments is a critical modeling choice. If a primitive is to be shaped into a custom model, the density of Segments determines how well that model will resemble the real object. For example, you can't bend a Box that has only one Segment. If you want the end result to be curved, you need to add more Segments.

CONSTRUCTION PLANES AND THE HOME GRID

When creating and manipulating objects in a 3D program, you are usually restricted to working with only two axes at a time. This is because the mouse or graphics tablet is a 2D pointing device. Also, a standard computer monitor is a flat, 2D display that does not show true 3D depth.

To stick to drawing in only two dimensions at a time, 3D programs enforce certain conditions during the object creation process. For example, suppose you're creating a cylinder. You have chosen to create in the Top view, so you are looking down on the XY plane. In this case, the Z axis is pointing right at you. When you create the cylinder in the Top viewport, you have no way of specifying where on the Z axis the cylinder will sit.

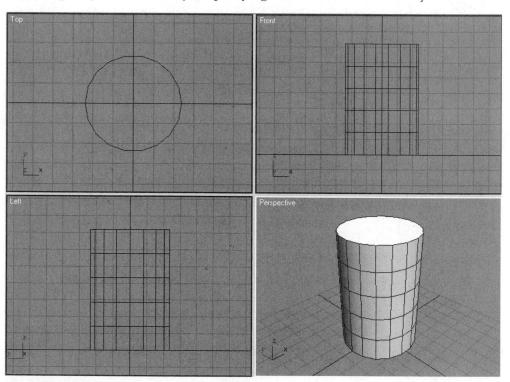

*Figure 2-28: Cylinder created on the **Home Grid***

When an object is created in a viewport, it is placed on a *construction plane*. The construction plane is designated by the grid lines running through a viewport. The default construction plane is the **Home Grid**, which is comprised of the world coordinate XY, YZ, and ZX planes. The intersection of the three construction planes marks the origin of world space. In the Top viewport, a cylinder would be created so that it sits on the XY construction plane. The base of the cylinder has a Z coordinate of zero.

Since the Home Grid is the master reference coordinate system for the scene, it cannot be moved. However, other grids can be created and activated as needed.

AUTOGRID

3ds Max has a very useful feature that enables quick and easy creation of objects on the surfaces of other objects. It is called **AutoGrid**, and it can be found on the Create Panel directly below the pulldown menu.

With AutoGrid enabled, as you move your cursor over existing objects in your scene, you will see an axis tripod. The Z axis of this axis tripod will always remain at right angles to the surface of the existing object as you move your cursor.

Click an existing object, and you will see a grid pop up. This is a temporary construction plane, allowing you to easily create objects off the Home Grid.

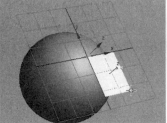

Figure 2-29: **AutoGrid**

OBJECT NAME AND COLOR

Every object you create in 3ds Max is given a default **Object Name** based on the type of object and the order in which is was created. For example, the first Teapot you create will be called Teapot01, the second one will be called Teapot02, and so on.

TIP: When creating scenes in 3ds Max, you will find it much easier to do your work if all objects in the scene are named intelligently. Don't settle for the default names given to you by 3ds Max. Use names that make sense to you and will allow you to identify objects in long lists of names. Take care not to use duplicate names; this is allowed by 3ds Max, but it will make your work difficult.

Each object is also assigned a random color at creation time. When you view an object in wireframe mode, it will be displayed in its **Object Color**. This makes it easier to identify objects in complex scenes with many wireframe objects. Until you assign a material to an object, its Object Color will also be seen in shaded viewports.

Object Name and Color can be found in the Create Panel under the aptly labeled **Name and Color** rollout. To change an Object Name, simply click in the text field and type in a new name. To change the Object Color, click the square *color swatch* to the right of the object name. This will launch a simple color picker that allows you to choose a new color.

After creation time, Object Name and Color can be changed from the Command Panels. The name and color of the currently selected object appear at the top of every Command Panel except Create, which has a Name and Color rollout.

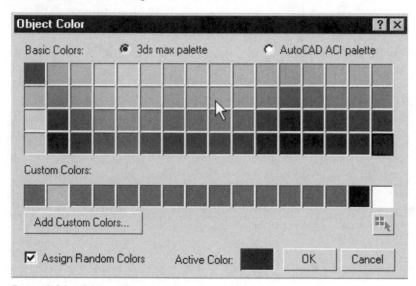

Figure 2-30: **Object Color** *dialog*

WARNING: Don't use white as an object color. White is the default color for a selected object in 3ds Max. If you have wireframe objects that are always white, it will be nearly impossible to tell which objects are currently selected and which aren't.

MODIFIERS

Modifiers are functions that can be applied to an object, primarily to change its shape. To add a modifier to a selected object, click the drop-down **Modifier List**. You will see a long list of modifiers arranged in alphabetical order. Select a modifier from the list, and the Modify panel will update to reflect the added modifier.

Figure 2-31: **Modifier List**

The Modifier List is context-sensitive. When an object is selected, 3ds Max knows which modifiers can be applied to it and which cannot. The modifiers that are not possible for the currently selected object are not displayed in the list. For example, the **Extrude** modifier is unavailable when a 3D primitive is selected, because this modifier can't be applied to primitives.

You can also add modifiers from the **Modifiers** menu, on the Main Menu. In this case, modifiers that cannot be applied to the currently selected object will be grayed out.

As you add a modifier to an object, its name appears near the top of the Modify panel in an area called the **Modifier Stack**. You can apply multiple modifiers to an object, and all modifiers for the current object are listed in the Stack.

Figure 2-32: **Modifier Stack**

EXERCISE 2.1: Bend Modifier

1. Reset 3ds Max. Maximize the **Perspective** viewport.

2. In the Perspective viewport, create a **Box** with a **Length** and **Width** of about **30** units, and a **Height** of about **100** units.

3. Click Zoom Extents.

4. Press the **<F4>** key to display **Edged Faces** in the viewport.

5. Increase the Box's number of **Height Segments** to **10**.

6. Go to the **Modify** panel. Open the **Modifier List** and click **Bend** to add the modifier.

7. On the Modify panel, increase the **Angle** parameter to **90**. The Box bends.

8. For the **Direction** parameter, drag the spinner until the parameter displays a value of **90**. This bends the object at an angle **90** degrees away from the original bend direction.

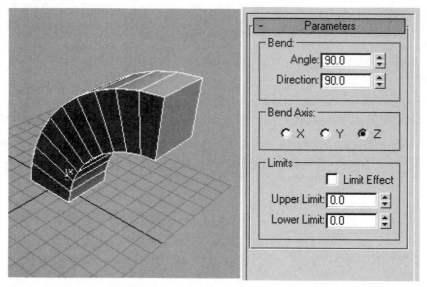

Figure 2-33: **Box** *with* **Bend** *modifier applied*

9. Save your scene as **BendyBox.max** in your folder.

THE MODIFIER STACK AND PARAMETRIC MODELING

The Modifier Stack processes the effects of each modifier in turn, moving from bottom to top. The result of one modifier may have a profound effect on the next modifier higher in the Stack. For this reason, the *order of operations* is a major consideration when using the Modifier Stack. Two objects with identical modifiers and identical parameters may turn out completely different if their modifiers are stacked in a different order. That's because the modifiers high in the Stack are *dependent* on the ones below.

In the following illustration, the two objects are the same except for the order of modifier operations. They are both Boxes with a **Bend** modifier and a **Taper** modifier applied. In the Box on the left, the Taper modifier is lower in the stack, so it is tapered first, then bent. In the Box on the right, the Bend modifier is lower in the Stack, so the Box is bent first, then tapered.

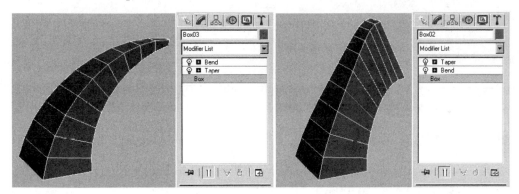

*Figure 2-34: **Modifier Stack** order*

The order of operations is not related to the chronological order in which you add the modifiers. In other words, you can add a modifier the next day, but place it lower in the Stack. In fact, you can add modifiers anywhere in the Stack, or drag modifiers around in the Stack to reorder them. You can even cut, copy, and paste modifiers among objects. Right-click a modifier to access a pop-up menu from which you can choose to cut, copy, paste, or perform other operations.

To access the parameters for a modifier, simply select it from the Stack. The parameters are displayed in the Modify panel below. You will see the end result of all modifiers, no matter which modifier is currently selected.

Sometimes it is useful or necessary to see the results of the modifiers only up to a certain point in the Stack. To do this, turn off the **Show End Result** button. If this is off, modifiers up to the currently selected Stack level still affect the object, but all modifiers higher in the Stack are ignored.

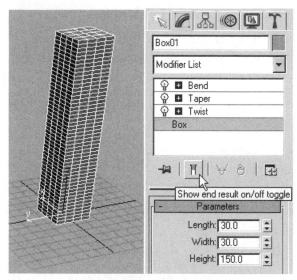

Figure 2-35: **Show End Result** *is off*

In addition, you may disable or enable any modifiers by clicking the little light bulb icons in the Stack. If the light bulb is dark, the modifier is turned off.

TIP: Using primitives and modifiers is called *parametric modeling,* because the shape of the object is defined by parameters. As you will see in the following exercise, parametric modeling gives you a great deal of freedom to make changes to a model at any time in the modeling process.

EXERCISE 2.2: Modifier Stack

1. Open the file **BendyBox.max** from your folder or from the disc. Maximize the Perspective viewport and press **<F4>** to view Edged Faces.

2. Select the Box and open the **Modify** panel. Select the Box creation level parameters by clicking the word **Box** in the Modifier Stack.

3. Click the **Modifier List** to open the drop-down list of modifiers. Apply a **Taper** modifier to the Box. Set **Amount** to **-1**.

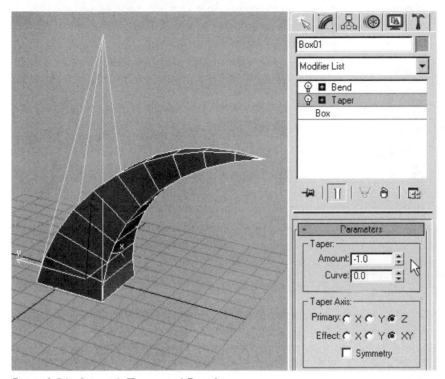

*Figure 2-36: Box with **Taper** and **Bend***

4. With the Taper modifier selected, turn off **Show End Result** [icon] to see the Box without the Bend applied. Experiment with the Taper Amount, then set it back to **-1**. Turn Show End Result back on.

5.   Now you will change the order of the modifiers. Select the Bend modifier in the Stack. Click and drag the Bend modifier downward. You will see a transparent box with the word **Bend** in it. When your cursor comes near another modifier, you will see a blue line appear between modifiers. Drag until the blue line is between Box and Taper in the Stack. Release the mouse button to complete the operation.

Figure 2-37: Changing modifier order

6.   The object has changed shape because the Bend is being applied earlier in the order of operations. Increase the **Angle** parameter to **180** degrees.

7.   Go back to the Taper modifier and adjust the Amount by dragging the spinner. Notice the result of applying the Taper modifier later in the order of operations. The Taper is now being applied to a Box that has already been bent.

     Restore the Taper Amount to **-1**.

8.   Drag the Bend modifier to the top of the Stack, so the order of operations is restored. The Stack should now read in the following order from bottom to top: Box, Taper, Bend.

9.   Select the Box creation level in the Stack. Go to the Main Menu and select **Modifiers > Parametric Deformers > Twist**. This is the same as adding a modifier from the Modifier List on the Modify panel.

10. Increase the Twist **Angle** to **360** degrees. The object is starting to resemble a horn.

11. Drag the Twist modifier up so that it is at the top of the Stack. The object deforms in a most unpleasant manner.

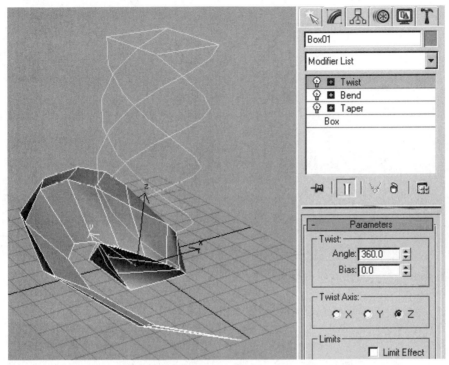

*Figure 2-38: Bizarre effect of placing the **Twist** modifier at the top of the Stack*

12. Drag the Twist modifier back down to its original location in the Stack, between the Box object and the Taper modifier.

13. In the Box object creation parameters, increase the number of **Height Segments** to **50**. This gives a much more pleasant curvature to the horn. Increase the Box **Height** to **150**. Increasing the Height results in a taller box, but the Twist, Taper, and Bend parameters haven't changed.

14. Zoom in on the object using the **Zoom** tool. Get in close enough to see the twists of the horn. Increase the number of **Length Segments** to **4**. Notice the more pleasant curvature of the horn.

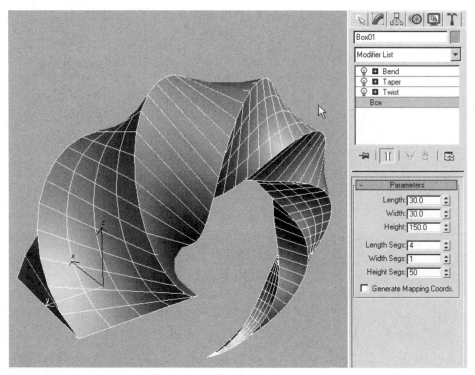

*Figure 2-39: Increasing the Box **Length Segments***

15. Increase the Length Segments to **10**. Notice that the curvature does not improve by increasing the Length Segments parameter from **4** to **10**, as it did when you increased the parameter from **1** to **4**. Additional segments did not create a better shape, but merely made the model needlessly complex. It is important that you learn to create only as much detail as needed for a particular model, otherwise your scenes will bog down and your render times will be excessively long. Reduce the Length Segments back to **4**, and increase the **Width Segments** to **4** also.

16. Click **Zoom Extents** to frame the horn in the viewport.

17. Save the file as **Horn.max** in your folder.

TRANSFORMS

In 3D graphics, the operations of position, rotation, and scale are collectively referred to as transformations, or *transforms* for short. There are three transform tools found on the 3ds Max **Main Toolbar** for changing an object's location, orientation, or size.

 Select and Move moves objects

 Select and Rotate turns objects

 Select and Scale changes the size of objects

When an object is transformed, the amount of transformation is displayed in the **Transform Type-In** area at the bottom of the screen. For example, when you rotate an object, the degree of rotation appears on the Status Bar as you rotate, updating interactively during the process.

In some places in the 3ds Max interface, you may see the acronym **PRS**, which stands for Position, Rotation, and Scale. PRS is a common expression throughout the computer graphics industry, and it is often used interchangeably with the term *transforms*. Also, the term *translation* is often used when referring to the position transform. To translate an object is to move it in a straight line.

Transforms are fundamentally different from modifiers. When you move, rotate, or scale an object, you don't see the operation appear in the Modifier Stack. Also, transforms don't have parameters. As a result, you can't go back and change a transform later the way you can with modifiers. The only thing you can do is apply a new transform operation to the object.

TRANSFORM GIZMOS

As discussed previously, it's only possible to predictably transform an object in one or two dimensions at a time, because the mouse is a 2D pointing device. To make 3D transforms easier, each transform tool in 3ds Max has its own 3D icon for manipulating objects in viewports. These icons are known as **Transform Gizmos**.

Whenever you select an object and activate a transform tool, the appropriate Move, Rotate, or Scale Gizmo appears onscreen. If this doesn't happen, the Transform Gizmos are probably hidden. Choose **Views > Show Transform Gizmo** from the Main Menu, and/or press the <X> key. Pressing the <X> key is the same as disabling the Transform Gizmos in the **Customize > Preferences > Gizmos** dialog box. If the Gizmos are disabled in the Preference Settings dialog, you will still see an object axis tripod. If the Gizmos are hidden by turning them off in **Views > Show Transform Gizmo**, you won't see any axis tripod. Both options must be on for Transform Gizmos to work.

NOTE: You may be wondering why anyone would want to hide or disable the Transform Gizmos. In fact, the Transform Gizmos are not required to transform objects. They merely make it easier to select which axis or axes you want. This is called choosing **Axis Restrictions**, and can also be done using the hotkeys **<F5>** through **<F8>**. In some cases it is easier to work with the Transform Gizmo hidden, and choose the transform axes with hotkeys. In the absence of a Transform Gizmo, a simple axis tripod appears, with the currently active axis or axes highlighted in red.

Like most things in 3ds Max, the Transform Gizmos are color-coded. The X axis is always represented by the color red. Y is always green, and Z is always blue. Currently selected Transform Gizmo axes are highlighted in yellow.

Move Gizmo

The **Move Gizmo** is probably the most straightforward and simple of the three. It resembles a simple axis tripod, such as the one seen at the bottom left of each viewport, except that the Move Gizmo has additional features.

As you move your cursor over an object, your cursor changes to a **Move** icon. When the cursor is over one of the Move Gizmo axes, the axis is highlighted in yellow. Click and drag an axis to move the object in that direction.

In the center of the Move Gizmo are three intersecting planes. These are the XY, YZ, and ZX planes. Place your cursor over one of the planes, and it is highlighted in yellow. The appropriate axis lines also light up in yellow. Click and drag one of these planes to move an object in two axes at once.

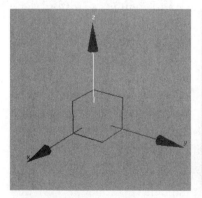

Figure 2-40: **Move Gizmo** *with Z axis active and with XY plane active*

Rotate Gizmo

The **Rotate Gizmo** is designed to act like a trackball. Instead of an axis tripod, the Rotate Gizmo looks like a transparent sphere decorated with circles. A small axis tripod appears in the center of the sphere by default.

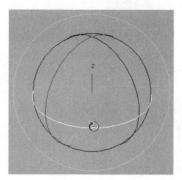

Figure 2-41: **Rotate Gizmo** with Z axis activated

Place your cursor over one of the circles on the "surface" of the Rotate Gizmo to select an axis for rotation. The circle highlights in yellow, and the appropriate axis of the Gizmo's axis tripod also highlights in red, green, or blue, depending on which axis you have chosen.

By default, when you click a rotation axis, a small line appears. This is called a **tangent vector**, and it shows you which way to drag your mouse to turn the object. The direction of rotation is also indicated by the tangent vector. One side of the tangent vector highlights in yellow to show you which way the object is turning. In addition, the rotation angle appears in a small numeric display, and a translucent slice icon appears on the plane of rotation. As you drag in the direction of the tangent vector, the slice icon gives you visual feedback on the amount of rotation.

This is the default behavior of the Rotate Gizmo, and it is referred to as the **Linear Roll Rotation Method**. There are other methods which can be chosen in the Preference Settings dialog, but Linear Roll is the default, because it is probably the easiest to use. This book always uses the Linear Roll method.

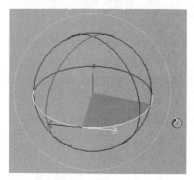

Figure 2-42: **Rotate Gizmo** in Linear Roll mode

If you click in the interior of the Rotate Gizmo or on the edge of the transparent sphere, the edge is highlighted. As you drag the mouse, the object rotates as if you were controlling a trackball. This is referred to as *free rotation*. It is much more difficult to control rotation with precision using free rotation, but it is good for quickly and roughly orienting objects.

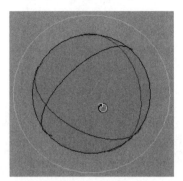

Figure 2-43: Free rotation

Finally, the Rotate Gizmo has a large circle surrounding the sphere icon. This is called the **Screen Handle**. Clicking and dragging this outermost circle rotates an object in the plane of the viewport. Since Perspective views are not aligned to world space, Screen rotation in a Perspective view can give unpredictable results.

EXERCISE 2.3: Moving and Rotating

In this exercise, you'll use the transforms on the Main Toolbar in conjunction with Transform Gizmo axis restrictions.

1. Reset 3ds Max and maximize the Perspective viewport. Create several objects, using Standard Primitives such as Cylinders, Boxes, and Teapots. Don't use Spheres, because it's hard to tell how a Sphere is rotated.

2. Save the scene as a new file called **LotsOfObjects.max** in your folder.

3. On the Main Toolbar, click **Select and Move**. Click any object. The **Move Gizmo** appears.

4. Move the cursor over the object's X axis until the cursor changes to the Move icon. Click and drag the Move Gizmo's X axis. The object's movement is restricted to the X axis.

5. Move the cursor to the Move Gizmo's Y axis. Click and drag; the object's motion is restricted to movement in the Y axis.

6. Select another object. Click the XY plane of the Move Gizmo, and drag in circles in the viewport. The object's movement is restricted to the XY plane. It moves around the scene, but never goes above or below the groundplane.

7. On the Main Toolbar, click **Select and Rotate**. The **Rotate Gizmo** appears. Click another object.

8. Move your cursor over the three circles on the "surface" of the Rotate Gizmo. Note that each circle is at right angles to one of the axes in the center of the Rotate Gizmo's sphere. Click and drag one of the circles to observe the result. The object spins around the axis that is perpendicular to the circle you have selected. The Rotate Gizmo's axis tripod shows the current axis highlighted in red.

9. Click in the center of the Rotate Gizmo, taking care not to click on one of the X, Y, or Z circles. Drag in circles to observe the effects of trackball-style free rotation. Save the scene again.

Scale Gizmo

One way of altering the size or shape of an object in 3ds Max is to use the **Scale** transform. Selecting an object with the **Select and Scale** tool activates the Scale Gizmo. As you move your cursor around the **Scale Gizmo**, notice that your cursor changes. Depending on the Scale option you have chosen and where your cursor is, you will see one of three different cursor icons.

Clicking and dragging in the center of the Scale Gizmo will change the size of an object. This is called Uniform Scale, meaning that the object is being scaled the same amount in all axes.

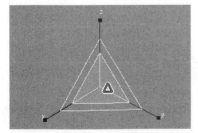

Figure 2-44: ***Uniform Scale***

If you select a single axis of the Scale Gizmo, you can scale an object in just that one axis. The result is that the object becomes elongated. This is called **Non-Uniform Scale**. As you scale an object up in one axis, it becomes longer in that axis, while the other two axes remain as they were.

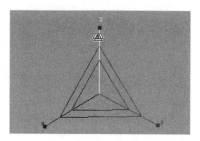

*Figure 2-45: **Non-Uniform Scale** in one axis*

WARNING: Select and Scale is not always the best way to change the size of an object. For Primitives such as a Box, it is better to change the parameters of the object on the Command Panel. This is because using the Scale transform often leads to problems with animated objects, especially if you are performing a Non-Uniform Scale. There are several solutions to these problems, which you will learn about in Chapter 6, *Keyframe Animation*.

Selecting a diagonal edge of the Scale Gizmo activates Non-Uniform scaling in two axes. By default, as you click and drag your cursor around the screen, you scale each of the two axes equally. If desired, you can set the behavior of the tool so that dragging up/down will scale in one axis, while dragging left/right will scale in the other axis. If you prefer to scale in this manner, you can disable the **Uniform 2-Axis Scaling** option in **Customize > Preferences > Gizmos**.

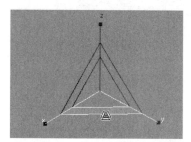

*Figure 2-46: **Non-Uniform Scale** in two axes*

Performing a Non-Uniform Scale always increases or decreases the size and volume of an object. In some cases, you may wish to change the scale of an object while maintaining a constant volume. One example would be the classic "squash and stretch" effect seen in cartoons. To achieve this effect with the Scale transform, you must select the **Select and Squash** tool from the Scale flyout.

With the Squash button active, the Scale Gizmo operates differently. When you click and drag one axis, the other two axes scale in inverse proportion. That is, if you increase the scale in the Z axis, the scale in the X and Y axes is automatically decreased. The result is that the volume of the object always remains constant.

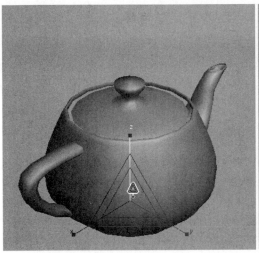

*Figure 2-47: **Squash** in the Z axis*

NOTE: You've probably noticed that the Scale flyout has three buttons, **Select and Uniform Scale**, **Select and Non-Uniform Scale**, and **Select and Squash**. The differences between the Uniform and Non-Uniform Scale buttons are minor. It doesn't really matter which of those two buttons is active, because you can perform uniform or non-uniform scale operations using the Scale Gizmo. Only the Squash button has a significant effect on the operation of the Scale Gizmo.

TRANSFORM TYPE-IN

Any transform that you can perform with a mouse operation can also be done with precision by entering numbers in X, Y, and Z fields of the **Transform Type-In** area at the bottom of the screen. Simply type in a number to move, rotate, or scale an object by that amount. The position of an object is measured in 3ds Max units, chosen in the **Customize > Units Setup** dialog. Rotations are measured in degrees, and scale is measured in percentages: 100% scale means that the object is the same scale as when it was created, and no scale transformation has been applied.

Transform Type-In has two modes: **Absolute** and **Offset**. The modes are chosen from the button on the left of the type-in fields.

 Absolute mode always displays the current amount of transformation to an object. For example, the position of any object can be seen by selecting the object with the Move tool active and viewing the readout in the Transform Type-In area. Changing the values in the Transform Type-In fields will send the object to that location.

Offset mode allows you to change a transform value by a certain amount, relative to the current transform value. For example, if you want to move an object 30 units away along the X axis, you don't need to know the current X position value. You simply type **30** in the X field, and 3ds Max adds 30 units to the current X position.

Offset mode always displays neutral values; zero for position, zero for rotation, and **100%** for scale. As soon as you type a number and hit **<ENTER>**, the object is transformed and the offset values return to neutral. Once you've transformed the object, the amount of offset is zero.

COORDINATE SYSTEMS

As we saw earlier in this chapter, a *coordinate system* is a way of tracking objects in space. It is absolutely essential that you master the techniques of working with coordinate systems in order to function in a 3D graphics environment.

You can see an example of a coordinate system by looking at the axis tripod at the lower left of each viewport. These axes are locked to the **World** coordinate system, in which the XY plane is always the "ground" and Z always points "up." This is so you can orient yourself in space no matter where your objects are located or how they're rotated.

However, for working with the objects themselves, you can choose one of many different coordinate systems. In fact, the default coordinate system for moving objects is not the World coordinate system. In this section of the book, you will learn more about the various coordinate systems.

REFERENCE COORDINATE SYSTEMS

The coordinate system to be used for transforms is chosen from the **Reference Coordinate System** pulldown menu on the Main Toolbar. By default, the **View** coordinate system is selected.

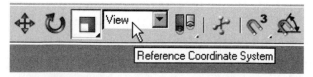

Figure 2-48: **Reference Coordinate System** *pulldown*

 **WARNING:** The **Render Type** pulldown menu also has a View item. Render Type is located at the far right of the Main Toolbar; Reference Coordinate System is found to the immediate right of the transform tools. The Main Toolbar pulldown menus are not labeled, but if you hover your cursor over a pulldown menu, you will see a ToolTip pop up.

When you choose a Reference Coordinate System, the Transform Gizmo or axis tripod usually changes orientation. In this way, you can turn the Transform Gizmo in the direction that is most convenient for your current task.

World Coordinate System

As previously discussed, the **World** coordinate system is the master reference for the location of all objects in the scene. World space is constant and cannot be changed. The axes are set in a certain orientation and stay that way, no matter which viewport is active. The default construction planes, known as the Home Grid, are identical to the XY, YZ, and ZX planes of the World coordinate system.

If the World coordinate system is active, the Transform Gizmo will always be aligned with the world. The Transform Gizmo's orientation will be identical to the axis tripod at the lower right of each viewport.

Users accustomed to technical work often prefer to use this coordinate system for most of their work. The fixed axes are helpful in work that requires accurate placement of objects. However, it imposes severe limitations on rotations, as described later in this section.

Screen Coordinate System

When the **Reference Coordinate System** is set to **Screen**, the X axis always points to the right of the screen, and the Y axis always points toward the top of the screen. With the Screen coordinate system, the XY plane is always parallel to the active viewport. The Z axis always points out toward the viewer.

The purpose of this coordinate system is to make it easy for you to move and rotate objects, primarily in the orthographic viewports. When **Select and Move** is selected and the Screen coordinate system is active, all movements in orthographic views take place parallel to the plane of the screen. The Z axis is disabled in this case.

WARNING: Working in the Screen coordinate system is intuitive for many users. However, the Screen coordinate system is almost never appropriate in a Perspective viewport. The Screen coordinate system is locked to the current viewport. Since the Perspective view can be positioned and rotated in any way, transformations in the screen space of the Perspective view are confusing and unpredictable.

For example, moving an object in the Y axis of the Perspective view may seem to be similar to moving in the World Z axis. In fact, if you move an object in the Y axis of the Perspective view, you end up moving the object in *all three* World axes at once!

View Coordinate System

The **View** coordinate system is a hybrid of **World** and **Screen** coordinate systems. In orthographic viewports, the **Screen** coordinate system is used. In Perspective, Camera, or User viewports, the **World** coordinate system is used.

This coordinate system is the default, because it gives the best of both worlds. With **View**, you have the intuitive use of the **Screen** coordinate system while working in orthographic views, while maintaining the predictability of the **World** coordinate system in the **Perspective** view.

Local Coordinate System

Every object in a 3D scene has its own individual coordinate system, called the **Local** coordinate system. The Local coordinate system enables transformation of an object relative to itself, rather than relative to the world. This is absolutely crucial for animation. For example, if there were no Local coordinate system, it would be impossible to create character animation with naturalistic joint rotations. We'd never be able to simulate animal motion, because all of the character's body parts would be restricted to rotating around the world axes.

To give another example, without Local movement, it would be difficult or impossible to make a spaceship fly in a straight line, unless the ship's path was lined up with the world axes. If you wanted the airplane to fly diagonally through the scene, you would have serious trouble. With the Local coordinate system, it is a simple matter to rotate the airplane out of alignment with the world axes, then move the ship along one of its local axes.

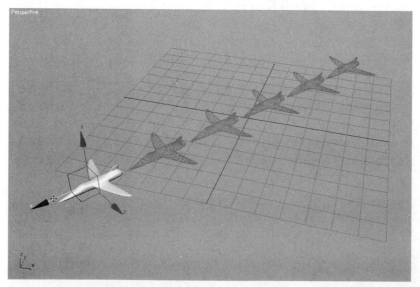

*Figure 2-49: Movement in the **Local** coordinate system*

To see a demonstration of how these Reference Coordinate Systems work in 3ds Max, view the file **ReferenceCoords.avi** on the disc that comes with this book.

PIVOT POINT

When transforming an object, a point of reference for the object is needed. If you want to rotate a Teapot, you need a point of rotation. Should you rotate around the handle or the spout? The reference point for a transform has a profound influence on the result.

An object is transformed relative to its **Pivot Point**. Each and every object has a single Pivot Point, which is the origin of the object's Local coordinate system. This is the true "center" of an object. All transforms of the object are made using this Pivot Point, or *local origin*, as the point of reference.

If the Transform Type-In area reports the location of a Box as **(0,30,0)**, this means the Pivot Point of the Box is located at these coordinates. The corners of the box are located at other coordinates.

By default, an object's axis tripod or Transform Gizmo marks the *location* of the Pivot Point. However, the object's axis tripod or Transform Gizmo does not necessarily point in the directions of the local axes. The orientation of the axis tripod is determined by the currently selected coordinate system. Only when the **Local** coordinate system is active can you see the true orientation of the object's local axes.

When an object is rotated, its Pivot Point rotates also. If the Local coordinate system is active, you can see this happening as you rotate the object. The Rotate Gizmo itself will rotate if the Local coordinate system is active. If the World coordinate system is active, the Transform Gizmo will remain locked to the Home Grid, and the Gizmo won't rotate.

At creation time, an object's Pivot Point is initially oriented according to the current construction grid. For example, the base of a Cylinder is always created on the current construction grid. When a Cylinder is created in an orthographic view, the Cylinder's local Z axis always points up out of the current viewport. So if you create a Cylinder in the Top view, the Cylinder's base is placed on the World XY plane, and the Z axis (height) of the Cylinder will be aligned with the World Z axis. However, if you create a Cylinder in the Front view, the base will be created on the World ZX plane, and the Cylinder's Z axis will be aligned with the World Y axis.

However, you have control over the rotation of the local axes, because the position and orientation of a Pivot Point can be changed. This is necessary for animation. For example, to make an object rotate around a certain feature of an object such as the Teapot handle, the Pivot Point must be moved to that location.

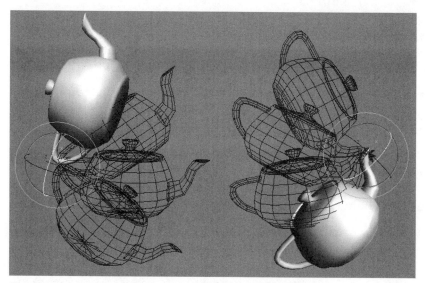

Figure 2-50: *Teapots rotating around different **Pivot Points***

To move or rotate the Pivot Point, open the **Hierarchy Panel**. In the **Pivot** subpanel, you will find controls for transforming the Pivot Point. If you choose **Affect Pivot Only**, you will see a new type of axis tripod. This is the Pivot Point icon. If you move or rotate this icon, the Pivot Point will be transformed.

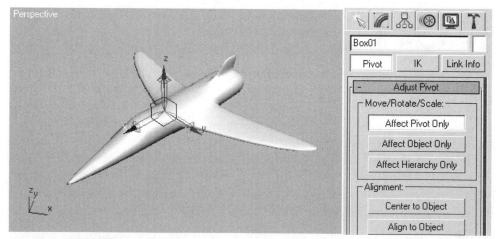

Figure 2-51: ***Affect Pivot Only***

The transformation of the Pivot Point is one of the few things in 3ds Max that cannot be animated. This means that you must set your Pivot Points before creating any animation. Also, there are simple methods for achieving the effect of an animated pivot, as you will see in Chapter 10, *Advanced Animation*.

OTHER COORDINATE SYSTEMS

While the **View, World,** and **Local** coordinate systems will work for most of your modeling needs, other coordinate systems are available. Occasionally, there will be times when other coordinate systems will come in handy. These will be discussed later in this book.

EXERCISE 2.4: Local Coordinate System

1. Open 3ds Max, or if it is already open, select **File > Reset**.

2. In the **Top** viewport, create a **Tube** at approximate XY coordinates **(66,-66)**. Give the Tube a **Radius 2** parameter of **13** units, and a **Height** of **40**.

3. In the Top viewport, create a **Teapot** near the center (*origin*) of the world, at XY coordinates **(0,0)**. Give the Teapot a Radius of about **45** units.

4. On the Main Toolbar, click on **Select and Rotate**. Position the cursor over the teapot in the Top viewport. Select the outermost circle of the Rotate Gizmo, and rotate the Teapot in the viewport's Z axis. Rotate the Teapot until its spout is pointing at the Tube object.

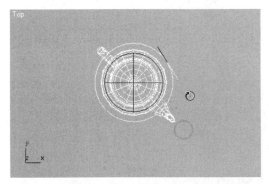

Figure 2-52: Rotate the Teapot to point the spout at the Tube

5. Click **Select and Move** in the Main Toolbar. Right-click an empty area of the **Front** viewport, then select the Y axis of the teapot's Move Gizmo. Move the Teapot about **50** units up, so the Teapot is hovering in the air just above the Tube.

6. Right-click in an empty area of the **Perspective** viewport to select it. Click **Zoom Extents**.

7. Select the **Arc Rotate** tool, so you can orbit around the scene. Click and drag in the center of the yellow circle, not outside it. Click-dragging outside the circle tilts the Perspective viewpoint rather than orbiting it.

8. Click **Select and Rotate** on the Main Toolbar. Select the Rotate Gizmo of the Teapot in the Perspective viewport. Attempt to rotate the Teapot as if you were to pour tea into the Tube. You are using the default **View** coordinate system, which, in the Perspective viewport, actually defaults to the World coordinate system. In the World coordinate system, it is impossible to rotate the Teapot to get the desired effect. The Teapot's spout always misses the Tube. You might be able to get it into a static position by making several rotations in various axes, but you can't simulate a pouring motion.

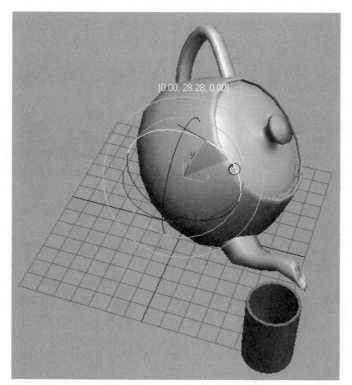

Figure 2-53: Teapot misses the Tube when rotating in world space

9. Hold down the **<CTRL>** key, and press **<Z>** to undo the rotations. You may need to perform multiple Undo operations to restore the teapot to the upright position seen in step 7.

10. With the teapot still selected, choose **Local** from the **Reference Coordinate System** drop-down list in the Main Toolbar. Observe how the Rotate Gizmo changes to indicate a different orientation of the Teapot's XYZ axes. Click and drag to rotate the teapot around its local Y axis. Pouring into the Tube is easily accomplished.

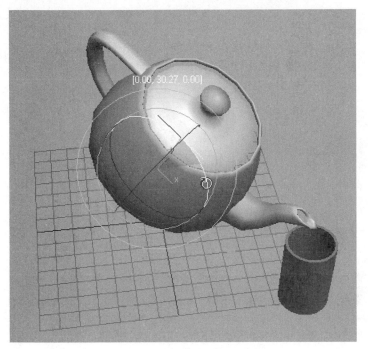

*Figure 2-54: Pouring tea using the **Local** coordinate system*

11. As you interactively rotate the Teapot, notice how unnatural the movement seems. This is because the **Pivot Point** is at the bottom of the Teapot. In the real world, a point of rotation might be near the object's center of gravity, or at a joint or connection. For a teapot, the handle is an appropriate point of rotation.

12. With the Teapot still selected and hovering in the pouring position, go to the **Hierarchy** panel. Press the **Affect Pivot Only** button; it turns blue to indicate that it is active. The Pivot Point tripod instantly appears, superimposed over the Move Gizmo.

13. Click **Select and Move**, and choose the **Local** coordinate system from the dropdown list in the Main Toolbar. In the Perspective viewport, select the ZX plane of the Move Gizmo. The Z and X axes of the gizmo turn yellow. Click on the ZX plane handle and drag the Move Gizmo and Pivot Point until they are located in the loop of the teapot's handle. Observe the movement of the gizmo and Pivot Point in the other viewports. Make sure the Pivot Point is inside the Teapot handle in all viewports. See the following illustration.

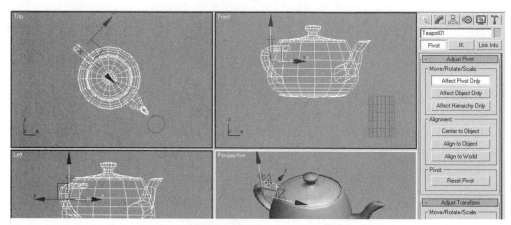

Figure 2-55: Moving the Teapot's **Pivot Point**

14. In the **Hierarchy** panel, click **Affect Pivot Only** again to turn off Pivot Point transforms. The Pivot Point tripod disappears, leaving only the Move Gizmo.

15. Click **Min/Max Toggle** to maximize the Perspective viewport. Click **Select and Rotate**, select the local Y axis of the Teapot once more, and rotate. With the Pivot Point in its new position, the Teapot now spins around its handle for a more convincing tea party.

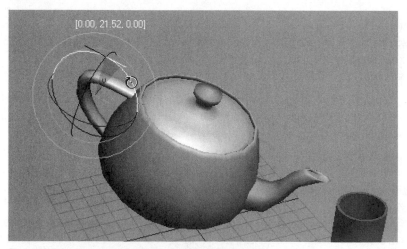

Figure 2-56: Rotating the Teapot around its handle

16. Save the file as **LocalCoords.max** in your folder.

SELECTING AND DISPLAYING OBJECTS

So far you have learned how to select one object at a time. When you need to select more than one object, different methods are required.

SELECT BY CLICK

To select one object, click the **Select Object** button on the Main Toolbar. Move the cursor over the object you wish to select. The cursor changes to a small white cross. Click the object you wish to select. The wireframe of the selected object turns white, and the object name appears on the Command Panel. In shaded views, white **Selection Brackets** appear around the object.

To select more than one object, hold down the **<CTRL>** key and click on additional objects. The objects turn white as they are selected, indicating that they have been added to the selection set.

To unselect a selected object, hold down the **<CTRL>** or **<ALT>** key and click the object. To unselect all objects, click an area of a viewport where no objects are present.

SELECT BY NAME

You can also select objects by clicking **Select by Name** on the **Main Toolbar**. A list of all objects in the scene appears in the **Select Objects** dialog.

Highlight an object name to select it. You can select additional objects by holding down the **<CTRL>** key while highlighting object names. Use the **<SHIFT>** key to select a sequence of objects from the list.

When you have finished selecting objects, click **Select** to select the objects and exit the dialog. You must exit this dialog before you can continue working with your scene.

This method of selection is very handy when a scene has many objects. Objects can be very difficult to select when they overlap in viewports.

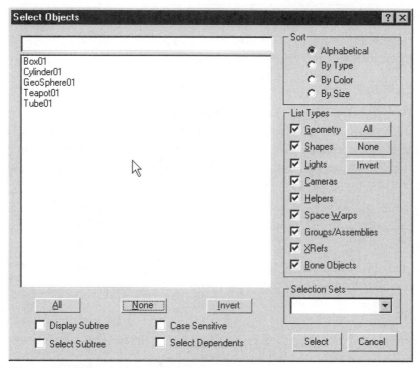

Figure 2-57: **Select Objects** *dialog*

In using the Select Objects dialog, the importance of naming objects intelligently becomes quite clear. If there are more than a few objects in the scene, it will become very difficult to select by name if all objects still have their default names.

Selection Floater

To select by name, you can also use a floater. Choose **Tools > Selection Floater** from the Main Menu.

The Selection Floater is identical to the **Select Objects** dialog. The only difference is that you don't have to exit the **Selection Floater** dialog before continuing to work with objects in the scene. This dialog can be left on the screen as long as you like to allow further selection of objects.

SELECT BY REGION

Another way to select objects is to draw around them. Move the cursor to an area of a viewport where no objects are present. Click and drag diagonally. A box appears, defined by dotted lines. This is called a *selection region* or *bounding area*. When the mouse is released, objects inside the area are selected.

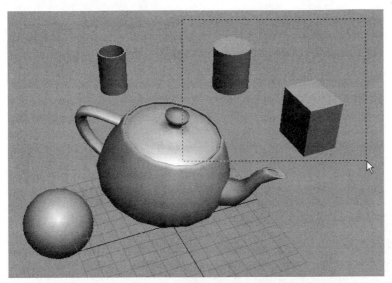

*Figure 2-58: Selecting with a **Rectangular Selection Region***

Window and Crossing Selection

How the bounding area works is determined by the status of the **Window/Crossing** button on the Main Toolbar. When **Crossing** selection is active, an object will be selected if any part of it is inside the bounding area. This is the default. When **Window** selection is active, an object must be completely enclosed within the bounding area in order to be selected.

Crossing   **Window**

*Figure 2-66: **Window/Crossing** selection icons*

Selection Region

In addition to a rectangular bounding area, there are four other types of regions you can use to select objects.

To change the shape of the bounding area, click and hold the **Rectangular Selection Region** button on the Main Toolbar at the upper left of the screen. A flyout with four additional icons appears.

 Circular Selection Region creates a circular bounding area. Click to define the center of the circle, and drag to set a radius.

Fence Selection Region allows you to draw an area with straight line segments. To draw a Fence, click and drag to start the fence, then move the cursor in another direction and click to set the next corner of the Fence. Move the cursor and click repeatedly to set the shape of the Fence bounding area. When the shape is complete, move the cursor near the start of the Fence shape until a crosshair cursor appears. Click to close the Fence.

 Lasso Selection Region is simpler and more intuitive than Fence. Just draw a curve of any shape, and release the mouse button. The Lasso is always closed, so there is no need to move the cursor close to the starting point to close the Selection Region.

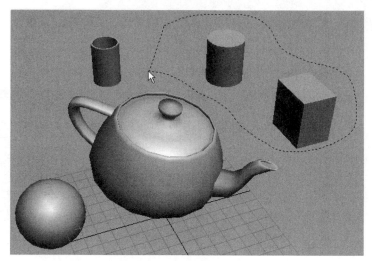

Figure 2-59: **Lasso Selection Region**

 Paint Selection Region is perhaps the simplest of all methods of selecting objects. Simply hold down the left mouse button and drag your cursor over objects to select them.

The <**CTRL**> and <**ALT**> keys can also be used in conjunction with bounding areas to add or subtract from a selection set. To add objects to the selection, hold down <**CTRL**> before beginning to draw the selection area, and keep it held down until the selection area is complete. To remove objects from the selection, use the <**ALT**> key in the same way.

WORKING WITH SELECTIONS

Once you have made a selection of objects, there are a number of ways to work with the selection as a whole.

Locking a Selection

After making a selection, left-clicking in a viewport will unselect all objects in the selection. To retain a selection while changing viewports, right-click in the new viewport.

When working with a complex scene, you might find that after you select an object, you have trouble retaining the selection as you work. Click the **Selection Lock Toggle** button ![lock icon] at the bottom center of the screen, or press **<SPACEBAR>** on the keyboard. The button turns yellow when the lock is on. As long as the selection is locked, you cannot deselect the currently selected objects nor select any other objects.

When you have finished with the locked selection and want to make another selection, be sure to unlock the selection by clicking Selection Lock Toggle or pressing **<SPACEBAR>** again.

WARNING: It is very easy to accidentally bump the spacebar and lock a selection unintentionally. If you can't select something, check the Selection Lock Toggle icon at the bottom of the screen.

Named Selection Sets

A selection can be stored so it can be accessed later on; 3ds Max calls this a **Named Selection Set.** This feature is very important when you have a number of objects that frequently need to be selected together.

To create a Named Selection Set, first select all objects you want to include. Then locate the **Named Selection Sets** pulldown list on the Main Toolbar.

Click in the entry field, type a name for the selection set, then press **<ENTER>**.

Later on, you can click the Named Selection Sets pulldown and choose the name from the list. The objects in the set will be selected automatically.

Figure 2-60: **Named Selection Sets** *pulldown list*

In addition, 3ds Max offers a **Named Selection Sets Floater** ![icon] that lets you manage selection sets. In this floater, you can create or delete selection sets. You can add or remove from selection sets, and even copy and paste objects among existing selection sets.

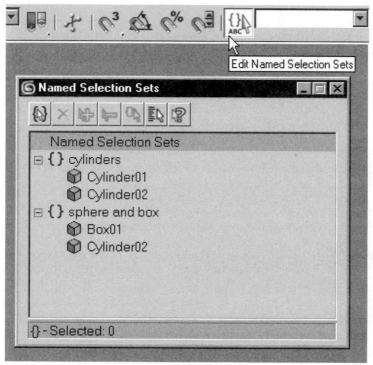

Figure 2-61: **Named Selection Sets** floater

EXERCISE 2.5: Selection Sets

1. Reset 3ds Max.

2. Create three Cylinders. Also create a few other objects, such as a Teapot, a Pyramid, and a Box.

3. With the Perspective viewport selected, hold down the **<ALT>** key and press **<W>** to maximize the viewport. Select the three Cylinders by holding down the **<CTRL>** key as you click each one.

4. In the **Named Selection Sets** entry field, enter the name **CylObjects** and press **<ENTER>**.

5. Select any other object on the screen.

6. From the Named Selection Sets pulldown, select the name **CylObjects**. The cylinders are now selected.

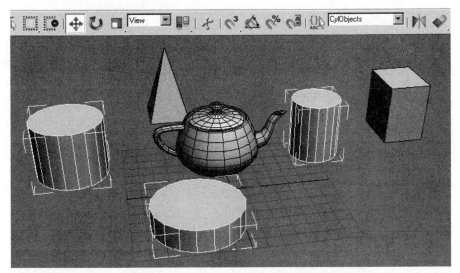

Figure 2-62: Select `CylObjects` *from* **Named Selection Sets** *pulldown*

7. Deselect the Cylinders by clicking an empty area in the viewport. Now no objects are selected. Open the **Named Selection Sets Floater** . Click the **Create New Set** button . Type `OtherObjects` as the name of the new selection set.

8. In the viewport, select all of the objects in the scene that are not Cylinders.

9. In the Named Selection Sets Floater, select the empty `OtherObjects` selection set. Click the **Add Selected Objects** button. All of the objects that are not Cylinders are added to the `OtherObjects` selection set. Click the plus sign next to the `OtherObjects` selection set to view the names of the objects in that set.

10. Leave the Named Selection Sets Floater open. Create a fourth Cylinder.

11. With the fourth Cylinder selected, go to the Named Selection Sets Floater. Use the Add Selected Objects button to add the Cylinder to the `OtherObjects` selection set.

12. In the Named Selection Sets Floater, select the name of the Cylinder which is in the `OtherObjects` set. Left-click the Cylinder's name, and drag to the `CylObjects` set. The new Cylinder is moved into the `CylObjects` set.

14. Click the **Select Objects In Set** button. All four cylinders are selected in the scene.

15. Save the file as `SelectionSets.max` in your folder.

CURSORS

In 3ds Max the cursor appearance changes frequently to match the current activity. It's essential that you take note of the current state of the cursor to ensure you're performing the operation you want to do.

During certain operations, 3ds Max turns off the selection tools while it performs other tasks. While creating primitives, for example, 3ds Max changes the cursor to crosshairs. If you switch to the Modify panel when you're done creating primitives, 3ds Max reactivates the last selection tool used. If you think about it, this makes a lot of sense. When you're creating objects, you don't want to be selecting objects. When you're done creating objects and are ready to modify them, you most likely want to select one or more objects to apply a modifier.

While working with 3ds Max, you can tell whether a selection tool is active by checking the appearance of the cursor. When an object type is selected from the Command Panel, the cursor appears as a set of crosshairs. When a selection tool such as **Select Object** or **Select and Move** is chosen, the cursor changes as shown below:

 An arrow when not over a selectable object

 A crosshair when over a selectable object

 A transform cursor when a transform tool is clicked and the cursor is over a selected object

A number of other cursors can appear depending on the current operation. The creation and selection cursors are the ones most frequently seen.

EXERCISE 2.6: Clown Head

In this exercise, you'll use what you've learned about creating primitives and moving them into place to create your own scene.

1. Reset 3ds Max.

2. Create a clown head with primitives. Make the head from a sphere, and the hat from a cone. For the rest of the head, be as creative as you like with the primitives. Use modifiers as often as necessary.

3. Save the scene with the filename **Clown01.max** in your folder.

Figure 2-64: Sample clown heads

Hints

- Use **AutoGrid** to create objects on the head, then move and rotate them into position.

- When using the Bend modifier, try each of the axes under the **Bend Axis** section. Also try the **Direction** parameter to achieve the bend you want.

HIDING AND FREEZING OBJECTS

There may be times when you want to hide objects to make work easier in a complex scene. To hide an object, select it and go to the **Display** panel . Under the **Hide** rollout, click **Hide Selected**. The object disappears from the screen. The object has not been deleted. It is still part of the file, but cannot be seen in viewports and will not render. To unhide the object, click **Unhide All** or **Unhide by Name**.

You can also keep an object on the screen but prevent it from being selected or altered. This is called *freezing* an object. Select the object and click **Freeze Selected**. The object turns a dark gray color and can no longer be selected. To unfreeze the object, click **Unfreeze All** or **Unfreeze by Name**.

Most of the Hide and Freeze commands can also be found in the Quad Menu. Select an object, then right-click anywhere in the viewport to access the Quad Menu. This is generally faster than going to the Display panel.

Display Floater

From the **Tools** menu, you can open the **Display Floater,** which can be used to hide or freeze objects. The floater can then remain on the screen while you continue your work. The Display Floater is useful for times when you need to hide and freeze objects frequently. Its controls work the same as those on the Display panel.

In addition, objects may be hidden, unhidden, frozen, or unfrozen from the Quad Menu.

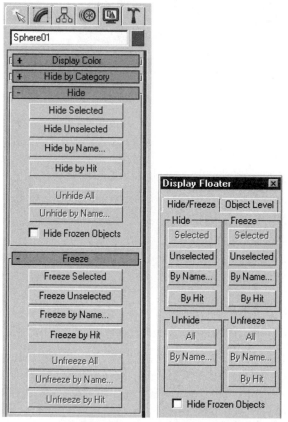

Figure 2-65: *Display panel and **Display Floater***

MATERIALS

3ds Max has a complex and powerful system for creating materials to put on your objects. A material can be thought of as paint or wallpaper, where colors and patterns can be mixed and matched. However, materials are also used to emulate bumpy patterns and to designate the shininess and transparency of an object. In short, the surface appearance of an object depends entirely on its material.

Materials are created and edited in the **Material Editor**. To access it, click the Material Editor button on the Main Toolbar, or choose **Rendering > Material Editor** from the Main Menu, or use the **<M>** keyboard shortcut.

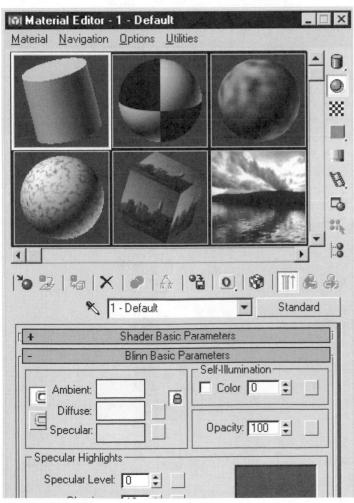

Figure 2-66: **Material Editor**

The Material Editor consists of three areas. At the top are sample slots that display materials as they would appear on a sample object, such as a sphere. The appearance of a material on a sample object gives you an idea of how it will look in the rendered scene. Only six sample slots appear by default, but you can access up to 24 sample slots by scrolling the sample slot display.

Below the spheres and along the right edge is the Material Editor toolbar. The toolbar contains options for assigning materials and customizing the sample display. These options can also be found on the menu at the top of the Material Editor panel.

Below the toolbar are rollouts with many options for setting colors and other parameters that make up a material. When you click a sample sphere, the parameters for the corresponding material are displayed on the rollouts.

The Material Editor also has its own menu. There are only a few commands in the Material Editor menu that are not found on the Material Editor toolbar.

PREMADE MATERIALS

Although you can create your own materials, 3ds Max comes with many premade materials. These materials are stored in a *material library*. A material library is simply a collection of materials saved in a file with the extension `.MAT`. When you load the Material Editor, the built-in library `3dsmax.mat` is loaded by default.

To get a material from the default material library, click the **Get Material** button in the Material Editor toolbar. The **Material/Map Browser** appears.

At the upper left of the Material/Map Browser is a section labeled **Browse From**. Choose **Mtl Library**. A list of materials appears.

The materials are listed by name. To see a sample of each material, click **View List + Icons** at the top of the Material/Map Browser. You can also click **View Small Icons** or **View**

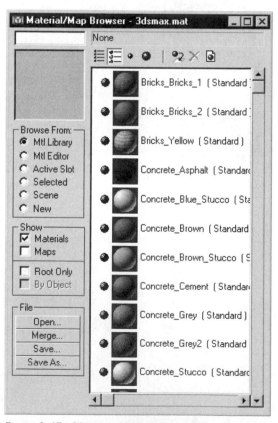

Figure 2-67: *Material/Map Browser*

Large Icons to see larger versions of the material samples. These methods show larger samples that make it easier for you to see the material, but can take longer to display. If you need to see more or fewer materials in the window, you can resize the Material/Map Browser by dragging its edges.

To choose a material, scroll down the display and locate a material with a blue sphere next to it. Click once to see the material display at the upper-left corner of the Material/Map Browser. Double-click the material name or sample picture to make the material appear in the currently selected sample slot on the Material Editor.

EXERCISE 2.7: Getting Materials

In this exercise, you'll get three materials from the default material library and put them in the Material Editor.

1. Load the file **LotsOfObjects.max** from your folder, or from the disc. Alternately, you can reset 3ds Max and create at least five primitives in the Top viewport.

2. On the Main Toolbar, click the **Material Editor** button . The Material Editor is opened, and the sample slot at the upper left is selected by default.

3. On the Material Editor toolbar, click the **Get Material** button . The **Material/ Map Browser** appears.

4. Under the **Browse From** section at the upper left of the Material/Map Browser, choose **Mtl Library**.

5. Click the **View List + Icons** button at the top of the Material/Map Browser to see the materials listed with small pictures.

6. Scroll down the list of materials. Find a material that looks interesting to you. Make sure it's a material with a blue sphere next to it. Click on the material once to see a larger representation of it at the upper left of the Material/Map Browser. If you like the material, double-click it to put it in the currently selected sample slot in the Material Editor.

7. Once you have selected one material, click a different sample slot in the Material Editor. Locate another material you like on the Material/Map Browser, and double-click it to put it in the sample slot.

8. Click another sample slot, and choose one more material for it.

9. Close the Material/Map Browser by clicking on the X at the upper-right corner of the dialog.

10. Save the file as **LotsOfObjects02.max**.

ASSIGNING A MATERIAL TO AN OBJECT

So far you have only loaded materials into the Material Editor to preview them. However, just because a material is loaded in the Material Editor does not mean that it is present in the scene. Materials must be assigned to objects in the scene in order for the objects to render with those materials.

To assign a material to an object, you can use one of three methods:

- Click and drag the material from the sample slot to the object.

- Select the object and click the **Assign Material to Selection** button on the toolbar.

- Select the object and choose **Material > Assign to Selection** from the Material Editor menu.

When a material is assigned to an object in the scene, small triangles appear in the corners of the sample slot. This indicates that the material is "hot," which means it is present in the scene. The triangles make it easy for you to tell at a glance which materials have been assigned to objects, and which have not.

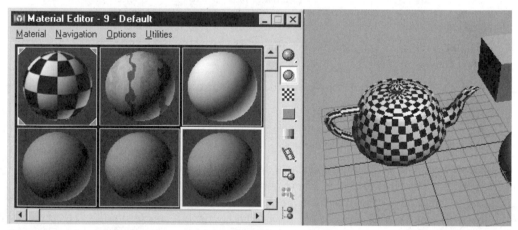

Figure 2-68: Triangles at corners of sample slot indicate a "hot" material

When you assign a material to an object, the object's appearance in shaded viewports will change. The material overrides the Object Color in shaded views. Wireframe views do not show materials in any way.

Some materials use maps, which are images applied to the surfaces of objects. By default, maps are not displayed in viewports. To see a map on an object in a viewport, select the appropriate sample slot, and click the **Show Map in Viewport** button on the Material Editor toolbar.

WARNING: The way an object looks in a shaded viewport does not indicate how it will look in the rendered scene. Real-time viewports can only provide an approximation of the quality level 3ds Max can achieve when the scene is rendered. The only way to see how an object will look when rendered is to render it. You will learn how to do this later in this chapter.

EXERCISE 2.8: Assigning Materials

In this tutorial, you'll assign materials to objects in a scene.

1. Load the file **LotsOfObjects02.max** from your folder or from the disc.

2. If the Material Editor is not already open, click the Material Editor button on the Main Toolbar. The three materials you got from the Material/Map Browser are displayed in sample slots in the Material Editor.

3. Click one of the sample slots, and drag the material to an object in the scene. Release the mouse when the cursor is over an object. The material is assigned to the object.

4. The complete material might not appear in the **Perspective** view, but if triangles appear at the corners of the sample slot, then you know you have successfully assigned the material. If the material doesn't appear in the Perspective view, don't be concerned. The material will appear when the scene is rendered. To see maps in viewports, enable

    Show Map in Viewport 🔳 for each material.

    To render the scene, activate the Perspective view and click the **Quick Render** button on the Main Toolbar.

    Note that 3ds Max renders the scene in the Perspective view, displaying all materials that have been applied. You will learn more about rendering in the next section of this book.

5. On the Main Toolbar, click the Select Object button. Click on a different object to select it.

6. Select another sample slot with a different material in it.

7. On the Material Editor toolbar, click **Assign Material to Selection** 🔳. The material is assigned to the selected object.

8. Assign another material to another object in the scene either by dragging the material from the sample slot to the object, or by selecting the object and clicking Assign Material to Selection.

9. Save the scene in your folder as **LotsOfObjects03.max**.

EXERCISE 2.9: Materials for Clown Head

In this exercise, you'll use what you've learned about materials to assign materials to the clown head.

1. Load the file **Clown01.max** from your folder or from the disc.

2. Get premade materials from the material library and put them in the Material Editor. Then assign them to objects in the scene.

3. In the Perspective view, use the Zoom buttons and Arc Rotate to get a good view of the clown. The newly assigned materials might not look the way you expect in the Perspective view. You can only see the finished result if you render the scene.

4. Activate the Perspective view and click the **Quick Render** button on the Main Toolbar.

5. Continue to assign materials and render until you're satisfied with the results.

6. Save the scene with the filename **Clown02.max** in your folder.

Hints

● In the Material/Map Browser, be sure to choose a material and not a *map*. A map is a pattern or image that can be included in a material. Materials are preceded by a blue sphere ● while maps are preceded with a green diamond ◆ . Materials can be assigned directly to objects, but maps cannot.

● Shiny materials like metals and chromes work best on rounded surfaces.

RENDERING

When you're ready to take a look at the final scene, you can render it. When you render, 3ds Max takes the view of the scene in one viewport and calculates what the scene should look like based on the objects, materials, and lighting from that view.

In general, you will want to render the view in the **Perspective** viewport or the view from a camera. You will learn how to set up a camera view in Chapter 5, *Cameras and Lights*.

To make the final image look exactly the way you want it to, you will most likely need to render it several times. Each time the image is rendered, you will need to look closely at it to see how it can be improved, then make the changes and render again. It's not unusual to render an image 20 or 30 times to get it to look exactly right.

RENDER SCENE DIALOG

To render the scene, activate the viewport you wish to render. Choose **Rendering >**

Render from the Main Menu, or click the **Render Scene** button 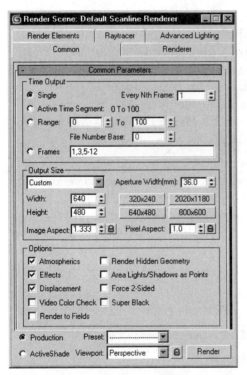 on the Main Toolbar. The Render Scene dialog appears.

There are many options in this dialog. Here we are concerned with three of them. They are all found on the **Common** tab of the dialog.

*Figure 2-69: **Render Scene** dialog*

Time. Do you want to render a single frame or an animated sequence? To render a single frame, make sure the **Single** option is selected under the **Time Output** section. To render an animated sequence, choose one of the three remaining options.

Resolution. There are several resolutions displayed under the **Output Size** section. For test renders, low resolutions such as 320x240 work well. You can click on any of the resolutions displayed or type in your own. The resolution you choose depends on how the image will be used. For example, if you plan to put your image on DV tape, you would choose the DV preset of 720x480 pixels. Or, if you want to print the image to 10x8 inches at 300 ppi, you would render to 3000x2400, a much higher resolution. Of course, the higher the resolution, the longer it will take to render the image.

Output. To cause the rendered image to be saved to a file, you must click the **Files** button under the **Render Output** section. You will be prompted for a filename and file type. When performing test renderings of still images, it is not necessary to save each test. For animation, a filename must be entered for you to be able to view the animation when rendering is complete. If you fail to provide a filename and file type, 3ds Max will render to the screen without saving anything!

To render the scene, make your selections, then click the **Render** button at the bottom of the dialog.

WARNING: It's important to understand the distinction between saving a rendered *image*, and saving a *scene*. A rendered image or animation is saved in a bitmap file format such as .BMP, .TGA, or .TIF. Bitmap files contain 2D picture information only. A 3ds Max scene file, on the other hand, has information about 3D objects. The extension .MAX is used for 3ds Max scene files. Scene files with the .MAX extension can only be saved and opened with 3ds Max, but 3ds Max is capable of importing and exporting other types of scene files, such as .3DS.

The .MAX file format is *forward compatible* with other versions of 3ds Max, but not *backward compatible*. This means that you can open a 3ds Max 7 scene in version 8, but you cannot open a 3ds Max 8 scene in version 7 or earlier. If you need to load your scene in an earlier version of 3ds Max, you should **Export** it to a .3DS file. Unfortunately, the .3DS format has limitations, and you will lose many features in the export process, such as modifiers.

EXERCISE 2.10: Render the Clown Head

In this exercise, you'll render the clown head with a custom resolution.

1.   Load the file **Clown02.max** if it's not already loaded.

2.   Adjust the Perspective view so you have a good view of the clown head.

3.   Click the **Render Scene** button ⬚ on the Main Toolbar, or choose **Rendering > Render** from the menu. The Render Scene dialog appears.

4.   Under the **Output Size** section, click the **320x240** button.

5.   Click the **Render** button on the Render Scene dialog to render the scene at this new resolution.

6.   Close the Render Scene dialog. Save the scene with the filename **Clown03.max** in your folder.

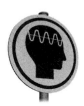

TIP: You can increment the number at the end of a filename quickly and easily by choosing **File > Save As** from the menu, then clicking the **+** button in the **Save File As** dialog.

ANIMATION

In 3ds Max, animation is accomplished with the use of *keyframes*. You give 3ds Max the object information at key (important) frames, and it figures out the in-betweens for you. This is called *keyframe interpolation*.

Keyframes are set with the help of the **Auto Key** button at the lower right of the screen. When clicked, this button turns red to show that it is in animation mode. When this button is red, keyframes are automatically set as you make changes to objects.

Figure 2-70: **Auto Key** *button*

3ds Max also allows you to set keyframes manually, using the **Set Key** feature. This topic is covered in Chapter 10, *Advanced Animation*.

As you create keyframes, they appear as small rectangles in the **Track Bar**. You can perform simple animation editing by selecting keyframes in the Track Bar and dragging them to a new time. Track Bar shows only the keyframes for the currently selected objects. Selected keyframes are highlighted in white.

AUTOMATICALLY SETTING KEYFRAMES

To automatically set keyframes, do the following:

- Click the **Auto Key** button to turn it on. The button turns red. The border around the current viewport and the Track Bar also turn red.
- Move the **Time Slider** so it is at the desired time for the new keyframe.
- Move, rotate, or scale an object to set a key for the object.
- Move the **Time Slider** again and move, rotate, or scale objects again. Repeat as many times as necessary to complete the animated sequence.
- When finished setting keyframes, click the Auto Key button to turn it off.

Once an animated sequence has been created, you can view it right in the viewport by clicking the **Play Animation** button in the **Time Controls**.

The Auto Key button must be on (highlighted in red) for the keyframe to be set. It doesn't matter whether you turn on the Auto Key button first or move to the frame first. A keyframe is set only when a change is made to an object.

After you have created keyframes, you can move them around in time by using the Track Bar.

MODIFIER ANIMATION

You can also animate any parameters associated with the object, including the following:

- **Radius** of a Sphere
- **Height** of a Cylinder
- The **Angle** parameter of a **Bend** modifier
- The **Amount** and **Curve** parameters on a **Taper** modifier

To animate a modifier, move to a frame other than zero and turn on the **Auto Key** button. Change one of the modifier's parameters. The parameter will now animate from its original value to the new value over a series of frames. Animated parameters are indicated by a red border around the parameter spinner.

SETTING THE CURRENT TIME SEGMENT

By default, 3ds Max gives you 100 frames over which to animate your scene. You can change the number of frames by clicking the **Time Configuration** button in the Time Controls area. The **Time Configuration** dialog appears.

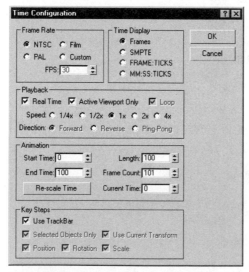

Figure 2-71: ***Time Configuration*** *dialog*

Enter a value for the **Length** parameter to set the new length of the animation.

EXERCISE 2.11: Animated Clown Head

In this exercise, you'll animate the clown head.

1. Load the file **Clown03.max** from your folder or from the disc.

2. Turn on the **Auto Key** button.

3. Go to a frame other than zero. Make a change to the clown head such as moving some part of it or changing a parameter on the **Modify** panel.

4. Continue to animate other parts of the clown head as you like. Move, rotate, or scale the eyes, nose, mouth, or hat.

5. To see your animation, click the **Play Animation** button in the Time Controls.

6. Turn off the Auto Key button when you are satisfied with your animation.

7. Save the scene with the filename **Clown04.max** in your folder.

8. Next, you'll render the animation to an **.AVI** file so you can play it onscreen. Click the **Render Scene** button ⬚ on the Main Toolbar. The Render Scene dialog appears.

9. In the Render Scene dialog, under the **Time Output** section, choose **Active Time Segment**. This will cause the entire animation to be rendered.

10. In the **Output Size** section of the Render Scene dialog, choose **320x240**.

11. In the Render Scene dialog, click the **Files** button. In the **Render Output File** dialog, click the **Save As Type** pulldown menu and choose **AVI File (*.avi)**. Enter the filename **ClownAnimation** in the **File Name** field. Be sure to note which folder the animation file is being saved in.

12. Click the **Save** button. If you are prompted to choose a video compression method, select **Cinepak** or whichever compressor you desire.

13. Click **Render** in the Render Scene dialog to begin rendering.

    Wait a few moments while the animation renders. You can watch the progress on the **Rendering** dialog that appears. In the **Rendering** dialog, the **Time Remaining** will tell you roughly how long the rendering will take.

14. When the file has finished rendering, you can play it. To do this, go to the the Main Menu and choose **File > View Image File**. Browse within this dialog to find the **.AVI** file. When Windows Media Player appears, click the **Play** button to view the animation. Close Windows Media Player when you have finished viewing the file.

15. Save the file again as **Clown04.max**.

Hints

- Don't forget to turn on the **Auto Key** button before animating, and turn it off when you're finished animating.

- When keyframes are less than three frames apart, the resulting animation can be rough and jerky. To avoid this, set keyframes at specific intervals, such as every 10 or 15 frames. This will also make it easy for you to remember where the keyframes are.

- Always stop playing the animation before making further changes to the scene.

EXERCISE 2.12: Advanced Exercise

Animate the hat or mouth bending back and forth.

END-OF-CHAPTER EXERCISE

EXERCISE 2.13: Table and Chairs

In this exercise, you'll create an interior scene.

1. Reset 3ds Max.

2. In the center of the **Top** viewport, create a **Cylinder** with a **Radius** of about **120** units and a **Height** of **-15**.

   Creating a Cylinder with a negative height places the top surface of the Cylinder at the construction plane. This means that when you create other objects in the Top viewport, they will automatically sit right on the Cylinder's top surface.

3. In the center of the Top viewport, create a **Teapot** with a Radius of about **35** units.

   Note that the Teapot is placed on the construction plane, which coincides with the top of the Cylinder.

4. On the **Main Toolbar**, click the **Select Object** button. Click the Cylinder to select it.

5. Go to the **Modify** panel. Highlight the name `Cylinder01`. Type the name `Table` to rename the object. On the keyboard, press **<ENTER>**.

6. Click the Teapot to select it. On the Command Panel, enter the name `YellowTeapot` to replace `Teapot01`.

7. Add to the scene, using more primitives to create a scene similar to the one shown in the following image.

Figure 2-72: Table and chairs

8. Assign materials to the objects in the scene. Remember to enable the **Show Map in Viewport**  option for each material that has a map..

9. Render the Perspective view and save the rendered image to a `.TGA` file with a resolution of 640 x 480. Give it the name `TableScene01.tga`.

10. Save the scene with the filename `TableScene01.max` in your folder.

Hints

• Use large, flat boxes for the walls and floor.

• You need only two walls and a floor. The other walls are behind your point of view in the Perspective view, so they do not need to be placed in the scene.

• The drinking glasses can be made of Tube primitives with Taper modifiers applied to them. You can't see the bottoms of the glasses, so you can't tell they don't have bottoms.

• Zoom in and out of viewports as needed.

• To do a Zoom Extents on all viewports except the Perspective view, hold down the **<CTRL>** key and click Zoom Extents All.

• Do as many test renders as you like. Save the rendered image to a file only when you feel you have completed it.

SUMMARY

In this chapter, you learned that the 3ds Max interface consists of **menus**, **toolbars**, the **Status Bar**, **Command Panel**, and **Viewports**. Object creation is done in viewports. Command Panels have *rollouts* with *parameters* and *spinners* for controlling the various aspects of objects and modifers.

The simplest types of objects, such as spheres and boxes, are called **Primitives**. **Modifiers** can be applied to objects, primarily to change their shape. Because primitives and modifers have parameters, modeling with these tools is called *parametric modeling*.

Multiple modifiers can be applied to a single object. They are placed in a Modifier Stack. The Modifier Stack flows from bottom to top, and this *order of operations* determines the end result of the Stack. You can go to previously applied modifiers and change their parameters, change the order of modifers, and cut and paste modifers among objects. As a result, parametric modeling with modifers is a very flexible method.

Objects can be manipulated in viewports using the transformation tools on the Main Toolbar: **Move**, **Rotate**, and **Scale**. Each transform tool has its own **Transform Gizmo**, which is a viewport icon that aids in transforming objects in virtual 3D space using a 2D pointing device, such as a mouse.

The result of a transform operation depends on the current **Reference Coordinate System**. Each object in a scene has a **Local** coordinate system that is unique to that object. An object is transformed relative to the origin of its Local coordinate system, known as the **Pivot Point**. Pivot Points can be moved and rotated to allow rotation and scaling around specific locations on a model.

The ability to select one or more objects is an important skill. There are many ways to select an object in 3ds Max, including clicking on the object or drawing a bounding area around it. Selections can be saved in **Named Selection Sets**, and these sets can be managed using the Named Selection Sets Floater.

Materials are selected and applied to objects via the **Material Editor**. Materials can be stored in files called *material libraries*.

The *rendering* process interprets the scene and creates a picture from it. 3ds Max can save rendered images in a variety of formats. Rendered images are 2D files, whereas 3ds Max scene files contain 3D data.

To animate objects, animators create *keyframes*, which are the most important or extreme frames in the animation. Computer graphics software calculates the in-between frames in a process called *interpolation*. When the 3ds Max **Auto Key** button is turned on, any changes made to objects become animation keyframes. You can use the **Time Slider** to go to various frames in the animation. Simple animation editing can be accomplished by moving keyframe icons within the **Track Bar**.

REVIEW QUESTIONS

1. What are *primitives?*

2. What are the three *transforms* in 3D computer graphics?

3. What is *parametric modeling?*

4. What is the main advantage of parametric modeling?

5. Where can *modifiers* be found?

6. Suppose you create a Cylinder and add a **Bend** modifier to it. Then later you want to go back to the original Cylinder and change the **Radius** parameter. How would you do this?

7. What is a *Pivot Point?*

8. Name three types of *coordinate systems.*

9. What is the basic purpose of coordinate systems?

10. Name three ways to select objects.

11. What is a *material?*

12. How is a material applied to an object?

13. What are the three main things to concern yourself with when rendering?

14. When rendering an animation, what happens if you don't assign a file output?

15. What are *keyframes?*

16. How are keyframes automatically created in 3ds Max?

Chapter 3
Modeling

OBJECTIVES

In this chapter, you will learn about:

- Types of models
- Modeling with modifiers
- Cloning objects and the 3ds Max data pipeline
- Modeling with 2D shapes
- Lofting
- Polygonal mesh objects
- Advanced transforms

ABOUT THE EXERCISES IN THIS CHAPTER

Previously, you learned how to zoom and pan viewports with the viewport controls at the lower right of the 3ds Max interface. For the tutorials in this chapter, it is expected that you zoom and pan whenever necessary, without specific instructions to do so.

In each tutorial, you will not be specifically instructed to change object names to appropriate names, but you are expected to do so on your own. Naming objects is an important practice that will help you work more efficiently in 3ds Max.

ABOUT MODELING

3D modeling can be done in a variety of ways. Becoming adept at modeling is a matter of learning all the tools available so you can choose the right tool for the job. Modeling an object often requires many different tools, used in the correct order for that particular object.

For any given model, there are two or three different approaches that will yield good results. When working on the job, 3D artists tend to use the tools they know well. However, all successful artists take time to explore functions that are less familiar to them so they'll be ready to tackle more complex models later on.

When learning the tools in this chapter, practice each one to gain familiarity with it. As your modeling skills improve, you can return to the more complex options and work on getting more adept at using them.

TYPES OF MODELS

Although there are many techniques for building geometric models, any model you build in 3ds Max will fall into one of three main categories: **Mesh**, **Bezier Patch**, or **NURBS**. These three categories define the most fundamental properties of models. Each type of model has its own strengths and weaknesses.

MESH MODELING

So far, all of the objects you have worked with have been *polygonal mesh* models. Sometimes this type of model is referred to as a polygon or poly model, or as a wire mesh, or simply a mesh.

The most distinguishing characteristic of a mesh model is that it has no true curvature. It is composed entirely of perfectly flat surfaces called *polygons*. These polygons are built from triangles, called *faces*. Faces, in turn, are built from straight lines, called *edges*. At either end of each edge is a point, known in 3D graphics terminology as a *vertex*. The plural of vertex is *vertices*.

The approximation of curvature is achieved by increasing the number of polygons. In the following illustration, the two spheres are identical, except that the one on the left has a higher value in the **Segments** parameter, resulting in more polygons, edges, and vertices. Therefore, it gives a more convincing illusion of curvature. The sphere with more Segments is said to have a higher mesh *density*, or a higher mesh resolution.

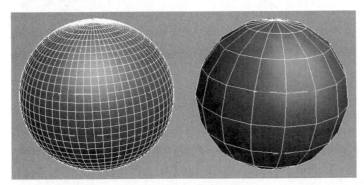

Figure 3-1: High- and low-density mesh objects

Increasing the density of a mesh object gives it a better approximation of curvature, but this comes at the price of increased render times. For this reason, it is important to know the intended use of the model before building it. If you are creating a model for a real-time game, then you will be forced to model at a lower *level of detail* than if you were creating a model for a feature film. Or, if a model is very significant in the scene, very close to the camera, or otherwise an object of close scrutiny, then it will need to be modeled to a high level of detail. Smaller, more distant, or less important objects should be modeled with less detail, to decrease render times.

As a general rule of thumb for non-real-time, pre-rendered animations, a single mesh polygon should never be smaller than a single pixel of the rendered image. Otherwise, the high detail of the mesh cannot be seen in the final image, and render times are increased unnecessarily. Models built for real-time applications, such as games, will have different considerations. In general, the density of real-time meshes will be dictated by the intended platform, such as a particular brand of game console.

In any 3D graphics application, fast render times are of paramount importance. One of the most efficient methods for speeding up render times is to disregard any polygons that are not pointing toward the camera. If the camera can't see it, it should be ignored during the rendering process.

By default, the polygons of mesh objects are only renderable on one side. The renderable side of a polygon is indicated by a line drawn at right angles to its surface. This line is called a *normal*. For any closed surface such as a cube or sphere, half of the normals are pointing away from the camera, and are therefore ignored at render time. This effectively cuts the number of polygons that must be calculated in half.

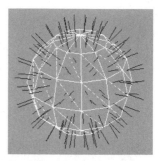

Figure 3-2: Face normals

Because mesh models are based on very simple building blocks, they tend to update and render more quickly than other types of models. For this reason, all 3D games use mesh modeling, although this will change in the near future as game consoles become faster.

With mesh objects, each vertex, edge, and polygon is explicitly defined, which is both a blessing and a curse. Polygonal meshes are great for hard surfaces such as architecture. They are the only choice when it is necessary to exert precise, low-level control over individual vertices and faces. However, raw mesh objects are not always the best choice for organic surfaces, due to the high number of polygons required for the illusion of smooth curves. Finally, it is difficult to change the level of detail of a raw polygonal model without a great deal of effort.

BEZIER PATCH MODELING

Another type of model category is the **Bezier Patch**. It is named after Pierre Bezier, one of the mathematicians who developed a type of curve used in computer graphics. This type of curve is called a Bezier spline, and you will learn about it later in this chapter.

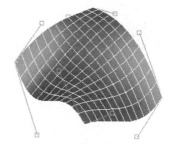

*Figure 3-3: A **Bezier Patch** with control vertices*

The advantage of modeling with Bezier Patches is that it is easy to achieve naturalistic curvature. The curves of Patch objects are controlled by a relatively small number of *control vertices*, or CVs. By manipulating a few CVs, an artist can quickly define the contours of a Patch model. This would be very time-consuming and difficult to achieve by transforming the vertices, edges, and faces of a mesh model.

In 3ds Max, all objects, even Bezier Patches, are ultimately converted to polygons when an image is rendered. So many of the same rules that apply to mesh objects also apply to Patches.

Level of detail is a major consideration in character modeling; the beauty of Patch modeling is that it is very easy to change the density of the model. The complexity of a Patch surface can be altered instantly using a parameter called **Steps**. This is simply not possible with a standard mesh model.

However, Bezier Patches have disadvantages. Because of their semiparametric nature, they take longer to calculate than mesh objects. Working with complex Patch models can slow down viewport interactivity dramatically compared to similarly complex mesh objects.

NURBS MODELING

NURBS is undoubtably the most challenging and complex form of modeling. NURBS is an acronym that stands for *Non-uniform Rational Basis Spline*. Objects made with this type of curve have more functionality than Bezier Patches. For example, using NURBS, you can trim and blend surfaces. In the following image, a chopped sphere is easily blended with a flat plane with a circular hole in it. The shape of the blended surface can be quickly altered with parameters found in the NURBS Modify panel.

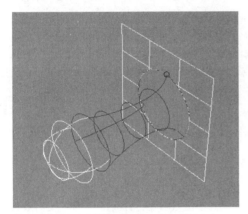

*Figure 3-4: A **NURBS** Blend Surface*

NURBS is a very valuable tool under some circumstances. Industrial and automotive designers rely heavily on NURBS modeling to produce complex surfaces with smooth curves. However, it is a difficult tool to master, and is only suitable for rounded surfaces. The NURBS in 3ds Max are a welcome addition to the modeling toolset, but in most cases you can achieve good results using other techniques.

SUBDIVISION SURFACE MODELING

Today, the most popular form of modeling for organic objects is *subdivision surfaces*. This is a technique for producing high-density polygonal meshes from low-resolution base meshes. The polygons of the original base mesh are divided into smaller polygons, and the angles between these polygons are smoothed to produce the effect of curvature.

Subdivision surfaces are a form of mesh modeling and are often described as mixture of polygons and NURBS. This isn't really accurate; from a technical standpoint, subdivision surfaces don't have anything do with NURBS. But in a way, subdivision surfaces combine the best properties of polygons with the best properties of semiparametric models such as NURBS.

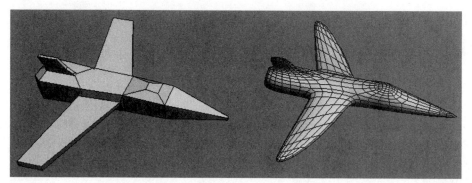

*Figure 3-5: A Subdivision Surface model before and after **TurboSmooth***

Creating organic detail with subdivision surfaces is very easy and flexible, as opposed to the performance problems inherent in Bezier Patch and NURBS modeling. For this reason, 3ds Max's subdivision surfaces, implemented through the MeshSmooth and TurboSmooth modifiers, are used extensively for characters and other curved surfaces.

PARAMETRIC PRIMITIVES

Earlier, you learned the basics of making primitives. Here you will learn a few additional methods for modeling with primitives.

STANDARD PRIMITIVES

You've already worked with several of the Standard Primitives, which are the default objects in the **Create > Geometry** panel. Each primitive has its own parameters that can be adjusted to change the shape of the object. Two more Standard Primitives merit special attention: **Geosphere** and **Plane**.

Geosphere

A Geosphere is a CG representation of a geodesic dome. Geodesic domes are very inexpensive and stable structures, invented by the famous architect and designer R. Buckminster Fuller.

The 3ds Max Geosphere has certain advantages over a traditional Sphere object. Because it does not have a polar structure like the regular Sphere, the polygons of a Geosphere are the same size all over. As a result, the Geosphere has a more uniform mesh density. This means that a Geosphere at the same approximate level of detail will actually have fewer polygons. In the following image, the Sphere has 960 polygons, and the Geosphere has only 720, which is only 75% as many polys. The polygon savings can be very important in many cases, such as in real-time games, or when there are many spherical objects in a scene.

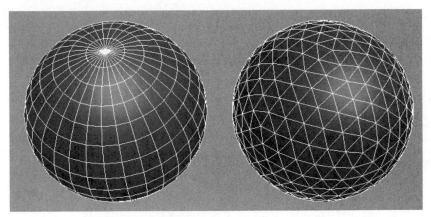

Figure 3-6: **Sphere** *and* **Geosphere**

Plane

Another important Standard Primitive is the **Plane**. It is a flat surface that is divided into **Length** and **Width Segments**. Its primary purpose is to act as a ground plane on which to place a scene. It is the only Standard Primitive which does not enclose a 3D volume. Only one side of a Plane is renderable.

As you will learn later, you can apply modifiers to a plane to produce terrain effects. Take special note of the **Render Multipliers** parameters of the Plane. This allows you to render the Plane with higher mesh density than what you see in the viewport. This comes in very handy when making terrains.

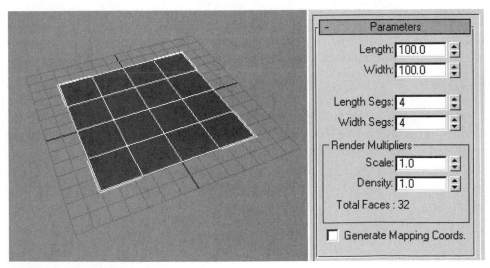

Figure 3-7: **Plane** *Primitive*

EXTENDED PRIMITIVES

By default, the **Create > Geometry** panel pulldown menu is set to **Standard Primitives**. Pull down the menu and choose **Extended Primitives**.

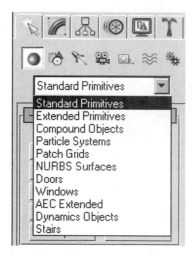

Figure 3-8: **Geometry** *pulldown menu*

The buttons on the Extended Primitives command panel can be used to create a number of different objects. Many of these primitives aren't needed very often, but you will find them useful from time to time for special cases. Experiment with these objects and their parameters to see what they can do.

ChamferBox and ChamferCyl

Some Extended Primitives are very useful in a variety of situations. Some in particular feature an important parameter: a rounded edge, known as a **Chamfer** or **Fillet**.

In the real world, there are no perfectly sharp edges on objects. All objects have thickness, and no two surfaces can meet at a perfect joint. So the Boxes and Cylinders you create in the Standard Primitives panel are not terribly realistic. If you are striving for realism, and you can afford to render the extra polygons, then you should create objects with rounded edges.

3ds Max provides the **ChamferBox** and **ChamferCyl** objects for just this purpose. A ChamferBox is a box with rounded edges, while a ChamferCyl is a cylinder with rounded edges at the caps.

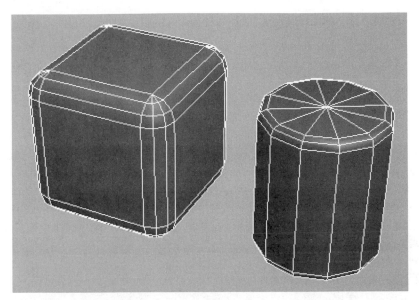

*Figure 3-9: **ChamferBox** and **ChamferCyl***

Fillets

Many Extended Primitives have an option for making a *fillet*. A fillet is a rounded edge on an object. In 3ds Max, the words *fillet* and a *chamfer* are often used interchangeably. However, in some places in 3ds Max, a chamfer is a flat joint between surfaces, and a fillet is a rounded joint.

If an Extended Primitive has a **Fillet** parameter, then the size of the fillet is set after the main part of the object has been drawn. After setting the dimensions of the object with two or more mouse clicks, move your mouse in an upward motion. The fillet size is set according to how far the mouse is moved up. If no fillet is desired, leave the mouse where it is and click again. You must click the mouse for the fillet before you can move on.

CANCELING PRIMITIVE CREATION

You can cancel the creation of a primitive at any time during the creation process by right-clicking. If you have already finished the creation process, you can delete the primitive by pressing the **<DELETE>** key on the keyboard.

MODELING WITH MODIFIERS

As you learned in Chapter 2, the shapes of objects can be changed with *modifiers*. Now you will learn more about controlling the effects that modifiers have on objects and parts of objects.

SUB-OBJECTS

You can manipulate geometry and modifiers at two basic levels: the **object** level and the **sub-object** level. If you simply select a primitive or modifier, then you are working at the object level, and you can alter parameters for the entire primitive or modifier. This is what you have been doing so far. If you want to work with just a *part* of an object or modifier, then you must enter sub-object mode.

Figure 3-10: **Bend** modifier sub-objects

NOTE: The use of the term "modifier sub-object" might be confusing, because modifiers are not renderable geometric objects, in the sense that primitives are. But in a more abstract sense, modifiers are like objects that you can manipulate.

A modifier has parameters just like a primitive, and as you will see, modifiers can be affected by transforms just as geometric objects can. In this way, you can move a modifier around to achieve different effects.

To access the transforms for a modifier, you must enter sub-object mode. This is accomplished by going into the Modifier Stack and clicking the plus sign next to the name of the modifier. Then you click the desired sub-object, and it highlights in yellow.

Gizmo and Center

When a modifier such as **Bend** or **Taper** is applied to an object, a box appears around the object. This is the **Gizmo** for the modifier. A modifier Gizmo is a box inside which the modification takes place. The Gizmo appears as an orange border around the object.

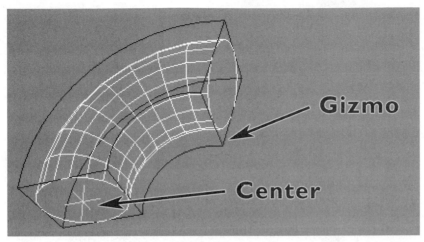

*Figure 3-11: Bend **Center** and **Gizmo***

The base point of a modifier is called its **Center**. For a Deformer such as Bend, the Center is the reference point around which the deformation takes place. For example, a Bend modifier applied to a Cylinder will cause the polygons of the Cylinder to bend more if they are farther away from the Bend Center.

By default, the Center of a modifier is placed at the object's **Pivot Point**, which is the origin of the object's local coordinate system. However, sometimes you will need the Center to be placed somewhere else. It is possible to move the Center of a modifier without moving the object's Pivot Point. To do this, choose **Center** from the list of modifier sub-objects. The Center of the modifier appears as a yellow cross. To move the Center, click **Select and Move**, then click and drag the Center to the desired location.

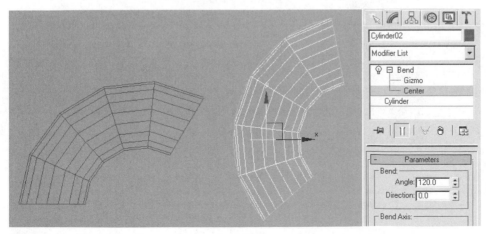

Figure 3-12: Moving the Center of a Bend modifier

The Cylinder will appear to rotate as you move the modifier Center. However, this is not the case. If you choose the **Select and Rotate** tool and look at the **Transform Type-In** values in the Status Bar, you will see that the Cylinder has not really rotated at all. As you change the Bend **Angle**, it will be clear that the point around which the Cylinder bends has moved.

NOTE: This procedure affects only the center of the Bend modifier, and has no effect on the object's Pivot Point.

The **Gizmo** sub-object can also be manipulated. In many cases, moving the Gizmo will produce the same results as moving the Center. This is because the modifier Center is actually the origin of the local coordinate system for the modifier Gizmo. It works the same as the Pivot Point of an object. When you move the Center, you offset the origin of the Gizmo's local coordinate system. When you move the Gizmo, you move it and its Center together.

This will come in handy later when you work with the **XForm** modifier.

EXERCISE 3.1: Modifier Center

In this exercise you'll change the Center of a **Bend** modifier.

1. Reset 3ds Max.

2. Create a cylinder in the **Top** viewport with a **Radius** of 40 and a **Height** of 200. Make sure the **Height Segments** parameter is set to at least 5.

3. Apply a **Bend** modifier to the cylinder. Change the **Angle** parameter to 90 to bend the cylinder.

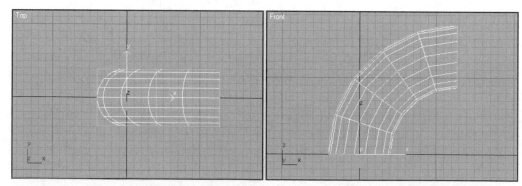

*Figure 3-13: Cylinder with **Bend** modifier applied*

4.   In the **Front** viewport, create a **Clone** copy of the Cylinder to the right of the first cylinder. To do this, click **Select and Move**, then hold down the **<SHIFT>** key while moving the Cylinder. On the **Clone Options** dialog, choose the **Copy** option and click **OK**.

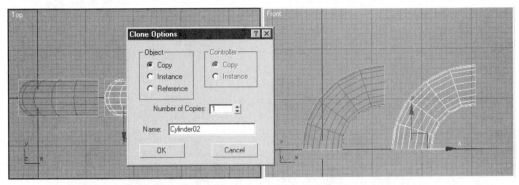

Figure 3-14: Cloned cylinder

The new Cylinder is automatically selected, and the Modify Panel shows the parameters for the **Bend** modifier.

5.   On the Modify panel, look in the Modifier Stack. Click the plus sign next to the Bend modifier. Choose the **Center** sub-object.

6.   In the **Front** viewport, notice the yellow Center crosshair icon. Move the Center in the Y axis to place it at the top of the Cylinder. The Cylinder will appear to rotate as you drag the Center.

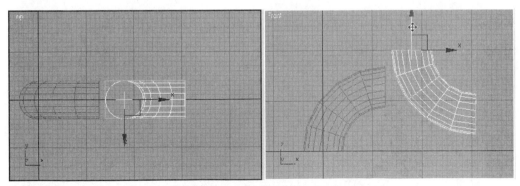

*Figure 3-15: **Center** of Bend moved to top of the Cylinder*

7.   On the Modify panel, click and drag on the **Angle** spinner while watching the **Front** viewport. The cloned Cylinder bends from its top instead of its bottom.

WARNING: When in sub-object mode, you can't select any other objects. If you can't select anything, check to see if you're in sub-object mode. This is a very common problem with users who are new to 3ds Max.

To exit sub-object mode, click the sub-object in the Modifer Stack, or click any other modifier or base object in the Stack. You will know if you're out of sub-object mode when nothing in the Stack is highlighted in yellow.

NOISE MODIFIER

The **Noise** modifier pushes and pulls vertices in a random fashion to give an object bulges and dents. It can be used on a Geosphere to make a lumpy rock or asteroid, or on a Plane to make natural-looking terrain.

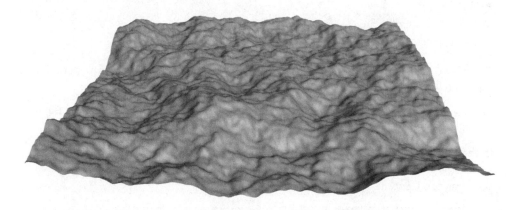

Figure 3-16: **Noise** *Modifier applied to* **Plane** *to make terrain*

Noise works by applying a variable push in the direction of the object's local X, Y, and Z axes. For Noise to work, you must enter a value under **Strength** for **X**, **Y**, and/or **Z**.

If the Noise modifier appears to have no effect on the object despite your having increased the Strength values, it is possible that the default size of the effect is too large for the object. You can remedy this by reducing the **Scale** parameter to a lower value such as 10 or 50.

The Noise modifier has a **Phase** parameter that determines which "phase" the noise is in. To see how the Phase parameter works, increase the Phase spinner while watching in the viewport. As the Phase is increased or decreased, the Noise pattern gradually shifts its shape. The **Frequency** parameter determines how fast the shape-shifting will occur. Higher values for Frequency make for quicker shifting.

The Noise modifier also has a **Seed** parameter. A *seed* is a number used to generate properties that appear to be random, but in fact are generated by the seed and an equation. When the Seed parameter is the same for the Noise on two objects, the pattern of bumps will be the same on the two objects. When the Seed is different for two objects, the bumps will be arranged differently.

You might think that changing the Seed incrementally would make a gradual change in the bump pattern. This is not so. Each time the Seed is changed, the equation that generates the bumps comes up with a completely different pattern of bumps. For this reason, there is little advantage to animating the Seed value.

EXERCISE 3.2: Asteroids

In this exercise, you'll use the **Noise** modifier to create an asteroid from a Geosphere.

1. Reset 3ds Max.

2. In the **Top** viewport, create a **Sphere** with a **Radius** of about **70** units. Set **Segments** to **32**.

3. Go to the **Modify** panel. Apply a **Noise** modifier to the sphere. Set the following parameters for the Noise modifier:

Scale	30
Strength X	25
Strength Y	25
Strength Z	25

    This creates an asteroid-like object.

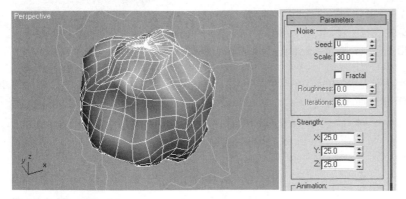

Figure 3-17: Asteroid

4. In the **Top** viewport, create a **Geosphere** next to the **Sphere** with a **Radius** of about **70**. Under **Geodesic Base Type**, choose **Icosa**, and set the Geosphere **Segments** to **6**.

5.   Select the Sphere. In the Modifier Stack, click the Noise modifer with the right mouse button. From the pop-up menu, choose **Copy**. Then select the Geosphere. In the Geosphere's Modifier Stack, right-click and choose **Paste** from the pop-up menu.

Maximize the Perspective viewport. The new asteroid looks nearly identical to the first one, with subtle differences at the tops and bottoms of the spheres. This is because the Geosphere has uniform mesh density all over the object, while the mesh of the Sphere is denser at its poles.

6.   With the Geosphere selected, access the object creation parameters by selecting the word **Geosphere** at the bottom of the Modifier Stack. Increase the number of Segments to **12**. The level of detail of the Geosphere is increased.

7.   Increase the number of Segments in the Sphere to **64**. Notice that the Geosphere still looks a little smoother. Select the Geosphere object and press the **<7>** key. A polygon/face counter appears at the top left of the viewport. The number of faces in the currently selected object appears. The Geosphere has **2880** faces. Select the Sphere; it has **3968** faces. So the Geosphere looks smoother, but has fewer polygons.

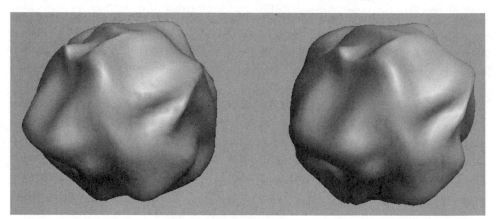

*Figure 3-18: Asteroids from **Sphere** and **Geosphere***

8.   For the second asteroid, change the **Seed** parameter. Note that the bumps now appear in different places.

9.   Save the scene with the filename **Asteroids.max** in your folder.

SUB-OBJECT MODELING

You can also apply a modifier to just part of an object. This is accomplished with a special category of modifiers called **Selection Modifiers**. A Selection Modifer doesn't actually change an object. It is used solely to select parts of an object so that the selected parts can be affected by other modifiers.

Mesh Select Modifier

One of the most commonly used Selection Modifiers is **Mesh Select**. When this Modifier is applied to a mesh object, you can select specific sub-objects within the object. The **Vertex**, **Face**, and **Polygon** sub-objects are the ones you'll use the most.

To use the **Mesh Select** modifier, expand the list of sub-objects in the Modifier Stack. Activate one of the mesh sub-object types, such as **Polygons**. As you do so, notice that one of the red icons in the Modify panel is now highlighted in yellow. These icons are merely shortcuts to the sub-object type in the Modifier Stack.

Sub-objects are selected with the same tools used for objects, such as clicking or drawing a bounding area around the sub-objects you wish to select. Selected sub-objects are highlighted in red by default.

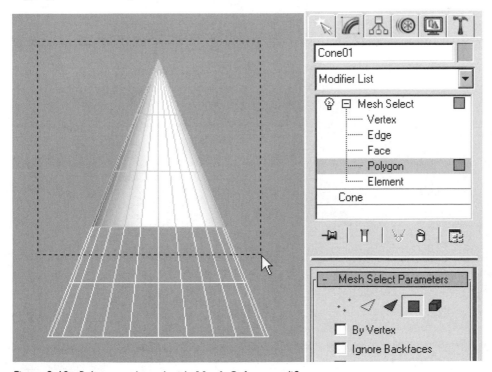

*Figure 3-19: Polygons selected with **Mesh Select** modifier*

When a sub-object mode is activated, the Mesh Select modifier is displayed on the Modifier Stack with a sub-object icon next to it. This indicates that the modifier is at a sub-object level and that a sub-object selection is expected.

Once a sub-object selection has been made, apply another modifier, such as **Bend**, above the Mesh Select. The sub-object selection made in Mesh Select is *passed up the Stack* to modifiers above it. The effect of the Bend is applied only to the sub-object selection chosen in Mesh Select.

A modifier placed in the Stack above a Mesh Select will also appear with a sub-object icon, indicating that the modifier is applied only to a sub-object selection. Each modifier higher in the Stack will affect only the selected sub-objects. The sub-object selection *continues up the Stack* until it is overridden by another Selection Modifier.

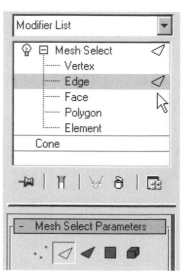

Figure 3-20:
Edge sub-object *mode*

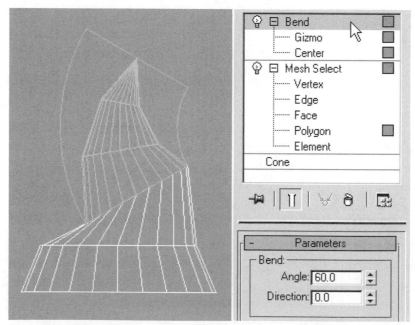

Figure 3-21: Bend applied to polygon selection

The **Poly Select** and **Patch Select** modifiers function in exactly the same manner. The only difference is the types of sub-objects they work with.

Dependencies

When you move to a lower level within the Modifier Stack, you might receive a warning message similar to the following:

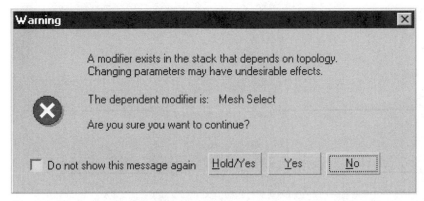

Figure 3-22: Dependent Modifier warning message

This message is telling you that making changes to the creation level of the object might cause subsequent modifiers to behave in unpredictable ways. This message most often occurs when a **Mesh Select**, **Poly Select**, or **Patch Select** modifier exists in the Stack.

If you are returning to the creation level to change the object's Radius or Height, or a modifier that doesn't affect the number of vertices, then you can click **Yes** on the warning dialog and proceed confidently. However, if you do change the number of vertices by altering the **Segments** or **Sides** parameters, you could run into serious problems.

The result of the Mesh Select depends on the information it gets from lower in the Stack, so it is called a *dependent modifier*. Any modifier depends on the object parameters and modifiers lower in the Stack, but sub-object selections are especially sensitive.

When an object is created, it is given a certain number of vertices based on the number of Segments or Sides, whether it is sliced or chopped, and so on. Each vertex is assigned an identification number.

When you select faces or vertices with Mesh Select, the selection is recorded according to the vertex numbers, *not* by the location of vertices on the object. Descending to the creation level and changing the number of Segments or Sides will cause the dependent Mesh Select modifier to select what appears to be an entirely different set of faces or vertices.

In fact, the same vertex identification numbers are still selected, but changing the number of Segments causes the vertices to move around on the object. Modifiers higher in the Stack will then operate on a selection which looks different from the one you originally made.

If you are planning to change a parameter that affects the number of vertices, you could still answer **Yes** to the dialog and change the parameter that affects the number of vertices. However, your object will probably look very strange.

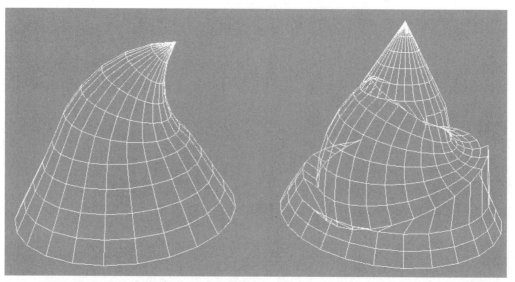

Figure 3-23: Changing number of Segments makes a mess if modifiers are applied to sub-object selections in the Stack!

To solve this problem, you could go to the **Mesh Select** modifier on the stack and reselect the correct faces or vertices. The object should return to its original appearance, but even this is not guaranteed to work.

A better solution to this problem is to plan ahead. Create the object so that it has the appropriate number of vertices for the job. Figure out how important the object is in the scene, and how much detail is needed. Dial in the correct number of Segments, and proceed to making your Mesh Select selections. If you are confident that you have the right level of detail for the mesh, you can even *collapse the Stack* before making a sub-object selection. Collapsing the Stack is explained later in this chapter.

Volume Select Modifier

If you know you'll really need to change the number of vertices in a model, you can use the **Volume Select** modifier instead of Mesh Select. Instead of selecting based on vertex numbers, Volume Select creates a Gizmo that encloses space. Anything inside the Gizmo is selected and passed up the Stack. You can choose Box, Sphere, Cylinder, or even a separate mesh object to use as the Volume Select Gizmo.

Volume Select is an extremely useful tool. Using it in conjunction with Soft Selection, as described later in this chapter, gives the artist a great deal of flexibility in modeling at the sub-object level. In many cases it is preferable to the other selection modifiers.

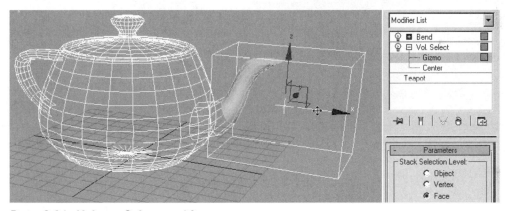

Figure 3-24: **Volume Select** *modifier*

Soft Selection

One very handy feature of 3ds Max is called **Soft Selection**. This feature allows you to define an area of influence based on a sub-object selection. You select sub-objects in the usual way, and if Soft Selection is turned on, the nearby sub-objects are *partially* selected.

To use this feature, open the **Soft Selection** rollout and enable the **Use Soft Selection** check box. If any sub-objects are selected, they are highlighted in red as usual. Other sub-objects display in different colors depending on the strength of the influence in that area.

- Fully selected sub-objects are displayed in red ("hard" selection).
- Partially selected sub-objects are displayed in orange to yellow.
- Unselected sub-objects are displayed in blue.

The warmer the color is, the stronger the selection is in that area. To increase or decrease the area of influence, adjust the **Falloff** parameter. See the following illustration.

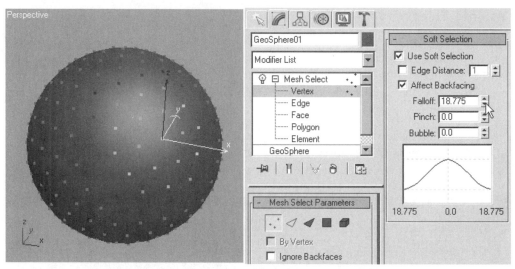

Figure 3-25: One vertex selected, surrounding vertices partially selected

Soft Selection allows you to create more natural shapes. If a you want to apply a modifier to only part of a mesh object and you don't use Soft Selection, then the transition between the modified and nonmodified areas of the object is harsh and looks unnatural. Soft Selection is an easy fix to this problem. It allows the modifier to have a decreasing amount of influence (Falloff) at the perimeter of a "hard" selection.

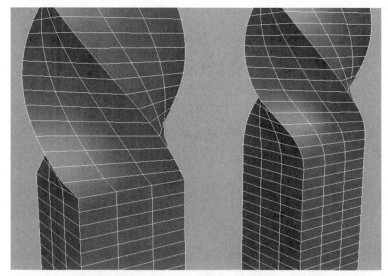

Figure 3-26: Twist applied to hard and Soft Selections

EXERCISE 3.3: Passing Modifiers up the Stack

In this exercise, you'll use multiple modifiers to alter the clown hat that you made in the last chapter.

1. Load the file **Clown03.max** from the disc.

2. Select the clown hat.

3. Go to the Modify panel.

4. Increase the number of **Height Segments** to **10**.

5. Apply the **Mesh Select** modifier to the hat.

6. Access the **Vertex** sub-object level. You can do this by clicking the **Vertex** button
    on the **Selection** rollout, or by selecting **Vertex** from the Modifier Stack.

7. In the **Front** viewport, select the vertices in the top half of the hat. To do this, draw a
    selection region around the vertices.

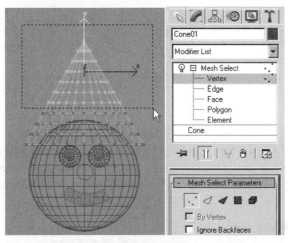

Figure 3-27: Vertices at top half of hat selected

You might want to use the **Top** or **Left** viewport to make the selection. Use whichever one works best.

8. Apply the **Bend** modifier to the hat.

9. Click the plus sign next to the Bend modifier. Choose **Center**.

10. On the **Main Toolbar,** click **Select and Move.** Move the center of the **Bend** to the base of the selection.

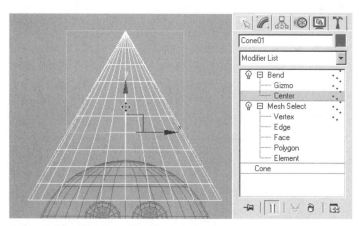

Figure 3-28: **Center** *of Bend moved to bottom of selection*

11. Increase the **Angle** parameter to bend the top of the hat.

12. Notice the sharp, abrupt transition area between the bent and nonbent areas of the hat. Go back to the **Vertex** sub-object level of the Mesh Select modifier, and enable **Use Soft Selection**. You see the vertices of the hat displayed in different colors. Increase or decrease the **Falloff** parameter to adjust the amount of influence the Bend modifier will have.

13. Turn on **Show End Result** . Now you can see a "before and after" picture of the Bend modification. The state of the object at the current level in the Stack (Mesh Select) is shown in orange. The final state of the object, after all modifiers have been applied, is shown in white. Adjust the **Falloff** parameter until you are happy with the result.

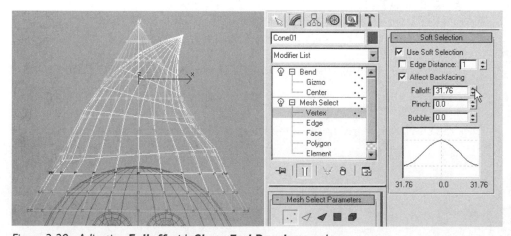

Figure 3-29: Adjusting **Falloff** *with* **Show End Result** *turned on*

14. Go back to the Bend modifier and adjust the Center again. **Do not exit sub-object mode in the Mesh Select modifier!** If you do, the Bend will be applied to the entire object.

    Turning on Soft Selection caused the Bend Gizmo to become bigger, and the Center is no longer at the base of the Gizmo. Move the Center of the Bend until it looks right to you. Notice that even vertices outside of the Gizmo are affected when you move the Center.

15. Apply a **Taper** modifier above the Bend modifier in the Stack. In the Taper modifier, choose the **Center** sub-object. Move the Center of the Taper to the base of the selection.

16. Adjust the Taper **Amount** and **Curve** parameters to produce a hat with an interesting kink at the top.

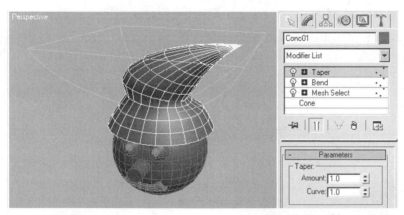

Figure 3-30: Kinked hat

17. Save the scene with the filename **KinkedHat.max** in your folder.

Stacking Selection Modifiers

Several selection modifiers can be applied to a single object. Each time you make a different selection with a modifier, it passes this new selection up the stack to any modifiers above it.

For example, after performing the Bend in the previous tutorial, you could choose a different set of faces and apply the Taper modifier to them. To do this, simply apply another selection modifier after the Bend. Make a different sub-object selection, and apply the Taper modifier on top of this selection. The Taper modifier is applied to the new selection. The new Mesh Select modifier sets a new selection, which it passes up the stack to the Taper modifier.

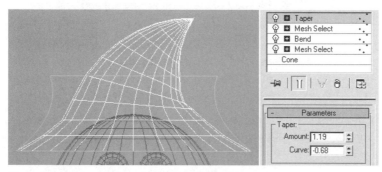

Figure 3-31: **Taper** applied to new selection

Applying Additional Modifiers to the Entire Object

If you create sub-object selections in the Stack, you can't apply modifiers to the entire object unless you reset the selection back to object level. To do this, apply another **Mesh Select** modifier, but don't activate sub-object mode. No sub-object type is highlighted in yellow, and no sub-object icon appears next to the Mesh Select. Modifiers higher in the Stack are applied to the entire object.

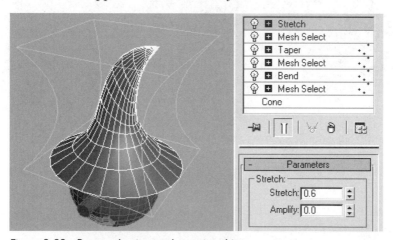

Figure 3-32:   Reset selection to the entire object

EXERCISE 3.4: Resetting a Passed Selection

In this tutorial, you'll reset the selection on the Stack so a modifier can be applied to the entire object.

1. Load the file **KinkedHat.max** if it is not already on your screen.

2. Select the hat and go to the **Modify** panel. The last modifier to be applied, **Taper**, may appear in the Stack highlighted in yellow, indicating that a there is a sub-object mode active somewhere lower in the Stack.

3. Apply the **Mesh Select** modifier to the hat. Don't click any sub-object type. The selection has now been reset to the entire object.

4. Apply a **Noise** modifier to the hat. Change the **Scale** parameter to **60**. Increase the **X**, **Y**, and **Z** parameters under the **Strength** section until the hat starts to wrinkle. Depending on the dimensions of the hat, you might have to change the **Scale** parameter to get good results.

5. Change the **Seed** parameter of the Noise modifier to randomize the vertices of the hat. Experiment with the **Scale** parameter and the **X**, **Y**, and **Z** parameters until the hat looks the way you want it to.

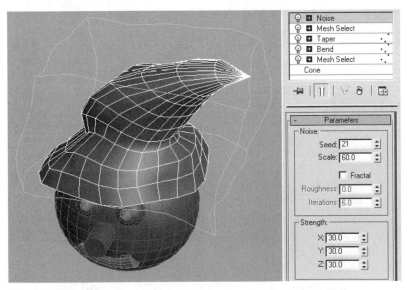

Figure 3-33: Lumpy hat

6. Save the scene with the file name **LumpyHat.max** in your folder.

CLONES

In 3ds Max, copied objects are called **Clones**. To clone an object, hold down the <**SHIFT**> key while transforming the object. In most cases, you will use the Move Gizmo to create Clones. You can also select the object and choose **Edit > Clone** from the Main Menu. Regardless of the method used to create the Clone, the **Clone Options** dialog box appears.

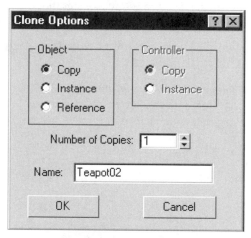

Figure 3-34: ***Clone Options*** *dialog*

There are three types of Clones: **Copy, Instance**, and **Reference**. A **Copy** is a Clone that has no relationship to the original object. Any changes to the original object have no effect on the new object.

An **Instance** is an object that is always identical to another object in the scene. The original object and the Clone are both considered Instances. An instanced Clone retains a two-way relationship with the original object. Any changes to one object's creation parameters are applied to the other object. In addition, modifiers applied to one object are applied to the other as well. However, a transform applied to one object is not applied to the other. Instance objects are useful when you want two or more objects to look the same, and you want to retain the freedom to change them later on without having to modify each object separately. Examples of this situation are a row of columns in front of a building, a school of fish, or a series of lights with the same intensity.

A **Reference** is a specialized type of Instance object. A Reference object can have one-way and/or two-way relationships with the original object, which is called the **Master** object. Modifiers applied to the Master object are also applied to the Reference object. However, a modifier applied to a Reference object is not necessarily applied to the Master, depending on where in the Stack the new modifier is applied.

If all of this sounds confusing, don't worry, because you'll get some hands-on experience with cloning objects in the next exercise.

You can tell if a modifier is from an Instance object by looking at the Modifier Stack. Modifiers from instanced objects are displayed in boldface type, and appear below a solid gray line. Modifiers that are unique to the Reference object appear in plain type above the line. If you add a modifier below the line, it will appear in the Master object and in any other References based on that Master.

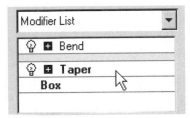

*Figure 3-35: **Reference** object with an instanced Taper and a unique Bend modifier*

EXERCISE 3.5: Copies, Instances, and References

This is a simple exercise to help you understand how Copies, Instances, and References work.

1. Reset 3ds Max.

2. Maximize the Perspective viewport. Press the **<F4>** key to see Edged Faces. Create a Box with a **Length** and **Width** of 20 and a **Height** of 100. Set **Height Segments** to 10. Change the name of the object to **BoxMaster**.

3. Click **Select and Move**. Hold down the **<SHIFT>** key and move the Box in the positive X axis to create a clone. When the **Clone Options** dialog appears, choose the **Copy** option and enter **BoxCopy** in the **Name** field. Click **OK**. Change the Object Color of **BoxCopy**, so that you can tell which is the Master and which is the Copy.

4. Select the **BoxMaster** object. Make another Clone by holding down **<SHIFT>** and dragging in the positive X axis. Place the new clone on the far side of **BoxCopy**, to the right. This time, when the Clone Options dialog appears, choose the **Instance** option, enter **BoxInstance** in the Name field, and click OK. Change the Object Color of the **BoxInstance** object.

You now have three objects on the screen: **BoxMaster**, **BoxCopy**, and **BoxInstance**.

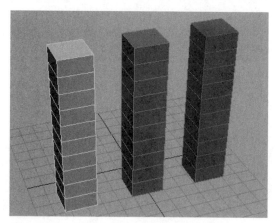

Figure 3-36: Master, Copy, and Instance

5.   Select **BoxMaster** and go to the Modify panel. Decrease the **Height** and observe what happens to the other two cylinders. **BoxInstance** changes, but **BoxCopy** does not.

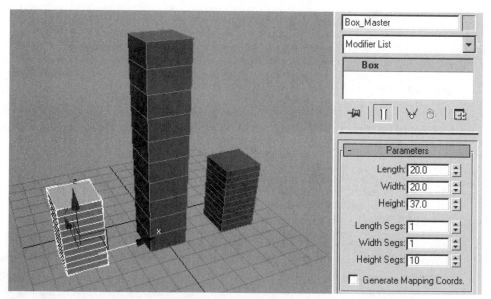

Figure 3-37: Instanced Box changes with Master; Copy does not

6.   Change the Height of **BoxMaster** back to **100**. Both of the instanced Boxes change back to their original heights.

7. Apply a **Taper** modifier to **BoxMaster**. A gizmo appears around both of the instanced Boxes. Change the **Amount** parameter to **-0.75** and observe the result.

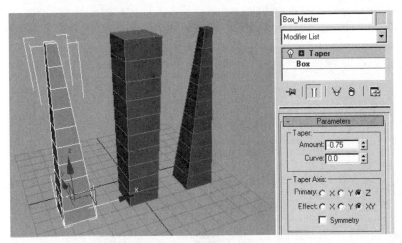

Figure 3-38: ***Taper*** *operates on both Instances*

8. With the Taper modifier still highlighted, click **Remove Modifier from the Stack**. The Taper modifier is removed from both instances.

9. Select **BoxInstance** and apply a **Bend** modifier. Change the **Angle** parameter to **30** degrees, and observe the result. Note that either the original or the clone can be used to control the other instanced object.

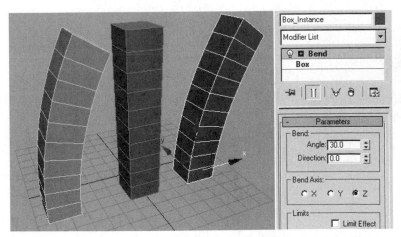

Figure 3-39: Bend modifier applied to **BoxInstance** *also bends* **BoxMaster**

10. Select **BoxMaster**. Make a clone of the object in the positive X axis, to the right of all the objects. When the Clone Options dialog appears, choose the **Reference** option, enter **BoxReference** in the Name field, and click **OK**. Change the Object Color of **BoxReference** to distinguish it from the other objects.

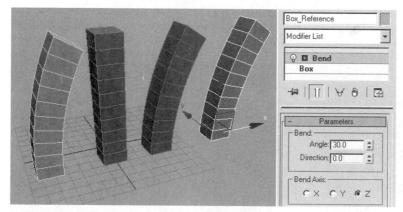

Figure 3-40: Master, Copy, Instance, and Reference

11. Select **BoxMaster**. In the Modifier Stack, select the Box creation parameters. Apply a **Twist** modifier to it. The Twist modifier appears above the Box and below the Bend modifier. Change the Twist **Angle** parameter to **180**. **BoxInstance** and **BoxReference** receive the **Twist** modifier as well.

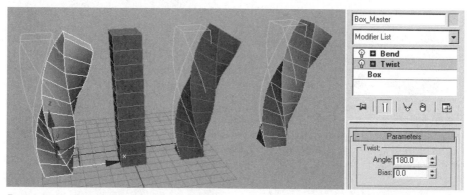

Figure 3-41: Twisty Boxes

12. Select **BoxReference**. Apply a **Noise** modifier, change the **Scale** parameter to 5, and change the **X Strength** and **Y Strength** to 10. The **Noise** is applied only to **BoxReference**.

Figure 3-42: Only **BoxReference** *is lumpy*

13. In the Modifier Stack for **BoxReference**, choose the **Bend** modifier. You can tell that it is present in another instanced object, because it is in boldface type and below the gray line. Drag the **Angle** spinner to change the parameter to approximately 90 degrees. **BoxMaster** and **BoxInstance** also reflect this change, because the same Bend modifier is present in all three objects.

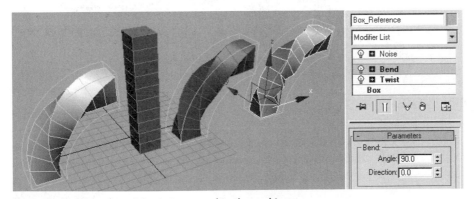

Figure 3-43: **Bend** *modifier is instanced in three objects*

14. As you dragged the Bend Angle spinner, you probably noticed that the polygons of **BoxReference** danced and wiggled. This is because the Noise modifier is applied above the Bend. If you were to animate the Bend Angle, you would have unpleasant results. The solution to this is to place the Noise modifier below the Bend modifier. Click the Noise modifier and drag it below the Bend modifier. Notice that the Noise modifier is now highlighted in boldface, and **BoxReference** and **BoxInstance** also receive the Noise.

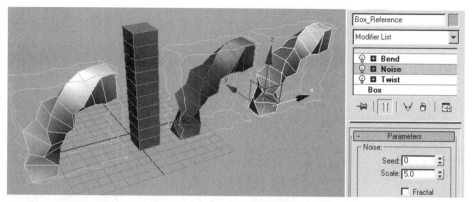

*Figure 3-44: Dragging the **Noise** modifier below the line*

15. Select **BoxInstance** and change the Bend Angle to **180** degrees. **BoxMaster** and **BoxReference** also change. The objects deform smoothly, without the dancing polygons.

16. Select **BoxReference**. You only want this object to be lumpy, not the others. Drag the Noise and Bend modifiers to the top of the Stack, above the line. Now the Bend modifier is no longer present in any other object; it is unique to the **BoxReference** object.

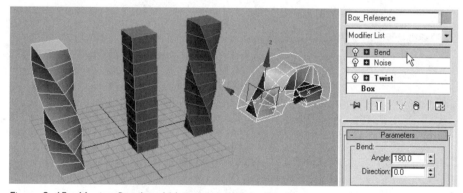

Figure 3-45: Moving Bend and Noise to top of the Stack

If you want the other objects to Bend, you have to add a new Bend modifier to them. However, if you do that, then **BoxReference** will have two Bend modifiers, and you will also have the problem of the dancing polygons again. We have discovered a limitation in the Modifier Stack. It is not possible to place a modifier that is present in an Instance object above a unique modifier in the Stack. There is a solution to this problem involving *instanced modifiers*, which we will explore in the next section.

17. Save the scene as **Instances.max** in your folder.

3DS MAX DATA FLOW

In Chapter 2, you learned that the order of modifiers in the Stack had a profound influence on the resulting object. This is an example of a *data flow*. In a logical process such as a computer program, the order of operations is crucial. Changing the order in which operations are performed changes the outcome.

Remember that in the Modifier Stack, the chronological order in which you add or alter modifiers is not important; all that matters is the resulting structure of the Stack. That is to say, the Modifier Stack has no notion of time, and it doesn't care *when* you add or change modifiers. It only cares *how* the modifiers are arranged in the Stack. Modifiers are evaluated from bottom to top in the Stack. This is the data flow within the Stack.

This concept of data flow is found throughout any 3D graphics program. In 3ds Max, every object you see on the screen is the result of a data flow, even if there are no modifiers applied to the object.

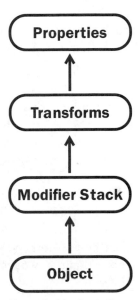

Figure 3-46: Data flow of a typical object with modifiers applied

The diagram illustrates the data flow for a simple object. It shows the data flowing from bottom to top, in the same way as the Modifier Stack. Various aspects of the object are evaluated in an internal order of operations. Even within each major stage in the process, there is an order of operations.

First, the object type and its parameters are processed. For example, if it is a Sphere object, it has parameters such as Radius and Segments.

Next, the object transforms are evaluated. Transforms are performed with the Move, Rotate, and Scale buttons on the Main Toolbar, and don't appear on the Modifier Stack. The transforms also have an internal order of operations. An object is always scaled, then moved, and finally rotated. This cannot be changed in 3ds Max, but there is a clever workaround using a modifier called XForm, which you will learn about in the next section.

If modifiers are applied to an object, then they are evaluated before any transforms. The entire Modifier Stack is always processed before the transforms.

Finally, the object **Properties** are processed. There are many object Properties such as its name, color, and material. Properties don't appear in the Modifier Stack. You can view and edit them by selecting **Edit > Object Properties** from the Main Menu or by right-clicking an object and choosing **Properties** from the Quad Menu.

When you create a Copy Clone of an object, the entire data structure of the object is recreated, with no shared information.

However, when you create an Instance of an object, the original object and the new Instance Clone share the same object creation parameters and modifiers, but they have their own transforms and object Properties. An Instance object shares part of its data flow with another object in the scene. This is an example of a *branching* data flow. Another way of saying this is that the two instances share the same data *pipeline*.

The fact that transforms are applied after modifiers is very important. This is what allows us to create Instances of an object and put them in different places in the scene. Instanced objects share object creation parameters and modifiers in common, but they have different transforms and Properties.

In a Reference Clone, the data pipeline is more complex, because some modifiers are shared and some are not.

As you see from this diagram, a Reference object also shares part of its data flow with another object in the scene. The difference is where the data flow branches. In an Instance object, the fork where the pipeline separates is after the entire Modifier Stack.

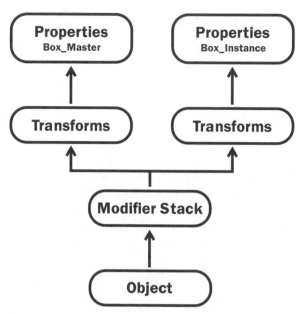

Figure 3-47: Data flow of Instance objects

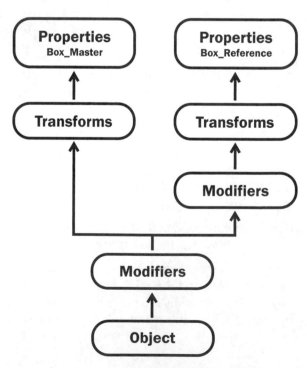

Figure 3-48: Data flow of a Reference object

In a Reference, the pipeline branches in the middle of the Modifier Stack. Some modifiers are shared, and some are not. The shared modifiers are shown in the Modifier Stack in boldface type, below the gray line. Modifiers that are not shared and are unique to the Reference object, are shown in plain type, above the gray line.

As you saw in Exercise 3.5, there are limitations to working with References. If you want to apply a modifier to a Reference object without affecting other objects in the scene, you have to place that modifier higher in the Stack, above the gray line. However, this not may produce the result you desire.

In the exercise, the Bend modifier worked better if it was higher in the Stack than the Noise. Unfortunately, moving the Bend higher in the Stack made the Bend unique to the Reference object. You could no longer change the Bend Angle parameter for all three objects simultaneously.

The solution to this problem lies in the idea of *instanced modifiers*. Instead of creating a Reference object, you can create a Copy and then apply an identical modifier to both objects. With an instanced modifer, the two objects have different data pipelines, but they share the same modifier.

This allows you to place shared modifiers anywhere in the Stack. There is no restriction on modifier placement, as there is with Reference objects. You can place an instanced modifier higher in the stack than a unique modifier, which is something you can't do if you're working with Reference Clones.

To create an instanced modifier, add a modifier to an object. Right-click the modifier in the Stack, and select **Copy** from the pop-up menu. Then select a different object. The other object doesn't have to be a Clone; it can be any other object, of any type. You can paste a modifier from a Box to a Cylinder.

Right-click in the Stack below where you want the instanced modifier to be added. From the pop-up menu, choose **Paste Instanced**. The modifier from the other object is instanced into the current object. Instanced modifiers appear in italic type in the Modifier Stack.

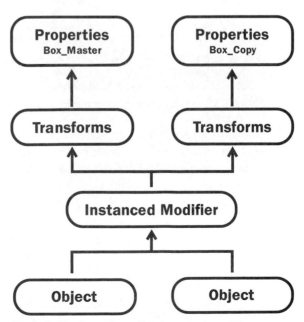

Figure 3-49: Data flow of an instanced modifier

You can also add an instanced modifier by selecting multiple objects and adding a modifier. The difference is that if you select multiple objects, the modifier Gizmo encloses all of the objects, and changes to the modifier affect the objects as a group. Pasting an instanced modifier results in the modifier affecting each object separately.

EXERCISE 3.6: Instanced Modifier

In this exercise you will fix the problem found in Exercise 3.5.

1. Open the scene file **Instances.max** from your folder or from the disc.

2 Delete the objects **BoxInstance** and **BoxReference**.

3. Select the object **BoxCopy**. Add a **Bend** modifier, and change the **Angle** to 40 degrees. Right-click the Bend modifier, and choose **Copy** from the pop-up menu.

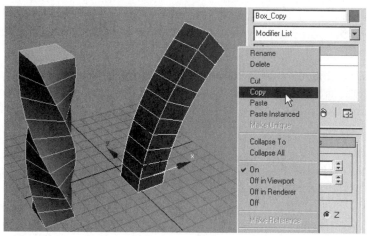

*Figure 3-50: Right-click the **Bend** modifier and choose **Copy***

4. Select the object **BoxMaster**. Right-click the **Twist** modifier and select **Paste Instanced** from the pop-up menu. The **Bend** modifier is added, and it is displayed in italic type. Drag the Bend Angle spinner to about **-60** degrees. The **BoxCopy** object also bends.

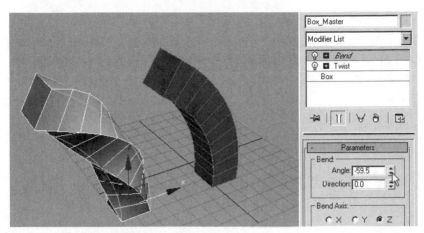

*Figure 3-51: Instanced **Bend** affects both objects*

5.   Select the object **BoxCopy** again. Select the **Box** creation parameters in the Stack and add a **Noise** modifier. The Noise is added above the Box and below the instanced Bend. Change the Noise **Scale** to 5, and the **X Strength** and **Y Strength** to **20**.

6.   Select the Bend modifier and adjust the Angle. Both objects bend equally. One of the objects has a unique Twist modifier; the other has a unique Noise modifier. The unique modifiers are lower in the stack than the instanced Bend, and the objects deform smoothly and predictably.

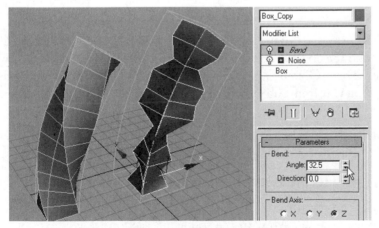

*Figure 3-52: **Noise** modifier below instanced Bend in the Stack*

THE XFORM MODIFIER

Earlier, you learned that transforms (position, rotation, and scale) are different from modifiers. One difference is that transforms don't go into the Modifier Stack. Another is that Instance and Reference Clones share modifiers, but don't share transforms.

When a change is made to an object directly from a transform on the Main Toolbar, the change to the object is always applied to it after all modifiers have been applied. In addition, you can't change the internal order of transformations. This sometimes yields unusual results.

To solve this problem, there is a modifier that simulates a transform. It is called **XForm**, because in engineering jargon, the letter "X" often stands as an abbreviation for "trans."

The **XForm** modifier moves, rotates, or scales an object in the same way a transform does. However, because **XForm** is a modifier, it has all the qualities of a modifier, such as going into the Stack and being shared by Instance objects. You can place an XForm anywhere in the Stack to accomplish various tasks, such as changing transform order.

XForm works by placing a Gizmo around the object and then transforming the Gizmo rather than the object. The Pivot Point of the object is not affected, but the object geometry is transformed.

Transforming the object with XForm puts the change in the Stack, where it will be evaluated according to its order on the stack. Compare this with transforms applied directly to the object, which are always applied after all modifiers in the stack.

When you add an XForm, the **Gizmo** sub-object level is selected by default. A yellow Gizmo appears around the object. You then choose a transform from the Main Toolbar and transform the XForm Gizmo. The transformation of the Gizmo is then passed on to the object. If you exit from sub-object mode, you will see that the Pivot Point is not affected.

XForm is particularly useful when positioning and animating a series of instanced objects. It is also very handy for scaling, as discussed later in this chapter.

WORKING WITH THE MODIFIER STACK

The Modifier Stack is a very powerful tool. Understanding how it works is crucial to your success in modeling in 3ds Max. In the following section we will explore some of the common pitfalls you may encounter when using the Stack, and you will learn how to avoid them.

Modifier Stack and Non-uniform Scale

Because of the order of operations in 3ds Max, performing a Scale Transform, especially a **Non-uniform Scale** or **Squash** transform, can have adverse effects on objects later on. Animated objects can skew and change shape unexpectedly. We will explore this in detail in Chapter 6, *Keyframe Animation*.

Whenever possible, avoid scaling objects. Instead, change the object's creation parameters to make it larger or smaller. If you absolutely must scale an object, use the **XForm** modifier to do so. Since **XForm** is a modifier and not a transform, this will place the scale factor in the Modifier Stack, and not in the Scale transform. This avoids all of the problems associated with the Scale transform.

If you exit sub-object level, select the **Select and Scale** tool in the Main Toolbar, and look in the **Transform Type-In** area of the Status Bar, you can see the Scale Transform values for the object. The X, Y, and Z scale values should all read **100%**. If they do, you won't have any skewing problems with animated objects.

Stack Depth and Collapsing the Stack

The advantage to working with modifier parameters is their incredible flexibility. But this flexibility comes at a price. As you add modifiers to objects, 3ds Max must hold all of that complex information in system memory. This is one reason why it is important for your computer to have lots of RAM — at least 512 megabytes.

As you make changes to modifiers, 3ds Max must reevaluate and recalculate the Stack. This can cause your system to run slowly. Even if you have a fast 3D accelerator card, Modifier Stack changes must be evaluated by the system processor. This is one reason why you must have at least a processor with a clock speed of at least 1 gigahertz.

Even with the fastest computer, you will eventually reach a point at which the scene bogs down. This happens if you have a lot of objects in the scene, or if you have very complex objects with many modifiers. If the Stack is very deep, with many modifiers layered over one another, you can experience system slowdowns.

Sometimes it is necessary to have many objects, or many modifiers, in a scene to achieve the effect you desire. One way of speeding up the process is to *collapse the Modifier Stack*. Collapsing the Stack fixes the object in its current state and destroys all creation parameters and modifiers. You are left with a raw polygonal mesh, in the state it was in when you collapsed the Stack.

If you are familiar with a 2D bitmap program such as Photoshop, you may relate to the following analogy. Collapsing the Modifier Stack is like flattening a Photoshop image. All layers are "burned in" to the image, and you are left with a single bitmap with no parameters. Collapsing the Modifier Stack works the same way; the vertices and faces of the object are "burned" into their present state.

There are several ways to collapse the Modifier Stack. One is to right-click the object and select **Convert To > Convert to Editable Mesh** from the Quad Menu. You can also choose **Convert to Editable Poly**, which will result in a different type of object.

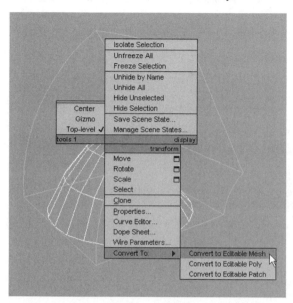

Figure 3-53: Right-click to access Quad Menu, **Convert To > Convert to Editable Mesh**

You can also collapse the Stack by right-clicking the Stack and selecting **Collapse All** from the pop-up menu. This is the same as **Convert to Editable Mesh**.

In addition, you can partially collapse the Stack, leaving modifiers in the upper part of the Stack intact. Right-click a modifier and choose **Collapse To** from the pop-up menu. Object creation parameters and modifiers up to and including the selected modifier will be destroyed, and you will have an Editable Mesh object at the bottom of the Stack. Modifiers above the current level will not be affected.

Collapsing the Stack results in a new type of object, which is *editable*. This is a fundamentally different way of modeling than parametric modeling with modifiers. Editable objects allow you to alter the shape of the model at the individual sub-object level, by directly transforming vertices, faces, and so on. You will learn more about editable objects later.

When you collapse the Stack, you gain an increase in system performance, and the ability to directly manipulate sub-objects. You lose the ability to instantly alter creation parameters, such as the number of Segments. For this reason, it is important that you only collapse the Stack if you are confident that you won't need to change the object creation parameters at a later time.

TIP: A good technique is to model an object with modifier parameters until you are ready to collapse the Stack. Then make a copy of the object and **Hide** the copy. Collapse the Stack of the unhidden model. If you need to go back to the parametric version of the model, it is still present in your scene, just hidden from view.

MODELING WITH SHAPES

In 3ds Max, you can use **Shapes** to make 3D objects. A Shape is a type of geometry, but it is different from the solid objects you've worked with so far. A Shape has vertices that control its curvature, but by default it has no faces or polygons. Since only faces are renderable, a Shape will not render by default. Shapes can be made to render by themselves, but more often, Shapes are used with various modeling tools to create solid 3D objects.

CREATING SHAPES

Shapes can be found on the **Create** panel. Click the **Shapes** icon to access the available Shapes. The default type of Shape in the pulldown list is **Splines**.

2D Primitives

Most Shapes are 2D primitives. These Shapes are created in a manner similar to solid 3D primitives. Click a Shape button, then click and drag in a viewport to create the Shape. As with 3D primitives, different Shapes are created in different ways.

Figure 3-54: **Shapes** *panel*

Before you create a Shape, you can specify how you prefer to draw the shape in the **Creation Method** rollout. For simple 2D primitives, you can choose **Edge** or **Center**. If you choose Edge, then click and drag to define the extents of the Shape. If you choose Center, click to define the center of the shape, and drag to define the radius.

After a Shape has been created, note that the Shape does not appear as shaded in the **Perspective** viewport. This is because a Shape has no polygons by default, and thus cannot be shaded. However, you can make a Shape renderable by changing its parameters in the **Rendering** rollout of the Command Panel.

EXERCISE 3.7: 2D Primitives

In this exercise, you'll practice creating 2D primitive Shapes.

1. Reset 3ds Max. Maximize the **Top** viewport.

2. On the **Create** Panel, click the **Shapes** button.

3. Click the **Circle** button to enter Circle creation mode. Click and drag to create a circle with a **Radius** of about 40 units.

4. Click **Rectangle**. Click and drag to create a rectangle with a **Length** of about **130** and a **Width** of about **50**.

5. Click **Ellipse**. Create an ellipse with a **Length** of about **30** and a **Width** of about **90**.

6. Click **Rectangle** again. In the Creation Method rollout, chose **Center**. Hold down the **<CTRL>** key while clicking and dragging. The rectangle is created as a square, with an equal **Length** and **Width**.

7. Click **NGon**. In the Creation Method rollout, choose **Center**. Create an NGon with a **Radius** of about **40**. Change the number of **Sides** to **5** to create a pentagon.

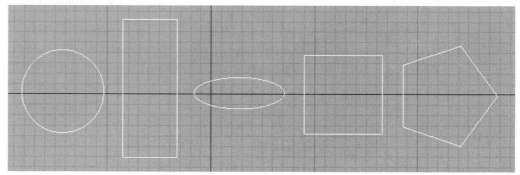

Figure 3-55: Circle, Rectangle, Ellipse, square Rectangle, NGon

8. Save the scene as **2dPrimitives.max** in your folder.

Fillets

Some Shapes can have a *fillet*, which is a rounded edge. In the 3ds Max interface, a fillet on a 2D primitive is created using the **Corner Radius** or **Fillet Radius** parameters.

To make a Shape with a parametric fillet, create a Rectangle, NGon, or Star object. Then increase the Corner Radius or Fillet Radius parameters.

Renderable Shapes

As mentioned previously, you can't render a Shape object by default, because it has no polygons. However, you can make Shapes renderable by adjusting their parameters in the **Rendering** rollout. This technique can be used to create simple objects that are not available as 3D primitives. Using the Rendering rollout, it's very easy to create ropes and cables with the **Line** or **Editable Spline** object, which we will discuss later in this chapter.

Turning on the **Enable In Renderer** check box allows any Shape to render, by creating polygons for the object. However, you won't see the results in the Viewports unless you also turn on the **Enable in Viewport** option. Check the **Use Viewport Settings** option to use the same settings for both the viewports and the renderer. Increase the **Thickness** parameter to see the mesh in the viewports. Change the number of **Sides** to achieve a faceted or a rounded look. Experiment with the **Radial** or **Rectangular** parameters to achieve different results.

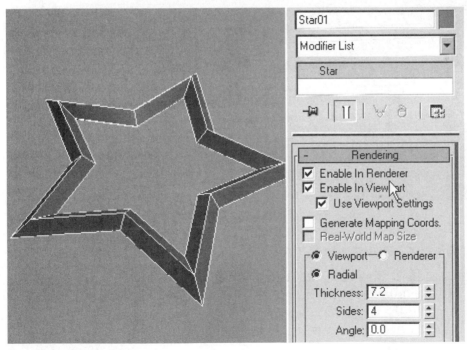

Figure 3-56: A renderable Shape

SHAPES AND THE EXTRUDE MODIFIER

The most basic way to turn a 2D Shape into a solid object is to apply the **Extrude** modifier. When a Shape is selected, the Extrude modifier is available, but when a 3D object such as a solid primitive is selected, the Extrude modifier is not available. After adding the Extrude modifier, the Shape is shaded in the **Perspective** viewport. At first, it has no visible thickness. Change the Extrude **Amount** value on the Modifiy Panel to give the object a thickness. Negative values will extrude in the opposite direction.

If you wish to deform the object with a modifier such as Bend, you should increase the number of Segments in the Extrude modifier. This will create more divisions along the thickness of the extrusion.

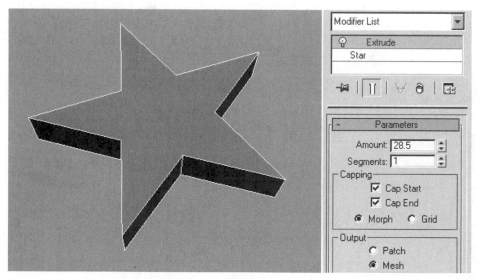

*Figure 3-57: Shape with **Extrude** modifier applied*

OPEN AND CLOSED SHAPES

Some Shapes are automatically joined, or *welded*, at the ends to form a loop. This is a *closed* Shape. If the end vertices are not welded, as in an **Arc** object, it is an *open* Shape.

Open Shapes don't work well with the Extrude modifier because there are no face normals on one side of the extruded Shape.

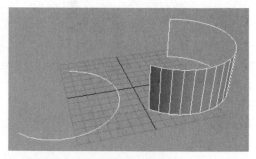

*Figure 3-58: Extruding an **Arc***

EXERCISE 3.8: Rendering and Extruding Shapes

1. Open the scene **2dPrimitives.max** from your folder or from the disc.

2. Maximize the Perspective viewport and click **Zoom Extents All**.

3. Select the **Rectangle** and open the Modify Panel. In the **Rendering** rollout, turn on the **Enable in Renderer** check box. Also turn on the **Enable in Viewport** and **Use Viewport Settings** check boxes. Increase the **Thickness** parameter to 10, and set the number of **Sides** to 7. You could use this object as a picture frame.

4. Select the **NGon** and apply an **Extrude** modifier to it. Change the **Amount** parameter to **20** to give the object some thickness.

5. Create a **Star** Shape with **5 Points**. Increase the **Fillet Radius 1** and **Fillet Radius 2** parameters to make a rounded Star. Add an **Extrude** modifier and increase the **Amount** of thickness.

6. Save the scene as **RenderableShapes.max** in your folder.

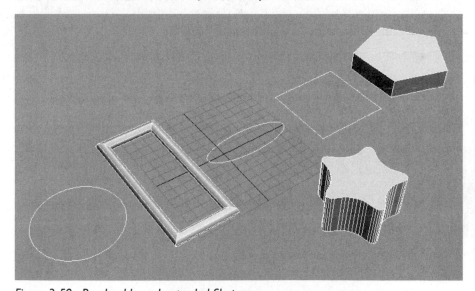

Figure 3-59: Renderable and extruded Shapes

MULTI-SPLINE SHAPES

A Shape is a type of object that is composed of *splines*. A spline is a computer graphics term that essentially means "curved line," although a spline can also be straight.

So far, the Shapes you have worked with have all consisted of a single, closed spline. However, Shapes often contain more than one spline. An example is the **Donut** Shape primitive, which consists of two concentric circles. The Donut is a single Shape object, with two splines within it.

Multi-spline Shapes, when extruded, can create many kinds of objects. When one spline is completely enclosed or *nested* inside the other, the extruded mesh object will be hollow. When two or more splines within one Shape are not nested, they can be extruded together.

There are two rules to remember when extruding multi-spline Shapes. First, all of the splines must be contained within the same Shape object. (You'll learn how to combine splines in the next section.) Second, the splines must not overlap or touch one another. If either of these rules is broken, you'll get unexpected results, such as missing hollow spaces or polygons intersecting one another.

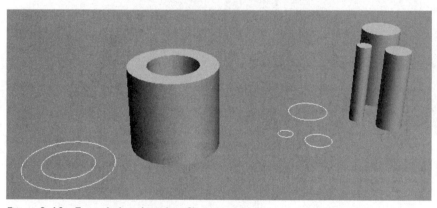

Figure 3-60: Extruded multi-spline Shapes

EDITABLE SPLINE OBJECTS

Earlier you learned about collapsing the Modifier Stack to convert parametric primitives to raw mesh objects. Now you'll see a similar technique in action with respect to 2D Shape primitives.

The curvature of a Shape is determined by points called *control vertices*, abbreviated *CV*. In 3ds Max, standard Shape CVs are usually simply called *vertices*. However, Shape vertices are different from the vertices in a mesh object. In a Shape, the segments between vertices can be curved, but in a mesh, the edges between vertices are always straight lines.

You can't directly edit the CVs of a Shape primitive. In fact, you can't even see them, but they are always there.

To change the form of a Shape object, you need access to its CVs at the sub-object level. To do this, you can make it *editable* by collapsing its Stack. Note that you don't need to have modifiers in the Stack to collapse it. You can convert a 2D Shape primitive directly to an **Editable Spline**. Right-click a Shape, and choose **Convert to Editable Spline** from the Quad Menu.

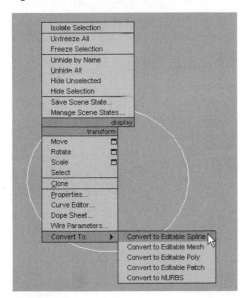

Figure 3-61: ***Convert to Editable Spline***

When you convert a Shape to an Editable Spline, the object type is changed. You no longer have a Circle or a Rectangle or an NGon; you have an Editable Spline object. If you open the Modify panel, you will see new sub-object types and new editing tools. The parameters of the 2D primitive are permanently discarded, but you gain the ability to edit sub-objects directly.

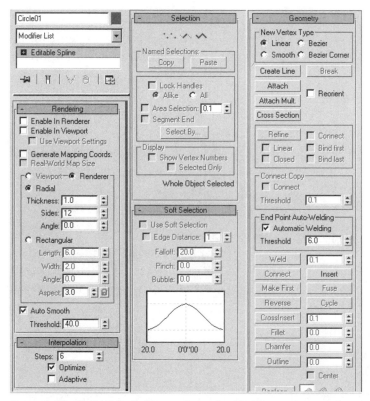

Figure 3-62: ***Editable Spline*** *tools on the Modify panel*

VERTICES, SEGMENTS, AND SPLINES

Shapes are made up of the following sub-objects: **vertices**, **segments**, and **splines**. As you learned earlier, there can be more than one Spline sub-object in a Shape. The **Donut** primitive is a perfect example. If you convert a Donut to an Editable Spline, you can enter Spline sub-object mode and transform the two circles independently.

WARNING: The fact that the **Editable Spline** object has a sub-object type called **Spline** is confusing. There are Spline sub-objects inside Editable Spline objects. To avoid confusion, in this book we will always make an explicit distinction between a Spline sub-object and an Editable Spline object .

Vertices, or control vertices, are points used to define the curvature of segments. **Segments** are the lines connecting the vertices. Segments can be straight or curved. The curvature of a segment is determined by the vertices at either end.

To edit the curvature of an Editable Spline, select the object and open the Modify Panel. Choose a sub-object type from the Modifier Stack or from the **Selection** rollout. Use the Transform tools, such as Select and Move, to transform vertices, segments, or Spline sub-objects.

Just as when you transform a modifier Gizmo, editing sub-objects does not affect the Transforms of the object itself. The Pivot Point of the object is not affected when you are in sub-object mode. You can move, rotate, or scale sub-objects, and the Transform values at the object level will never change.

Vertex Types

By default, each shape's vertices are of a certain type. For example, a converted rectangle has vertices that cause the segments around them to be straight. A converted circle has vertices that cause the segments to be curved.

To change a vertex type, right-click a vertex or a selection of vertices. From the Quad Menu, select a vertex type.

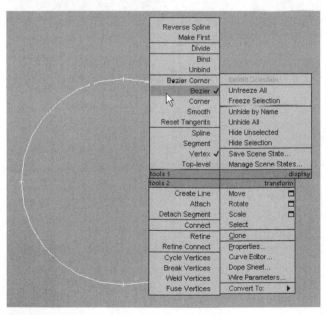

Figure 3-63: Right-click to convert vertex

The four available vertex types for Editable Spline objects are listed as follows.

Corner makes straight segments between two Corner vertices.

Smooth makes curved segments.

Bezier makes curved segments with handles for changing curvature of both adjoining segments together.

Bezier Corner makes curved segments with handles for changing the direction of adjoining segments separately.

Bezier and Bezier Corner vertices have small green boxes attached to them. These green boxes are called **Vector Handles**. The handles can be moved to change the curvature of the segments around the vertex. In other software, they are often called Bezier handles.

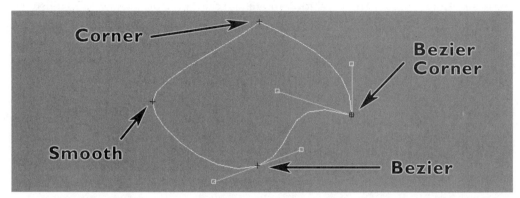

Figure 3-64: Vertex types

You can also convert a Bezier vertex to a Bezier Corner by holding the **<SHIFT>** key while moving a vector handle. Remember that if you're at the object level, the **<SHIFT>** key will create a Clone.

Disabling the Transform Gizmo

Transforming vertices is easy; simply use Select and Move as usual. However, editing vector handles can be tricky. When you select a Handle, the Transform Gizmo does not move to the handle. Instead, the Transform Gizmo stays behind with the vertex. Since you are only able to move the handle in the currently active transform axes, you will probably have to activate the desired axis *before* selecting the handle. Click the Transform Gizmo, and the axis or axes is highlighted in yellow. Then you can move the Handle in the desired direction.

Sometimes the Transform Gizmo is superimposed directly over a Vector Handle, making it hard or impossible to move the Handle in the desired direction. To solve this problem, you can disable the Transform Gizmo and *constrain* (restrict) the transform with keyboard shortcuts, or **Hotkeys**.

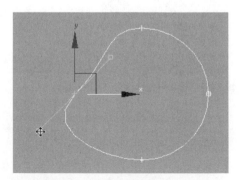

Figure 3-65: Handle selection does not affect Transform Gizmo

To disable the Transform Gizmo, press the **<X>** key on the keyboard. This is the same as turning the Gizmo off in **Customize > Preferences > Gizmos**. With the Transform Gizmo disabled, an object axis tripod is still visible on the screen, showing the position of the Pivot Point by default. You can't click an axis to select it, but the object tripod will highlight the active axis or axes in red.

Then, choose the axis you wish by pressing the appropriate Hotkey.

<F5> – **Restrict to X axis**

<F6> – **Restrict to Y axis**

<F7> – **Restrict to Z axis**

<F8> – Cycles among **Restrict to XY Plane**, **Restrict to YZ Plane**, and **Restrict to ZX Plane**. Press the **<F8>** key repeatedly to select the desired plane for transform restriction.

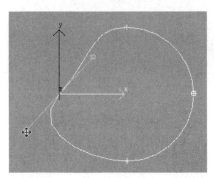

Figure 3-66: Disable Transform Gizmo, Restrict to X axis with **<F5>**

You can use the axis constraint hotkeys even if the Transform Gizmo is enabled, but this is not usually necessary.

The axis constraint controls are also as icons found on a toolbar. Right-click on an empty area of the Main Toolbar and choose **Axis Constraints** from the pop-up menu. A floating toolbar opens that allows you to click the desired axis. Constraining to a plane is done by choosing XY, YZ, or ZX from the flyout.

WARNING: There is also an item in the Main Menu, **Views > Show Transform Gizmo**, which, if turned off, will *completely hide the Transform Gizmo and the object axis tripod*. If you have an object selected, yet can't see the Transform Gizmo or axis tripod anywhere, make sure that Show Transform Gizmo is enabled in the Views menu.

First Vertex

Each Shape has a vertex that is different from the others: it has a box around it. This is the **First Vertex**.

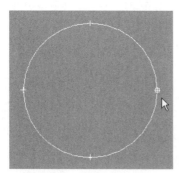

Figure 3-67: **First Vertex**

For some modeling functions, 3ds Max uses this vertex to line up Shapes. You can change the First Vertex by selecting a vertex and clicking the **Make First** button on the Modify panel. This function will become important later when you learn how to create Loft objects.

Editing Segments

To transform a segment, choose **Segment** as the sub-object, then select one or more segments. Like all sub-objects, segments turn red when selected. You can then transform the segment as a unit. Vertices on either end of the segment are also transformed.

You can also change a segment to a curve or a straight line. Right-click a selected segment, and the Quad Menu appears. Select **Curve** or **Line**, and the segment is converted. When a segment type is changed, the vertices at either end of the segment must usually change to a different type in accord with the new segment type.

Editing Spline Sub-Objects

As mentioned earlier, a spline is a single, continuous line or curve, while a Shape is an object made up of one or more splines. The **Editable Spline** object type has a **Spline** sub-object type, which allows editing of individual curves within the Shape object. As with all sub-objects, selected splines can be transformed or otherwise edited.

In the Editable Spline panel, you will see many new tools. The first ones we will explore operate at the Spline sub-object level. To use these tools, select one or more Spline sub-objects, and then click the tool button in the Modify panel.

Mirror

The **Mirror** tool button mirrors a Spline sub-object. You can find this button on the **Geometry** rollout of the Editable Spline object's Modify panel. Mirror is particularly useful for making symmetrical shapes. You can mirror and copy an open spline to create a new spline. The new, mirrored spline can then be welded to the original to form a closed spline.

To use the **Mirror** option, select the Spline sub-object to be mirrored. Check the **Copy** check box to make a mirrored copy of the selected spline. Click the **Mirror Horizontally** , **Mirror Vertically** or **Mirror Both** button. These buttons work with the currently active viewport to determine the direction of the mirror.

Outline

The **Outline** tool duplicates a spline and makes the new spline slightly larger or smaller by an even increment all the way around the original spline. This option is much more effective than scaling if you wish to retain the overall form of the original spline.

To appreciate this point, see the accompanying illustration. The kidney-shaped spline on the left was scaled with **Select and Uniform Scale**. On the right, the same spline was enlarged with the **Outline** tool. The Outline spline keeps the original form better than the scaled spline.

The Outline tool is not perfect. Sometimes you will need to edit the new spline manually to get the look you want.

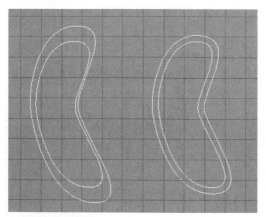

Figure 3-68: Spline copied and enlarged with scaling (left) and **Outline** *(right*

EXERCISE 3.9: Valentine

In this tutorial, you'll create a heart-shaped valentine candy.

Figure 3-69: Valentine candy

1. Reset 3ds Max. Turn on **3D Snaps** from the Main Toolbar. Now when you use a creation or transform tool, your cursor displays as a blue crosshair as it snaps to Grid points.

2. In the **Front** viewport, create a circle. Click at the origin to snap the center of the circle to the world origin. Drag to define the Radius. Then turn off 3D Snaps.

3. To make the heart, you'll start by deleting half of the circle. Right-click the circle and choose **Convert To > Convert to Editable Spline** from the Quad Menu.

4. Choose **Segment** as the sub-object level. Select the two segments that make up the left side of the circle.

5. Press **<DELETE>** on the keyboard to delete the selected segments.

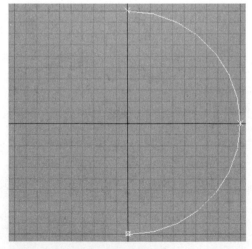

Figure 3-70: Segments deleted

6. Next, you'll shape the half circle into half a heart. Choose **Vertex** as the sub-object level. Click **Select and Move**. In the Front viewport, select the bottom vertex. The vertices of a circle are Bezier vertices by default, so they have handles.

7. In the Front viewport, move the bottom vertex downward in the Front view's Y axis. Do not move it left to right in the view's X axis.

8. Select the vector handle of the bottom vertex. Move the handle in the X and Y axes to make a slanted bottom to the spline. If you can't select the handle, turn off the Transform Gizmo with the **<X>** key, and choose the XY plane with **<F8>**.

9. Move the topmost vertex downward in the Y axis a little. Do not move it in the X axis. Move its vector handle so the segment points upward, as shown in the illustration.

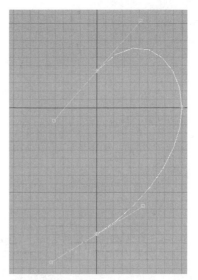

Figure 3-71: Half a heart

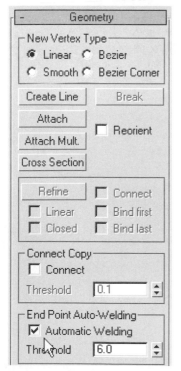

Figure 3-72: ***Automatic Welding*** *option*

10. Next, you'll mirror the spline to make the rest of the heart shape. Choose **Spline** as the sub-object level. Select the Spline sub-object in the viewport.

11. On the Modify Panel, locate the **Geometry** rollout. Look for the option labeled **Automatic Welding**. Turn it on by checking the box. With Automatic Welding turned on, the vertices will be welded if they come closer together than the **Threshold** parameter.

12. Also in the Geometry rollout, look for the **Mirror** tool button.

WARNING: Do not use the Mirror icon button on the Main Toolbar for this operation! Make sure you are using the text button labeled "Mirror" on the Editable Spline **Geometry** rollout.

13. Check the **Copy** check box located under the Mirror tool button on the Geometry rollout.

14. Click **Mirror**. The selected spline is mirrored in the viewport.

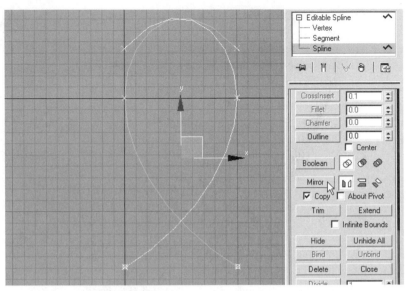

Figure 3-73: Mirrored spline

15. Move the new spline to the left in the Front viewport so its endpoints are placed over the other spline's endpoints. The vertices are automatically welded. Before welding, there were two open splines with a total of six vertices. Now there is one closed spline with four vertices. Click the spline, and both sides of the heart are selected.

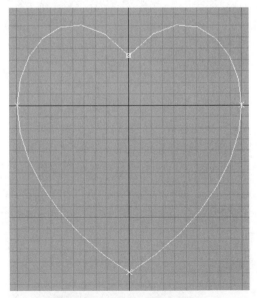

Figure 3-74: Moved and welded spline

16. Apply the **Extrude** modifier to the Shape. Increase the **Amount** spinner until the object looks like a valentine candy.

    Note that **Extrude** is applied to the entire object regardless of any sub-object selection still in effect. The Extrude modifier is different from most. Selections are not passed up the stack to the Extrude modifier, nor are they passed beyond it.

17. Change the color of the object to pink.

18. Save the file as **Valentine.max** in your folder.

LINES

You can draw a free-form Shape by using the **Line** tool. Actually, you create an **Editable Spline** object from scratch, without needing to collapse the Stack. The Line object and the Editable Spline object are identical.

To draw a Line, click and move the cursor to draw the points of the Line, then right-click when finished. You can also click over the first point to close the shape.

When you click to create a point, a **Corner** type vertex is created. To create a **Bezier** vertex, click and drag, then release the mouse when the Line is sufficiently curved.

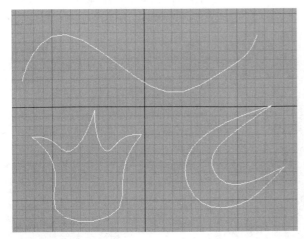

Figure 3-75: Lines

To edit a Line, you don't have to collapse the Stack. Just go to the **Modify** panel. There you can access the Vertex, Segment, or Spline sub-object levels.

Adding and Removing Line Sub-Objects

To remove vertices from a Line or Editable Spline, select the vertex and press the <**DELETE**> key. You can safely delete vertices in the middle of a spline. This will not break a closed spline or split a spline into two. However, if you use the <**DELETE**> key while in Segment or Spline sub-object mode, segments or splines will be deleted. Deleting a segment causes closed splines to be opened, and open splines to be split into two spline sub-objects.

To add vertices, use the **Refine** tool button in the **Geometry** rollout. You must be in Vertex or Spline sub-object mode. Press the Refine button and click to add a vertex in the middle of a spline. The curvature of the spline does not change, but the new vertex might not appear exactly where you clicked. Right-click to exit the Refine tool.

Creating a new Spline sub-object is easy. Press the **Create Line** tool button in the Geometry rollout, and click in the viewport to make the line. Right-click to exit the tool, as usual.

If **Automatic Welding** is enabled, then the Create Line tool will allow you to extend the end of a spline. Enable Automatic Welding, click Create Line, and then click an endpoint in the viewport. The next place you click will create a new vertex. Continue clicking to make more vertices.

The **Break** tool button lets you split segments and open splines by adding sub-objects instead of deleting them. Select a vertex and click **Break**; the segment is split, and there are now two vertices instead of one. In Segment sub-object mode, you can click anywhere in a segment to split it at that location and create two new vertices.

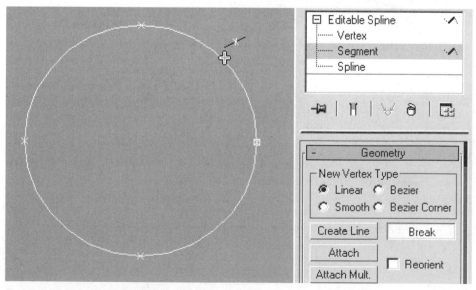

*Figure 3-76: Using the **Break** tool*

Here is a quick summary of other Editable Spline tools that can add or remove sub-objects:

Attach imports another Shape object into the current Editable Spline. The attached Shape is no longer a separate object; it is a new Spline sub-object within the current Editable Spline.

Connect creates a new segment between two vertices. Enter Vertex sub-object mode and click Connect. Click a vertex, hold the mouse button down, then drag to another vertex. A new straight line segment is created.

Insert is similar to Refine, but changes the shape of the curve. In any sub-object mode, activate the **Insert** tool, and click a spline to create and move new vertices.

Divide is also similar to refine, but adds vertices where needed to evenly divide a segment. Select a segment, and find the **Divide** button near the bottom of the Geometry rollout. Click the Divide button to create evenly spaced new vertices on the current segment. The curvature of the segment does not change. The number next to the Divide button defines the number of new vertices to create.

Fillet and **Chamfer** convert vertices into segments. These tools only work in Vertex sub-object mode. Activate the **Fillet** tool, and click and drag a vertex to create a smooth, curved segment. The **Chamfer** tool creates straight line segments.

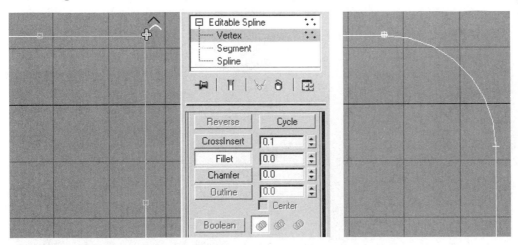

*Figure 3-77: Before and after using the **Fillet** tool*

EXERCISE 3.10: Lines

Being able to make lines quickly and easily is an important modeling skill. In this exercise, you'll practice making lines.

1. Reset 3ds Max.

2. On the **Create** panel, click **Shapes**. Click the **Line** button.

3. Maximize the **Front** viewport. Click and drag to start creating a Line. Move the cursor and click to create more of the Line. Right-click when the Line is complete.

4. Create more Lines. When you click to set a vertex, click and drag to make a curve in the Line.

5. When you have made several Lines, select one of them and go to the **Modify** panel.

6. Go to the **Vertex** sub-object level. You can click the word Vertex in the Modifier Stack, or you can click the **Vertex** button on the **Selection** rollout.

7. Select a vertex on the Line. If the vertex is not a **Bezier** vertex, right-click on it to access the pop-up menu, and choose **Bezier**.

8. Use **Select and Move** to adjust the vertex handles.

9. Continue to practice creating and editing lines until you are comfortable doing so.

10. Experiment with some of the sub-object tools, such as **Refine**, **Break**, and **Fillet**.

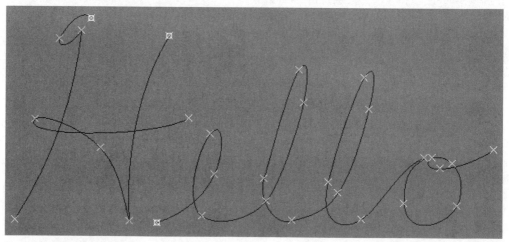

*Figure 3-78: A **Line** object*

LATHE

The **Lathe** modifier can be used to make a round object from a shape. Lathe spins a shape around an axis to form a 3D object. This is sometimes called a "revolve" or a "surface of revolution," because the Shape revolves around an axis to create the surface. The Lathe modifier is available only for 2D shapes and not for 3D objects.

Lathe works best on open 2D shapes with the axis of revolution placed at the open side of the Shape. You will have best results if the endpoints of the Shape are vertically aligned.

To make a lathed object, maximize the Front viewport and turn on **3D Snaps**. Create and edit your Shape in the Front viewport. Snap the Shape's endpoints to the vertical axis of the viewport. Do not edit vertices in the other viewports, or parts of the curve might move off the construction plane, resulting in problems with the Lathe modifier.

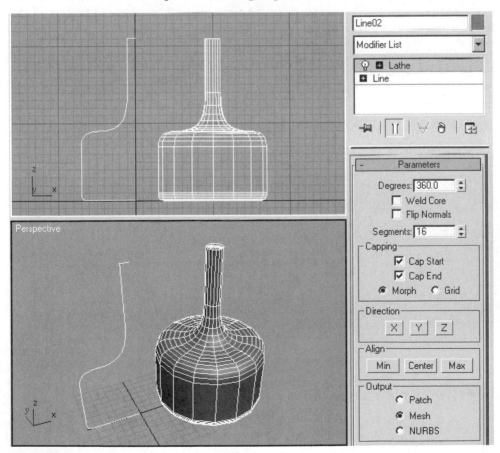

Figure 3-79: Shape lathed to create 3D object

The **X**, **Y**, and **Z** axis settings under the **Direction** section can be used to orient the axis of the lathe. The **Min**, **Center**, and **Max** buttons set the location of the Shape's local axis in relationship to the lathe axis. The easiest way to use these settings is to try each one until the 3D object looks the way you want it to.

The resulting lathed object might appear to be inside out. The easiest way to check for this problem is to look at the object in a shaded viewport such as the Perspective view.

If the object appears to be inside out, check the **Flip Normals** check box on the panel. If you're not sure if the object is inside out, check and uncheck the Flip Normals check box a few times and look at the object closely in a shaded viewport. You will usually be able to tell whether the check box should be on or off.

In most cases you will want the points at the center of the Lathe to be welded to prevent rendering errors. You may see this as flickering around near the top and bottom of the object in a shaded view. Turn on the **Weld Core** option to fix this.

Another common problem with Lathes is that the axis is not properly aligned. The Lathe modifier has a single sub-object type: the **Axis**. Sub-object Axis mode lets you move the Lathe Axis. In this way you, can edit the resulting 3D object and correct errors such as faces crashing through one another. This is often the problem when the object appears to be partially inside out.

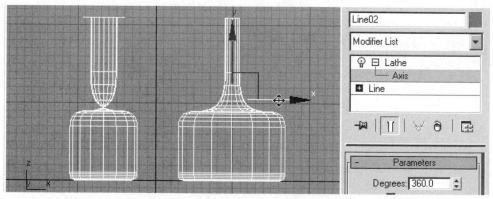

*Figure 3-80: Moving the Lathe **Axis***

EXERCISE 3.11: Lathed Bottle

In this exercise, you'll use the **Lathe** modifier to make a bottle.

1. Reset 3ds Max.

2. Turn on **3D Snap** 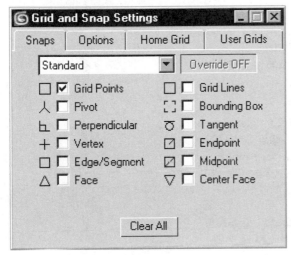. Right-click the 3D Snap button to access the **Grid and Snap Settings** dialog. In the dialog, make certain that **Grid Points** is the only item checked. Close the dialog.

*Exercise 3.81: **Grid and Snap Settings** dialog*

3. Maximize the Front view. Create a **Line** as shown in the following illustration. It can be either open or closed. The important thing is that the CVs at the Lathe Axis must be perfectly aligned vertically, in the world Z axis. That's why we used Snaps. Make sure you draw the Line on the right side of the screen, in the world's positive X axis.

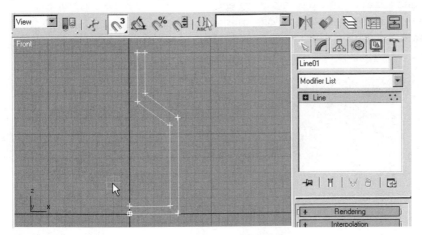

*Figure 3.82: Create a **Line***

4. Once you've set those CVs at the Lathe Axis, you can do whatever you want to the rest of them to shape the object. Your goal is to create a pleasing shape with a thin wall of revolved glass. Turn off 3D Snap with the **<S>** hotkey. Customize your bottle a little by adding CVs with the **Refine** tool if desired, then moving points around as usual. Only move points in the Front view. If you use any other viewport you will move the points off the same plane, which will yield strange results.

    To convert among the different CV types, such as Bezier and Corner, use the right-click Quad Menu. To reset the length of Bezier vector handles, convert the CV to **Smooth** and then back to **Bezier**.

TIP: The Move Gizmo is not necessarily your friend. Sometimes it gets in the way, especially when moving Bezier handles. In this case, use the **<X>** key to hide the Gizmo. Then press the **<F8>** key until the X and Y axes of the object tripod are both red. Now you only move the CVs in the XY plane of the Front view.

5. Pay careful attention to the points at the ends. The Bezier handles might be too long, causing the Line to cross itself. This must be corrected to prevent an illegal Lathe.

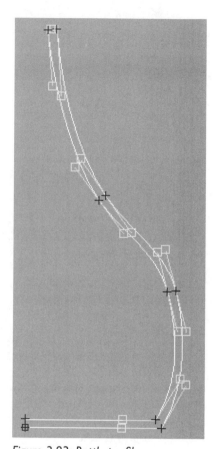

Figure 3.83: Bottle profile

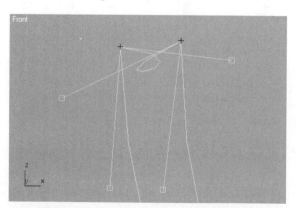

Figure 3.84: Problem: vector handles are too long

6. Experiment with some of the tools you haven't used yet. Try the **Fillet** tool on a corner CV. This converts it to a Segment. In the **Geometry** rollout, click the **Fillet** button, then click, hold, and drag the mouse over one or more control vertices. Release the mouse button when you've set the Fillet size you wish. You can also select vertices, click the Fillet button, and enter a size into the spinner field to the right of the button.

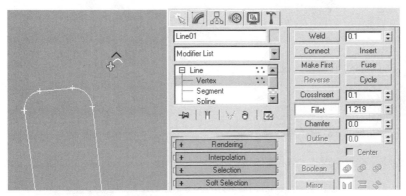

*Figure 3.85: Using the **Fillet** tool on two Bezier Corner vertices*

7. Before adding the Lathe modifier, double-check the vertices located at the Lathe Axis. In this case they are at the bottom left of the profile Line. They must be aligned along the world Z axis, or the resulting Lathe object will have holes in it.

8. When you're happy with the bottle profile, add the **Lathe** modifier. The first result is unsettling; the object is twisted inside out. To fix it, locate the **Align** section of the Lathe parameters. Click the **Min** button to set the Lathe Axis. The axis of revolution is sent to the world Z axis. See the following illustration.

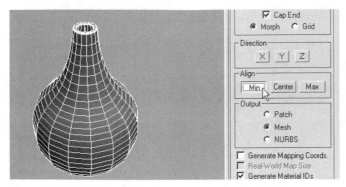

*Figure 3.86: Click the **Min** button to set the Lathe Axis*

9. Increase the number of **Segments** in the Lathe parameters. The object has smoother radial symmetry. You can also adjust the settings in the **Interpolation** rollout of the Line.

10. Use the **<ALT> + middle mouse** hotkey to Arc Rotate. Look at the bottom of the bottle in the Perspective view. If you see this problem, enable the **Weld Core** option in the Lathe parameters. It's probably a good idea to enable it anyway, because that should also lower the polygon count.

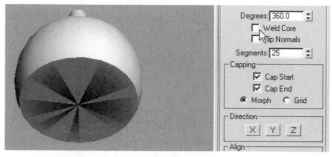

Figure 3.87: **Weld Core** *corrects a problem*

11. Go back to the Line level in the Modifier Stack, and resume editing CVs. Only move the points in the Front view. If you enable **Show End Result** 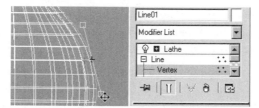, you can see the results on the 3D object in real time.

Figure 3.88: Use **Show End Result** *to view changes to the 3D object*

12. Save the file as **lathed_bottle.max** in your folder.

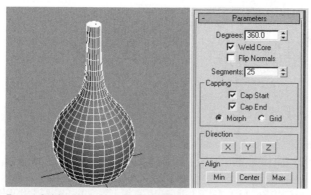

Figure 3.89: Completed Lathe object

SHAPE INTERPOLATION

When working with shapes and Editable Spline objects, sometimes you see straight lines where you expect to see curves. This is easily fixed by changing the parameters on the **Interpolation** rollout.

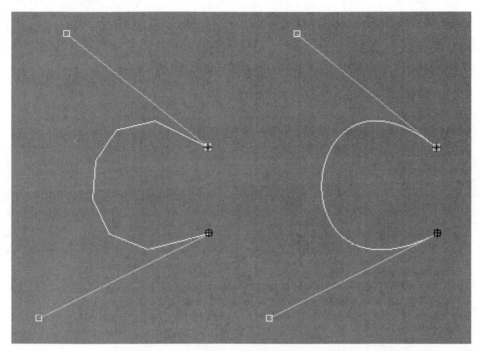

*Figure 3-90: Two identical lines with different **Interpolation** settings*

Interpolation is the automatic calculation of in-between values. The curvature of each segment is calculated from the positions of the CVs and from the Interpolation parameters. This means that the shape of the curve can change even if the control points don't move. Changing Interpolation values alters the smoothness of the curve.

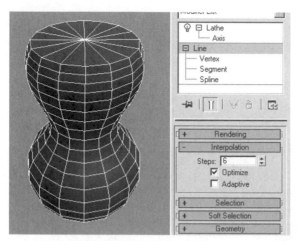

Figure 3-91: A Lathe object with default Shape Interpolation

The **Steps** parameter is the level of detail for Line segments. A higher Steps value creates more divisions along the length of the segment. At a Steps value of zero, the segments are always straight lines. A low Steps setting may be sufficient for objects that are rendered very small or far away from the camera.

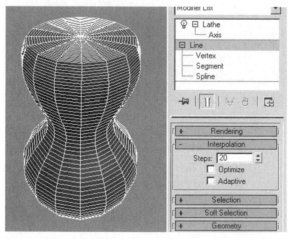

*Figure 3-92: Increasing the Line's **Steps** parameter to **20***

The Steps setting is global to all segments in the Shape, meaning that you have just one Steps setting for the whole object. If you have segments with vertices close together, and other segments with vertices far apart, you will see that one global Steps setting causes the longer segments to be more blocky and the shorter segments to be too dense.

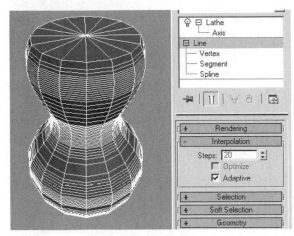

Figure 3-93: **Adaptive** *Interpolation*

The problem is easily solved with the **Adaptive** option in the Interpolation rollout. With Adaptive on, 3ds Max does a very good job of "filling in the blanks" for you, adding extra steps in extremely curved segments and removing steps where they aren't needed. Manual control of curvature via the Steps parameter is disabled.

If Adaptive is off, you have the **Optimize** option available. Optimize sets the step value to zero for segments that have Corner vertices at both ends. Optimize only affects linear segments, and leaves curved segments as they are defined by the Steps value. When Adaptive is on, then optimization is always enabled, no matter whether the Optimize box is checked or not.

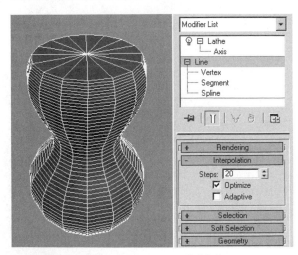

Figure 3-94: **Optimize** *option with* **20** *Steps*

You can see the effects of the Interpolation settings quite easily if you create a Lathe object with a mix of curved and straight line segments. View the Lathe in a shaded Perspective view with **Edged Faces** turned on. Use **<F4>** to turn on Edged Faces. Also remember to have **Show End Results** enabled in the Modify panel.

TIP: In some circumstances, Adaptive Interpolation is good enough. However, sometimes you will need more control over fine details of the curve. In this case, use manual interpolation and add or remove control vertices using the **Refine**, **Divide**, and **Delete** tools in the **Geometry** rollout of Editable Spline.

TEXT

Text objects are 2D parametric shapes that produce typography. They are easily turned into 3D flying logos using the **Extrude** or **Bevel** modifiers.

In the **Text** box under the **Parameters** rollout at the bottom of the Command Panel, replace the text **MAX Text** with your own text.

Click on the screen to place the text. After placing the text, change the **Size** as desired on the Command Panel.

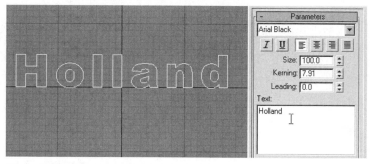

Figure 3-95: **Text**

You can create the text in any font available on your Windows system by choosing it from the pulldown at the top of the Parameters rollout. Since the Text object is parametric, you can change your text in the Modify panel. Basic typesetting features are available, such as **Alignment, Kerning, Leading**, and **Italics**. However, keep in mind that direct editing of the geometry is sometimes necessary to achieve precise spacing. Convert the object to an Editable Spline using the Quad Menus, and move the Spline sub-objects as desired.

MODIFIERS FOR SHAPES

In addition to **Extrude** and **Lathe**, several additional modifiers can be used with shapes.

Bevel

The **Bevel** modifier extrudes an object, and also creates an outline of the shape as it extrudes. This modifier can be used to create a 3D object with angled edges, sometimes called a chamfer. It can also produce curved surfaces, similar to a fillet.

Up to three extrusions of the shape can be created by a single Bevel modifier. In the **Bevel Values** rollout, change the **Height** of Level 1, which is the first extrusion. Negative values are better, because the shape extrudes back into the viewport if you're in the default viewing position.

The angle of the bevel is controlled by the extrusion distance (Height) and the **Outline** parameter. Outline works similarly to the Outline tool in Editable Spline.

When Outline is zero, the shape remains the same as it extrudes. When Outline is increased, the shape becomes larger. Outline can also be decreased to a negative number to make the shape smaller. Bevel works best when Outline values are set to small increments such as **2.0** or **-1.5**.

WARNING: Some shapes, when beveled, result in 3D objects with sharp spikes or jagged edges sticking unexpectedly out of the object. This is because the **Bevel** modifier enlarges and shrinks the shape as it is extruded. If the shape has sharp corners or curves, the beveling process causes the shape to cross back over itself, resulting in spikes in the 3D object. The **Keep Lines from Crossing** option can help prevent the spikes, but you might need to edit the shapes manually.

EXERCISE 3.12: Beveled Text

In this exercise, you'll create beveled text suitable for a flying logo.

1. Reset 3ds Max.

2. On the Create panel, choose **Shapes**. Click the **Text** button.

3. In the text entry area, the text currently reads **MAX Text**. Erase this text and enter your first name.

4. Click in the **Front** viewport to place the text.

5. Select the **Arial Black** font from the pulldown list. Leave the **Size** parameter at **100**.

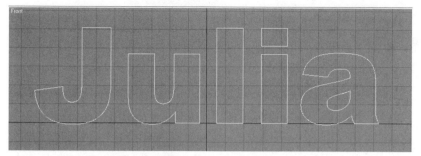

Figure 3-96: Text in Front viewport

6.   With the text selected, go to the **Modify** panel. Choose **Bevel** from the Modifier List.
      The Bevel modifier has been applied to the text. The text now has a visible surface in
      shaded viewports.

7.   Scroll down the panel to the **Bevel Values** rollout.These values will be used to
      extrude and outline the text at the same time.

8.   Leave **Start Outline** at zero. Under **Level I**, set **Height** to -4. Leave **Outline** at
      zero. Note that the text has been extruded by **4** units.

9.   On the **Bevel Values** rollout, check the box next to **Level 2**.

10.   Under **Level 2**, set **Height** to **-20**. The text has been extruded by another **20** units,
      which you can clearly see in the Left and Front viewports.

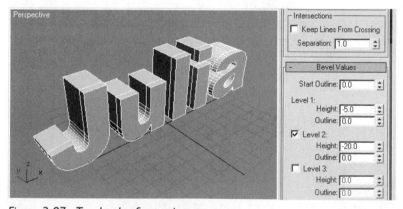

Figure 3-97: Two levels of extrusion

11.   Click and drag on the **Outline** spinner for Level I. Watch in all four viewports as the
      Outline value changes. If you drag it too far, you will get very unpleasant spikes in the
      geometry. Type a value of **2** for the Outline of Level I.

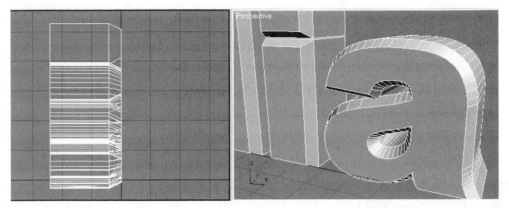

Figure 3-98: **Bevel** *in Left viewport*

The new extrusion has been beveled away from the center of each spline by **2** units, This beveled edge is visible in the **Front** and **Perspective** viewports.

12. You may notice that with an Outline value of **2**, some of the geometry in the letters is crashing into itself. This is called *self-intersecting geometry*, and it can lead to rendering errors. Fix it by reducing the **Start Outline** value to **-2**.

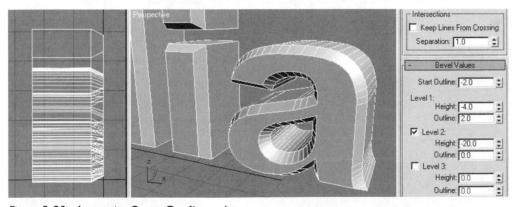

Figure 3-99: A negative **Start Outline** *value*

13. The bevel you just created is a straight edge. Next, you'll make a curved bevel.

    On the **Parameters** rollout, select the **Curved Sides** option. Change **Segments** to **3**. The increased number of segments makes it possible for a curved bevel to be created. When **Segments** is **1**, the bevel is forced to a straight edge regardless of whether Linear Sides or Curved Sides is selected.

14. If the beveled object has spikes sticking out of it or any other problems, adjust the Outline values. Since the object is parametric, you can change the font of the Text object on the Modifiy panel. This may require you to change the Bevel modifier parameters. You can also change the Interpolation settings of the Text object. **Adaptive Interpolation** works well with the Bevel modifier.

15. Experiment with other options and values on the **Bevel** panel until your text looks the way you want it to.

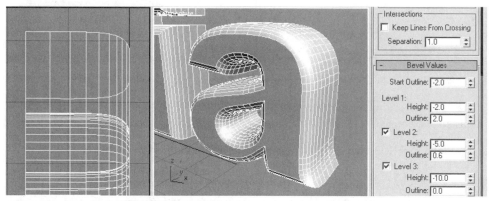

Figure 3-100: **Curved Sides**

16. Save the scene with the file name **Bevel.max** in your folder.

Bevel Profile

The **Bevel Profile** modifier uses a shape to set the profile of the beveled object. This modifier works best when the shape used as the profile is small in comparison to the shape being beveled.

The profile shape can be as fancy as you like. However, intricate edges or curls on the shape being beveled might result in jagged edges or spikes in the resulting 3D object. For example, using **Bevel Profile** on text in a font with *serifs* (small "tails" at the tops and bottoms of letters) might cause unwanted spikes in the beveled object.

EXERCISE 3.13: Bevel Profile Modifier

In this tutorial, you'll create beveled text using a profile shape and the **Bevel Profile** modifier. You'll use the same text you created in the last tutorial.

1. Load the file **Bevel.max** that you created in the previous tutorial. If you did not do the previous tutorial, load **Bevel.max** from the disc that came with this book.

2. Select the beveled text.

3. Go to the **Modify** panel.

4. Remove the **Bevel** modifier by clicking the **Remove Modifier from the Stack** button. The text becomes a shape again, rather than a 3D object.

   Next you'll create the profile for the bevel, using a rectangle shape as a base.

5. Go to the **Create** panel and click **Shapes.** Click the **Rectangle** button.

6. In the **Top** viewport, draw a small rectangle, and set the following values for it.

Length	15
Width	3
Corner Radius	1

7. In the **Top** viewport, zoom in on the rectangle.

8. Right-click the Rectangle and select **Convert To > Convert to Editable Spline**.

9. Go to the **Segment** sub-object level. Select and delete the long lengthwise segment on the left of the rectangle. Exit sub-object mode.

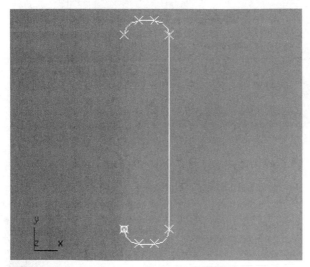

Figure 3-101: Delete one segment of profile object

10. Select the text. Choose **Bevel Profile** from the Modifier List.

11. On the Modify panel, click the **Pick Profile** button, then click the rectangle profile shape.

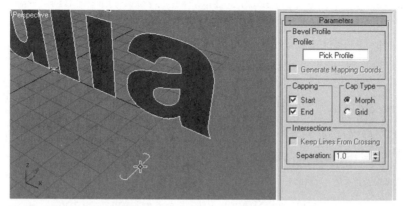

Figure 3-102: **Pick Profile**

12. You may notice self-intersecting geometry in the Bevel Profile object. In the following illustration, the geometry of the letter "A" is self-intersecting, and there is not enough space between the letter "I" and its dot.

Figure 3-103: Beveled text

13. In the Modifier Stack, activate the **Profile Gizmo** sub-object of the Bevel Profile modifier. In the Top viewport, use Select and Move to move the Profile in the X axis. The position of the Profile sub-object affects the thickness of the lettering. Watch the shaded Perspective view closely as you move the Profile. Adjust it so that there are no self-intersecting polygons or holes in the model.

*Figure 3-104: Move the **Profile Gizmo***

14. Save the scene with the filename **BevelProfile.max** in your folder.

Sweep

Using an Extrude or Bevel modifier lets you create a 3D object from a 2D shape by extending the shape along a straight line. You can't use these modifiers if you want to extrude a shape along any other path, such as a curved line or a closed shape. The **Sweep** modifier, new to 3ds Max 8, is a quick and easy way to extrude shapes along a complex path. In prior versions of the program, the **Loft** Compound Object was the way to do this. We'll take a look at advanced Loft techniques later, but first let's examine Sweep, which is quicker and easier to use.

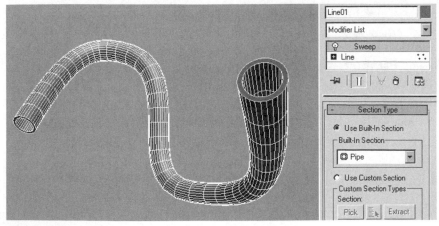

*Figure 3-105: Line object with **Sweep** modifier applied*

To create a Sweep object, start with any 2D primitive or Editable Spline. This is the path, to which you apply the Sweep modifier. You can choose from a number of preset shapes, called Sections, to extrude along the path. You can select a custom Section that you have created. In this way, you can create a wide variety of interesting objects. The most common use of a Sweep is to create architectural details, such as steel girders or picture frames.

EXERCISE 3.14: Picture Frames with Sweep

In this exercise you'll use the Sweep modifier to create two picture frames: one modern, metal frame, and one carved oval frame.

1.   Open or reset 3ds Max. In the Front view, create a 2D primitive **Rectangle** Shape with a **Length** of 120 and a **Width** of 240 units.

2.   Apply the **Sweep** modifier. In the **Section Type** rollout, choose **Wide Flange** from the **Built-In Section** pulldown list.

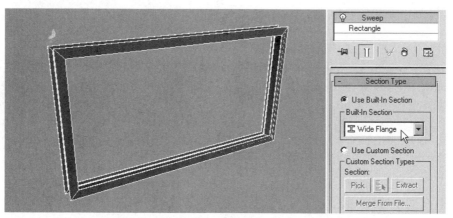

*Figure 3-106: Choose **Wide Flange** from the **Built-In Section** pulldown list*

3.   In the **Sweep Parameters** rollout, set the **Angle** to **90** degrees. The Section rotates around the path, producing a different form to the picture frame.

4.   In the **Parameters** rollout, adjust the **Length**, **Width**, **Thickness**, and **Corner Radius** to your liking.

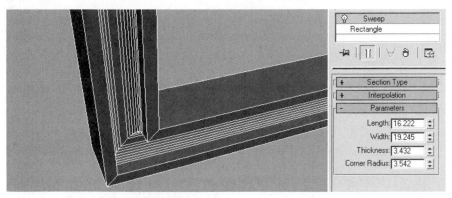

*Figure 3-107: Adjust the Sweep **Parameters***

5. Create a 2D primitive **Ellipse** Shape in the Front viewport. Give it a Length of **150** and a Width of **100** units.

6. Zoom in the Front view, so the Ellipse is much larger than the available viewport area. Create a closed Line object similar to the one shown in the following illustration. Snap to the Grid so that two sides of the Line make a right angle, as shown.

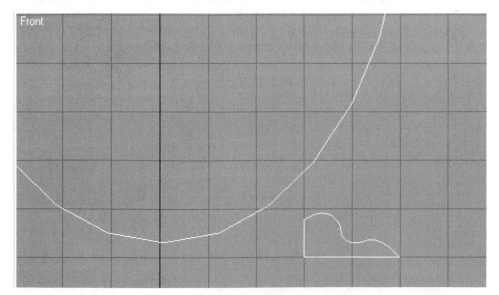

*Figure 3-108: Create an **Ellipse** and a closed **Line***

7. Apply a Sweep modifier to the Ellipse. In the **Sweep Parameters** rollout, set the **Angle** to zero. In the Section Type rollout, choose the Use Custom Section option. Then click the **Pick** button and click the Line object in the viewport. The Line is swept along the Ellipse, creating a picture frame that looks carved from wood. See the following illustration.

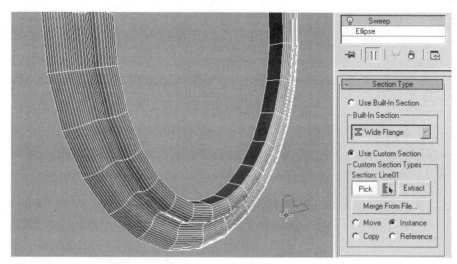

*Figure 3-109: Choose the Line as the Sweep **Section***

8.   Notice how the carved frame looks rounded along its profile, because the Section is rounded. But the frame doesn't have enough detail around its radius. To make it look better, select the Ellipse in the Modifier Stack. Make sure **Show End Result** is enabled. Open the **Interpolation** rollout and enable the **Adaptive** check box. Now the picture frame has much more detail around its radius. The number of polygons in that direction is determined by the curvature of the Ellipse.

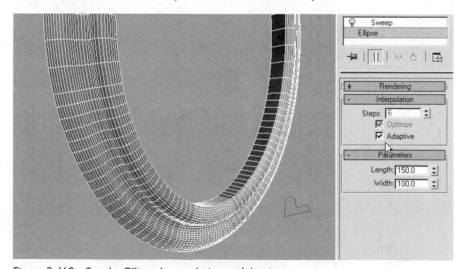

Figure 3-110: Set the Ellipse Interpolation to Adaptive

9.   Save the scene as `sweep_frames.max` in your folder.

CUSTOM MULTI-SPLINE SHAPES

A shape can consist of more than two embedded splines. In the following illustration, Spline sub-objects are enclosed or *nested* within one another. When this shape is extruded, the outermost spline defines a solid area. The next spline inward defines an open area, the next spline defines a solid area, and so on.

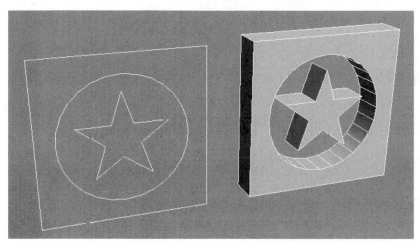

Figure 3-111: Nested splines

To extrude a multi-spline shape, any spline inside another spline must be completely inside or outside with no overlapping segments.

Earlier, you saw how to make a multi-spline shape with a 2D primitive, a **Donut**. You can also make a custom multi-spline shape from several individual shapes.

Start New Shape Check Box

One way you can create your own multi-spline shape is by turning off the **Start New Shape** check box at the top of the Shapes Create panel. When Start New Shape is disabled, and a shape is selected, all new shapes are created as Spline sub-objects within the current shape. Note that 2D primitives such as Circles lose their parameters and are automatically converted to Editable Splines.

Editable Spline Tools

There are several tools in the **Geometry** rollout of Editable Spline that let you create multi-spline shapes. Using the **Attach** tool, you can attach one or more Shape objects to the currently selected Editable Spline. In the Geometry rollout, click **Attach**, then click each shape that you wish to be part of the current object. The Modifier Stacks of the attached shapes are collapsed, destroying all parameters. Attached shapes become Spline sub-objects in the current Editable Spline.

A multi-spline shape can also be created with the **Outline** tool. Outline always creates a copy of a Spline sub-object.

Individual splines in a multi-spline shape can be moved by accessing the Spline sub-object level. Holding down the **<SHIFT>** key while moving a Spline sub-object creates a copy, just as the **<SHIFT>** key is used to create Clones of objects.

The **Create Line** tool lets you draw new Spline sub-objects directly within the current Editable Spline. If **Automatic Welding** is enabled, you can use Create Line to easily extend or close Spline sub-objects that are open.

Edit Spline Modifier

The **Edit Spline** modifier performs all of the functions of an Editable Spline object. At first this might seem to be a major advantage over collapsing the stack to an Editable Spline, because the object creation parameters are preserved when you use the Edit Spline modifier. Unfortunately, if you change any shape parameters that alter the vertex count, such as the number of sides of an NGon or the font of a Text object, the vertices will be scrambled.

WARNING: If you try to change the creation parameters of an object that has an Edit Spline modifier in the stack, you will get a warning about undesirable effects. This is the same dependency warning that appears when you try to change the Segments or Sides parameters of a 3D primitive while a Mesh Select is present in the stack. If you proceed with changing the creation parameters and end up scrambling the geometry, you'll be forced to remove the Edit Spline modifier, add a new Edit Spline, and do all of your edits over again.

The Edit Spline modifier has limited usefulness unless the parameters you wish to change don't affect the vertex count, such as font **Kerning** in a Text object. Also, Edit Spline modifiers use up RAM, so you're better off converting to an Editable Spline.

EXERCISE 3.15: Custom Multi-spline Shapes

In this exercise, you'll learn to attach shapes to make a multi-spline shape.

1. Reset 3ds Max. Press **<S>** to activate **3D Snap**.

2. On the Create panel, click **Shapes**.

3. Click **NGon**. In the **Top** viewport, click at the origin and drag to create an NGon with a **Radius** of about **100** units.

4. On the **Object Type** rollout, uncheck the **Start New Shape** check box.

5. Click **Circle**. In the **Top** viewport, click at the origin and drag to create a circle inside the NGon. Make sure the circle stays completely inside the NGon.

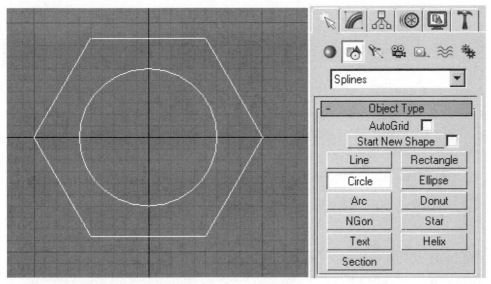

Figure 3-112: **Circle** *inside* **NGon**

You have created a multi-spline shape. When the **Start New Shape** check box is off, all shapes drawn become part of the same shape.

6.   Go to the **Modify** panel. Note that the object type shown in the Modifier Stack is not an NGon or Circle, but an Editable Spline.

7.   Apply the **Extrude** modifier to the shape. Change the **Amount** parameter to **30**.

The two shapes are extruded together, with the circle creating a hole in the rectangle.

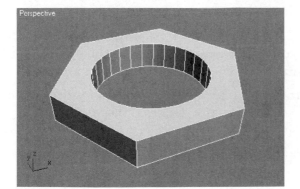

Figure 3-113: Extruded Editable Spline

Next, you'll use the Attach tool to make a multi-spline shape.

8.   In the **Top** viewport, next to the NGon and Circle, create a second NGon of dimensions similar to the first one.

9. Make sure the **Start New Shap**e check box is enabled. Create a circle inside the new NGon, similar to the first circle.

   You have just created two separate shapes. Now you'll use the **Edit Spline** modifier to attach them into one shape.

10. Select the NGon. Go to the **Modify** panel and apply the **Edit Spline** modifier.

11. On the **Geometry** rollout, click **Attach**. Click the circle to attach it to the NGon. Click the **Attach** button again to turn it off.

   The shapes are now attached into one multi-spline shape.

12. Apply the **Extrude** modifier to the shape.

   The shapes that were attached with the **Edit Spline** modifier result in exactly the same object as those created with the **Start New Shape** check box turned off.

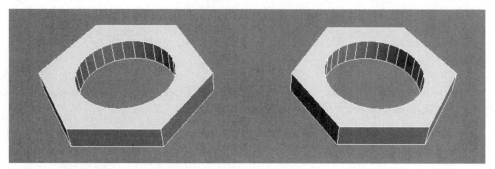

Figure 3-114: Identical objects created two different ways

13. Select the **NGon** creation parameters at the bottom of the stack. The Dependent Modifier warning pops up. Click **Yes** to continue to the NGon parameters.

14. Make sure **Show End Result** is enabled. Adjust the **Radius** parameter of the NGon. Since the Radius doesn't change the number of vertices in the shape, this parameter is safe to change. It is also OK to adjust the settings in the **Interpolation** rollout.

15. Try adjusting the **Sides** or **Corner Radius** parameters. The geometry is scrambled because the number of vertices in the NGon have changed. Change the Sides and Corner Radius parameters back to their defaults of **6** and **0**.

16. Save the file as `MultiSpline.max` in your folder.

Boolean Operations on Splines

The **Boolean** tool within the Editable Spline object can be used to join two splines, cut one spline from another spline, or extract the overlapping area of two splines. A Boolean operation works on two splines at a time, both of which must be part of the same shape. The Boolean tool only works in Spline sub-object mode. In addition, the two splines being joined, subtracted, or intersected must be closed, and they must overlap one another.

To get both splines into the same shape, use any of the methods described in the previous section. To perform a Boolean operation on splines, access the **Spline** sub-object level. Look for the **Boolean** button on the **Geometry** rollout. Choose a Boolean operation, **Union** ⬡, **Subtraction** ◉, or **Intersection** ◉ . Select a spline, then click the **Boolean** button and click the other spline. The Boolean operation will be performed, leaving a single spline.

EXERCISE 3.16: Bubble Blower

In this exercise, you'll create a bubble blower object with a multi-spline shape using several tools, including Boolean operations.

Figure 3-115: Bubble blower

1.   Reset 3ds Max.

2.   In the **Front** viewport, create a circle with a **Radius** of about 75 units.

3.   At the top of the **Shapes** Command Panel, turn off the **Start New Shape** check box.

4. In the **Front** viewport, create two more circles overlapping the first circle.

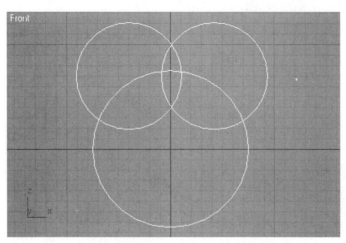

Figure 3-116: Overlapping circles

5. Go to the **Modify** panel. Activate **Spline** sub-object level.

6. Select one of the splines. Scroll down to the **Geometry** rollout and locate the

    Boolean button. Click **Union** [⊗].

7. Click the **Boolean** button, and click on the other two splines. The intersecting areas are removed and the three splines are joined together.

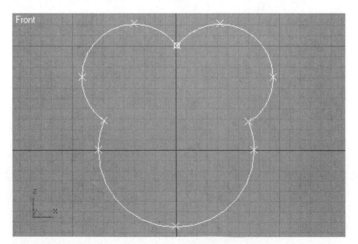

*Figure 3-117: Circles combined with Boolean **Union***

8. Right-click in the viewport to turn off the **Boolean** button.

9.   Next, you'll create an outline spline for the shape. Just above **Boolean**, next to **Out-line,** enter a value of **10** and press the **<ENTER>** key. An outline is created outside the original spline.

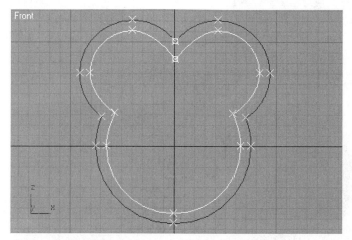

Figure 3-118: Outline spline

10.   Zoom out the **Front** viewport. Create a long Rectangle shape that slightly overlaps the outside spline, but extends below it, as shown in the following image.

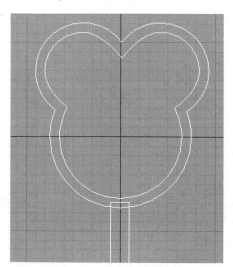

Figure 3-119: Rectangle overlapping spline, close-up view

11.   Select the first shape and go to the Modify panel. Click the **Attach** button, then click the Rectangle. The rectangle is now attached to the Editable Spline object. Click the **Attach** button again to turn it off.

12. Go to the **Spline** sub-object level. Select the rectangle, click **Boolean**, and click the outline spline. The rectangle is joined to the spline.

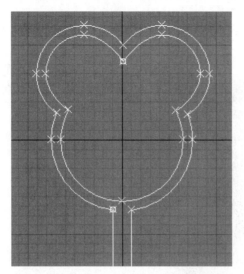

Figure 3-120: Rectangle and outer spline joined with Boolean Union

13. Apply the **Extrude** modifier to the shape, with an **Amount** of about **15** units.

14. The bubble blower is now complete. Save the scene with the file name **BubbleBlower. max** in your folder.

LOFT COMPOUND OBJECT

A Loft Compound Object is very similar to the Sweep modifier, but it has many more options. A Loft can make a variety of objects by extruding a cross-section shape along a path. You can use any shape for the cross-section. Lofting an open shape gives you an object with no normals on one side. The path is always a single open or closed spline, such as a Circle primitive or a Line with only one Spline sub-object.

The main advantage of a Loft over a Sweep modifier is the ability to use more than one cross-section shape along the path. You can also deform the cross-section shapes as they are extruded down the path. As a result, the variety of objects you can create with lofting is much greater than is possible with the Sweep modifier.

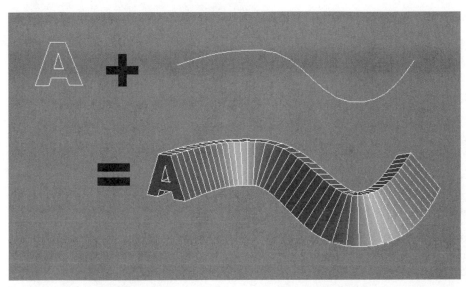

*Figure 3-121: Cross-section shape, Loft path, and resulting **Loft** object*

BASIC LOFTING INSTRUCTIONS

When lofting, you will always follow the same basic procedure.

1. Create two shapes, a Loft path and the shape to loft.

2. Select the path.

3. On the **Create** panel, click **Geometry** , and choose **Compound Objects** from the pulldown list.

4. Click on the **Loft** button.

5. Click **Get Shape**, then click the cross-section shape to loft.

Get Shape or Get Path?

You can also start by selecting the shape and clicking **Get Path**, but the procedure described above gives more intuitive results.

If you select the path before clicking Loft, then click Get Shape, the shape jumps to the path. If you select the cross-section before clicking Loft, then click Get Path, the path jumps to the shape.

CHANGING THE LOFT

When you create a Loft object, the **Instance** option on the **Creation Method** rollout is on by default. Remember that instances allow you to change one object and have the other follow suit.

In the case of Loft objects, this means that when you change the original shape or path, the Loft object changes accordingly. To best see this, move the Loft object away from the original path. Select the path and access the **Vertex** sub-object level. Move any vertex on the path. The Loft object changes according to the new path.

Earlier, you learned that Instance objects are affected by modifiers, but not transforms. Changes to modifiers or sub-objects are observed across instances, but instances have thier own transforms.

The same is true for Lofts. If you move a shape object as a whole, the instanced Loft object is not affected. If you move a vertex on the shape, however, the Loft is changed.

PATHS AND CORNER VERTICES

When creating a path for lofting, pay special attention to the vertices at both ends of the path. Sometimes the ends of the resulting Loft might have a hook or crease instead of a flat cap.

To avoid this situation, use **Corner** and **Smooth** type vertices at the ends of the path. These vertex types are the most reliable, but may not give you the form you need. **Bezier Corner** vertices give you more control, but can also give you trouble. Check the vector handles on the end vertices and correct them if necessary.

A related but less serious problem can also occur when a Bezier vertex is at the end of a Loft. The Loft object segments can be irregularly distributed. This can cause problems later on when you attempt to put more than one shape on the path.

To convert a vertex, right-click it and select a vertex type from the Quad Menu. Changing the vertex type can be done either before or after the Loft is created, as long as the Loft was created with the **Instance** option enabled.

TIP: Sometimes it is necessary to reset a Bezier vertex, so that its vector handles are the default length and position. To reset a Bezier vertex back to its default vector handles, convert it to a Smooth vertex and then back to Bezier.

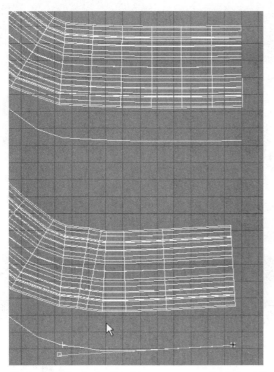

*Figure 3-122: Loft with **Corner** vertex (top) and **Bezier** vertex (bottom). Notice the irregular spacing of the bottom Loft.*

PATH AND SHAPE DETAIL

Earlier, you learned that changing the Interpolation settings for a Shape will change the level of detail for a dependent Lathe. However, if you change the Interpolation settings of the shape or path of a Loft object, the resulting compound object is not affected. This is one of the quirks of working with Lofts.

For a Loft, the level of detail is set not on the shape or path object, but on the Loft object's **Skin Parameters** rollout. Look in the **Options** section. **Path Steps** sets the number of steps between each vertex on the path. **Shape Steps** sets the number of steps between each vertex on the shape. There are other controls in the Skin Parameters rollout, such as **Optimize Shapes** and **Adaptive Path Steps**. These are similar to the controls in the Interpolation rollout of a Shape object.

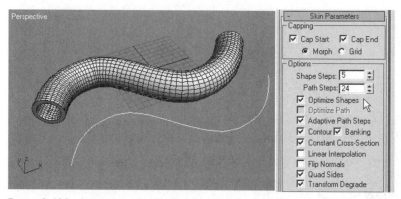

*Figure 3-123: Increasing the **Path Steps** parameter*

To keep the number of faces and vertices in your scene to a minimum, reduce **Path Steps** and **Shape Steps** to the smallest number possible while still retaining enough detail to make the Loft object look smooth.

TIP: If certain areas of your Loft need more or less detail, remember that you can edit the vertices of the instanced shape and path. Use the **Delete, Refine,** or **Divide** tools in Editable Spline.

LOFTING TWO OR MORE SHAPES

Several shapes can also be lofted sequentially along one path. To do this, select the path, and use **Get Shape** to put the first shape at one end of the path.

Under the **Path Parameters** rollout, change the **Path** value. Your current path location is marked with a small yellow X that you can see on the path. If **Percentage** is on, the Path value is the percent down the path where the next shape will be placed. If **Distance** is on, the Path value is the distance in units where the next shape will be placed.

When the Path value has been set to the desired percentage or distance, click **Get Shape** and choose the next shape. The shape is placed wherever you have positioned the yellow X.

You can place as many shapes as you like on the path in this manner. Shapes placed on a path don't have to have the same number of vertices, but they do have to have the same number of splines. For example, you cannot place a **Circle** and a **Donut** on the same path because the Circle has one spline, while the Donut has two.

LOFT SUB-OBJECTS

If you create a Loft with the **Instance** option selected, you can edit the sub-objects of the original shapes to make changes. However, sometimes it's faster or easier to make those changes directly to the Loft using its sub-objects.

A Loft has two sub-object types: **Path** and **Shape**. The Path sub-object only allows you to create a new instance of the Path, in case the original one is missing. The Shape sub-object, however, gives you a lot of control over the Loft. You can transform shapes in their own local coordinate systems. For example, moving a Shape in its Z axis will move it down the Path. In this way you can control the transitions between Shapes more precisely.

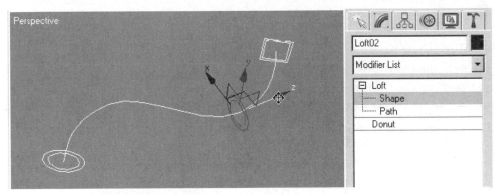

*Figure 3-124: Moving a **Shape** sub-object*

Choose Shape sub-object level, then select the shape directly on the Loft object in the viewport. Because shapes and wireframe edges are both displayed in white, seeing Shape sub-objects can be difficult in wireframe views. Uncheck the **Display > Skin** option at the bottom of the Skin Parameters rollout, and the wireframe is hidden, leaving only the Path and Shapes. If you need to see the mesh skin, you can also use a shaded viewport with Edged Faces turned off, allowing you to see the Loft Shapes. Selected Shapes are highlighted in red.

While in Shape sub-object mode, you can use the Select and Move tool on the Main Toolbar to move shapes along the path. To delete a shape within a Loft, simply select it and press <DELETE>. You can also clone a Shape to another location on the Path. Hold down the <SHIFT> key and move a shape along the Z axis to copy it. Rotating a Shape sub-object around its Z axis is one way of fixing a twisted Loft.

LOFT DEFORMATION

After you create a Loft, you can go to the Modify panel and access the **Deformations** rollout. The five kinds of deformation listed there can be used to further adjust the Loft object.

Scale scales the shape as it goes along the path.

Twist rotates the shape around its Z axis. The shape turns around the path.

Teeter turns the shape along its X or Y axis.

Bevel makes beveled edges, usually at the start and/or end of the path.

Fit uses two profile shapes and fits the Loft object to them, to make smooth, complex shapes.

Deformation Windows

When you choose a Loft deformation, a window appears. Here you can edit Bezier curves that control the deformation effect. Each deformation window is slightly different, but for the most part they all work the same way.

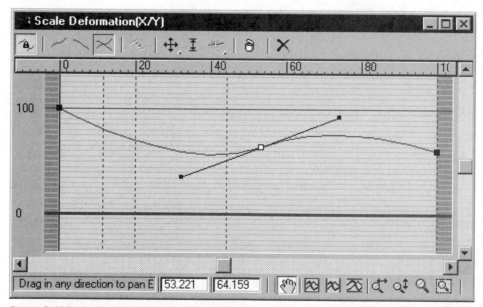

*Figure 3-125: **Scale Deformation** window*

The curves inside the deformation window are the *deformation curves*. These are similar to animation function curves in the Curve Editor. The height of a deformation curve determines the amount of deformation.

By default, a deformation curve is straight with two endpoints. The points on the **Scale** deformation curve default to **100** at each end, meaning the Loft shape is **100%** of its original size.

The numbers across the top of the window refer to a percentage of the path distance. The left end of the deformation curve affects the Loft object at the beginning of the path, while the right end affects the end of the path.

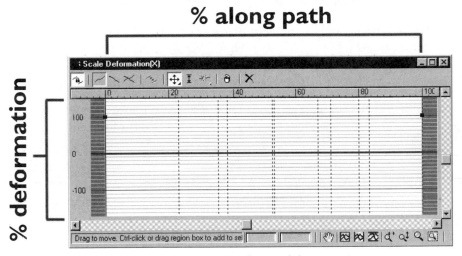

*Figure 3-126: Loft **Deformation** window numbers and their meanings*

Deformation Control Points

A deformation curve is changed by adding points to it and moving the points.

To change a deformation curve, add points to the curve by selecting **Insert Corner Point** from the Deformation window toolbar. You can also click and hold this button to choose the **Insert Bezier Point** button from the flyout. Then click the deformation curve to add a control point.

To select a point on the deformation curve, click **Move Control Point** and select the point. Points can also be selected by holding down the **<CTRL>** key while clicking each point, or by drawing a bounding box around points. Click and drag the point to move it.

When a point is selected, its percentage and value are displayed in entry boxes at the bottom of the deformation window. You can type values in these entry boxes.

A Loft Deformation can be performed in two dimensions: the X and Y axes of the Loft shape. You can choose to edit X and Y deformation curves together or separately.

Make Symmetrical causes the same curve to be used for both axes.

Display X Axis displays the deformation curve for the X axis as a red line.

Display Y Axis displays the deformation curve for the Y axis as a green line.

Display X/Y Axes displays the deformation curve for both the X and Y axis.

EXERCISE 3.17: Lofted Chili Pepper

In this exercise, you will use Scale and Twist Deformations to create a Lofted chili pepper. You will also add Noise to make the pepper look like it has been dried.

Loft the Pepper

1. Open 3ds Max, or if it is already open, select **File > Reset**.

2. In the Top viewport, create a **Circle** shape. The size and location of the circle are not important. Right-click to complete the circle creation.

3. Also in the Top viewport, create a **Line** with a simple curve as shown in the following illustration. Click to create a Corner point, then click-drag to create the second point (Bezier), and click to create the last Corner point. You are creating a Path for a lofted chili pepper. Make the Line approximately **200** units in length.

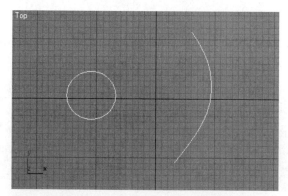

Figure 3-127: Shape and Path for chili pepper

4. With the Line object selected, open the Create panel to **Geometry > Compound Objects**. Activate the **Loft** command. Under the **Creation Method** rollout, make sure the **Instance** option is selected, and click the **Get Shape** button. Click the Circle to select it. Right-click to complete the operation.

5. Move the new **Loft01** object over a little so you can easily select and edit the original Line. Select the Line, open the **Modify** Panel, and enter sub-object **Vertex** editing mode. Adjust the Line's control points in the viewports to edit the shape of the curve. Note that the Loft object updates as you edit the control vertices; this is because the Loft Path is an instance of the Line.

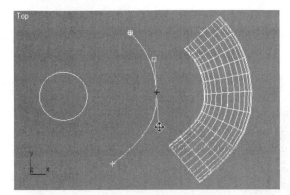

Figure 3-128: Edit the control points on the Path

6. Likewise, you can adjust the Loft Shape by editing the instanced object– in this case, a Circle. Reduce the **Radius** parameter of the Circle to about **5** units.

   A tubular shape will only work for the stem of a chili pepper. The fruit itself should have a cross section like a rounded hexagon. A Loft may have only one Path, but it can have more than one Shape. Therefore we can use a second closed spline object as another Loft Shape.

7. In the Top viewport, create an **NGon** shape. Give the NGon six sides. Set the **Radius** to about **30** units. Increase the **Corner Radius** amount to give the NGon soft, filleted corners.

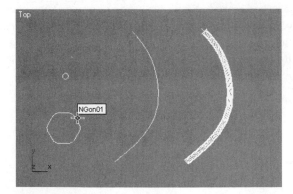

*Figure 3-129: Create an **NGon** with filleted corners*

Now you will add additional Shapes to the Loft.

8. Select the Loft object and maximize the Perspective view. Zoom and Rotate the view so the Loft is oriented diagonally on the screen. On the Modify panel, find the **Path Parameters** rollout. Click and drag the **Path** spinner to move the yellow cross on the Loft Path. Assign the Path parameter a value of about 7; this places the yellow cross at 7% of the total length of the Path.

    To add another Shape to the Loft, click the **Get Shape** text button. Then select the NGon object.

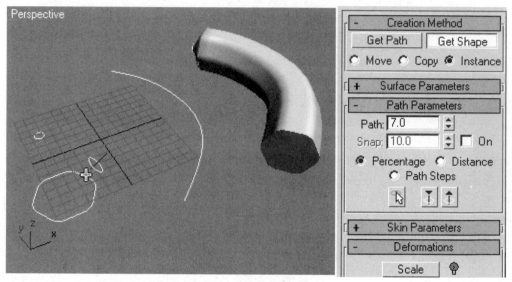

Figure 3-130: Add the NGon as a new Shape in the Loft object

9. You may adjust the position of a Shape after it has been added to the Loft. In the Modifier Stack, open up the Loft object, and activate **Shape** sub-object mode. Then click on the Loft Shape you wish to move, in this case, the instanced NGon. It highlights in red to indicate that it is selected. Use the Select and Move tool to translate the Shape along its Z axis. You can also do this by adjusting the **Path Level** parameter. Set the NGon Shape at around **20%** of the Path distance.

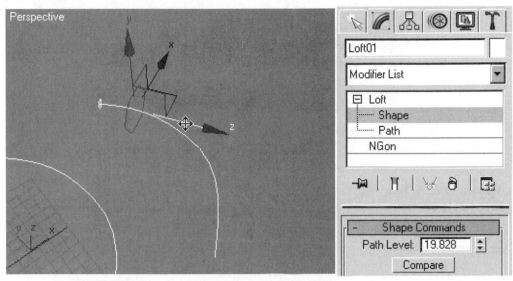

*Figure 3-131: Move the **Shape** sub-object in its Z axis*

10. To create a stem, we can reuse the Circle by placing it as a new Shape. Exit sub-object mode. Set the **Path** parameter to **18**. Click **Get Shape**, then click the Circle.

     This creates a tubular stem. The cross-section of the Loft is now perfectly circular up to **18%**, then makes a sharp transition from a circle to a rounded hexagon.

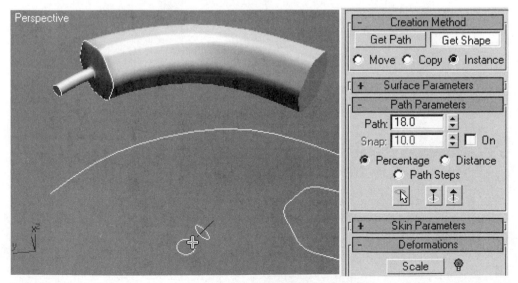

*Figure 3-132: Add the Circle again as a new Shape, at **18%** along the Path*

11. Select the Circle and reduce its Radius to create a more realistic stem. Both Circle Shapes in the Loft update simultaneously, since they are both instances of the Circle object. If you wish, go back to the Shape Sub-Object of the Loft and adjust the positions of the Shapes. The second Circle might look better at **10%** along the Path.

Deform the Loft

1. We could create more NGons to add as additional Shapes to complete the model, but there is an easier and better way. Exit sub-object mode if necessary, and open the **Deformations** rollout at the bottom of the Modify panel. Click the **Scale** button, and a new **Scale Deformation** window opens.

   Activate the **Move Control Point** tool on the Scale Deformation toolbar and drag the far-right control point down to about **20**. Observe the result in the Perspective viewport.

2. Add more points to the Scale curve using the **Insert Bezier Point** tool. Click and hold the **Insert Corner Point** flyout to access the tool. Then click on the curve to add a Bezier point at around **60%** along the path. Add another Bezier point at around **95%**. Then use the Move Control Point tool to adjust the Scale curve points.

   You may also right-click a control point to convert it among three vertex types: Corner, Bezier-Smooth, and Bezier-Corner.

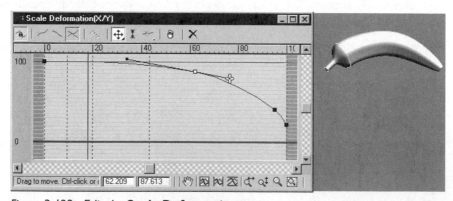

*Figure 3-133: Edit the **Scale Deformation** curve*

3. To make the chili pepper more realistic, we can make the scale deformation asymmetrical. To apply separate deformation curves to the X and Y axes of the Loft, disable the **Make Symmetrical** lock. Then choose to **Display X Axis** (red curve), **Display Y Axis** (green curve), or both. At first, both curves will be identical, so they will be superimposed. It's easiest to display one curve, edit it, and then display both curves to compare them. The orthographic viewports will show the results most effectively.

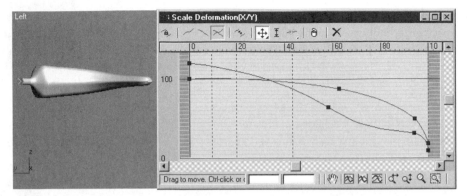

Figure 3-134: Separate X and Y Deformation curves result in a flattened pepper

4. Close the Scale Deformation window. Use the same deformation techniques to add a twist to the chili pepper. Click the **Twist** button in the Deformations rollout. Edit the Twist curve to produce the result you desire.

5. View the Perspective viewport in wireframe mode, using the **<F3>** keyboard shortcut. Open the **Skin Parameters** rollout in the Modify panel. Experiment with the settings for **Shape Steps** and **Path Steps**. More steps result in a denser, more complex mesh; fewer steps result in a coarser, simpler mesh object. Restore the Shape Steps and Path Steps back to their default values of **5**.

6. Try the **Optimize Shapes** option; the straight line segments from the NGon Shape are optimized and the overall mesh density is reduced. Disable Optimize Shapes to return the Loft to its original form.

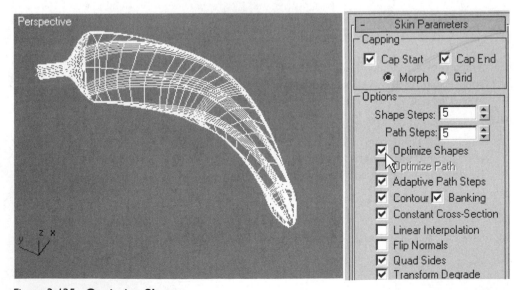

*Figure 3-135: **Optimize Shapes***

7. Disable the **Adaptive Path Steps** option. Now the density of the Loft along its Path is no longer influenced by the curvature of the original Path object or by the Deformation curves. With Adaptive Path Steps turned off, the Path steps are spaced at regular intervals. This is helpful in situations in which the Loft will be deformed by modifiers higher in the Stack.

In this exercise, we will use the **Noise** modifier to add random surface detail to the Loft. For the Noise modifier to work most effectively, the polygons of the Loft mesh should be laid out in a fairly regular grid pattern. Increase the **Path Steps** parameter to **25**.

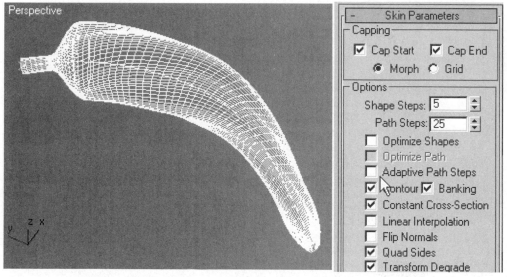

*Figure 3-136: **Adaptive Path Steps** option disabled; **Path Steps** parameter set to 25*

8. Add a **Noise** modifier. Reduce the **Scale** parameter to about **50**. Increase the X, Y, and Z **Strength** to about **10**. Enable the **Fractal** option and reduce the number of **Iterations** to **3**.

Experiment with the Noise Parameters. **Seed** will randomize the Noise, producing different random forms for the chili pepper. Fractal **Roughness** will amplify the effect of the Fractal option, giving more pronounced slopes to the noise. More Fractal **Iterations** will produce finer and finer surface details.

9. Increase the **Path Steps** in the Loft object to **50**. Increasing the Path Steps produces a higher level of detail, which in turn results in a better Noise effect.

10. Save the file as **LoftedChili.max** in your folder.

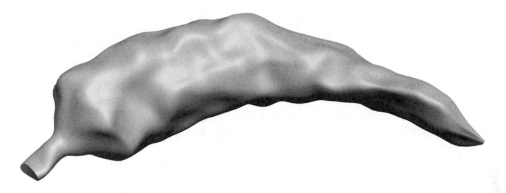

Figure 3-137: Completed chili pepper

POLYGONAL MESH OBJECTS

Earlier, you learned a little about **Editable Poly** and **Editable Mesh** objects. Now we will go into more detail about these powerful modeling tools.

EDITABLE MESH

A Editable Mesh object is essentially the most basic form of 3D model in 3ds Max. It has no parameters, such as Radius, which are found in primitives. Instead, Editable Mesh objects have sub-object components that can be directly edited. In this way, an Editable Mesh is similar to an Editable Spline.

The most fundamental method of working with an Editable Mesh is to transform its sub-objects. Moving, rotating, and scaling sub-objects changes the shape of the model. The sub-object types are explained as follows.

Vertex – A point on a mesh object. Vertices are not renderable. They are joined to one another by edges.

Edge – A line between two vertices. Some edges are visible in viewports, some are invisible by default.

Face – A renderable triangle with exactly three vertices and three edges. Each face has a renderable surface on one side only; the side that can be rendered is indicated by a face normal.

Polygon – A collection of one or more faces that all lie on the same plane. Faces within a polygon are bounded by invisible edges.

Element – A collection of polygons that usually defines a section or piece of the overall model.

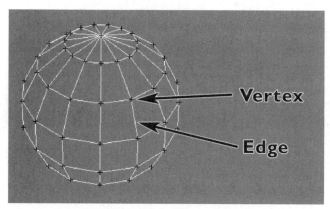

Figure 3-138: Vertices and Edges

Faces vs. Polygons

The distinction between faces and polygons is a subtle one. In some 3D programs, they are one and the same. However, 3ds Max is very precise, and faces and polygons are actually two different things.

A face is always a triangle. A triangle is the simplest possible 2D object, and the three points of a triangle always lie in the same geometric plane. Therefore, it makes sense to use triangles as the basic building blocks of renderable 3D objects.

In fact, when a scene is rendered in 3ds Max, every geometric model is rendered using triangles. Even nonmesh objects such as Bezier Patches and NURBS surfaces are converted to triangles in a process called *tessellation*. Tessellation is dividing a surface up into tiles. When a nonmesh object is rendered in 3ds Max, the object is automatically divided into triangular faces.

Polygons are collections of triangular faces. Polygons exist because, in general, they are easier to work with than faces. For example, it's easier to select a single polygon than it is to select the whole collection of faces that make up the polygon.

Strictly speaking, all of the faces within a polygon should lie on the same plane. This is a requirement of many 3D programs and game engines. In 3ds Max, you are allowed to create polygons that are not coplanar; in this case, all of the faces within a polygon do not lie on the same plane. In general practice, this is not a good idea, especially if you need to export your model to a different application.

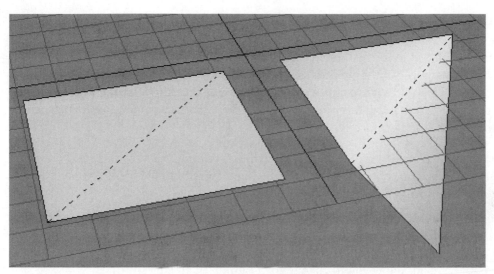

Figure 3-139: Coplanar polygon on left, non-coplanar polygon on right

Visible vs. Invisible Edges

Ordinarily, the triangular faces inside a polygon are not visible. 3ds Max hides the interior edges within polygons in order to speed up the display and make it less cluttered. These hidden edges are called *invisible* edges.

To see the invisible edges of an Editable Mesh object (or a primitive), select the object and go to the **Display** panel. Scroll down to the **Display Properties** rollout and uncheck the **Edges Only** option. The invisible edges are now displayed as dashed lines.

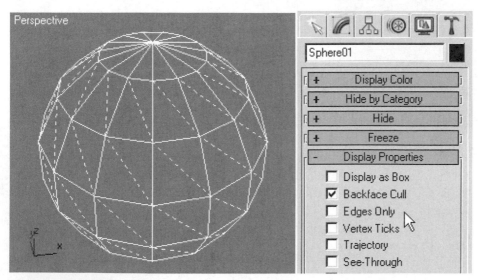

*Figure 3-140: Disable **Edges Only** on the Display Panel to see invisible edges*

WARNING: If you use the Direct3D display driver, all edges, including invisible ones, are drawn as solid lines by default. You will only see dashed edges if you select an Editable Mesh object and open the Modify panel. To make the Direct3D driver hide invisible edges in the traditional 3ds Max fashion, go to **Customize > Preferences > Viewports > Configure Driver** and uncheck the **Display All Triangle Edges** option. Disabling this option will negatively impact viewport performance on complex scenes, therefore the option is on by default.

The visibility of an edge is closely related to the distinction between faces and polygons. Visible edges define boundaries among polygons. Invisible edges define the boundaries among faces within a polygon.

You have the power to decide how 3ds Max interprets an edge. An invisible edge can be made visible, or a visible edge can be made invisible. Enter Edge sub-object mode, select an edge, and scroll down to the **Surface Properties** rollout. The **Visible** and **Invisible** buttons allow you to change the visibility of an edge.

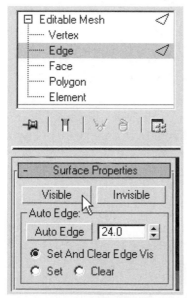

Figure 3-141: Edge visibility

Most polygons in 3ds Max have four sides and are composed of two triangular faces. In this case, the edge shared by the two faces is invisible. However, if you convert the invisible edge to a visible edge, you end up with two triangles, each with three visible edges. In this special situation, the triangles are considered to be polygons, since all of their edges are visible.

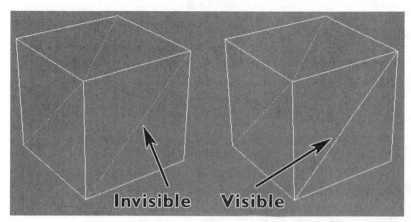

Figure 3-142: In the box on the left, all interior edges are invisible, so each side is a single polygon composed of two faces. In the box on the right, one of the edges has been made visible. There are now two polygons on that side, instead of one.

In low-polygon modeling, it is often necessary to make an edge visible to avoid creating polygons that don't all lie on the same plane. You might need to do this if you are exporting to another program or game engine that does not support non-coplanar polygons. The visibility of edges is also a major consideration when modeling with subdivision surfaces. You will learn more about this in Chapter 9, *Advanced Modeling*.

Edge Turn

The placement and orientation of edges on a mesh model has profound influence over how the model responds to changes in its structure. Even moving a single vertex can have unexpected or undesired effects unless you clearly understand how edges work.

In the following illustration, the center vertex at the top of a cube has been moved. Notice how the faces on the top of the object are not symmetrical.

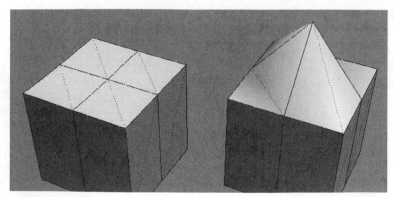

Figure 3-143: Moving a vertex creates an asymmetrical pyramid

This situation is easily fixed by using the **Turn** tool. In Edge sub-object mode, look in the **Edit Geometry** rollout. Click the Turn button to activate the tool, then click an edge. The edge is turned so that it joins the opposite pair of vertices on the polygon.

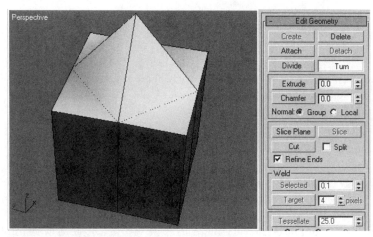

*Figure 3-144: Edge **Turn***

Soft Selection

Previously, you used the **Soft Selection** feature of Mesh Select to create a falloff effect for modifiers. With Soft Selection, the influence of a modifier decreases with distance away from selected sub-objects. In this way, you can create a more natural transition between areas of the model that are modified and those that are not.

Soft Selection is found in many places in 3ds Max. One of them is Editable Mesh. In this context, Soft Selection is typically used to create smooth curves on a mesh object. This is very difficult to achieve by transforming hard-selected mesh sub-objects, but Soft Selection makes it easy.

Soft Selection works by establishing a zone of influence around selected sub-objects. As you transform a sub-object, nearby sub-objects are transformed also. The farther away a soft selected sub-object is, the less it will be influenced by the transform. The radius of the zone of influence is determined by the **Falloff** amount. The shape of the Falloff is controlled with the **Bubble** and **Pinch** parameters.

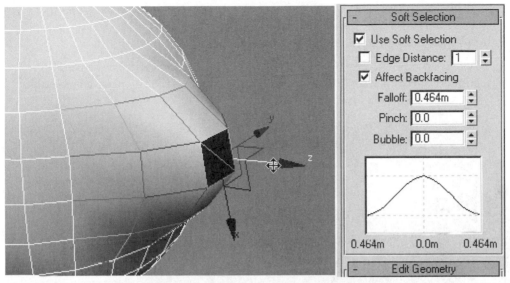

*Figure 3-145: Using **Soft Selection** to transform a polygon in the Local coordinate system*

Because the zone of influence is determined by distance measured in max units, the soft selection will usually change after you make a transform. For this reason, edits made with Soft Selection are generally nonreversible. For example, once you move a soft-selected region of polygons, it's nearly impossible to move them back so that the model is the same shape as before. This is one of the disadvantages of working with Soft Selection on mesh objects. If nondestructive editing of models is important to your project, you may decide to use patch modeling or subdivision surfaces. But for quick and simple organic curvature of mesh objects, Soft Selection is an extremely useful tool.

EXERCISE 3.18: Soft Selection

In this exercise you will use Soft Selection to create organic curvature on an Editable Mesh. You will move soft-selected vertices to create bumps on the bottom of the apple you created earlier.

1. Open the file **Apple.max** from the disc.
2. Select the apple in the Top viewport. Right-click the apple and choose **Convert To > Convert to Editable Mesh** from the Quad Menu.

3. Right-click the viewport label in the upper-left corner of the Top view. A pop-up menu appears. Select **Views > Bottom** from the viewport label menu. The viewport switches to the Bottom view.

4. Activate the Perspective view. Hold down the **<ALT>** key and press the **middle mouse button** to Arc Rotate around the scene. Rotate the Perspective view so that you can see the bottom of the apple.

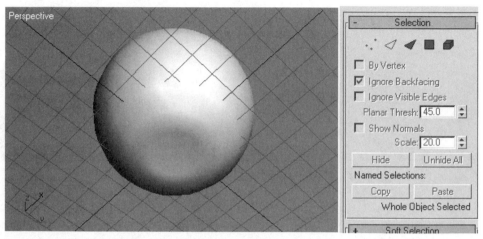

*Figure 3-146: Rotate the Perspective view with **<ALT> + middle mouse button***

WARNING: You need a three-button mouse to work effectively in 3ds Max. If you have a wheel mouse, press the wheel down to use the middle mouse button. You may need to change settings in the Windows **Mouse Control Panel**, so that the mouse wheel functions as the middle button when pressed.

5. On the Modify panel, select **Vertex** sub-object mode. Look in the **Selection** rollout. Check the **Ignore Backfacing** option. When this option is active, you can only select sub-objects that are facing toward you in the viewport. This prevents accidental selection of sub-objects on the side of the model that is facing away from you.

6. In the Bottom viewport, select a vertex as shown in the following illustration.

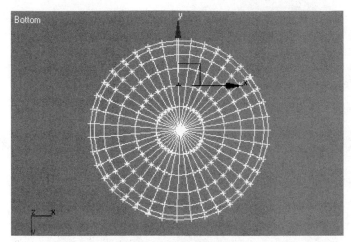

Figure 3-147: Select a vertex in the Bottom view

7.   Hold down the **<CTRL>** key to add to the current selection. Select three more vertices, for a total of four vertices in a cross configuration, as shown in the following illustration.

Figure 3-148: Four vertices selected

8.   On the Main Toolbar, locate the **Named Selection Sets** entry field. With the vertices still selected on the model, type the name **bottom_verts** in the Named Selection Sets field, and press the **<ENTER>** key. You have created a selection set of those four vertices. If you deselect the vertices, you can easily reselect them by choosing the **bottom_verts** selection set while in Vertex sub-object mode.

Figure 3-149: **Named Selection Sets** *entry field*

9.   In the Front view, move the four vertices down in the viewport's Y axis. Notice that this creates unpleasant sharp angles, unlike the smooth curvature of an apple.

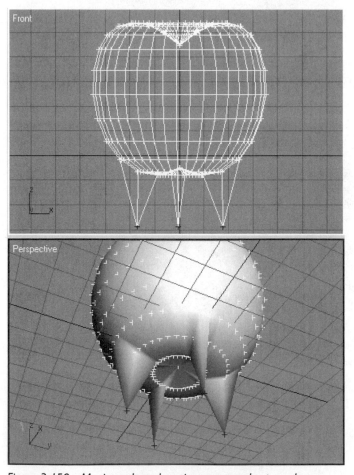

Figure 3-150: Moving selected vertices creates sharp angles

10. Hold down the **<CTRL>** key and press the **<Z>** key to undo the last operation. The apple is restored to its previous state.

11. Open the **Soft Selection** rollout. Activate **Use Soft Selection**. The vertices and edges of the model change color.

12. In the Front view, move the selected vertices down in the viewport's Y axis to create bumps on the apple. With the default Falloff value of **20**, the contours of the apple are still too sharp.

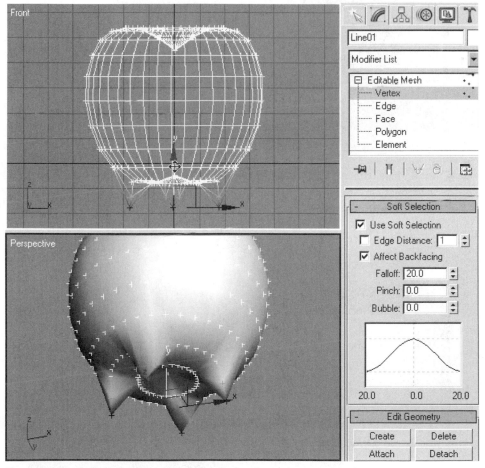

*Figure 3-151: Default Falloff value of **20** also produces sharp angles*

13. With the vertices still selected, attempt to move them back into their original position. It is not possible to restore the apple to its original form.

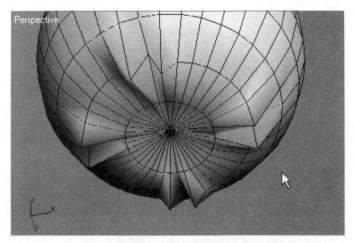

Figure 3-152: Moving vertices back to their original position creates a mess

14. On the Main Toolbar, right-click the **Undo** button. The history of most recent operations appears in a pop-up dialog. Select the second **Move** command in the list. When you click **Undo**, all recent operations up to and including the selected command are undone. The apple is restored to its initial state.

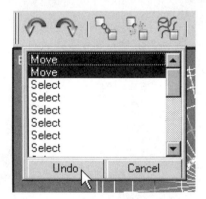

Figure 3-153: **Undo History**

15. Adjust the Soft Selection **Falloff** value to create a larger region of influence at the bottom of the apple; try a value of **50** units.

16. In the Front view, move the selected vertices down in the viewport's Y axis.

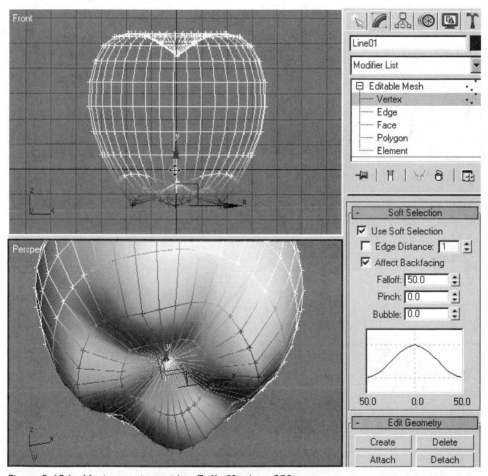

*Figure 3-154: Moving vertices with a **Falloff** value of **50***

The apple now has a much more natural shape.

17. Save the scene as **Apple02.max** in your folder.

EDITABLE POLY

Modeling with Editable Mesh is a very powerful technique. The Editable Mesh object has many useful tools for polygonal mesh editing. However, there is an object type that is even more advanced: **Editable Poly**.

The Editable Poly object shares much in common with Editable Mesh. It is essentially a new and improved version of Editable Mesh. In fact, Editable Poly is superior in many ways, and it is intended as a replacement for the Editable Mesh object type. However, there are situations in which Editable Mesh is still required, such as opening up a scene created in an earlier version of 3ds Max.

This book focuses on the tools within Editable Poly. Some of those tools can also be found in Editable Mesh. However, they work a little differently in Editable Poly. For example, Editable Poly has pop-up dialog boxes that allow you to precisely control the options for many tools.

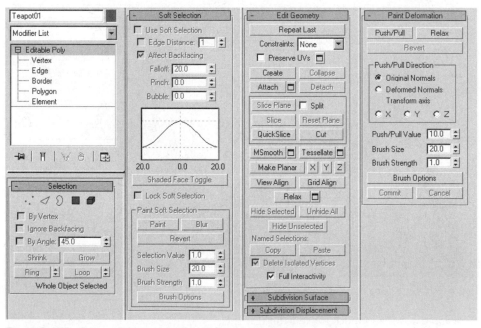

*Figure 3-155: Some of the tools within **Editable Poly***

Editable Poly has so many tools that you can make just about anything with it. Some artists use it exclusively.

Editable Poly Sub-objects

The most obvious difference between Editable Mesh and Editable Poly is the difference between their respective sub-object types. The Editable Poly object has a sub-object that Editable Mesh does not: the **Border** sub-object. A Border is a collection of edges. The edges of a Border form a ring around a hole in the mesh. Usually, holes in the mesh are undesirable, so the Border sub-object toolset includes a **Cap** tool that creates a polygon from the selected border.

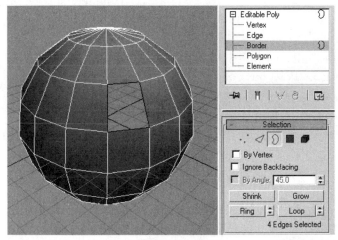

*Figure 3-156: Selecting a **Border** sub-object*

The most fundamental difference between Editable Mesh and Editable Poly is that Editable Poly objects don't have a **Face** sub-object type. This is sometimes confusing to new users, and there is a common misconception that Editable Poly objects don't have triangular faces at all. Even the 3ds Max documentation is misleading in this regard.

In fact, Editable Poly objects do have triangles with invisible edges. If you select an Editable Poly object and disable the **Edges Only** option in the Display panel, you will see the invisible edges. However, because there are no Face sub-objects in an Editable Poly, you can't select the triangles or their invisible edges.

In most cases, this is an advantage. The lack of a Face sub-object type means there is less for the artist to worry about. 3ds Max re-triangulates the polygons within an Editable Poly object in real time, so you are able to do cool things like **Remove** edges without punching a hole in your model. This function is not available with Editable Mesh objects.

Triangulation

It's not possible or desirable to completely do away with triangles in Editable Poly. The triangles are still there and can pose the same types of problems as they do in an Editable Mesh. Earlier, you saw how to **Turn** an Editable Mesh edge to control the structure of a low-density object. The same function is found in Editable Poly, but the procedure is different.

To change the orientation of invisible edges on an Editable Poly object, use the triangulation tools in the Polygon sub-object mode. You can retriangulate a polygon automatically or manually. To retriangulate automatically, select the polygon and open the Edit Polygons rollout. Click the **Retriangulate** button.

To perform a manual operation similar to an Editable Mesh edge **Turn**, click the **Edit Triangulation** button in the Edit Polygons rollout of Editable Poly. Click a vertex and drag to another vertex to define the new invisible edge. Click again to complete the operation.

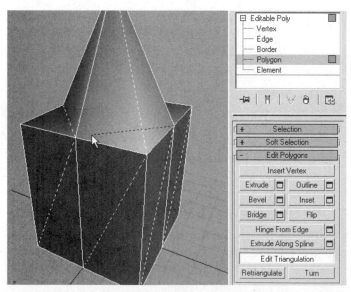

*Figure 3-157: Turning an Editable Poly edge with the **Edit Triangulation** tool*

In a situation such as the one shown in the previous illustration, it is a good idea to make the invisible edges on the top of the object visible. That way, all polygons will be coplanar. To do this, you must create new edges, since Editable Poly has no Visible or Invisible edge buttons.

Use the **Create** tool within Edge sub-object mode. In the **Edit Geometry** rollout, click **Create**. Click a vertex, then click another vertex to create a new edge, separating a polygon into two new polygons.

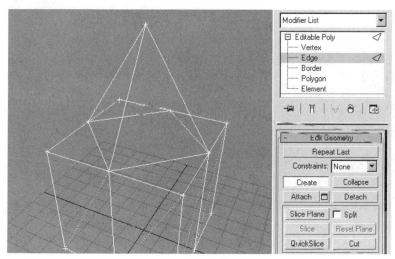

Figure 3-158: Edges created to prevent non-coplanar polygons

Extrude Tool

One of the most powerful features of Editable Poly is the **Extrude** tool. Not to be confused with the Extrude modifier, the Extrude tool is a button on the Editable Poly panel. It is also available within Editable Mesh.

Whereas the Extrude modifier operates on an entire Shape object, the Extrude tool in Editable Poly operates on sub-object selections. You can use the Extrude tool to quickly and easily create *branching* in your model. For example, using the Extrude tool you can create arms on a character with a minimum of effort and without visible seams.

Simply select a sub-object type such as Polygon, and activate the Extrude tool. Click a polygon and drag upward in the viewport to extrude the polygon, creating new polygons around the sides. The mesh remains closed, with no seams or holes.

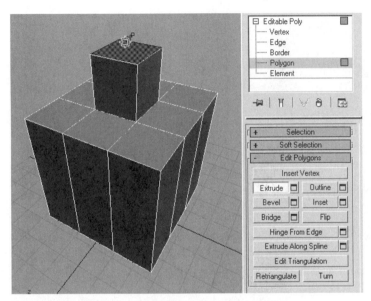

Figure 3-159: Extruding a single polygon

The Extrude tool can also be used to create recesses or concave surfaces. Simply click and drag the mouse downward in the viewport instead of up.

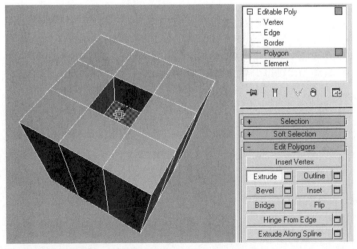

Figure 3-160: Extruding a polygon inward to create a concave surface

You can also extrude vertices and edges in a similar manner. Extruding vertices and edges creates pyramid-like structures, since the end of the extrusion is not a flat polygon. In addition, when extruding a vertex or edge, you must define an extrusion height and a size for the base of the pyramid. Click and drag a vertex or edge, holding down the mouse button. As you drag vertically in the viewport, the extrusion height changes. As you drag horizontally, the size of the base changes.

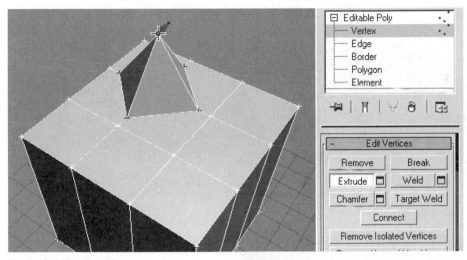

Figure 3-161: Extruding a single vertex

If you select multiple sub-objects, you can extrude them together in a single operation. When extruding multiple connected polygons, you have additional options to consider. Do you wish to extrude them as a unit, or do you wish to create separate extrusions for each polygon?

To access the options for a polygon extrusion, click the **Settings** button next to the Extrude tool. This opens a dialog box that allows you to change the options for that tool. Many of the tools in Editable Poly have this feature, but the tools in Editable Mesh do not. The tool options dialog box is modeless, meaning you can leave it open while you work.

When you open the dialog, the tool performs its function based on the most recent parameter values entered. If you wish to change these values, do so. If you like the result, click **OK** to accept the changes and close the dialog box. Clicking Cancel will override the changes and restore the model to its prior state.

The **Apply** button allows you to perform multiple operations without closing the dialog. Click Apply, then make another selection and/or change the tool parameters. A new operation is performed. When you are finished with the tool, click OK or Cancel.

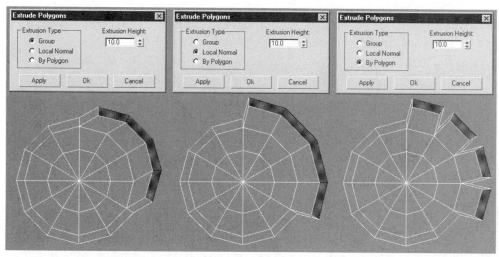

Figure 3-162: Options for extruding polygons

Bevel Tool

Most of the time you will be extruding polygons. Often you will wish to change the shape of the extruded polygons to create angled sides. The **Bevel** tool is a handy feature that lets you do both of these operations at once. It performs an Extrude, then performs an **Outline** command on the polygons at the end of the extrusion. This Outline is identical to the one found in Editable Spline, except that it works on polygons instead of shapes, and it works directly on the selected polygons, without creating a copy.

To use the Bevel tool, activate it, then click a polygon or collection of selected polygons. Click and drag to define the extrusion height, then release the mouse button and drag to define the outline amount. Click again to complete the operation. Of course, these controls are also available from the Bevel Polygons dialog.

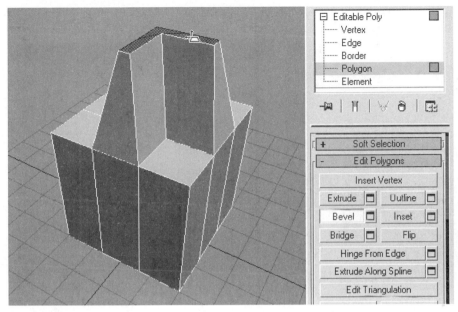

Figure 3-163: **Bevel** tool

Outline and Inset

Sometimes you will need to change the shape of a collection of polygons without extruding them. In this case, the **Outline** tool is helpful. Activate the Outline tool, and click and drag selected polygons. Just as with the Outline tool in Editable Spline, the edges move, maintaining a constant distance away from their original position.

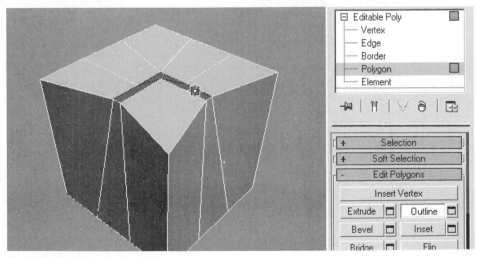

Figure 3-164: **Outline** tool

Even more useful than Outline is the **Inset** tool. This tool actually behaves more like the Editable Spline Outline, because it creates a copy. Use this tool to create detail in irregularly shaped areas of a polygon model.

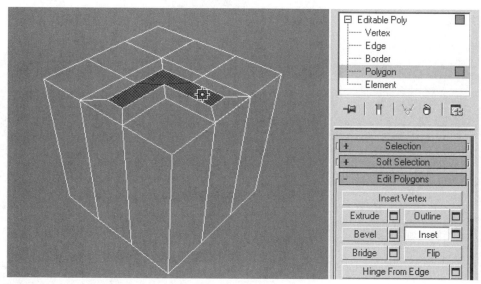

Figure 3-165: **Inset** *tool*

Slice

When modeling polygonal objects, you will often need to create detail by splitting polygons and creating new edges and vertices. One simple way to do this is with the **Slice** tool found in the **Edit Geometry** rollout.

Slice works by dividing selected polygons based on the position and rotation of a gizmo called the **Slice Plane**. The Slice Plane is available in Editable Poly from any sub-object mode.

Click the Slice Plane button, and you will see a yellow rectangle in the viewports. This is the Slice Plane gizmo. Anywhere the Slice Plane intersects with your model, new edges and vertices will be created. Use the transform tools to place the Slice Plane where you want it. You will see a real-time preview of the changes that will be made to your model. Then click the Slice button, and the operation will be performed.

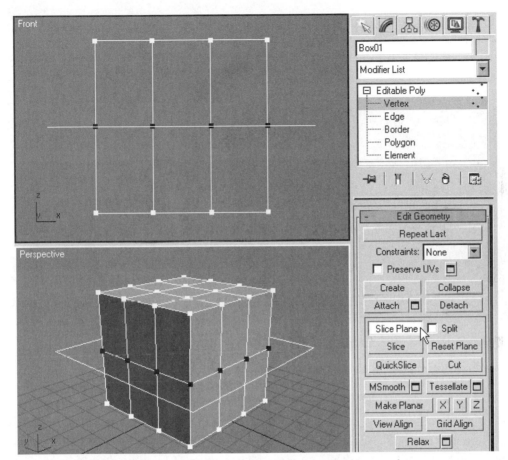

Figure 3-166: **Slice Plane** *applied to entire object in Vertex sub-object mode*

If the **Split** check box is enabled, a Slice operation will create a double row of edges and vertices. The result will be an invisible gap in the model, which will cause rendering errors. You should enable Split only if you wish to chop your model in half, or intentionally create a hole in the geometry.

WARNING: The scale of the Slice Plane does not affect the operation. More importantly, the influence of the Slice Plane extends infinitely, and does not stop at the borders of the gizmo. Geometry that is outside the boundaries of the Slice Plane gizmo will still be sliced.

In Vertex, Edge, or Border sub-object mode, any sub-object selections are ignored, and the Slice is applied to the entire object. In Polygon sub-object mode, you must select polygons to be sliced *before* activating the Slice Plane. Unselected polygons will not be sliced.

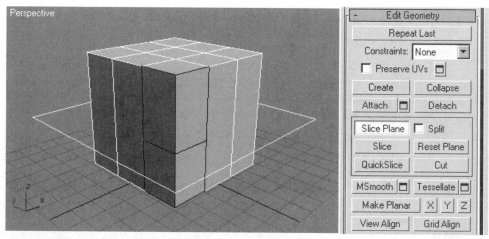

Figure 3-167: **Slice Plane** applied to selected polygons

The Slice tool is also available within Editable Mesh, but without the real-time preview of changes to the geometry.

QuickSlice

As the name implies, **QuickSlice** is a fast and easy way to split polygons and create edges and vertices. Instead of relying on a planar gizmo, QuickSlice performs a slice operation relative to the active viewport.

Just as with the Slice Plane, QuickSlice operates on the entire model unless you have previously made a polygon selection. To use this tool, click the QuickSlice button, then click in the viewport. Move the mouse in the viewport to define an angle for the slice. Click again to complete the operation.

Since QuickSlice uses the current viewport as its frame of reference, you will achieve the most predictable results if you use an orthographic view. It's a bad idea to use a Perspective view in conjunction with QuickSlice.

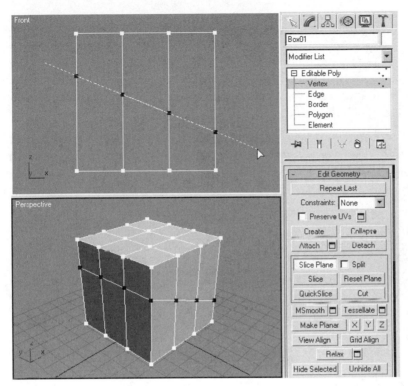

Figure 3-168: **QuickSlice** *tool*

TIP: You will find that it is much easier to make precision QuickSlice operations if you have **3D Snap** enabled. That way, your cursor will snap to Grid points or any other options you select in the **Customize > Grid and Snap Settings** dialog.

Cut

The **Cut** tool is perhaps the easiest and most intuitive way to split polygons if you are not concerned with precision. Unlike Slice and QuickSlice, Cut does not respect sub-object selections at all. Instead, the Cut cursor snaps to vertices and edges on an Editable Poly, making it unnecessary to make a prior selection merely to control which polygons get split and which do not.

Simply press the Cut button and move the cursor over the model. The cursor changes depending on whether it is over a vertex, an edge, or a polygon. Click anywhere on the model, drag the mouse, and click again to finish the Cut operation. You will see a real-time preview in the viewport.

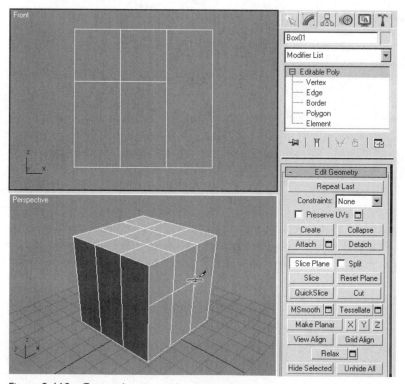

Figure 3-169: **Cut** *tool*

Unlike QuickSlice, the Cut tool tends to work best in a Perspective view.

WARNING: The real-time preview of modeling changes within Editable Poly can sometimes slow down your computer. If this happens, disable the **Full Interactivity** option at the bottom of the Edit Geometry panel.

EXERCISE 3.19: Flat Panel Monitor

In this exercise, you will create a model of a flat screen monitor from an ordinary Box object. Its low polygon count will make it suitable for use in a real-time 3D video game.

The Monitor Stand

1. Reset 3ds Max.

2. Press the **<S>** key to activate **3D Snap**. This causes the cursor to snap to Grid Points.

3. In the Top viewport, create a Box at the origin. Use 3D Snap to place the corners of the Box directly on Grid Points. Make the Box **80** units long, **80** units wide, and **5** units tall. The Box should have just one **Segment** in all axes.

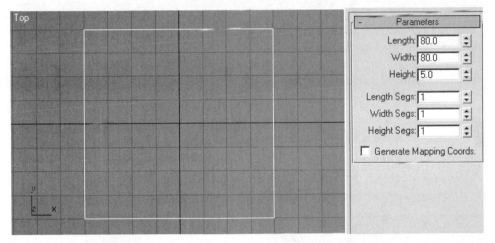

*Figure 3-170: Flat **Box***

4. Right-click the Box and choose **Convert To > Convert to Editable Poly** from the Quad Menu.

5. Press the **<A>** key to activate **Angle Snap**.

6. Now you will slice the top polygon of the box so you can extrude it later. Enter **Polygon** sub-object mode. In the Perspective view, select the polygon on the top of the box.

7. In the **Edit Geometry** rollout, activate the **Slice Plane**. In the Left view, rotate the Slice Plane **90** degrees, so it is at right angles to the selected polygon.

8. The Transform Gizmo can interfere with Snaps, so we'll turn it off for now. Press the **<X>** key to disable the Transform Gizmo.

9. Press the **<F5>** key to activate the X axis. Move the Slice Plane -30 units in the X axis, so it is on the left side of the Left viewport. Your screen should look like the following illustration.

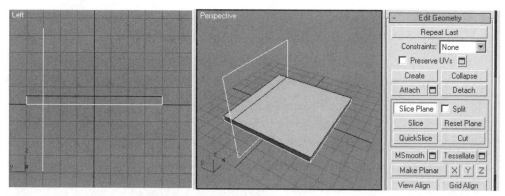

*Figure 3-171: Rotate and position the **Slice Plane***

10. Click **Slice** to perform the operation. Then click the Slice Plane button again to exit the tool. Click anywhere outside the object to deselect all polygons.

11. Press the **<S>** key to deactivate 3D Snap. Press the **<A>** key to deactivate Angle Snap. Then press the **<X>** key to enable the Transform Gizmo.

12. Click the **Extrude** button in the **Edit Polygons** rollout. In the Perspective view, select the new polygon that you have just created, and drag upward. Release the mouse when the extrusion is about the same height as the width of the box.

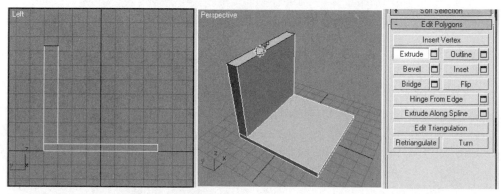

*Figure 3-172: **Extrude** the new polygon*

13. With the extruded polygon still selected, right-click in the Left viewport. Press the **<W>** key to activate the Move tool. Move the polygon in the viewport's XY plane, to angle the extrusion so it looks like the following illustration.

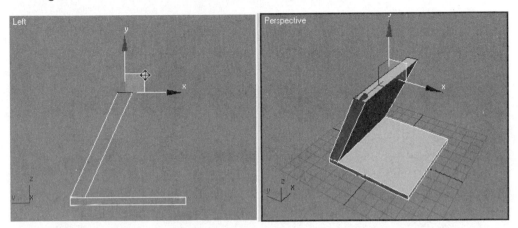

Figure 3-173: Move the selected polygon

The Screen-Mounting Hardware

1. With the polygon at the top of the object still selected, click the Extrude Settings button ⬚. In the pop-up dialog, enter a value of 5 units for the **Extrusion Height**. Click the **Apply** button three times, then click **OK**. You have now made four identical extrusions on the top of the object.

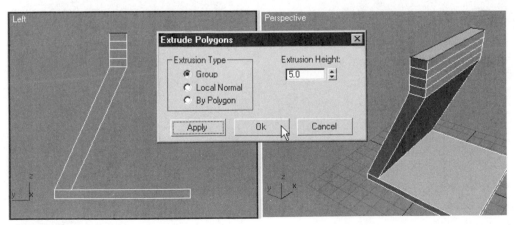

Figure 3-174: Four identical extrusions

2. On the Main Toolbar, click the **Window/Crossing Selection** button to switch it to the Window state. Now, when you make a Region Selection, only objects or sub-objects completely within the Region will be selected.

3. Now you will create some simple curvature. In the Left view, zoom in on the top of the object. Click and drag to create a window around the two extruded segments near the top of the object, as shown in the following illustration.

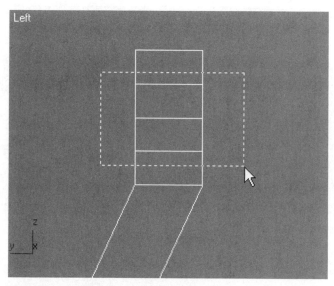

Figure 3-175: Select extruded polygons

4. In the Left view, move the selected polygons slightly to the right.

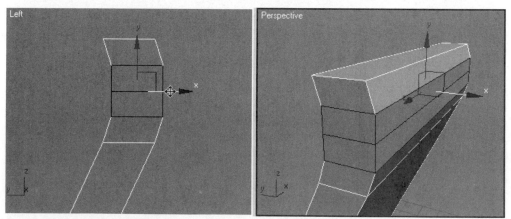

Figure 3-176: Move selected polygons

5. Switch to Vertex sub-object mode. In the Left view, draw a selection region around the vertices shown in the following illustration. Move the two vertices slightly to the right, to create a smooth curve.

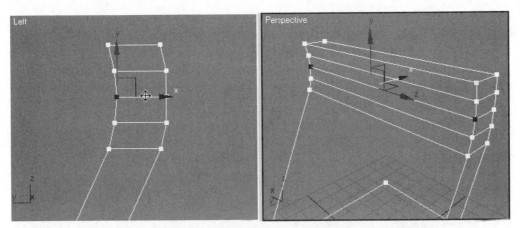

Figure 3-177: Move two vertices to create curvature

6. On the Main Toolbar, switch back to **Crossing Selection** mode . Enter Edge sub-object mode. In the Left view, select all of the horizontal edges on the interior of the object. The edges on the side facing toward you *and* the edges facing away from you are selected. If necessary, use the **<CTRL>** key to add to the selection until all the edges shown in the following illustration are selected. Remember to select the edges near the bottom of the object also.

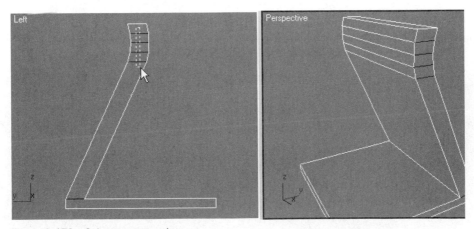

Figure 3-178: Select interior edges

7. Press the **<BACKSPACE>** key on the keyboard. This is the same as the **Remove** tool on the Edit Edges rollout. The extra edges are removed. Now each side of the object is a single polygon instead of six.

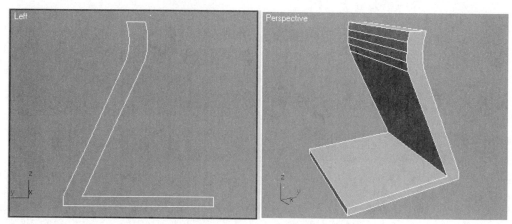

Figure 3-179: **Remove** *the selected edges with the* **<BACKSPACE>** *key*

WARNING: Do not press the **<DELETE>** key! If you delete edges, you also delete the polygons that share those edges. The **<BACKSPACE>** key, however, will remove edges and keep the polygons.

8. Select the edge on the front of the object, in the center of the curved area, as shown in the following illustration. Press the **<BACKSPACE>** key to remove it.

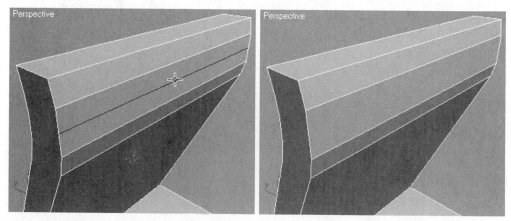

Figure 3-180: Before and after removal of edge

9.   Enter Polygon sub-object mode. Select the large polygon in the center of the curved area on the front of the object. Extrude the polygon by about 5 units.

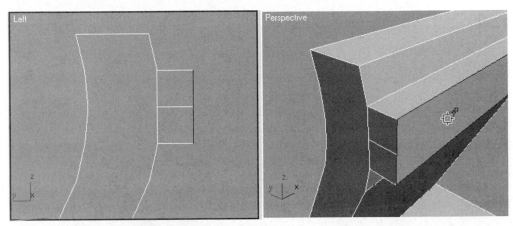

*Figure 3-181: **Extrude** the center polygon*

10.   Notice that there is an extra, unexpected edge on each side of the extrusion. This is because when you Remove edges, the vertices are not necessarily removed. There was an extra vertex left over on each side, and when you extruded the center polygon, new edges sprouted from those leftover vertices.

Go back into Edge sub-object mode and select the interior edges again. Press the **<BACKSPACE>** key to Remove them.

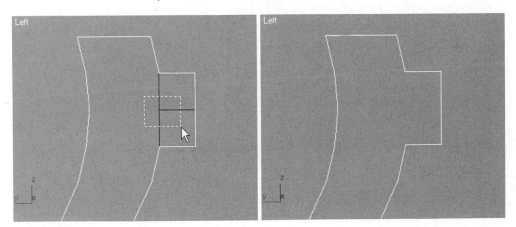

Figure 3-182: Before and after removing extra edges

11. In Vertex sub-object mode, check to make sure the leftover vertices have been removed. In this case they have been, because you removed the edges all around the leftover vertices.

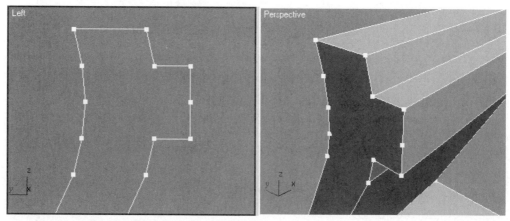

Figure 3-183: Some vertices removed automatically

12. There are still two vertices left over after deleting the edges. Enter Vertex sub-object mode, draw a selection region around the extra vertices.

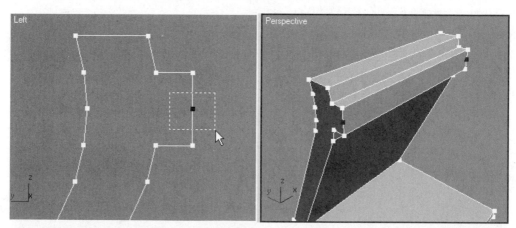

Figure 3-184: Select the remaining extra vertices

13. Press the **<BACKSPACE>** key to remove the extra vertices. The model is now clean, with no extra edges or vertices.

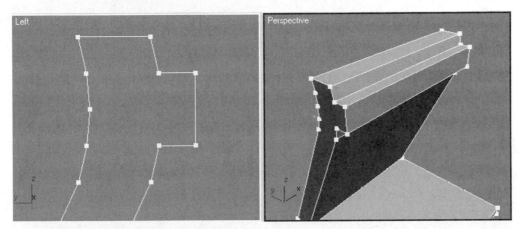

Figure 3-185: Extra vertices removed

The Screen

1. Switch to the Perspective view. Activate **Polygon** sub-object mode. Press the **Bevel** button in the Edit Polygons rollout. Click and drag the center polygon at the cap of the most recent extrusion. This creates a new extrusion. Release the mouse, drag to create the bevel, and click again to finish the operation. This creates a large, shallow Bevel as shown in the following image.

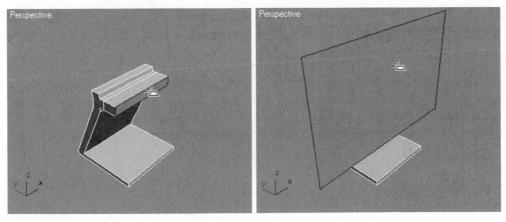

*Figure 3-186: Create a large, shallow **Bevel***

2.   At this stage, the model should look something like the following illustration.

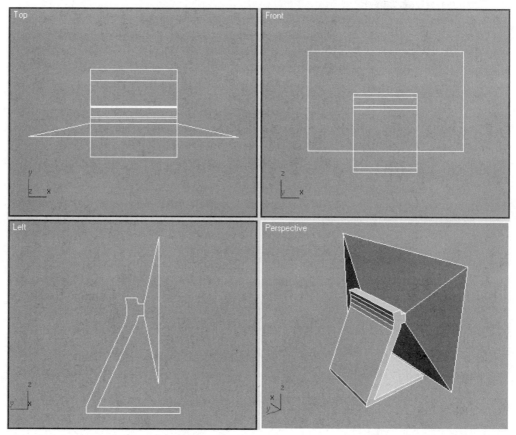

Figure 3-187: Monitor screen created by Bevel

Look in the Front viewport; does the monitor screen have the correct proportions? The height of the screen should be about three-fourths of its width. The *aspect ratio* (proportion of screen width to screen height) of a computer monitor is 4 to 3.

3. If the polygon at the end of the Bevel isn't still selected, select it. Activate the Front view. Select the Scale tool from the Main toolbar. Scale the polygon non-uniformly in the X or Y axes of the viewport to achieve the approximate proportions of a computer monitor.

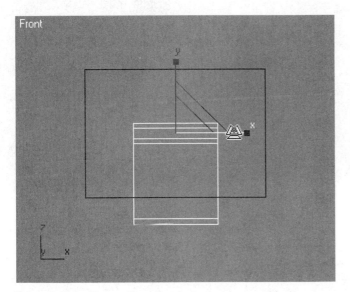

*Figure 3-188: Non-uniform **Scale** in the viewport X axis*

4. With the big polygon still selected, extrude it by about **10** units to create some thickness for the screen.

5.   Activate the **Inset** tool. Click and drag the big polygon to create an inset. Make the inset about 10 units. Your model should look something like the illustration that follows.

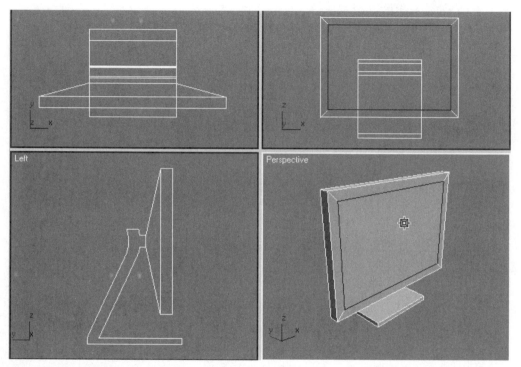

*Figure 3-189: **Inset***

6. With the inset polygon selected, click the Settings button 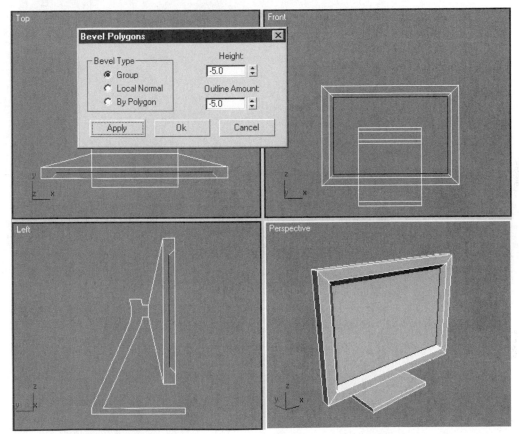 for the Bevel tool. Enter a value of **-5** for both the **Height** and the **Outline Amount**. This creates a Bevel with sides at exactly **45** degrees.

*Figure 3-190: Precision **Bevel** operation*

Finishing Touches

The model is nearly complete. Now you will add a few details to make it more realistic.

1. The screen should be slightly tilted to make it look like it's being used. Using Window

   Selection mode , select the polygons of the screen in the Left viewport. Make sure you don't select any of the polygons that connect the monitor stand to the screen.

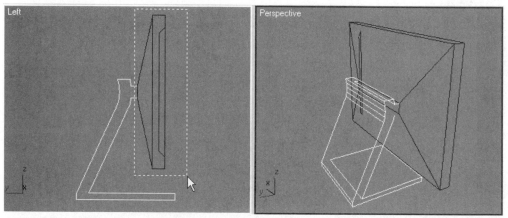

Figure 3-191: Select the polygons of the screen

2. Once the screen polygons have been selected, rotate them very slightly around the Z axis in the Left viewport.

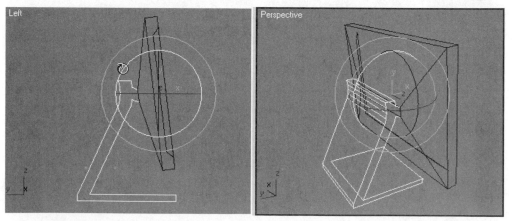

Figure 3-192: Rotate the monitor screen

3. Zoom in on the polygons that connect the base to the monitor screen. Notice that they have become skewed. With the polygons of the monitor screen still selected, move them up slightly in the Left viewport, so the connecting polygons are once again parallel to the viewport's X axis.

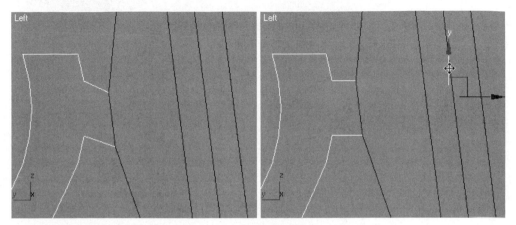

Figure 3-193: Before and after moving the monitor screen

4. Enter Edge sub-object mode. Activate the Perspective view. In the **Edit Edges** rollout, press the **Chamfer** button. Click and drag the edge at the inner joint of the monitor base. The edge is chamfered to create a new polygon.

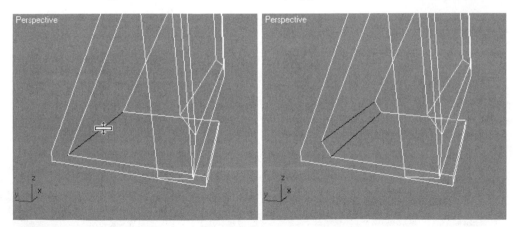

*Figure 3-194: Before and after using the **Chamfer** tool*

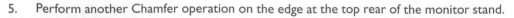

5.   Perform another Chamfer operation on the edge at the top rear of the monitor stand.

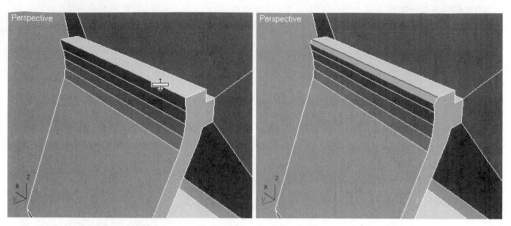

Figure 3-195: Before and after the second Chamfer operation

6.   The model is finished. Maximize the Perspective view and Arc Rotate around the scene to admire your work. Press the **<7>** key on the keyboard to see a polygon count for the selected object in the top left of the active viewport.

   The polygon counter reads only **36 Faces**. The term "faces" used here is inconsistent; what it really means is there are 36 polygons. If we counted the number of triangles, there would be 100.

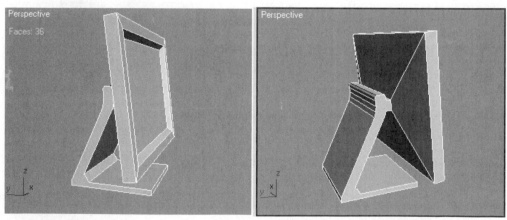

Figure 3-196: Completed model

7.   Save the file as **LowPolyMonitor.max** in your folder.

ADVANCED TRANSFORMS

So far you have learned to transform objects with the standard Move, Rotate, and Scale buttons on the Main Toolbar. However, these are not the only tools you will need for transforming objects. Occasionally you will need to align objects to one another, create arrays, space objects along a path, and so on. The tools you need to perform these tasks are collectively referred to as the **Transform Tools**. In the next section we will explore some of the most commonly used of these tools.

ALIGN

The **Align** command aligns one object with another, aligning by position or rotation. You can place one object directly on the surface of another object, or turn an object so its Pivot Point is oriented exactly with another object's Pivot.

To use Align, first select the object or objects to be aligned, then click the **Align** button

on the Main Toolbar. You can also choose **Tools** > **Align** from the Main Menu. The cursor changes to the align cursor. Move the cursor over the object to which you want to align the selected object, and click to pick it. This is the target object. The **Align Selection** dialog appears.

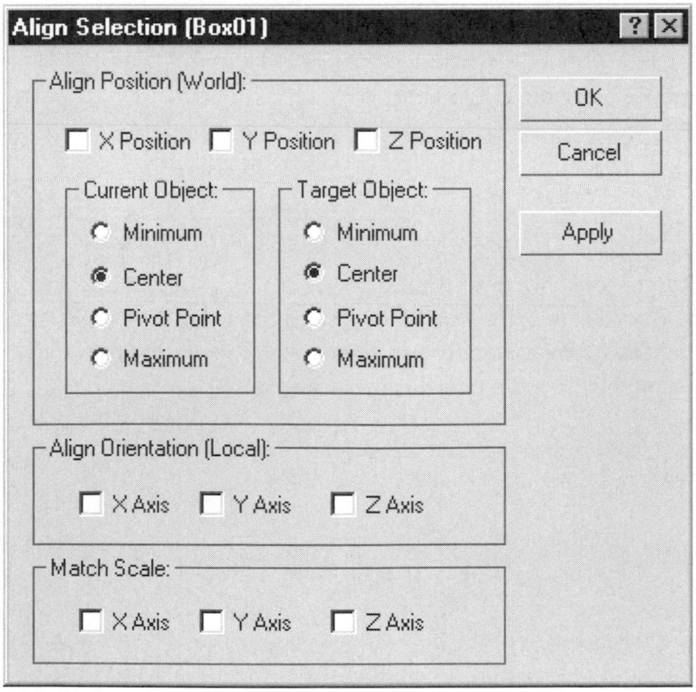

Figure 3-197: **Align Selection** *dialog*

The **Align** command uses the current Reference Coordinate System. Choose any or all of the **X Position**, **Y Position**, or **Z Position** check boxes to align the object along these axes.

You can choose to align the selected object's Pivot Point or geometric center with the picked object's Pivot Point or center. **Minimum** and **Maximum** align to the outermost edges of the object.

The options under the **Align Orientation** section will align the rotation of the selected object to the target object's Pivot Point.

Under **Match Scale**, you can align one object's scale to the other's. Note that this feature matches scale percentages, not actual sizes.

TIP: Sometimes you may find that you cannot perform all of the alignments you want in one operation. In that case, you'll need to execute the Align command more than once. You can do this without exiting the Align dialog by using the **Apply** button. After pressing Apply, you will see the axis tripod move and/or rotate, indicating that the operation has been performed. Then you can change the settings in the dialog box and press Apply again.

EXERCISE 3.20: Align

In this exercise, you'll practice using the **Align** tool.

1. Reset 3ds Max.

2. In the Top viewport, create a Box and a Teapot.

3. Select the Teapot in the Perspective viewport and move it up or down in the Z axis.

4. Choose the Select and Rotate tool. Switch to the **Local** Coordinate System by selecting it from the Reference Coordinate System pulldown list on the Main Toolbar. Rotate the Teapot in all three axes, so its Pivot Point is not aligned with the world in any axis. Your scene should look something like the following illustration.

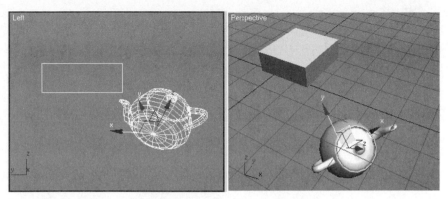

Figure 3-198: Teapot moved and rotated

5.   Switch back to the **View** coordinate system. With the Teapot selected, click the Align button on the Main Toolbar. The cursor changes to an Align cursor. Click on the Box to select it as the target object.

6.   The **Align Selection** dialog box opens. In the **Align Orientation** section, activate the **X Axis** check box. The teapot rotates so that its X axis lines up with the X axis of the Box. Click the Y or the Z Axis check box. The Teapot is rotated in alignment with the Box. Click the **Apply** button and leave the dialog box open.

    TIP: You need to select only two axes to force the object into total alignment.

7.   In the Align Selection dialog, in the **Current Object** section, select **Minimum**. Under **Target Object,** select **Maximum**. Activate the **Z Position** check box. The bottom (minimum) of the Teapot is aligned with the top (maximum) of the Box. You can see this clearly in the Front and Left views. Click the Apply button; the axis tripod of the Teapot moves to its new position. Leave the dialog box open.

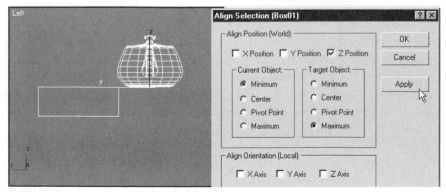

*Figure 3-199: Align **Minimum Z value** of Teapot to **Maximum Z value** of Box*

8. In the Align Selection dialog, select **Pivot Point** in both the Current Object and Target Object sections. Deactivate the Z Position check box, and activate both the **X Position** and **Y Position** check boxes.

The Teapot now sits precisely on the Box.

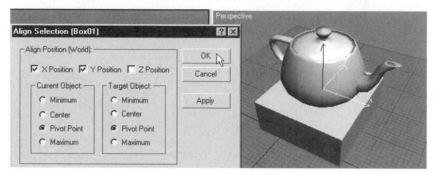

Figure 3-200: Teapot aligned with Box

MIRROR

The **Mirror Selected Objects** button  can be used to mirror and clone objects. Select an object to mirror, and click **Mirror Selected Objects**. Choose the axis over which to mirror the object. The axes are based on the current coordinate system, not the object's local axes. As you change axes, the mirrored object changes interactively in the viewport.

You can also choose to create a **Copy**, **Instance**, or **Reference** of the original object when mirroring.

TIP: For advanced modeling operations, such as character modeling, you should not use the Mirror Selected Objects command. Instead, you should use the **Symmetry** modifier. You may see tutorials in books or on the Internet that instruct you to use the Mirror Selected Objects command for character modeling; these tutorials are based on earlier versions of 3ds Max. The Symmetry modifier is far superior. You will learn how to use it in Chapter 9, *Advanced Modeling*.

ARRAY

The **Array** tool makes multiple clones of an object, arranged in a pattern. To use **Array**, select an object, then choose **Tools** > **Array** from the Main Menu. The **Array** dialog appears.

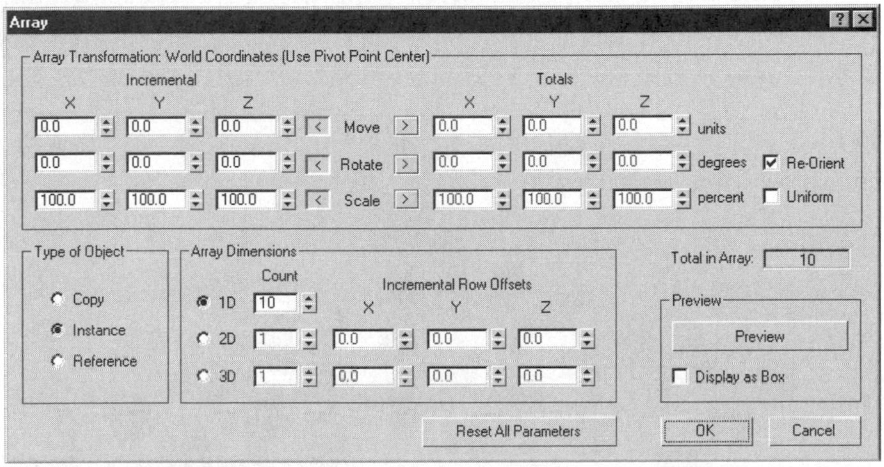

*Figure 3-201: **Array** dialog*

TIP: The Array dialog appears fairly complicated, because it has to be. If you want to clone a lot of objects in a pattern in space, there are many variables to consider. The Array dialog actually makes this very easy, once you get used to how it works. You can also experiment with different kinds of Arrays until you get familiar with the tool. Just create an Array, then Undo, and then create another Array with different parameters.

In the **Type of Object** section, you can specify whether the cloned objects created by the Array tool will be copies, instances, or references of the original object.

The **Array Transformation** section of the Array dialog pertains to moving, rotating, or scaling cloned objects while creating the array. To create a matrix of objects, use **Move**. To create a rotated array, use **Rotate**. To create a series of incrementally scaled objects, use **Scale**. You can use these in combination with one another to create complex structures.

The **Incremental** section is where you specify the increments between each array object. For Move, this is the distance between the Pivot Points of cloned objects in 3ds Max units. The Rotate Increment values indicate how much, in degrees, each clone of the object will be rotated. The Scale Increment fields determine the percentage of scaling applied to each clone, relative to the clone before it.

If you click the right arrow next to the transform name (**Move, Rotate, Scale**), then the **Totals** settings are used instead of the values in the Incremental fields. You enter the total distance, angle, or scale, and that value is divided by the **Total in Array** value to determine the increments between each object in the array. Total in Array indicates how many objects will be created, *including the original object*. Total in Array is automatically calculated by the numbers you enter in the **Array Dimensions** section.

The **Array Dimensions** section relates to moving the clones during array creation. You can choose to create a **1D, 2D** or **3D** Array. A 1D Array creates objects in a straight line. A 2D Array creates a series of objects with columns and rows, while a 3D Array creates columns, rows, and vertical levels.

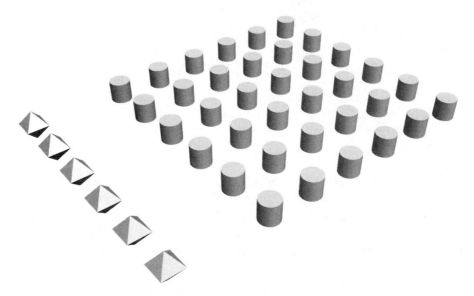

Figure 3-202: A one-dimensional Array of Hedras, and a two-dimensional Array of Cylinders

The **Count** parameter sets the number of objects along each dimension of the Array. For a 1D Array, the Count field determines the total number of objects in the linear Array, and the distance between them is what you entered in the Array Transformation section.

If you create a 2D Array, you'll have columns and rows. In the following illustration, the first Count field, next to the 1D button, determines the number of columns. The second Count field, next to the 2D button, determines the number of rows. The Total in Array value is the number of rows multiplied by the number of columns.

For a 2D Array, the **X, Y,** and **Z Incremental Row Offsets** determine the spacing between rows. For a 3D Array, they also determine the vertical spacing between levels.

Study the following illustration to get a sense of how the Array tool is designed to work. The value of **20** units in the **Incremental Move X** field means the objects will be 20 units apart in the world X dimension. In the **1D Count** field, a value of **5** is entered, meaning there will be five objects in each column.

The **2D Count** field has a value of **3**, so there will be three rows of objects. The resulting **Total in Array** value is 5 x 3 = 15. In the **Incremental Row Offsets**, a value of **50** means that each row will be spaced 50 units apart.

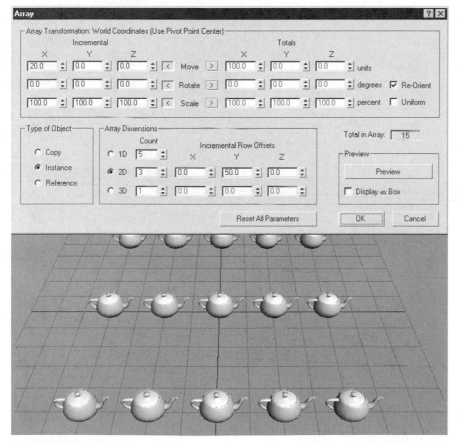

*Figure 3-203: An **Array** of objects*

WARNING: It takes practice to learn how to use the **Array** tool effectively. A common error is to leave all the Incremental and Total values at their default values, resulting in multiple objects piled on top of one other with exactly the same position, orientation, and size. Making this mistake will fill your scene with unnecessary objects, which can interfere with viewport performance, increase rendering time, and cause rendering errors. Before creating an Array, be sure to save your scene to avoid creating multiple objects in the wrong places.

SPACING TOOL

The **Spacing Tool** allows you to create clones of objects that are evenly spaced. You can specify the space between objects or let 3ds Max calculate the spacing based on the number of objects and the total distance. You can create a series of objects by selecting points in the viewport or by placing them on a path. The path creation method is most common. Using this technique, you can do things such as automatically place streetlights at regular increments on a curved road.

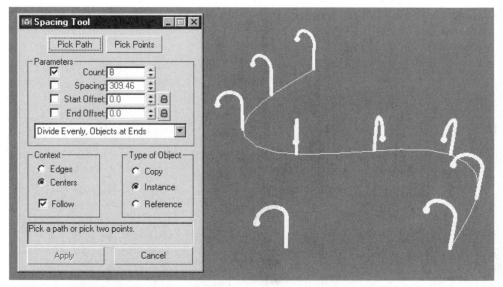

Figure 3-204: **Spacing Tool**

To use the Spacing Tool, select an object and choose **Tools > Spacing Tool** from the menu. The Spacing Tool dialog appears. To space objects on a path, click the **Pick Path** button and select an open or closed shape in the viewport. To space objects in a straight line, click the **Pick Points** button and draw a line in the viewport.

The Spacing tool has controls for the number of clones, the spacing between them, and other options. One important feature is the **Follow** check box. With this option enabled, objects will rotate to remain parallel to the path.

WARNING: If you deselect the dialog box of the original object, you will lose your real-time preview of the results of the Spacing Tool. Both the object and the dialog must be selected. When you get the result you want, click the **Apply** button. The Spacing Tool dialog will remain open if you wish, so you can continue creating more clones. Simply select another object and/or pick another path, and click Apply again.

TRANSFORM TYPE-IN

Earlier, you learned that you can use the **Transform Type-In** area of the Status Bar to enter transform values. You can also right-click on an active transform button on the Main Toolbar to type in specific values for the transform. The main advantage of using the Transform Type-In dialog instead of the Status Bar is that you can see the Absolute world value and an Offset value at the same time.

For example, suppose you want to position an object at the exact point of **20,16,23** in world space. Select the object, and click the **Select and Move** button. Choose the **World** coordinate system. Right-click the Select and Move button, and the **Move Transform Type-In** dialog appears.

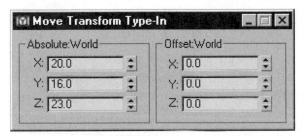

Figure 3-205: **Move Transform Type-In** *dialog*

Under the **Absolute** column, you can enter the exact **X**, **Y**, and **Z** coordinates to which you want to move the object. If you enter coordinates in the **Offset** column, the object is moved by that amount. Note that the coordinate system being used is displayed on the window title bar at the top of the dialog.

HELPER OBJECTS

As the name indicates, a **Helper** is a utilitarian object that helps you perform tasks. Helper objects are found on the Create panel, in the Helper subpanel.

Helpers do not render. Their sole purpose is to enable operations on other objects. For example, the Grid Helper object allows you to create objects on a customized construction plane, rather than on the default Home Grid of the world coordinate system. Create a Grid Helper object, move and rotate it as desired, and activate it. Right-click the Grid Helper and choose **Activate Grid** from the Quad Menu. Now, when you create an object, it will be placed on the active Grid Helper instead of the Home Grid.

Figure 3-206: **Helper** *objects on the Create panel*

The most commonly used Helper object is the Point Helper. A Point Helper is simply a non-rendering point in the scene. It can be used as a control handle for complex animation tasks, such as character animation. As you will see in the following pages, it can also be used for modeling tasks. In this chapter, we use a Point Helper as a temporary center for transforming objects. You'll learn more about Helper objects in Chapter 10, *Advanced Animation*.

SNAPS

Earlier, you used the **Snaps** on the Main Toolbar to precisely position the cursor on intersecting points of the Home Grid. Now let's look at Snaps in more detail. Four Snap buttons are available, each with a specialized function.

 3D Snap is used in conjunction with the Move transform. When 3D Snap is active, the cursor will snap to objects, sub-objects, or other viewport elements in 3D space. By default, 3D Snap will cause the cursor to snap to points on the active constuction grid. Other options are available from the **Customize > Grid and Snap Settings** dialog, such as snapping to vertices. Hotkey = <**S**>.

 Angle Snap. Rotations are constrained to precise angle increments. Rotating an object with Angle Snap enabled will cause it to snap to 5-degree increments of rotation. You can change the number of degrees in the Grid and Snap Settings dialog. Hotkey = <**A**>.

 Percent Snap. Any amount that is expressed in percentages, such as Scale, is rounded off to 10% increments. You can change the percentage amount in the Grid and Snap Settings dialog.

 Spinner Snap. All parameter spinners in 3ds Max will increase or decrease by a precise amount when you click a spinner arrow once. Clicking and dragging spinner arrows is not affected.

By default, Spinner Snap adds or subtracts a value of one to a parameter when you click a spinner arrow. The addition or subtraction is relative to the current value. So if a parameter value is **57.005**, and you click a spinner up arrow, the new value will be **58.005**. The incremental amount can be set in the General tab of the Customize > Preferences dialog.

Options for Snaps can be accessed from the Customize menu, or by right-clicking a Snap button. The **Grid and Snap Settings** dialog has many options. For example, you can choose to snap to Pivot Points instead of Grid Points. You can choose to snap to more than one viewport element. For example, snapping to both Grid Points and object Vertices is a common setup for polygonal modeling.

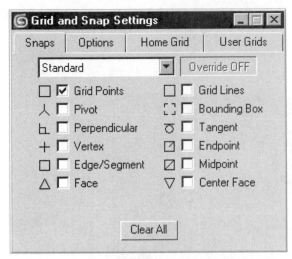

Figure 3-207: **Grid and Snap Settings** *dialog*

TRANSFORM CENTERS

By default, an object transforms relative its Pivot Point. For modeling tasks, you can also cause objects to transform around other points.

Choose a *transform center* from the flyout immediately to the right of the Reference Coordinate System pulldown on the Main Toolbar. This flyout has three buttons:

Figure 3-208: **Transform Center**

 Use Pivot Point Center transforms objects around local Pivot Points. This is the default when a single object is selected.

 Use Selection Center transforms objects around the geometric center of all selected objects. This is the default when multiple objects are selected.

 Use Transform Coordinate Center transforms objects around the origin of the currently active Reference Coordinate System. For example, if the active coordinate system is **World**, then selected objects are transformed relative to the world origin.

Selection Center

When you select a group of objects, the transform center is automatically set to **Use Selection Center**. This makes sense; when you make a selection, you will usually want to work with the center of the selection, especially when rotating the entire selection.

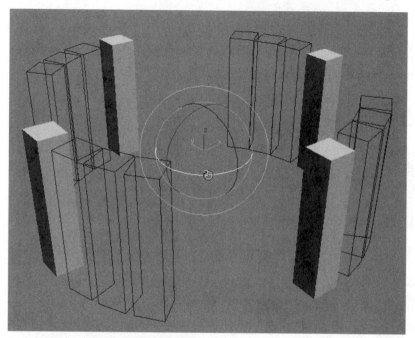

*Figure 3-209: Rotating multiple objects with **Use Selection Center***

However, you can choose instead to rotate around each object's Pivot Point. To try this out, select several objects. The transform center is set to **Use Selection Center** automatically. Click **Select and Rotate**. Rotate the objects in each viewport and observe the result.

Pivot Point Center

Next, click and hold the **Use Selection Center** button until the flyout appears, and choose **Use Pivot Point Center**. Rotate the objects in each viewport and observe the difference. Each object rotates independently around its own Pivot Point.

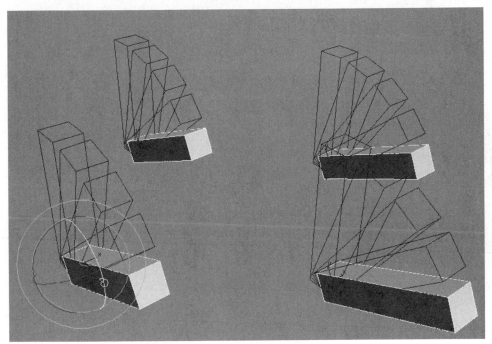

*Figure 3-210: Rotating multiple objects with **Use Pivot Point Center***

Transform Coordinate Center

There will be times when you need to transform an object around a point that is neither its Pivot Point nor the center of the current selection. When this happens, you could move the object's Pivot Point. However, this is bad practice, because the Pivot Point is the center for animated transforms. You should only move or rotate the Pivot Point as a preparation for animating the object(s), not for modeling.

Fortunately, 3ds Max makes it easy to transform an object around any point in your scene. This is accomplished with the help of the **Use Transform Coordinate Center** option.

When you choose Use Transform Coordinate Center, 3ds Max uses the origin of the currently active Reference Coordinate System as the center for transforms. So, if you choose **World** as your Reference Coordinate System, and choose Use Transform Coordinate Center from the flyout, then you can rotate one or more objects around the world origin.

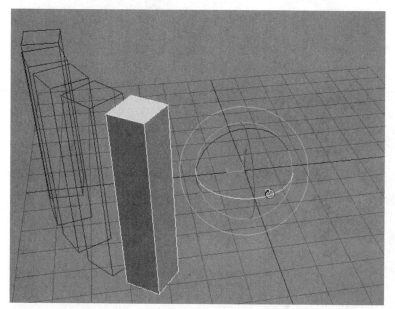

*Figure 3-211: Rotating an object around the world origin with **Use Transform Coordinate Center** and the **World** Reference Coordinate System*

Use Transform Coordinate Center is very versatile. The origin of any coordinate system you can choose from the drop-down list is available as a center for transforms. This is most useful in conjunction with the **Pick** coordinate system.

The Pick coordinate system lets you choose the Pivot Point of any object in your scene as the origin of the current **Reference Coordinate System**. So, you are not limited to rotating an object around its own Local Pivot Point or around the world origin. This opens many possibilities.

To set this up, first determine the point that you want the object to transform around. Place another object so that its Pivot Point lies at the spot around which you want to transform the object. If there isn't an object already there, you can create a Box primitive or a **Point** Helper to do the job.

Choose which transform you wish to use, such as Select and Rotate. Then, for the Reference Coordinate System, select **Pick** from the pulldown list, then immediately click the object you wish to transform around. The object you picked is now a new

Reference Coordinate System in the pulldown list, and its name is displayed on the Main Toolbar.

On the transform center flyout, choose the last button, **Use Transform Coordinate Center**. Select the object you wish to transform. Now the transform center has moved to the Pivot Point of the picked object. If the Transform Gizmo is enabled, you will see it located at the picked object, not at the Pivot of the object you wish to transform. Use the Transform Gizmo, or the Axis Restriction hotkeys **<F5>** through **<F8>,** to choose the axis for the transform, and proceed to transform the object.

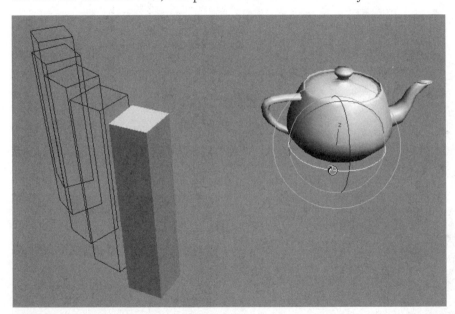

Figure 3-212: Rotating an object around another object with **Use Transform Coordinate Center** *and the* **Pick** *Reference Coordinate System*

In setting up the scene for this procedure, you are working from left to right on the Main Toolbar. First you select the transform, then the coordinate system, then the coordinate center. This is the necessary sequence. If you click buttons in any other order, the operation will not work. For example, if you choose the Pick coordinate system, then change the transform, the coordinate system will revert to the most recent coordinate system chosen for the transform. This is because 3ds Max "remembers" the coordinate systems and transforms centers you choose for each type of transformation.

NOTE: The transform center flyout options are only for modeling operations. If an animation mode is activated, the transform center flyout is set to Use Pivot Point Center, and the button is greyed out. You are unable to select any other option. However, there are ways to accomplish the same results with animated objects, using linking and Helper objects. This is covered in Chapter 10, *Advanced Animation*.

Transform Centers and Sub-object Selections

While modeling at the sub-object level, you will find many uses for the transform center options. The result of a rotation or scale operation performed on a selection of polygons or vertices will vary greatly depending on the active transform center.

In the following series of illustrations, different transform centers are used for the rotation of a polygon selection. In each case, the result is different.

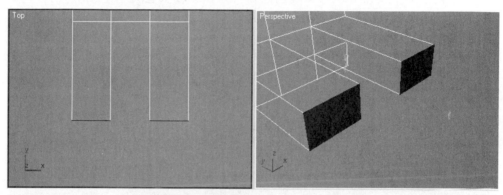

Figure 3-213: Polygon selection with no rotation

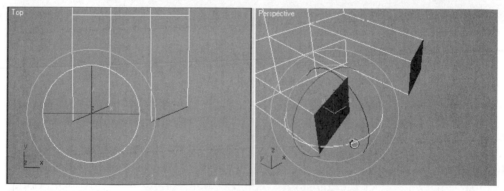

*Figure 3-214: Polygon selection rotated with **Use Pivot Point Center***

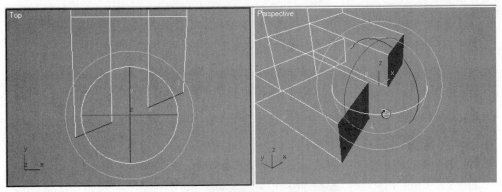

Figure 3-215: Polygon selection rotated with **Use Selection Center**

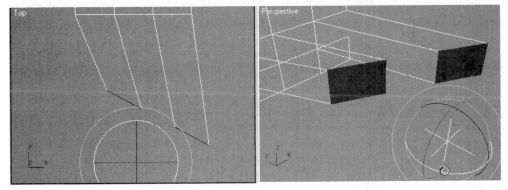

Figure 3-216: Polygon selection rotated with **Use Transform Coordinate Center** *and the* **Pick Reference Coordinate System**

In the final illustration in the series, a Point Helper object is employed to rotate the polygon selection around an arbitrary transform center. This means that you get to pick exactly where the center of the transform is.

You will find that as you become more experienced with polygon modeling, you will rely on the Pick coordinate system and Point Helpers to accomplish sub-object transforms that would otherwise be impossible.

Unfortunately, in 3ds Max there is no way to use a sub-object as a transform center. For example, you can't pick a vertex on a model and use that as the center for transforming other vertices. That is why it is often necessary to use other objects, such as Point Helpers, in conjunction with the Pick coordinate system.

EXERCISE 3.21: Flower

In this exercise, you'll use what you've learned about coordinate systems to make a flower.

1.   Reset 3ds Max.

2.   In the **Front** viewport, create a small, somewhat flat Cylinder.

3.   From the **Shapes** panel, create a tall ellipse in the **Front** viewport, and move it so it sits above the Cylinder.

4.   Use the **Align** tool 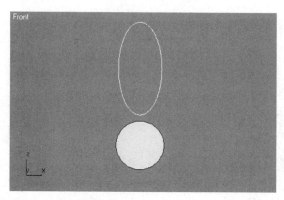 to line up the ellipse with the Cylinder vertically in the Front viewport.

Figure 3-217: Ellipse over Cylinder

5.   Extrude the ellipse to about the same height as the Cylinder. This object will be used as a flower petal.

     Next, you'll rotate and copy the petal around the Cylinder to form five more petals.

6.   Press the **<A>** key to activate **Angle Snap**. Click **Select and Rotate**. Hold down the **<SHIFT>** key and rotate the petal.

     The petal rotates around itself. This isn't what we want. We want to make the petal rotate around the Cylinder.

7.   Cancel the rotation, or click **Undo** to return the petal to its original orientation.

     To rotate the petal around the Cylinder, we'll use the **Pick** coordinate system, and follow the procedure described earlier for setting up and using the coordinate system. The transform is already selected, so we can go on to the next step, choosing the coordinate system.

8. Make sure **Select and Rotate** is still on. Choose the **Pick** coordinate system, and immediately click on the Cylinder in the active viewport.

9. From the flyout to the right of the coordinate system, choose the last option, **Use Transform Coordinate Center**.

    The axis tripod is now at the center of the Cylinder, even though the petal is still selected.

10. Hold down the **<SHIFT>** key, and click and drag at the center of the Cylinder to rotate the petal on the Z axis. Rotate the petal by 60 degrees. On the **Clone Options** dialog, enter 5 for the number of copies.

    A flower with six petals is created.

Figure 3-218: Flower

11. Save the scene with the filename **Flower.max** in your folder.

END-OF-CHAPTER EXERCISES

These exercises are designed to enable you to use some of the features and techniques you've learned in this chapter.

EXERCISE 3.22: Microphone

In this exercise, you'll use filleted cylinders to create a microphone. The exercise illustrates parametric modeling with primitives.

1. Reset 3ds Max.

2. On the **Create** panel, click **Geometry**. On the pulldown menu, choose **Extended Primitives**.

3. Click **ChamferCyl**. In the Top viewport, click and drag to begin creation of the ChamferCyl. Drag until the **Radius** is about **20** units. Release the mouse.

4. Move the cursor upward until the **Height** is about **200**. Click to set the Height.

5. Move the mouse upward again until the **Fillet** parameter is about **8** units. Click to set the fillet.

6. Increase the **Fillet Segs** parameter to **3**, and increase **Sides** to **24**.

    You have now created a smooth ChamferCyl. This will be the microphone handle.

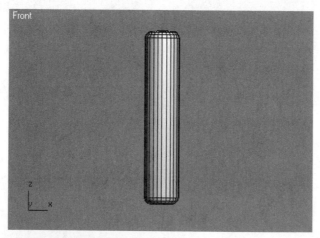

Figure 3-219: Microphone handle

7. Go to the Modify panel. Apply a **Taper** modifier to the ChamferCyl. Change the **Amount** parameter to **0.2**. The ChamferCyl is slightly tapered. Rename the object **Handle**.

8. In the Top viewport, create another ChamferCyl with the following dimensions:

Radius	38
Height	90
Fillet	15
Height Segs	1
Fillet Segs	3

9. Apply a Taper modifier to the new ChamferCyl. Change the **Amount** parameter to **-0.2**. This object will be the bulb of the microphone.

10. In the Front viewport, move the microphone bulb so it sits above the handle, leaving some space between the two objects, as shown in the following illustration.

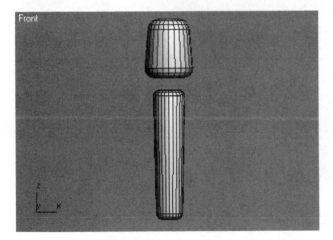

Figure 3-220: Bulb moved upward in Front viewport

11. On the Create panel pulldown menu, choose **Standard Primitives**. Click **Cone**.

12. In the Top viewport, create a **Cone** with the following dimensions:

Radius 1	24
Radius 2	38
Height	33
Sides	24

    This object will connect the microphone to the bulb.

13. In the Front viewport, move the Cone so its bottom edge covers the top fillet of the microphone handle.

14. In the Front viewport, move the bulb so about half of its fillet is inside the Cone.

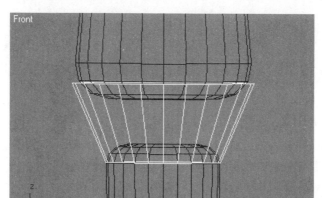

Figure 3-221: Bulb moved partially inside cone

Next, you'll make sure the objects are aligned by using the **Align** tool.

15. Select the cone. On the Main Toolbar, click **Align**, then click the microphone handle. The **Align Selection** dialog appears.

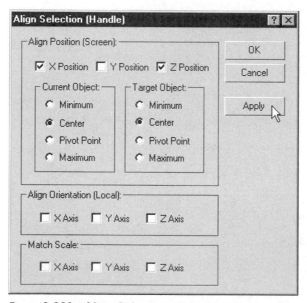

*Figure 3-222: **Align Selection** dialog*

16. Check the **X Position** and **Z Position** check boxes. As you do so, you will see the cone shift slightly as it aligns with the handle. Click **OK** to exit the dialog.

17. Align the bulb to the handle in the same way.

Be sure to select the bulb before clicking Align.

Next, you'll create two more ChamferCyl objects to be used as rings around the connecting parts.

18. On the Create panel pulldown menu, choose **Extended Primitives**. In the Top viewport, create a **ChamferCyl** with the following dimensions:

Radius	**38.5**
Height	**10**
Fillet	**1**
Fillet Segs	**1**

19. Move the new ChamferCyl in the Front viewport so it covers the area where the cone meets the bulb.

    To ensure the ring is covering the joint area, change the Front view to Smooth + Highlights display. To do this, right-click the Front viewport label and choose **Smooth+Highlights** from the pop-up menu.

    You can also activate the viewport and press the **<F3>** Hotkey.

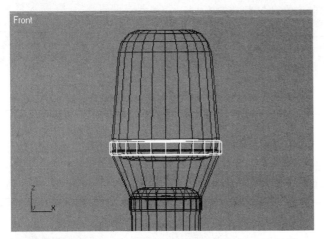

Figure 3-223: Ring covering area where Cone meets bulb

20. Use the Align tool to align the ring with the Cone, using the same technique you used earlier to align the Cone and bulb with the handle.

21. In the Front viewport, copy the ring downward to create another ring for where the Cone meets the handle.

22. On the Modify panel, change the new ring's **Radius** to **25**.

The microphone is now complete.

Figure 3-224: Completed microphone

23. Save the scene with the filename **Microphone.max** in your folder.

EXERCISE 3.23: Microphone Stand

In this exercise, you'll create a microphone stand for the microphone, complete with swiveling head.

1. Load the file **Microphone.max** you created earlier, if it is not already on your screen. If you don't have this file, you can load the file from the disc.

2. Create a Named Selection Set for the microphone. To do this, select all the objects. Go to the **Named Selection Sets** field on the **Main Toolbar**, and enter the name **Microphone**. Be sure to press **<ENTER>** after you have entered the selection set name.

3. Hide all the objects in the scene. To do this, select all objects, go to the **Display** panel and click **Hide Selected**.

Next, you'll create the tripod base of the microphone stand.

4. In the Front viewport, create a **Capsule** with a **Radius** of **20** and a **Height** of **500**. This object will be one leg of the tripod.

5. Press the **<A>** Hotkey to activate **Angle Snap**. In the Top viewport, hold down the **<SHIFT>** key and rotate the capsule by exactly **120** degrees. When the **Clone Options** dialog appears, choose the **Instance** option and set **Number of Copies** to **2**.

    The tripod has been created.

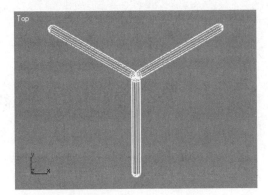

Figure 3-225: Microphone stand tripod

6. Select one of the tripod legs. Apply the **XForm** modifier to it.

    Because the legs are instanced, the XForm modifier is applied to all the objects.

7. Click Select and Rotate and change the **Reference Coordinate System** to **Local**. In any viewport, click and drag the leg's local X axis and rotate the leg by 25 degrees.

    Because the objects are instanced and the XForm modifier is used to rotate one of them, they all rotate together.

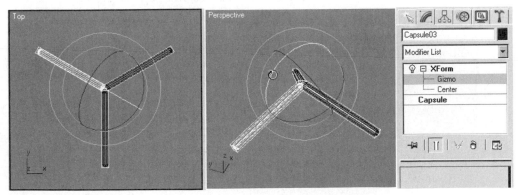

*Figure 3-226: Rotating **XForm Gizmo** of instanced tripod legs*

8. In the Top viewport, create a **Cylinder** with a **Radius** of **20** and a **Height** of **2000**. This object won't be bending, so you can set the **Height Segments** to **1**.

    This Cylinder will be used as the shaft of the microphone stand.

9. Move the Cylinder so its base sits at the center of the tripod.

10. In the Top viewport, create a **Sphere** with a **Radius** of about **70** units. Place the Sphere where the Cylinder meets the tripod.

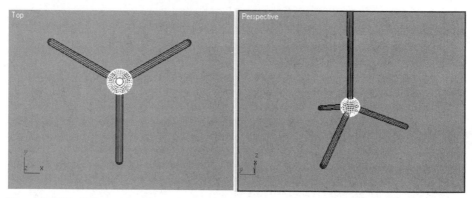

Figure 3-227: Sphere placed at tripod center

Next, you'll create the clip for the microphone.

11. In the Front viewport, create a **Tube** with the following dimensions:

Radius 1	32
Radius 2	25
Height	115

12. At the bottom of the **Parameters** rollout, check the **Slice On** check box. Change **Slice From** to **-80**, and **Slice To** to **80**.

The Tube has been sliced into a sleeve that can be used as a clip for the microphone. Many Primitives have similar slicing parameters.

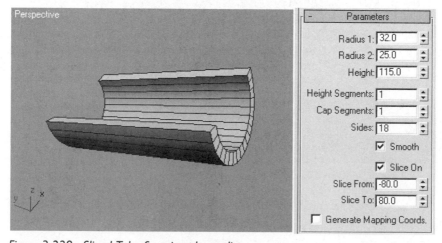

Figure 3-228: Sliced Tube for microphone clip

13.   Move the microphone clip near the top of the shaft.

Next, you'll make the swivel assembly for the microphone clip.

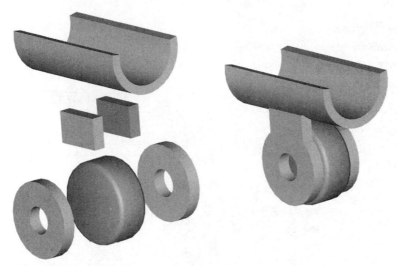

Figure 3-229: Swivel assembly

14.   The swivel assembly is made of several primitives. The previous illustration shows the parts of the assembly and how to fit them together. Use Standard Primitives and/or Extended Primitives to create the assembly shown above.

The attachment piece that connects the swivel with the shaft is a little more complex. This object will have to be made by applying a modifier to a sub-object selection.

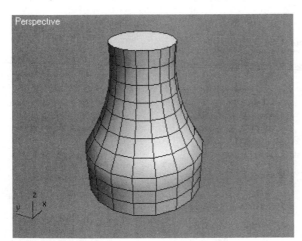

Figure 3-230: Attachment object

15. In the Top viewport, create a Cylinder with the following parameters:

   Radius   24
   Height   60
   Height Segments 10

16. Apply a **Mesh Select** modifier to the cylinder. Access the **Polygon** sub-object level. Select the seven top rows of polygons on the cylinder.

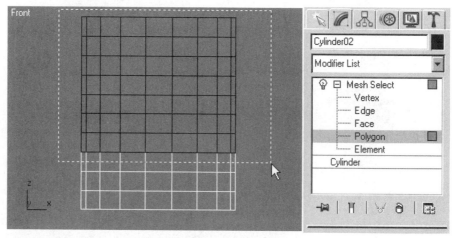

Figure 3-231: Top seven rows of polygons selected

17. Apply a **Taper** modifier to the selection. Change the **Amount** to **-0.5** and **Curve** to **-0.4**.

18. In the Front viewport, move the center of the Taper to the bottom of the selection. To do this, select the **Center** sub-object. In the Front viewport, move the Center down to the bottom of the selection.

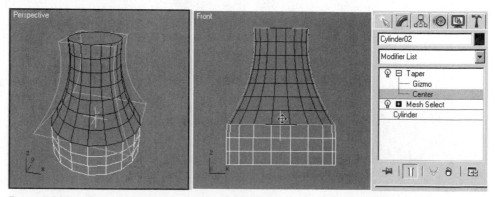

*Figure 3-232: Move the **Center** of the Taper modifier*

The attachment object is complete.

19. Position the attachment object at the bottom of the swivel, and move the entire assembly to fit over the top of the shaft.

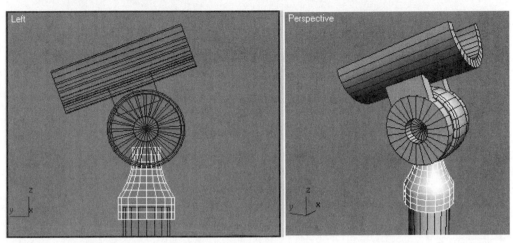

Figure 3-233: Assembly fitted to shaft

20. Unhide the microphone objects. Use the Named Selection Sets to select all of the microphone objects. Position the microphone in the clip.

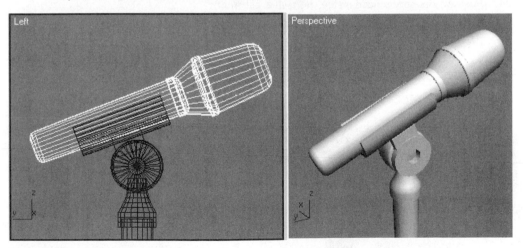

Figure 3-234: Microphone positioned in clip

Next, you'll create a screw head to fit into the assembly.

21. In the Left viewport, zoom in on the round cogs of the swivel assembly.

22. In the Left viewport, create a Sphere that fits inside the hole in the outside cogs.

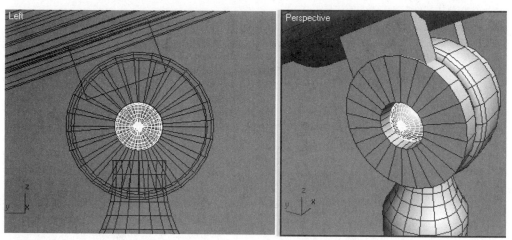

Figure 3-235: Sphere inside cog

23. On the Modify panel, change the **Hemisphere** parameter to **0.5**.

     This cuts the sphere in half and makes it resemble a screw head.

24. Position the screw head inside the swivel assembly.

25. Place a small cylinder in the hole on the other side of the swivel assembly to complete the screw.

     The scene is now complete.

Figure 3-236: Completed microphone and stand

26. Save the scene with the filename **MicStand.max** in your folder.

EXERCISE 3.24: Space Scene

In this exercise, you'll use the following techniques to create an alien landscape:

- Primitives
- Volume Select modifier
- Soft Selection
- Noise modifier

1. Reset 3ds Max.

2. In the Top viewport, create a **Plane** primitive with the following dimensions:

Length	250
Width	250
Length Segs	10
Width Segs	10

3. Go to the Modify panel and apply the **Noise** modifier to the Plane. Set **Scale** to **25**. Under the **Strength** section, increase the **Z** parameter to **20**.

    The Plane is now a low-density space terrain.

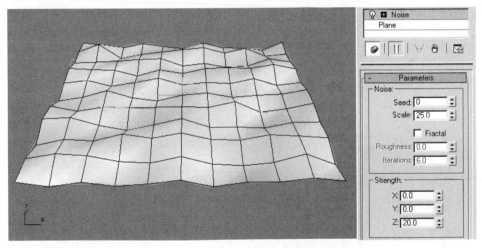

*Figure 3-237: Low-density **Plane** with **Noise***

4. Now we will increase the density of the Plane, but only when it is rendered. This will create a much more realistic terrain, without slowing down viewport performance.

    In the Modifier Stack, select the Plane creation parameters. In the **Render Multipliers** section, increase the **Density** parameter to **10**. The **Total Faces** read-out displays an amount of **20000**. This means that when the Plane renders, it will have 20,000 faces.

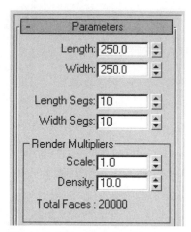

Figure 3-238: **Render Multipliers** *parameters*

5.   Adjust the Perspective view so it appears you are looking out over an alien landscape.

On the Main Toolbar, press the **Quick Render** button .

Figure 3-239: Alien landscape

The landscape renders with much higher polygonal density.

6.　Select the Noise modifier in the Stack. Enable the **Fractal** option. This causes the Noise to be calculated numerous times, producing a rougher terrain. Press the Quick Render button again to see the results.

Figure 3-240: **Fractal** Noise

7.　The landscape is more rugged, but now it appears that the geometry is not dense enough. You can see jagged edges in the Plane mesh.

Go back to the Plane in the Modifier Stack. Increase the density of the rendered mesh by entering the following values in the Plane creation parameters:

Length Segments	20
Width Segments	20
Density	20

Note that the number of Total Faces is now **320,000**.

8.   Press the Quick Render button to observe the results. Notice how much longer it takes to render the scene with 320,000 faces, instead of 20,000.

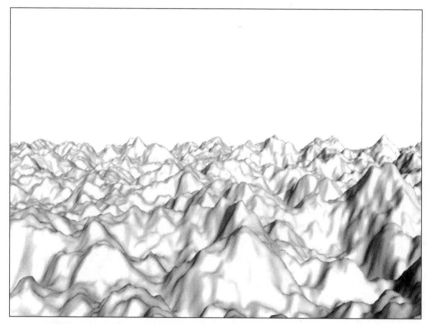

Figure 3-241: Fractal Noise with increased mesh density

The landscape looks fairly realistic, but is too rugged for our uses. We would like to have a more flat plane closer to the viewpoint, so we can place objects on the landscape later. To do this, we will use a Volume Select modifier to mask the effects of the Noise modifier.

9.   To reduce the time spent waiting for rendering, reduce the Plane Density to **5**. The Total Faces value now reads **20,000**.

10. With the Plane selected in the Modifier Stack, add a **Volume Select** modifier. Be certain not to add a Mesh Select or Poly Select, as they will not work for this procedure, due to the vertex ID problem discussed earlier.

11. In the Volume Select modifier, select **Vertex** in the **Stack Selection Level** section. This allows a Vertex sub-object selection to be passed up the stack. Turn on **Show End Result** ⎸⎸ in the Modifier Stack.

12. Enter **Gizmo** sub-object mode in the Volume Select modifier. In the Top viewport, move the Volume Select Gizmo up in the viewport's Y axis. Position the Gizmo so that about one half of the Plane object is selected. Your screen should look something like the following illustration.

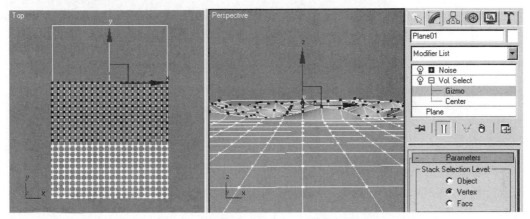

Figure 3-242: Moving the **Volume Select Gizmo**

13. Select the Perspective view and click the Quick Render button. The transition between the flat plane and the fractal noise is far too abrupt.

14. In the Volume Select modifier, scroll down to the **Soft Selection** rollout. Enable **Use Soft Selection** and increase the **Falloff** value.

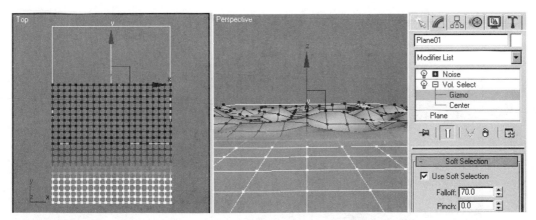

Figure 3-243: Volume Select with **Soft Selection**

15. Select the Perspective viewport and click Quick Render. Experiment with the Noise, Volume Select parameters to get a result you are happy with. Continue to do test renders as you go. Remember to select the Perspective viewport before clicking Quick Render.

16. In the Top viewport, create a Sphere with a **Radius** of about **30** units. Position the sphere to appear as a planet in the sky. Move it far back into the scene.

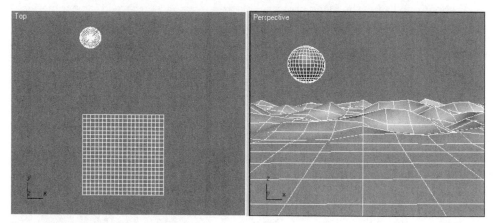

Figure 3-244: Planet over landscape

17. Create a **GeoSphere** with a **Radius** of about **5** units. Apply the **Noise** modifier to it, and experiment with the Noise settings until it looks like an asteroid. Place the asteroid so it looks as if it's about to land on the terrain.

18. Increase the Plane Density once more and do a few more test renders.

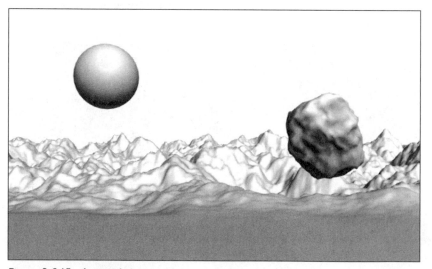

Figure 3-245: Asteroid on terrain

19. Save the scene with the filename **SpaceScene01.max** in your folder.

SUMMARY

In this chapter, you learned that there are several types of models, including *mesh*, *Bezier Patch*, *NURBS*, and *subdivision surfaces*.

Objects can be created from *primitives*, and then altered through the use of *modifiers*. This is called *parametric modeling*, and allows a great deal of flexibility in making overall changes to the object.

The **Modifier Stack** and its many options are an important part of modeling. Modifiers can be applied to entire objects or sub-object selections within objects.

Modifiers themselves have sub-objects that control their effects. For example, the **Center** of a Bend modifier specifies the location around which an object will be bent.

Multiple modifiers can be applied to an object. The order of modifiers in the Stack has a profound influence on the result.

An understanding of the *data flow* within 3ds Max is crucial to successful modeling. The data flow for an object with modifiers is: object parameters, then modifiers, then transforms, and finally object properties.

When objects are cloned in 3ds Max, you have the option of making the clone an independent **Copy**, an **Instance**, or a **Reference**. Instance objects are always identical, and share the same Modifier Stack. Reference objects have a branching data flow, in which some of the modifiers are shared, and additional unshared modifiers may be added.

The Modifier Stack can be collapsed, which makes the effects of modifiers permanent. This results in an editable object, which can be directly manipulated at the sub-object level. Editable objects include **Editable Spline**, **Editable Mesh**, and **Editable Poly**. Each of these objects has many specialized tools in its Modify panel.

2D shapes can be used in conjunction with several modifiers, including **Extrude**, **Lathe**, **Bevel**, and **Sweep**, to make 3D objects. Shapes are edited by changing the *vertices*, *segments*, or *splines* in the shape. The curvature of shape segments is controlled with the shape **Interpolation** settings.

A **Loft** is a compound object that uses two or more shapes to make a 3D object. A Loft always has exactly one path. One or more shapes can be placed on the Loft path. Many types of objects can be made with this process.

3ds Max offers many tools for controlling transforms. These include the **Align** and **Spacing** tools, the **Snap** options, and **Helper** objects.

The *transform coordinate center* plays an important part in transforming objects and sub-objects. The center of a selection, or any object in the scene, can be used as the center of a transform, instead of the object's Pivot Point.

REVIEW QUESTIONS

1. What are the different types of models in 3ds Max?

2. What is the difference between a **Copy** and an **Instance**?

3. What does collapsing the Modifier Stack do? How does it affect your ability to edit an object?

4. What is the difference between a transform and a modifier? Name as many differences as you can.

5. How do you control what parts of an object are affected by a modifier?

6. What is the main difference between **Editable Mesh** and **Editable Poly**?

7. What are vector handles on a vertex? What are they used for?

8. What are the vertex types on an **Editable Spline**? How do you switch from one type to another?

9. How do you make a multi-spline shape?

10. What is a *face normal*? What does it have to do with rendering?

11. On a **Loft** object, what can happen if you don't make sure the vertices at the ends of the path are the **Corner** type?

12. How do you fix a twisted Loft object?

13. What is the *transform center* used for?

14. What is the difference between the Extrude modifier and the Extrude tool in Editable Poly?

15. What is the purpose of **3D Snap**?

16. How do you control the smoothness of Shape segments?

17. What is an **XForm**?

18. What is the advantage of **Soft Selection**?

Chapter 4
Materials

OBJECTIVES

In this chapter, you will learn about:

- Core concepts of materials
- Working with the Material Editor
- Creating materials with maps
- Using maps to simulate environments
- Placing maps on objects
- Applying multiple materials per object
- Face smoothing
- How to organize materials and material libraries

ABOUT THE MATERIAL EDITOR

Materials are the "paint" or texture to put on your models. All of the surface properties of renderable objects such as color, bumpiness, shininess, and transparency are defined by materials.

Materials are created, edited, and assigned in the **Material Editor**. The Material Editor has many options. In this chapter, we will go over the options you are most likely to use when making materials. These are also the fundamental building blocks you will need to make more complex materials later in this book.

Some of the advanced options in the Material Editor, such as *raytracing*, can cause rendering times to increase dramatically, by as much as 100 times. For this reason, it is strongly recommended that you stick with the features described in this chapter until you gain more experience with 3ds Max. In Chapter 11, *Advanced Materials*, you'll learn how to use the advanced features efficiently.

MATERIAL EDITOR LAYOUT

To access the Material Editor, click the Material Editor button on the Main Toolbar. You can also access the Material Editor by choosing **Rendering > Material Editor** from the Main Menu, or by pressing the <**M**> key.

The Material Editor is a floating window that consists of four main areas:

Sample Slots are the large squares near the top of the Material Editor. In each slot is a *sample object*. The default sample object is a sphere, but you can also choose other objects. Only one sample slot can be active at any time. As you define or modify a material, a representation of the material appears on the sample object, to help you visualize how the material will look in the scene. When you are satisfied with how the material looks in the sample slot, you then assign it to an object in the scene.

The **rollout panel** displays parameters for defining the material. The rollout panel changes frequently to display the current options.

The buttons on the Material Editor **toolbars** give you access to Material Editor commands. Some of these buttons affect the display only, and not the materials themselves. For example, you can choose what type of sample object to use, specify how many sample slots you wish to view at once, or control material preview options.

In addition, the commands on the Material Editor toolbar are also available from a text-based **Menu Bar** at the top of the Material Editor window.

TIP: The Material Editor may be resized vertically, to make it taller or shorter, and fit better on your screen. Click and drag the top or bottom edge of the window to resize the Material Editor.

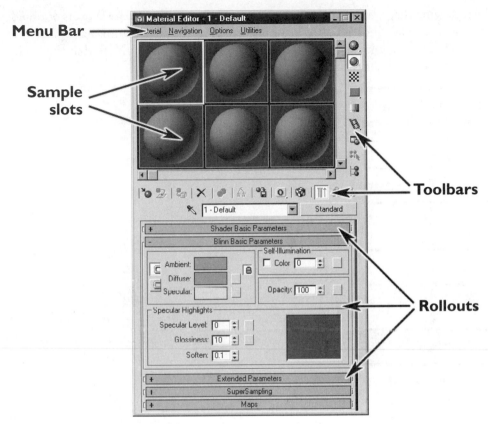

Figure 4-1: Material Editor layout

MATERIAL DEFINITIONS

A material is sometimes referred to as a *material definition*, which hints at the fact that a material is actually a whole host of attributes. In some 3D programs, the term *shader* is used instead of the term material. In either case, a material or shader controls how an object will render.

3ds Max comes with a number of ready-made materials that you can assign to your objects. You can also edit these materials or create new ones from scratch. Materials can be assigned to entire objects or portions of objects.

MATERIAL DATA FLOW

Like everything in 3D graphics, material definitions have an internal data flow. This is the priority of operations that are performed to achieve the end result. In the case of a material, there are four major levels of a material definition.

In the following diagram, the data flow of a material definition in 3ds Max is outlined. In this case, we have indicated the data flowing from top to bottom, because that is how it is visually organized within the 3ds Max Material Editor. Note that this is the opposite of the Modifier Stack, which flows from bottom to top.

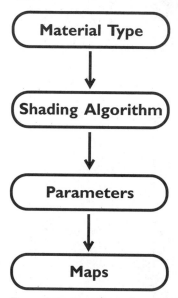

Figure 4-2: Data flow of a material definition

The four main levels of data within a material definition are: *material type*, *shading algorithm*, *parameters*, and *maps*. Each term is explained in the following sections.

MATERIAL TYPES

3ds Max has several different types of materials. A material type is the most fundamental property of a material definition in 3ds Max. Each material type has its own specific function. For example, the **Ink 'n Paint** material is designed to create renderings in a cartoon style, and the **Composite** material is designed for complex layerings of other materials.

To create a new material from scratch, you begin by clicking the **Get Material** button in the Material Editor. This launches the **Material/Map Browser**. If you choose **Browse From > New** in the Material/Map Browser, you will see a list of all of the currently available material types.

Figure 4-3: **Material/Map Browser** *displaying the available material types*

3ds Max offers many different material types for specialized uses. However, the vast majority of the materials you create will be of the **Standard** type. Despite their unimpressive name, Standard materials are highly customizable, and you will find only occasional situations in which you will need one of the other material types.

Material types are plug-ins in 3ds Max, so other material types can be created to extend the functionality of the Material Editor. You can purchase material plug-ins from software developers, and there are even some plug-ins available on the Internet for free.

SHADING ALGORITHMS

Shading is the process of rendering geometry so that it responds to simulated light in the CG environment. An *algorithm* is a set of logical instructions, such as a mathematical equation or a computer program. So, a *shading algorithm* is a method of rendering geometry to achieve a certain look. Some shading algorithms are designed to simulate particular surfaces, others are designed for a wide variety of applications.

The choice of shading algorithm can have a great influence over how a material looks when rendered. For example, the **Phong** shading algorithm has the appearance of plastic, while the **Strauss** algorithm is designed to simulate metal surfaces.

3ds Max refers to these algorithms as **shaders**. Unfortunately, this choice of terminology is at odds with the usage of the term within the larger 3D graphics community. As mentioned previously, the term *shader* is often synonymous with the term *material*. In this book, we explicitly refer to shading algorithms as such, to avoid confusion. A shading algorithm is not a material, it is only one of the attributes of a complete material definition.

In the 3ds Max Material Editor, the shading algorithm for a material are chosen from a pulldown list in the **Shader Basic Parameters** rollout. By default, the pulldown list reads **Blinn**.

Many of the shading algorithms in 3ds Max are also found in other programs. This is because the most popular algorithms were not

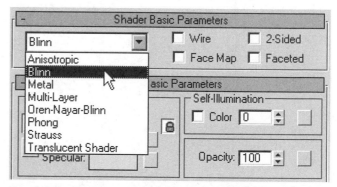

Figure 4-4: *Selecting a shading algorithm in the* **Shader** *pulldown*

written by the programmers at Autodesk or by their competitors. Instead, they were created by computer scientists at various institutions around the world. In most cases, the algorithms are named after their creators.

One of the most brilliant of these graphics programmers is James Blinn. Due to its versatility, Blinn's shading algorithm is used widely throughout 3D graphics, and it is the default shading algorithm in 3ds Max. You will find that, like the Standard material, the **Blinn** shading algorithm is all you'll usually need, and you won't use the others very often. You will learn more about the other shading algorithms in Chapter 11, *Advanced Materials*.

MATERIAL PARAMETERS

Like an object or a modifier, a shading algorithm has parameters that an artist controls to achieve the desired results. These include color, transparency, shininess, and many other attributes of a material.

In the Material Editor, the most fundamental material parameters are found in the rollout directly below the Shader Basic Parameters rollout. The name of this rollout changes depending on which shading algorithm is chosen. So, the default name of the rollout is **Blinn Basic Parameters**. If you pick the Strauss shading algorithm, then the rollout will read **Strauss Basic Parameters**.

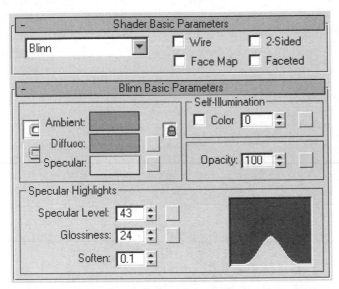

Figure 4-5: **Blinn Basic Parameters** *rollout*

Since the shading algorithms are all different, the parameters available on the rollout will vary depending on the current algorithm. If you choose a different algorithm, the rollout will change.

However, some of the parameters are commonly found in many or all of the different shading algorithms. For example, nearly all shaders have a **Diffuse** color, which is the color of the material where it is directly illuminated.

In addition, there are other material parameters that are not dependent on the type of shading algorithm chosen. These are found in the various other rollouts, such as **Extended Parameters**.

MAPS

Surfaces in the real world are very complex. They never look like the perfectly smooth, featureless surfaces you see in the default sample slots of the Material Editor. To make your materials really look like something, you need a way to vary the material parameters across the surface of your objects. That's where *maps* come in.

A map is a way to alter the parameters of a material. The most basic use of a map is to place a pattern or image on an object, to simulate changes in color across the object. This is commonly referred to as a *texture map*, although it is more accurately known as a **Diffuse Color** map. In this case, the image varies the color of the material, not its roughness or texture.

Imagine how difficult it would be to create a rendering of a wooden floor with a rug on it if you had to model every tiny detail. You'd have to create an incredibly complex scene that simulated the grain of the wood and the fibers of the rug. Practically speaking, this would be impossible. However, with mapping, it is a simple matter to create two low-polygon plane objects and assign appropriate materials to them.

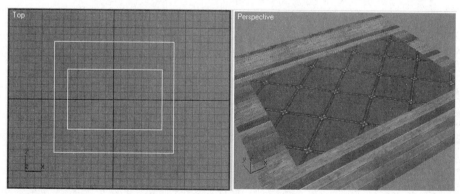

*Figure 4-6: Scene with only two polygons, each assigned a material with a **Diffuse Color** map*

There are many different types of mapping. You can build up realistic materials by assigning maps to vary the color, bumpiness, transparency, shininess, and many other attributes of a material. Maps can control most of the material parameters in 3ds Max.

Maps are assigned in the **Maps** rollout of the Material Editor. This rollout lists the different mappable parameters available to the current material type. In the following illustration, a map has been assigned to the Diffuse Color component of the material to create the illusion of a pattern on the rug.

TIP: Computer graphics are all about illusion. There is no need to create an accurate simulation of an object or an event. All that is necessary is to create an illusion convincing enough to fool the audience, or at least suspend their disbelief. Think of the shooting set of a movie, where only the facades of the buildings are built, or a stage set in a play, where wooden flats are painted to look like brick. Don't waste energy constructing something the audience will never see!

	Amount	Map
☐ Ambient Color . . .	100	None
☑ Diffuse Color	100	Map #2 (PAT0039.TGA)
☐ Specular Color . .	100	None
☐ Specular Level .	100	None
☐ Glossiness	100	None
☐ Self-Illumination .	100	None
☐ Opacity	100	None
☐ Filter Color	100	None
☐ Bump	30	None
☐ Reflection	100	None
☐ Refraction	100	None
☐ Displacement . .	100	None

Figure 4-7: **Maps** *rollout*

MATERIAL EDITOR BASICS

WORKFLOW

When making materials, you will always follow this procedure:

1. Select a sample slot.

2. Work with the parameters on the **Material Editor** rollouts to define the material. Watch the sample slot to see how the material is shaping up.

3. When you are satisfied with the material, assign the material to an object in the scene.

4. Render the scene to check how the material looks with the current scene setup.

5. If necessary, return to the Material Editor, change your materials, and re-render the scene.

In this chapter, you will use this procedure to create and assign custom materials.

SAMPLE SLOTS

The first step in setting up a material is activating a sample slot for the material to occupy. You can activate a sample slot simply by clicking in it. A white border appears around the sample slot to indicate that it is active.

By default, neutral gray materials are set up in the sample slots. You can change any and all attributes of these materials to make your own custom materials.

Lighting, the camera angle, and other objects in the scene can affect how a material looks in a scene. What you see in the sample slot will give you an idea of how a material will look in the rendering, but it won't always be accurate. You must render the scene to see how the material really looks.

Editor Samples vs. Scene Materials

A common misconception among beginners is that the sample slots show you which materials are in the scene. Nothing could be farther from the truth. The sample slots are merely places for previewing materials, which *may or may not be in the scene.*

Just because you see a material in the Material Editor does not mean it is assigned to any object in the scene. If a material *is* assigned to an object, then you will see white triangles at the corners of the sample slot. This is called a *hot* material.

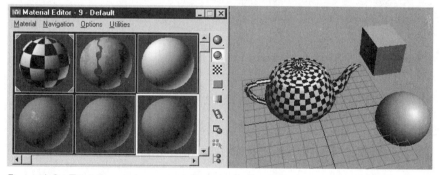

Figure 4-8: Triangles at the corner of a sample slot indicate a material that is in the scene

When the object to which the material has been assigned is selected in the viewport, the triangles are solid. When the object is not selected, the triangles become gray.

If the material isn't assigned to any objects in the scene, there are no triangles at the corners of the material slot. A material that has not been assigned in the scene is called a *cool* material.

You can assign the material to an object in the scene in two ways. One is to click and drag from the sample slot to an object or to multiple selected objects. The other way to assign materials is to select the object or objects in the scene, then click the **Assign Material to Selection** button on the Material Editor toolbar.

You can use the scrollbars at the bottom and right of the sample slot area to view more than the six sample slots you see initially. There are a total of 24 slots available, but there is no limit to the number of materials that you may have in a scene. If you run out of sample slots, you can just reuse the existing slots. You can clear sample slots without erasing any materials in the scene. And, as you will learn later, you can always reload materials from the scene into the Material Editor, using the Material/Map Browser.

BASIC CUSTOM MATERIALS

To begin making custom materials, you must first activate a sample slot. In the rollout area, locate the **Blinn Basic Parameters** rollout. You will use these parameters to change some of the material attributes.

Diffuse Color

On the **Blinn Basic Parameters** rollout, locate the **Diffuse** color swatch. This parameter sets the overall color of the material. To change it, click the **Diffuse** color swatch. The 3ds Max **Color Selector** appears.

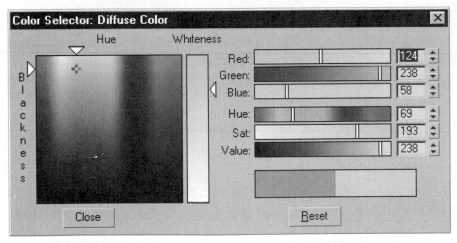

Figure 4-9: Color Selector

The Color Selector provides multiple ways to set a color. As you change parameters, the new color appears at the lower right of the window. The **Diffuse** color swatch on the **Blinn Basic Parameters** rollout updates each time you change the color. You can leave the Color Selector open as long as you need it. There is no need to close this window unless you need more room on your screen.

EXERCISE 4.1: Simple Custom Material

In this exercise you'll create three objects, and create custom materials for them.

1. Reset 3ds Max. In the Top viewport, create a teapot.

2. In the Top viewport, make two copies of the teapot to the right of the original teapot. To do this, click **Select and Move** on the Main Toolbar. Hold down the **<SHIFT>** key while moving the teapot to the right. When the **Clone Options** dialog appears, choose the **Copy** option, and enter 2 for **Number of Copies**. Click **OK** to create the two new teapots.

Figure 4-10: Three teapots in the Perspective view

3. Open the **Material Editor** by clicking the button on the Main Toolbar, or by choosing **Rendering > Material Editor** from the Main Menu, or by pressing the **<M>** key.

4. Select the first sample slot.

5. On the **Blinn Basic Parameters** rollout, click the **Diffuse** color swatch. The **Color Selector** opens. Change the color to a bright yellow by increasing the Red and Green values.

6. Click and drag from the sample slot to the first teapot. The material is assigned to the teapot.

    Note that four small triangles have appeared at the corners of the sample slot, indicating that the material has been assigned in the scene.

7. Click another sample slot to activate it. On the Blinn Basic Parameters rollout, click the Diffuse color swatch, and change the color to a bright blue. Assign this material to the second teapot.

8. Click a third sample slot to activate it. Change the Diffuse color to a bright red. Assign this material to the third teapot.

9. Save your work as **SimpleMaterials.max** in your folder.

NAMING MATERIALS

Each material is assigned a default name that appears on the Material Editor toolbar. The currently active material is also listed at the very top of the Material Editor window, in the window title bar. Default material names are **01 - Default, 02- Default,** and so on.

When a material is assigned to an object, it is linked to the object through its name. Names of materials assigned to objects have to be unique. Unlike objects, you can't have two materials with the same name in the scene.

As you create materials, it is important that you assign a unique name to each one. Otherwise, you will end up with a scene full of names like **5- Default** and **17 - Default**. When your scene has become populated with 20 or 30 objects, you will find that you can no longer remember which material is assigned to which object. This will cost you much confusion and wasted time.

You can change the material name by highlighting the existing name on the Material Editor toolbar and typing a new name.

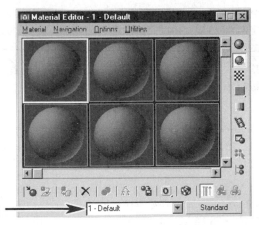

Material name

Figure 4-11: Material name

When assigning names, you can choose a name that describes either the material or the object, or a name that describes both. For example, suppose you are creating wood materials for a room. The room will have a wooden floor and wooden chairs, each a different type of wood. In this case, you could call one **Wood Floor** and the other **Wood Chairs**.

You can return to the **Material Editor** at any time to change material names. If you change the name of a material in the editor, the names of the materials in the scene are updated to reflect the change.

SHININESS

In life, a shiny object has highlight areas that reflect light. In the **Material Editor**, basic shininess qualities are simulated with four parameters, all of which are located in the **Blinn Basic Parameters** rollout.

The **Specular** color component sets the color of the highlight. Click the Specular color swatch to assign a color to the highlight. In the real world, the highlights on most objects will be the same color as the light source. In computer graphics, you will usually want to leave the Specular color as a near-white color, which will cause the highlights to reflect the color of lights in your scene.

The remainder of the shininess controls are found in the **Specular Highlights** section at the bottom of the Blinn Basic Parameters rollout.

The **Specular Level** parameter is a value from 0 to 999 that sets the overall brightness of the highlight. Higher values make a brighter highlight. Values above 100 will cause the highlights to be "blown out" or overexposed, resulting in loss of detail in the highlight area. However, this may be desirable for some materials, such as polished metal.

Glossiness is a value from 0 to 100 which sets the size of the highlight. Higher values make the highlight smaller. The more highly polished an object is, the smaller its highlights will be.

Soften is a value from 0.0 to 1.0 that controls the softness of the highlight edge. Higher values make the highlight edges softer.

The small graph on the right of the Specular Highlights section works in conjunction with the Specular Level and Glossiness parameters. This graph gives an indication of the size and brightness of the highlights. You can't edit the shape of the graph directly; it merely displays the results of the parameters.

The height of the curve within the graph is an indicator of the Specular Level of the material, while the width of the curve indicates the amount of Glossiness. For example, a high Specular Level makes the graph tall, indicating a bright highlight. A low Glossiness setting makes the graph narrow, making a small highlight.

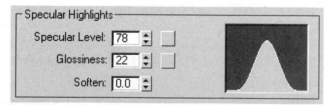

Figure 4-12: Shininess graph depicts highlight qualities

By default, sample slots are illuminated by two virtual light sources. Sometimes it is easier to evaluate the shininess of a material by turning off the **Backlight** option in the Material Editor toolbar.

Figure 4-13: ***Backlight*** *disabled*

Using the Specular Highlights controls and the default Blinn shading algorithm, you can create materials with a wide variety of shininess, from a dull stone to a shiny metal. In the following illustration, the materials are identical except for their Specular Highlights parameter values.

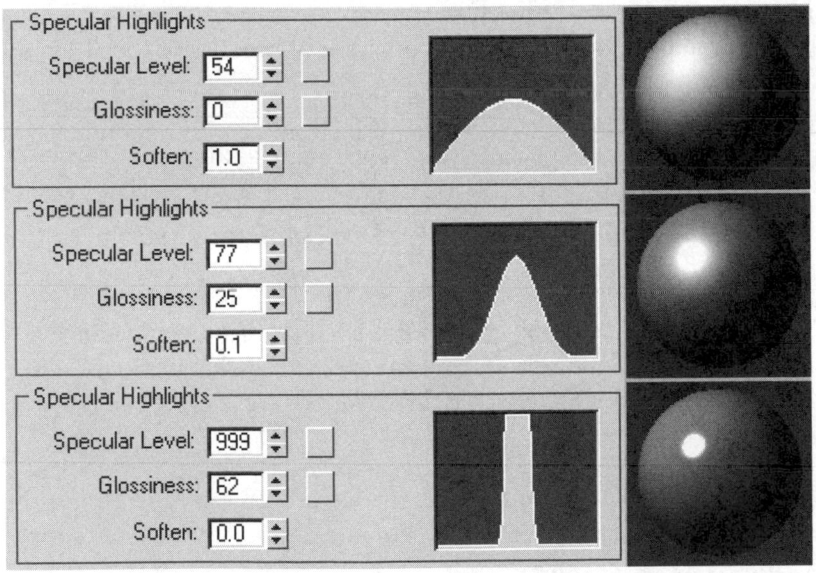

*Figure 4-14: Materials with varying **Specular Highlights** settings*

ADDITIONAL PARAMETERS

A few additional parameters appear on the **Blinn Basic Parameters** rollout. These parameters are not used as frequently as the ones you have already learned.

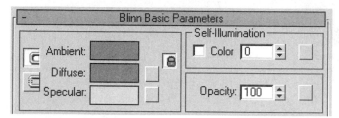

*Figure 4-15: Additional parameters on **Blinn Basic Parameters** rollout*

Locking Colors is accomplished with the buttons to the left of the color swatches. If colors are locked, changing one swatch will affect the other swatch as well.

The **Ambient** color component sets the color of the material when in shadow. By default, the Ambient color is locked to the Diffuse color. If you desire more contrast in your material, you may unlock the colors and reduce the luminance of the Ambient color with the **Value** slider in the Color Selector.

Self-Illumination makes an object look as if it were lit from the inside, like a light bulb. When **Color** is unchecked, a value from 0 to 100 determines the amount of self-illumination. When **Color** is checked, the self-illumination is tinted with the color. By default, the black **Color** makes no self-illumination. The **Color** must be changed to a lighter color to see any self-illumination.

NOTE: By default, a self-illuminated object does not shed any light in the scene. Self-illumination merely brightens the material in renderings. Self-illumination can only add light if you use **Radiosity**, which is an Advanced Lighting technique discussed in Chapter 12.

Opacity is a value from 0 to 100 that sets the degree of overall transparency of the object. 0 is completely transparent, while 100 is completely opaque. If you choose to make an object partially transparent, turn on the **Background** button ▓ on the **Material Editor** toolbar. This places a background in the sample slot so you can see how transparent the material has become.

Simply changing the Opacity alone usually doesn't give good results. Other parameters need to be used to make the material look realistic. The proper use of transparency is covered in Chapter 11, *Advanced Materials*.

Shader Basic Parameters

The **Shader Basic Parameters** rollout offers additional options to further define the material.

Shader Basic Parameters
Blinn ▼   ☐ Wire   ☐ 2-Sided   ☐ Face Map   ☐ Faceted

Figure 4-16: **Shader Basic Parameters** *rollout*

Wire makes a wireframe material, causing the object to render as a wireframe. The width of the wire is set on the **Extended Parameters** rollout under the **Wire** section, and can be set in **Pixels** or 3ds Max **Units**. This option is useful for objects made of wires or mesh, such as a basket. It can also be used for test rendering animations, because wireframe rendering is much faster than shaded rendering.

Face Map causes the maps in the material to be placed on each of the object's polygons. This option is used for special cases such as particle systems. You'll get a chance to use face mapping and particle systems in Chapter 8, *Special Effects*.

The **2-Sided** option causes both sides of each face to render, regardless of which way the face normals are pointing. This option is very handy for one-sided objects, but it increases rendering time, so use it only when you need it. Generally speaking, it is usually preferable to model your objects with interior and exterior surfaces, rather than relying on 2-Sided materials.

The **Faceted** option forces the object to render with visible seams between faces. This overrides any face smoothing applied to the object. Face smoothing is described later in this chapter.

EXERCISE 4.2: Additional Parameters

In this exercise, you'll use the additional parameters to make changes to the materials you created earlier.

1. Load the file **Shininess.max** from the disc.

2. In the **Material Editor**, select the sample slot for the **Teapot Blue** material. On the **Shader Basic Parameters** rollout, check the **Wire** check box. In the **Perspective** view, you can see that the teapot now appears as a wireframe.

3. Render the **Perspective** view. The blue teapot appears as a wireframe.

Figure 4-17: Teapot rendered as wireframe

4. In the rollouts area, scroll down to the **Extended Parameters** rollout. Under the **Wire** section, increase the **Size** parameter to 3.0.

   Changing the wire size in this way does not affect the display in the viewport.

5. Render the Perspective view. The teapot wireframe is now thicker.

6. On the Shader Basic Parameters rollout, check the **2-Sided** check box. Render the Perspective view. The other side of the wireframe teapot is now visible.

7. Select the sample slot for the **Teapot Yellow** material. On the Blinn Basic Parameters rollout, under the **Self-Illumination** section, change the **Color** parameter to 80.

   In the Perspective view, the yellow teapot now appears to glow, as if it were lit from the inside. Render the Perspective view to see the results.

Figure 4-18: Self-illuminated teapot at left

NOTE: Increasing the self-illumination of the teapot has caused it to become more uniform in color, eliminating much of the contrast between light and dark areas of the material. This can make the object look flat and two-dimensional. This type of look is appropriate only on objects that are lit from the inside, and not for ordinary 3D objects in a scene. Self-illumination should be used only when you want an object to appear illuminated, and not to compensate for bad lighting in a scene.

8. Select the sample slot for the **Teapot Red** material. On the Shader Basic Parameters rollout, check the **Faceted** check box. The sample slot changes to show a faceted sphere, and the teapot in the viewport appears faceted.

9. Render the **Perspective** view. The third teapot appears to be made up of flat rectangles.

Figure 4-19: Faceted teapot at right

10. Save the file with the filename **MaterialParameters.max** in your folder.

CREATING MATERIALS WITH MAPS

As discussed earlier, one of the basic building blocks of materials is a *map*. A map is a pattern or image that affects a material parameter. There are several types of maps available in 3ds Max. The ones you will use most often are discussed in this chapter.

BITMAPS

The most commonly used type of map is a *bitmap*, which is a file containing colored pixels. A bitmap can be any picture file, whether a scanned photograph or a picture drawn in a paint program such as Photoshop. Images rendered in 3ds Max are also bitmaps.

Figure 4-20: Bitmaps

Using a Bitmap for the Diffuse Color

One of the most common uses for a bitmap is to define the overall color of an object.

To create a material using a bitmap for color, select a sample slot. Expand the **Maps** rollout for the material. Locate the **Diffuse Color** label. Click on the button labeled **None** across from Diffuse Color. The **Material/Map Browser** appears.

Look at the **Browse From** section at the upper left, and make sure it is set to **New**. This will enable you to load a new map.

Select **Bitmap**. To do this, you can highlight Bitmap and click **OK**, or you can double-click Bitmap. A file selector appears, in which you can choose a bitmap.

Figure 4-21: Material/Map Browser

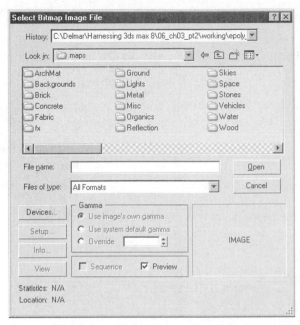

Figure 4-22: Bitmap file selector

The file selector remembers the folder you chose the last time the file selector was opened. If you navigate to the **3dsMax8\maps** folder, you will find a number of additional folders containing bitmaps.

When you click a filename, a thumbnail image appears at the lower right of the dialog. Highlight the desired bitmap, then click **Open** to choose the bitmap, or simply double-click the filename.

The sample object in the slot is now colored with the chosen bitmap.

Figure 4-23: Sample slot colored with bitmap

WARNING: The bitmap file browser displays files based on their format. If a file format such as .TIF is selected in the **Files of Type** pulldown, then the file browser will hide all files that are not .TIFs. This may mislead you to believe that your files are missing. Choose **All Formats** in the Files of Type pulldown to view all files in the current folder.

This material can now be assigned to an object. Click and drag from the sample slot to an object to assign the material.

The map you just chose won't automatically show in the viewport. To make it appear, click the **Show Map in Viewport** button 🔲 on the Material Editor toolbar.

Also note that the rollout area of the Material Editor has changed to display parameters for the **Bitmap** map type. You will explore some of these parameters later in this chapter.

EXERCISE 4.3: Diffuse Color Bitmap

In this exercise, you'll create a custom material with the **Diffuse** color defined by a bitmap.

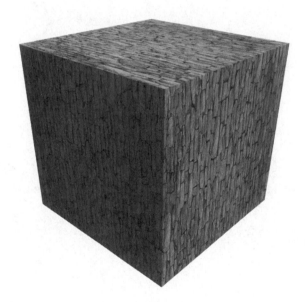

Figure 4-24: Box with custom material

1. Reset 3ds Max.

2. In the Top viewport, create a box with a **Length**, **Width** and **Height** of about **40** units.

3. Open the Material Editor. The first sample slot is active by default.

4. Expand the **Maps** rollout. Locate the **Diffuse Color** label. Click the box labeled **None** across from Diffuse Color. On the **Material/Map Browser**, choose **Bitmap**. The **Select Bitmap Image File** dialog appears.

5. In the Select Bitmap Image File dialog, navigate to the **Maps\Stones** folder.

6. Click the file **LIMEST01.JPG**. A thumbnail image appears at the lower right of the dialog, displaying an image of limestones. Click **Open** to load the file into the Diffuse Color map slot.

    The bitmap appears on the sphere in the sample slot. The rollouts have also changed to display the bitmap parameters.

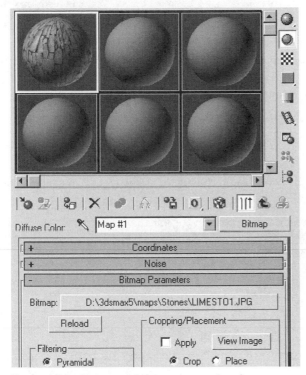

Figure 4-25: *Limestone bitmap on sample sphere*

7. Click and drag from the sample slot to the box to apply the material to the box.

    The material has been applied to the box, but the bitmap doesn't show in the Perspective viewport. Right now, the color in the Diffuse color swatch appears on the box.

8. On the **Material Editor** toolbar, click **Show Map in Viewport** .

    The **LIMEST01.JPG** bitmap appears on the box.

9. Save the scene with the filename **DiffuseMap.max** in your folder.

EXERCISE 4.4: Bitmap Materials for Clown

In this exercise, you'll create custom materials for the clown head you created earlier in this book.

1. Load the latest version of the clown scene from your folder. If you have done all the previous exercises in this book, this file is called **LumpyHat.max**. You can also load this file from the disc.

2. Open the Material Editor. Using the technique described in the previous exercise, create a custom material with a bitmap as the **Diffuse Color** map. Look through the bitmaps in the various folders under the **Maps** folder. Click several bitmaps to view their thumbnails. Choose one that is appropriate for the clown's hat.

3. Apply the material to the clown's hat. To do this, you can click and drag from the sample slot to the hat, or you can select the hat and click **Assign Material to Selection** on the Material Editor toolbar.

4. Select the next sample slot. Use the same technique to create a custom material for other parts of the clown head.

5. Save the scene in the file **ClownBitmaps.max** in your folder.

PROCEDURAL MAPS

The other main category of maps, besides bitmaps, are *procedural maps*. As the name implies, procedural maps are based on algorithms that generate patterns or images. Rather than using colored pixels to define colors and patterns, these maps use a series of equations or commands to generate a pattern. Procedural maps don't rely on external bitmap files, they are created internally by 3ds Max.

3ds Max has many procedural maps that define a pattern, such as **Noise**, **Gradient Ramp**, **Checker**, or **Smoke**. Additional procedural maps are available as plug-ins for 3ds Max.

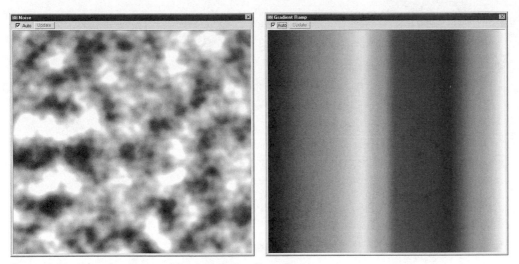

Figure 4-26: Procedural maps: **Noise** *and* **Gradient Ramp**

There are two types of procedural maps, **2D Maps** and **3D Maps**. A 2D map creates an image in two dimensions across the surface of an object, in much the same way that a bitmap does.

A 3D map creates a three-dimensional pattern in space. The intersection of the pattern with an object results in an image on the surface of the object. 3D maps achieve an effect as if the object were carved out of a block made of the map.

Using a 2D Procedural Map

To use a 2D map as a material's main color, select the material's sample slot. In the rollouts area, expand the Maps rollout. Across from Diffuse Color, click the box labeled None. The Material/Map Browser appears.

Locate the area at the lower left of the Material/Map Browser that lists the types of maps. Choose **2D Maps**. The display area of the browser changes to display only 2D map types.

Bitmap is the only type of 2D map that is not a procedural map.

Checker makes a checker pattern with two colors or maps.

Gradient makes a transition from one solid color to another, using three colors in all.

Tiles creates a regular pattern that can be used for bricks or tiles.

Figure 4-27: Checker and Gradient maps

Gradient Ramp creates transitions among colors, using any number of colors. It has many more options than the simple Gradient map.

Combustion accesses the Autodesk **combustion** software if it is installed on your system.

Swirl makes a swirling pattern from two colors.

Figure 4-28: Gradient Ramp and Swirl maps

To use one of these maps, select it from the Material/Map Browser. The rollout area changes to display the options for the map.

For some 2D maps, the bottom of the rollout area displays color swatches. Change these color swatches to change the colors in the map.

MATERIAL HIERARCHIES

Earlier, you learned that material definitions are built from an internal data flow. Another way of saying this is that materials have an internal *hierarchy*, or chain of command. You will find this concept throughout 3D graphics, and indeed, computers in general.

In a computer graphic hierarchy, certain objects or modules have precedence over others. Those with higher priority are generally earlier in the data flow, and they are often called *parents*. Those with lower priority are called *children*.

When you choose a map type, the rollouts on the Material Editor change to display the options for the chosen map type. At this point, you are at a *child* level of the material. The parent level can have many children.

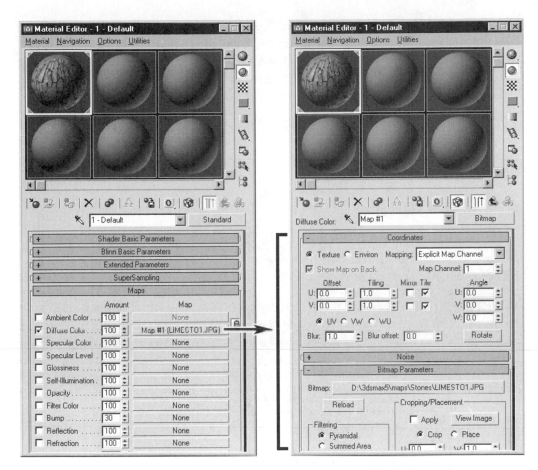

Parent Level   **Child Level**

Figure 4-29: Parent/child relationship

Go to Parent

You can tell if you are at a child level for a material by checking the status of the **Go to Parent** button 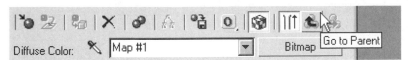 on the Material Editor toolbar. This button is available only when you are at a child level of a material.

Figure 4-30: **Go to Parent** *button available only at child level*

Once you are on the parent level, you can get back to the child level by going to the **Maps** rollout and clicking the button across from the map attribute name. When a map has been assigned to the map attribute, the button is no longer labeled **None**; it is now labeled with the map type you assigned to it. When you click this button, you are returned to the child level for the map type.

If you can't find the parameters you want on the rollout displayed, it's probably because you're not at the right parent or child level. To go to the parent level, click **Go to Parent** . To go to a child level, click across from the appropriate map attribute on the **Maps** rollout.

Material/Map Navigator

In addition, you can see the structure of a material, and move around within its hierarchy, by launching the **Material/Map Navigator** . Click the button on the Material Editor toolbar to launch the Navigator.

WARNING: Don't confuse the Material/Map **Navigator** with the Material/Map **Browser**. The Browser allows you to load new materials into sample slots. The Navigator is for visualizing the structure of a material hierarchy, and for navigating within it.

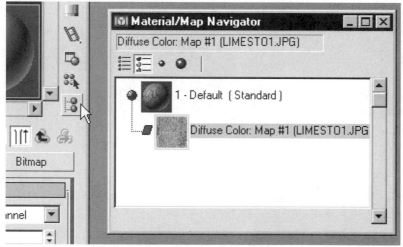

*Figure 4-31: Launching the **Material/Map Navigator***

Once the Material/Map Navigator is open, you can change the display within the Material Editor rollout panel by selecting the desired level within the Navigator. If you want to access the parameters for a map, just click on the map in the Navigator, and the Material Editor will open that map, and the rollouts will change.

CLEARING SAMPLE SLOTS

After you have created six materials, all the visible sample slots will have been used up, and you'll need more unused sample slots so you can make further materials. You can pan the sample slot area to unused slots.

To pan the slots over, use the scrollbars. You may also move your cursor over the boundary between two slots until the hand cursor appears. Click and drag to see more sample slots.

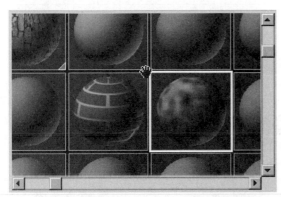

Figure 4-32: Panning sample slots up and to the left

Eventually, you may use up all 24 sample slots. In this case, you can reset a used sample slot so it holds the default gray material. To reset a sample slot, activate the slot, then go to the material's parent level. Click the **Reset Map/Mtl to Default Settings** button ☒ on the **Material Editor** toolbar. If the material is being used in the scene, a dialog will appear.

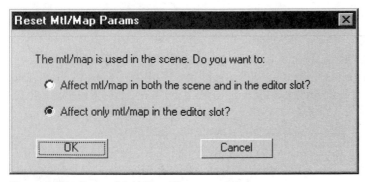

Figure 4-33: **Reset Mtl/Map Params** *dialog*

The first option will reset the material both in the scene and in the sample slot. The second option, the default setting, will reset only the material in the sample slot and will not affect the material applied to objects in the scene.

If the material has not been applied to an object in the scene, a different message appears.

It's important to understand that a material doesn't have to be present in a **Material Editor** sample slot to be present in the scene. The Material Editor sample slots are holding areas for materials so you can create and work on them. Once a material is applied to an object in the scene, it is stuck there permanently unless you specifically do something to change the material properties or replace the material entirely.

TILING AND MIRRORING MAPS

Tiling Maps

When the **Checker** map is used, the rollout area changes to display rollouts particular to this type of map. The first rollout is the **Coordinates** rollout.

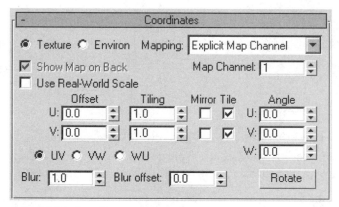

Figure 4-34: Coordinates rollout

The **Tiling** parameter sets the number of times the map will repeat, or *tile*, across the surface of the object. Increasing these values causes the map to tile more frequently, making it appear smaller on the object.

The number of tiles is determined by multiplying the **U** and **V Tiling** values. For example, if Tiling for U is **4** and V is **3**, the material has 12 tile areas. These tile areas are spread over the sample object in the material slot. They affect the number of tile areas both in the material slot and on the object to which the material is applied.

You can think of U as going across from left to right, and V going from bottom to top. When setting these parameters, simply play with them until the map looks the right size.

Figure 4-35: Bitmap with 1 x 1 Tiling and with 3 x 3 Tiling

If you choose the **Use Real-World Scale** option, then the **Tiling** field labels change to **Size**, and you can enter an absolute size for the map. The size is measured in whatever units are currently active, chosen in the **Customize > Units Setup** dialog. For Real-World Scale to work, you must also enable the **Real World Map Size** option for each object in the scene. This check box is found in the parameters for a primitive, or in the **UVW Map** modifier, which is discussed later in this chapter.

The Tiling parameter also appears on the **Coordinates** rollout for other map types. With 2D maps, a seam is visible where one tile meets the next tile. This is not a problem with the Checker map, because you would expect a seam between tiles, but it is very noticeable with some bitmaps.

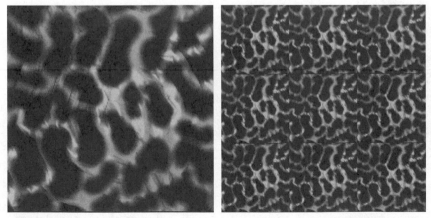

Figure 4-36: Bitmap with no tiling, and with 2 x 2 tiling and obvious seams

In general, if the bitmap has a random pattern, such as stones or grass, the tiling seams will not be very obvious. Seams are also not visible for bitmaps that have been created especially to be tiled, such as brick and fabric patterns. On background images and photographs, the tiling seams are usually visible unless the bitmap has been specially altered for tiling in an image editing program.

Mirroring Maps

If a map has obvious seams when tiled, you can sometimes make the tiling less obvious by checking the **Mirror** check box on the **Coordinates** rollout for both the **U** and **V** directions. This option flips the bitmap as it tiles it, so the right edge of one tile matches the left edge of the next tile.

Figure 4-37: Map with no tiling, 2 x 2 tiling, Mirror tiling

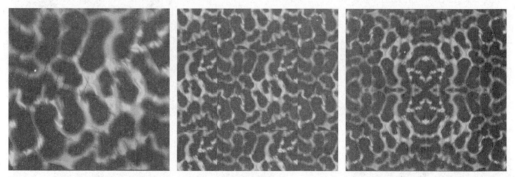

Figure 4-38: Map with no tiling, 2x2 tiling, Mirror tiling

When Mirror is checked, the Tile check box is automatically unchecked, and the Tiling values are doubled internally. For example, if both Tiling values are **3** and Mirror is checked, the effect is 6 x 6 tiling with each tile flipped to match the edges of the previous tile.

Mirroring sometimes eliminates or softens an obvious tiling effect, but sometimes makes it worse. Mirroring can also create accidental patterns that look like carvings or insignia. The only way to know if mirroring will make your tiles look better is to try it out.

EXERCISE 4.5: Tiling

In this exercise, you'll practice tiling with maps.

1. Load the scene **CheckerMap.max** from the disc.

2. Create another copy of the checkered box, and place it next to the two boxes.

3. Open the Material Editor.

4. Select the first sample slot to select the material created with the **LIMEST01.JPG** bitmap as the **Diffuse Color** map.

5. If the **Coordinates** rollout is displayed on the Material Editor, go on to the next step. If the **Blinn Basic Parameters** rollout is displayed, this means you are at the parent level of the material. In this case, click the button next to the **Diffuse** color with the letter **M** on it.

   This takes you directly to the child level of the material.

6. On the **Coordinates** rollout, under **Tiling**, change both the **U** and **V** values to **2.0**.

   The bitmap now has 2 x 2 tiling, with two tiles going across and two down, for a total of four tiles. The tiling appears in both the sample slot in the Material Editor, and on the box in the scene.

Figure 4-39: Box with 2 x 2 tiling on bitmap

Because the **LIMEST01.JPG** pattern is somewhat random, no seams are visible where the tiles meet. The stones simply appear to be smaller due to the higher tiling values.

7.   Activate the second sample slot to select the **Checker** material.

8.   If the Checker child level of the material is not already displayed, click the button next to the **Diffuse** color with the letter **M** on it to go to the child level.

9.   On the **Coordinates** rollout, under **Tiling**, change both the **U** and **V** values to **2.0**.

One tile of the Checker map contains four black or white squares, with two going in each direction. Changing the Checker map to 2 x 2 tiling means you now have four squares going in each direction.

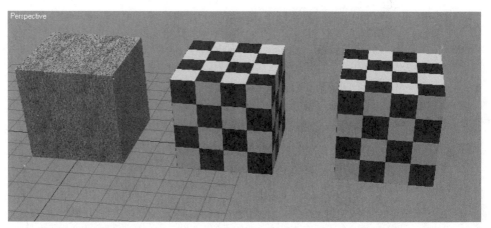

Figure 4-40: Checker tiling

Next, you'll create a third material using a tiled bitmap. This bitmap will have seams where the tiles meet.

10.   Activate the third sample slot. Click the button next to the **Diffuse** color. The **Material/Map Browser** appears.

11.   In the Material/Map Browser, choose **Bitmap**. When the **Select Bitmap Image File** dialog appears, open the **Backgrounds** folder. Select the bitmap **LAKE_MT.JPG**.

This bitmap is a photograph of a lake and mountain scene.

12.   On the **Coordinates** rollout, under **Tiling**, change both the **U** and **V** values to **2.0**.

13. Apply this material to the third box. Click **Show Map in Viewport**  to see the map on the box.

The image is tiled on the box, with one tile centered on each box face, and the other tiles arranged around it. Note that there is an obvious seam where one tile meets the next.

Figure 4-41: Box with tiling seams

14. Save the scene as **Tiling.max** in your folder.

2D MAP PARAMETERS

The remaining options on the **Coordinates** rollout can be used to further customize a map.

Offset can be used to shift the position of the map along the U and V directions. The Offset is expressed as a fraction of the map height or width. For example, an Offset value of **0.5** for U shifts the map along the U direction by half its width.

Tile causes the map to repeat, or tile, over the object when the **Tiling** values are above **1.0**. This check box is on by default. If Tile is unchecked, the map appears only once on the object even if the Tiling values are above **1.0**.

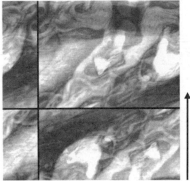

V Offset = 0.4

U Offset = 0.2

Figure 4-42: U and V Offset

Blur blurs the map based on its distance from the camera. Compare with **Blur Offset**, which blurs the map regardless of its distance from the camera.

Angle rotates the map along the U, V, or W axes. This feature is used for aligning a map on the surface of an object. It is handy for making a map look more organic or for varying a map that looks too similar to another map being used in the scene.

To rotate the map interactively, click the **Rotate** button to bring up a window where you can rotate the map using the mouse. Move the cursor anywhere on the window, then click and drag to rotate the mapping.

BITMAP PATHS

Sometimes when you load or render a scene, you will see a dialog that warns you that the bitmaps needed for rendering are not found. This is a common occurrence, and you may even experience it when loading the sample files from the disc that came with this book.

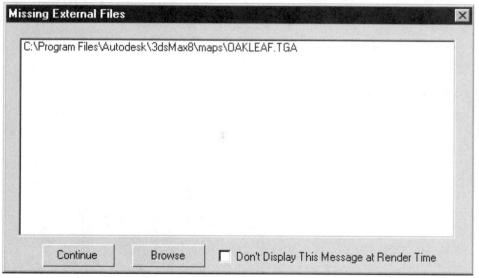

Figure 4-43: ***Missing External Files*** *dialog*

The reason this happens is that 3ds Max can't find the bitmaps that are used in the scene. 3ds Max stores the *absolute path* to bitmaps, meaning that it keeps an internal record of the complete directory or folder path where the maps are located. If the maps are not where 3ds Max thinks they should be, then you will see the warning dialog when you load or render the scene. There are many circumstances that will cause this problem, such as 3ds Max not being installed in the default directory.

It is easy to remedy this situation. When you see the **Missing External Files** dialog, click the **Browse** button. *Do not* press the **Continue** button, or you will not be able to add a new bitmap path, and your scene will not render correctly.

When you click Browse, the **Configure External File Paths** dialog pops up. This dialog can also be accessed from the Main Menu, by choosing **Customize > Configure User Paths** and clicking the **External Files** tab.

The current bitmap paths are listed in the Configure External File Paths window. Click the **Add** button to add a new path.

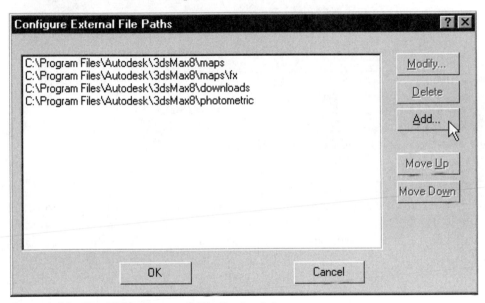

Figure 4-44: ***Configure External File Paths*** *dialog*

When you click Add, yet another dialog opens up, which reads **Choose New External Files Path**. Navigate to the folder where your bitmap files are located. Select the **Add Subpaths** check box if you wish to add the current folder and all folders currently visible in the window. Click the **Use Path** button, and the selected folders are added to the 3ds Max bitmap paths. You may now proceed to render the scene.

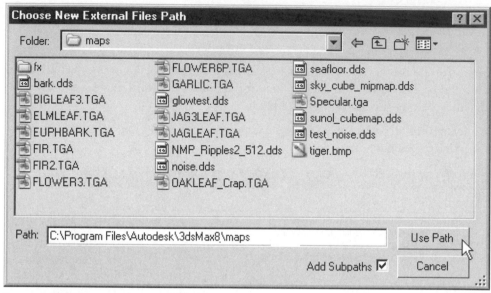

Figure 4-45: **Choose New External Files Path** *dialog*

3D MAPS

3D procedural maps are applied to materials in the same way as 2D maps. However, there are many more types of 3D maps, and they have more parameters and options. 3D maps also have a **Coordinates** rollout that works similarly to the Coordinates rollout for 2D maps.

Generally speaking, these maps don't look like much in their default states; they are merely the building blocks of more complex materials. It is only through careful parameter adjustment and layering of maps that these procedural algorithms become artistically valuable. As always, you must decide what colors, textures, and patterns will look right on an object, then design materials accordingly.

In the following illustrations, the 3D maps are shown with default settings, so you will become accustomed to them.

Cellular uses three colors to create a cell-like image. It is most useful as a bump map, to create a semiregular texture, such as scales or alien skin. Bump mapping will be covered later in this chapter.

Figure 4-46: **Cellular** *map*

Dent and **Stucco** make random patterns of dots from two colors. These are also very useful as bump maps. Dent was designed to add roughness to metallic objects. Stucco is mainly used for textured surfaces such as plaster walls.

Figure 4-47: **Dent** *and* **Stucco** *maps*

The **Falloff** map sets colors according to the directions of face normals. This map is used primarily to set opacity, transparency, and self-illumination. Use of the Falloff map is covered in Chapter 11, *Advanced Materials*.

Marble and **Perlin Marble** use colors to make a marble pattern.

Figure 4-48: **Marble** *and* **Perlin Marble** *maps*

Noise creates a random pattern from two colors. Noise is by far the most commonly used of all procedural maps. It is very versatile and can be used to achieve a variety of effects. A little bit of Noise will go a long way toward making your scenes look more realistic.

Figure 4-49: **Noise** *map*

Particle Age and **Particle MBlur** are used exclusively with particle systems. Their uses are covered in Chapter 8, *Special Effects*.

Planet makes a pattern from several colors that resembles land and water masses on a planet.

Figure 4-50: **Planet** *map*

Smoke is a map designed to be animated, simulating clouds of smoke. The effect of smoke can be more realistically depicted using particle systems, but the Smoke map is occasionally useful when a different type of chaotic noise is needed.

Figure 4-51: **Smoke** *map*

Speckle and **Splat** create random dotted patterns from two colors. As their names indicate, they are commonly used to add the illusion of dirt to a material.

Figure 4-52: **Speckle** *and* **Splat** *maps*

Water makes a pattern of waves from two colors. **Wood** uses two colors to make a pattern with veins and curves like that of wood.

Figure 4-53: **Waves** *and* **Wood** *maps*

Note that these 3D maps are named for their similarities to marble, wood, and so on, but this doesn't mean that making realistic materials is just a matter of using a particular map. For example, the Waves map doesn't instantly make realistic water without any further action from you.

EXERCISE 4.6: Something in the Bathtub

Somebody forgot to clean the bathtub, and now there's something growing in there. In this tutorial, you'll use the **Cellular** map to put a bacterial material on the slime in the bathtub.

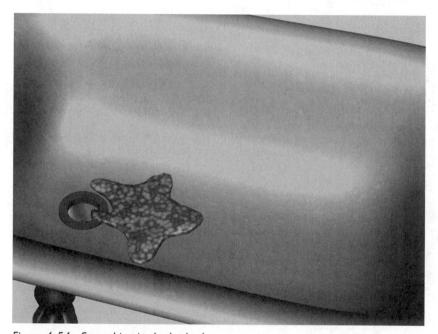

Figure 4-54: Something in the bathtub

1. Load the file **Bathtub.max** from the disc that comes with this book.

    The **Perspective** view shows some slime oozing down the drain.

2. Open the Material Editor.

3. Activate the third sample slot, which holds the material **Slime**. This material is currently a plain gray material.

4. On the **Blinn Basic Parameters** rollout, click the small box next to the **Diffuse** color swatch to assign a map. In the **Material/Map Browser,** choose the **Cellular** map type.

    The Material Editor rollout panel changes to show the parameters for the Cellular map type. A Cellular map is defined by three colors.

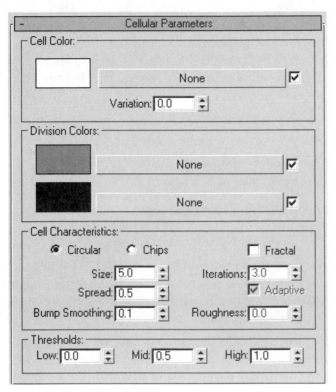

Figure 4-55: **Cellular** *map parameters*

5. On the **Cellular Parameters** rollout, change the colors to the following.

Cell Color	Yellow
Division Colors	Green and orange

    The material has changed, but the cellular map doesn't show in viewports. Click the

    Show Map in Viewport button to see a viewport preview.

6. Render the **Perspective** view. The slime now has a multicolored cellular surface.

    You can also make the slime a little transparent if you like.

7. Click **Go to Parent** to go to the top level of the material. On the **Blinn Basic Parameters** rollout, change the **Opacity** parameter to **50**.

    This makes the material 50% transparent.

8. Render the **Perspective** view.

    The slime is now transparent. Opacity and transparency are covered in more detail in Chapter 11, *Advanced Materials*.

9. Save the scene with the file name **BathtubSlime.max** in your folder.

3D Map Size

Every 3D map also has a parameter for setting the size of the effect. This size is an arbitrary number and doesn't refer to a particular number of blobs or splats or anything else of the sort. Setting the size is simply a matter of looking at the effect and deciding whether you want it to be larger or smaller.

On most 3D maps, this parameter is called **Size**. On some 3D maps, the parameter has a slightly different name.

Planet map size is set with the **Continent Size** parameter.

Waves map size is set with the **Wave Radius** parameter.

Wood map size is set with the **Grain Thickness** parameter.

3D Map Tiling

When you increase the **Tiling** values for a 3D map, you are not really repeating the map. You are actually decreasing or increasing its size. When the Tiling value is increased, no visible seams are generated on 3D maps. To get the approximate tile size in each direction, divide the map Size by the coordinate Tiling values.

You have independent control over Size and Tiling so that you have the option of using the same map in several materials. In this manner, you can control the global size of the map for all materials with the Size parameter, and you can alter the size of the map on each material individually using the Tiling parameters. This technique involves using instanced maps, which you will learn about later in this chapter.

Phase

Some 3D maps, such as Noise and Smoke, have a **Phase** parameter. These types of maps are designed to change over time. You can animate this parameter to achieve basic fluid or gaseous effects. You can also use Phase to create variations on 3D maps. In this way, you can customize the look of objects in your scene, so they all don't look the same.

For example, suppose you want to use the **Smoke** map to make a pattern on two different objects. If you create a material with **Smoke** as the **Diffuse Color** map and assign it to both objects, the pattern will look exactly the same on both objects, which is unrealistic. Even if you create two separate materials and use the same **Smoke** map on each, the two objects will still have exactly the same pattern.

Look closely at the two trunks in the following illustration. The patterns on both trunks are exactly the same. This is because the **Smoke** maps applied to each one have the same **Phase** values. This makes the scene look fake and computer-generated.

*Figure 4-56: Same **Phase** value on both objects*

To remedy this situation, you can create two materials, each of which uses a **Smoke** map as the **Diffuse** map. On one material, change the **Phase** value for the **Smoke** map to another number, such as 3.2. This will "flow" the smoke pattern through the object, giving you an entirely different pattern, but with the same-size smoke blobs.

*Figure 4-57: Different **Phase** values on each object*

SUBMAPS

Many 2D and 3D maps are defined by colors. Each of these plain colors can be replaced by a map. A map within a map is called a *submap*.

For example, the **Checker** map uses two colors. By default, these colors are black and white. Either of these colors can be replaced by a submap, so that the area usually covered by black or white is instead filled in with the submap.

*Figure 4-58: Colors within **Checker** map replaced with bitmaps*

To assign a submap, click the box labeled **None** next to the color swatch. The **Material/Map Browser** appears. Choose a map type, and the rollouts change to display the parameters for the selected map. Set the parameters as desired.

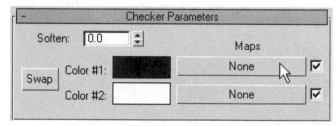

*Figure 4-59: Assign a submap within the Checker map by clicking the button labeled **None***

Parents, Children, Siblings

Assigning a submap within an existing map will take you down to yet another child level. Clicking **Go to Parent** will take you back up the material hierarchy to the parent map, and repeatedly clicking Go to Parent will eventually take you back to the top level of the material.

When two or more maps exist at the same child level, you can click **Go Forward to Sibling** to move to the next sibling. You can press Go Forward to Sibling as many times as necessary to cycle through all the siblings at the same level.

If the Material Editor is displaying a child level of a material, you can click the pulldown list on the Material Editor toolbar. This is where the name of the current material or map is displayed. Select a material or map from the pulldown list to move to that level. Note that you can only move up to the parent levels with this pulldown list; you can't move to children or siblings.

Material Tree

As discussed earlier, the various levels of a material make up the material hierarchy. This is also called the *material tree*. When you have multiple maps within a material, the **Material/Map Navigator** [image] becomes very useful. In this floater, you can view the structure of the material tree. Click a material attribute, such as a map, at any level within the material, and the parameters for that map are displayed in the Material Editor rollout panel.

In the following example, there are three levels to the material. The top level is a Standard material called "Checker Material." The second level is a Checker map applied to the Diffuse color. Within the Checker map are two submaps, which both happen to be bitmap files. The two bitmaps are at the third level of this material.

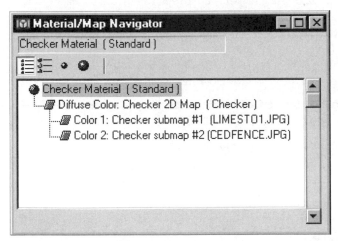

Figure 4-60: **Material/Map Navigator** *displaying a map with submaps*

Naming Maps

Like materials, maps can also be named. Maps with meaningful names can be helpful in organizing materials, and when using a map multiple times in different materials.

To name a map, simply navigate to the map's parameters in the Material Editor. The map name appears in the pulldown list on the Material Editor toolbar. This is the same area where the material name appears.

To enter a new map name, highlight the old name, type in a new name to replace it, and press <**ENTER**>. The new map name appears on the **Material/Map Navigator** and in other areas of the Material Editor.

EXERCISE 4.7: Submaps

In this exercise, you'll define the color of a **Checker** map with a bitmap.

1. Load the file **CheckerMap.max** from the disc.
2. Open the Material Editor.
3. Select the material that has a Checker map assigned as the Diffuse Color map.
4. Go to the **Diffuse Color** level of the material.
5. On the **Checker Parameters** rollout, click the box labeled **None** next to the **Color #1** color swatch. The **Material/Map Browser** appears. Choose **Bitmap**. Open the disc that came with this book, open the **maps** folder and choose the file **CASCADE.TGA**.

    In the sample slot, the black area of the checker map is replaced by the bitmap. Note that the bitmap does not appear in the viewport. Only one map per material can be displayed in viewports at a time. Right now, the Checker map is being displayed in the Perspective viewport.

6. Click **Go to Parent** ![icon] to return to the **Checker** map level of the material.
7. Click the box labeled **None** next to the **Color #2** swatch. On the **Material/Map Browser,** choose **Bitmap**. Select any bitmap you like.

    The white area of the **Checker** map is now defined by a bitmap.

8. Render the **Perspective** view to see the new material on the box.

    The checkers on the box are now defined by bitmaps.

Figure 4-61: Checker colors replaced by bitmaps

9. Save the file as **Submaps.max** in your folder.

BUMP MAP

In addition to Diffuse Color, the Maps rollout has many other map attributes. These settings allow you to assign a map to a number of material properties.

Diffuse Color is the most commonly used map attribute. Another frequently used attribute is **Bump**. The Bump map attribute uses the brightness values of the map to make the material appear bumpy. Lighter areas of the map set the high parts of the bumps, while darker areas make indentations.

A bump map uses the grayscale values from a map. All color information is discarded, and only the brightness and darkness of map areas are used to set the bumps.

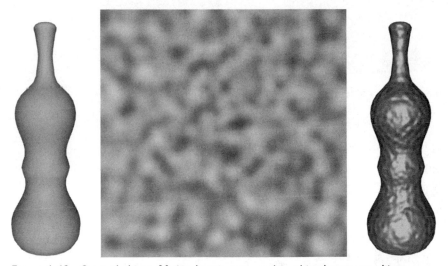

Figure 4-62: Original object, **Noise** *bump map, and resulting bumps on object*

The **Amount** value on the Maps rollout controls the depth of the bumps. Larger values make deeper bumps. The Amount value for the Bump map attribute can range from **-999** to **999**. In many cases, a value of about **30** is effective, and therefore that is the default.

A bump map doesn't change an object's geometry—it's just a rendering effect. It changes the way the material reacts to light, so that the object appears bumpy. You could create a similar effect by editing the object's geometry, but this is usually impractical from a modeling standpoint and might result in unacceptably long render times.

EXERCISE 4.8: One Bad Apple

In this tutorial, you'll make a delicious apple rotten and lumpy with a bump map.

Figure 4-63: Original apple, and lumpy version

1. Load the file **Apple.max** from the disc.

2. Open the Material Editor. Select any sample slot.

3. On the **Blinn Basic Parameters** rollout, change the **Diffuse** color to red or green. Assign the material to the apple.

   Higher shininess will make the bumps show up better.

4. Increase the **Specular Level** parameter to **75**.

5. Expand the **Maps** rollout.

6. Across from the **Bump** map attribute, click the button labeled **None**. On the **Material/Map Browser**, choose **Noise**.

   The Noise map has been applied as the Bump map. This map does not appear in viewports.

7. On the **Noise Parameters** rollout, change the **Size** parameter to **10**.

8. Render the Perspective view.

   The bumps could be deeper.

9. Click the **Go to Parent** button ![icon] on the Material Editor toolbar to return to the parent level of the material.

10. On the **Maps** rollout, increase the **Amount** parameter next to **Bump** to **50**.

11. Render the Perspective view.

Figure 4-64: Apple with Noise as bump map

You now have a lumpy rotten apple.

12. Save the file as **BumpMap.max** in your folder.

MAP AMOUNTS

On the Maps rollout is a column called **Amount**. By default, the value in the Amount column next to Diffuse Color is **100**. This indicates that the map assigned for the Diffuse Color determines 100% of the material's main color.

The map Amount can be changed to a lower number. This has the effect of reducing the map's influence over the Diffuse Color. As you reduce the map Amount percentage, the Diffuse color swatch is mixed with the map.

For example, suppose you assign a bitmap as the Diffuse Color map on the Maps rollout. You then change the Amount for this map on the Maps rollout to a lower number such as **60**. The Diffuse Color swatch on the Blinn Basic Parameters is red. Now 60% of the material's color comes from the bitmap, and 40% of it comes from the red color.

*Figure 4-65: Diffuse bitmap with **Amount** of 100 and **Amount** of 50*

You can use this technique to make subtle effects for a material. For example, you could use the Noise map as the Diffuse Color map, but set the Amount to a low number such as **10**. Then the Diffuse Color from the Blinn Basic Parameters rollout would make up most of the main color, and the Noise map would add a subtle dirty or worn effect.

OTHER MAPPED PARAMETERS

In addition to Diffuse Color and Bump, the Maps rollout has numerous map attributes. These settings allow you to assign a map to various material properties. For example, you could use a grayscale map for the **Specular Level** to vary the degree of shininess across the object.

When mapping certain material parameters, only the brightness of the map will affect the material parameter. For example, when a map is used for the Specular Level, all color information in the map is discarded, and only the map's light and dark areas are used to set the brightness of the highlight across the object.

You can easily tell whether a material parameter can accept colors or whether it can only accept grayscale values. If the parameter has a color swatch, it can accept a color map, complete with hue and saturation information. If the parameter is a single number, then it can only be mapped with a grayscale value, and color information will be discarded.

WARNING: It is not a good idea to use color bitmaps, or procedural maps with colors, to map parameters that can only accept grayscale values. The results will probably not be what you expect. For example, let's say you want to use a Checker as a bump map. You choose one color swatch in the Checker map and assign a color with red, green, blue amounts of 255, 0, 0. This gives a bright red color. Then for the other Checker swatch you choose a bright green color, with RGB amounts of 0, 255, 0. The resulting bump map will have *no bumps at all,* because when these bright red and green colors are converted to grayscale, they have the exact same brightness value.

WORKING WITH MAPS

Locking Maps and Colors

When a Diffuse Color map is assigned to a material, you will usually want the same map assigned as the Ambient Color map. This is so the material will appear to have a uniform color pattern in both lighted and dark areas of the object.

By default, whenever you assign a Diffuse map to a material, the same map is used for the Ambient map. In other words, the Ambient Color map attribute is locked to the Diffuse Color map. This setting can be seen on the Blinn Basic Parameters rollout and the Maps rollout. To the right of both map attributes is a small button called **Ambient and Diffuse Map Lock** 🔒, which is depressed by default.

You can also lock color swatches on the Blinn Basic Parameters rollout. To the left of the **Ambient**, **Diffuse**, and **Specular** labels are two lock buttons 🔳. The button between Ambient and Diffuse locks these two colors together, while the button between Diffuse and Specular locks those colors together. The difference between locking maps and locking color swatches is a subtle one. In general practice, it is best to have the Diffuse and Ambient color swatches locked, and also have the Diffuse and Ambient maps locked. This is the default in 3ds Max. Some artists prefer to use a dark ambient color instead of locking the Diffuse and Ambient colors. It's a matter of personal preference.

Changing a Map Type or Removing a Map

After maps have been assigned, there may be times when you want to replace a map with a different map type altogether. For example, you may have assigned a bitmap as the Diffuse Color map for a material, but now have decided to use a Noise map instead.

One way to change a map or material type after it has been assigned is to click the **Material/Map Type** button on the lower right of the Material Editor toolbar. This button is labeled with the currently active map or material type. Clicking this button launches the Material/Map Browser, and you can choose another map or material type.

*Figure 4-66: Click the **Material / Map Type** button to assign a new map type*

If you want to remove the map altogether, follow the same process, but choose **NONE** from the **Material/Map Browser**. If you choose a map that can have multiple map children, you will be asked if you want to discard the current map or use it as a submap within the new map.

Copying and Instancing Maps

When creating a material, you will sometimes find that you want to reuse a map that is already present in another material. This is particularly helpful when you have taken great care to set up a map's parameters in a certain way and want to use the map and its parametesrs again elsewhere.

A quick and easy way to copy a map and its parameters is to simply drag and drop the map. For example, if you have a Diffuse Color map you wish to reuse, open the Maps rollout for the material. Click and drag the Diffuse Color map button, and drop it in a sample slot. The map is loaded into the slot, and you see a 2D rendering of the bitmap or procedural map. Note that the sample slot does not display a sample object such as a sphere, because you are not displaying a material definition, but merely a map.

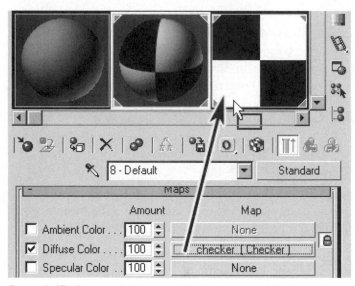

Figure 4-67: Drag and drop a map button to a sample slot

When you drag and drop a map, you will see a pop-up dialog asking if you wish the new map to be a **Copy** or an **Instance**. As with objects, a Copy of a map will be an independent entity that starts off with the same settings as the original map. An Instance will always be identical to the original map.

Working with instanced maps in sample slots is very convenient, even if you're not using the map in several different materials. You can see what the map looks like by itself, and also what it looks like once it has been applied to the material, at the same time. If the material is hot (present in the scene), then the instanced map used by that material will also be hot. In the following illustration, there are triangles in the corners of two sample slots: the Checker map and the material that uses it. This tells you that both the material and the map are hot.

If you wish to use the map in another material, select the second material and display its Maps rollout. Then simply drag the sample slot of the instanced map to the Maps rollout of the second material.

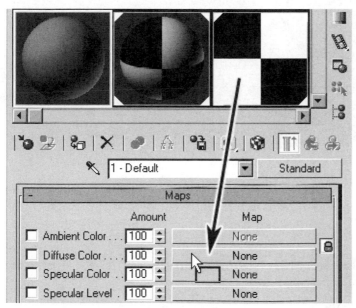

Figure 4-68: Drag and drop a map sample to the Maps rollout of another material

When dragging and dropping a sample slot, take care to press the mouse button, hold it down, then drag. If you just click in the sample slot, the Material Editor rollout panel will switch to display the parameters for that slot.

After releasing the mouse button, you will once again be asked if you wish to create a Copy or an Instance. If you choose Instance, then any changes to the map, or the instance of the map in either material, will be seen in all three sample slots.

Dragging and dropping maps can also be done within the Maps rollout. You might want to use the same map for the **Bump** attribute as well as the **Specular Level** attribute. In that case, drag and drop between the map buttons on the Maps rollout. You might want to make the new map a copy, so you can have independent control over the parameters of each.

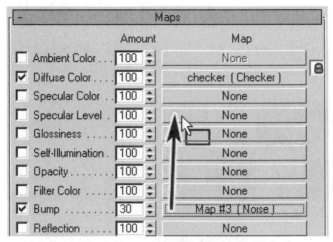

Figure 4-69: Drag one map button onto another

Another way to copy and instance maps is to use the **Material/Map Browser**. You can view all of the materials and maps in the current scene by selecting **Scene** in the **Browse From** section of the Material/Map Browser. Choose a map fron the list and load it into a sample slot.

If the **Root Only** check box is on, turn it off to display the maps, which are subordinate to the root (parent) level of materials.

In addition, you can browse the materials and maps in other ways:

Selected displays materials and maps from the objects in the scene that are currently selected. This option is handy for narrowing your search for a map.

Mtl Editor displays maps and materials that are currently in sample slots in the Material Editor.

Mtl Library displays maps and materials from the currently loaded material library file. You'll learn more about material libraries at the end of this chapter.

ENVIRONMENT MAPPING

Up to this point, all of the maps you've used have been applied directly to objects. As you move the object around in the scene, the map stays "stuck" to the surface of the model. This type of map is called a *UVW map*, and we will discuss it in detail later in this chapter.

There is another category of maps, called *environment maps*. This type of mapping applies an image or pattern to the whole scene, rather than directly to an object. One example of this is a background image. Another is a map designed to simulate reflections on objects.

REFLECTION MAP

In life, a shiny object reflects its environment to some degree. In a 3D rendering, shiny objects must show some sort of reflection in order to look realistic. If a shiny object doesn't reflect anything, it looks as though it is sitting in an empty environment.

Figure 4-70: *Teapot without reflections; teapot with reflections*

The **Reflection** map attribute simulates reflection in a 3D rendering. When a bitmap, 2D procedural or 3D map is assigned to the Reflection map attribute, the object appears to reflect the map.

Although reflection maps are assigned in the Maps rollout of the Material Editor, they are not applied directly to the object. Instead, 3ds Max places the reflection map on a large invisible sphere that encompasses the entire scene. When the scene is rendered, the reflection map is projected from the invisible sphere onto the object. If the object moves, rotates, or deforms, the reflection on the object's surface changes, just as it would if there were a genuine environment reflected on a shiny object.

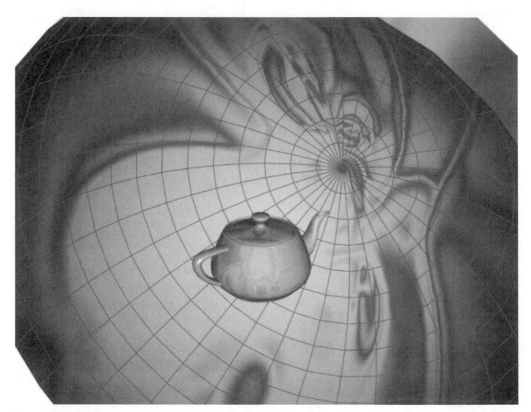

Figure 4-71: Reflection map on invisible sphere around scene

By default, the **Amount** for the Reflection map attribute is set to **100**. At an Amount of **100**, the reflection map tends to take over the material, showing only the reflection map and none of the other maps or colors on the material. For more realistic materials, try setting the Reflection Amount to a lower number. In addition, the **Reflection Dimming** controls in the **Extended Parameters** rollout can be used to achieve a more naturalistic look.

Reflection works best when the material has some shininess to it. Increase the material's **Specular Level** parameter to **50** or more.

The bitmaps in the `maps\Reflection` folder are created specifically to be used as reflection maps. However, as you progress in 3D graphics, you will want to create your own reflection maps to customize the look of your renderings.

Reflections can also be simulated based upon the surrounding objects in the scene; you'll learn about these techniques in Chapter 11, *Advanced Materials*.

STAND-ALONE MAPS

For a background image, a map alone can be used. To do this, select an unused sample slot. Click the **Get Material** button  and choose a map from the list. Maps are indicated by a green diamond. After you choose a map, its parameters appear on rollout panel of the Material Editor.

When a map is in a sample slot, the **Go to Parent** button is disabled. This is because there is no parent for this map. It is simply a map, with no material attributes whatsoever. 3ds Max will not allow you to assign this map to an object, because only materials can be assigned to objects, not stand-alone maps.

Setting Up a Map as a Rendered Background

Regardless of the map type used, it is assigned as a background in the same way. Choose **Rendering > Environment** from the Main Menu. The **Environment and Effects** dialog appears. Look in the **Background** section at the top of the dialog. Locate the **Environment Map** attribute button, which is currently labeled **None**. Click and drag the map from the Material Editor sample slot to the Environment Map attribute button. The **Instance (Copy) Map** dialog appears. Select **Instance** and click **OK**. Now, any changes made to the map in the Material Editor will also be made to the Environment Map.

*Figure 4-72: Drag a map from a Material Editor sample slot to the **Environment Map** button*

2D Maps and Backgrounds

When creating an Environment Map, you have several choices for how it is applied. First, you must specify that the map is to be applied to the scene, not to objects. On the **Coordinates** rollout of the Material Editor, click **Environ** to use the current map as a background. (The default type of mapping for a stand-alone map is **Texture**, which is for applying maps to objects.)

There are several options for how an Environment Map is applied to the scene. Choose the mapping type with the **Mapping** pulldown.

The most basic type of Environment Map uses **Screen** mapping. This places the map in the background of the scene. Objects in the scene always render over the top of the Screen Environment map. When you change the point of view with the viewport controls, the Screen map never changes.

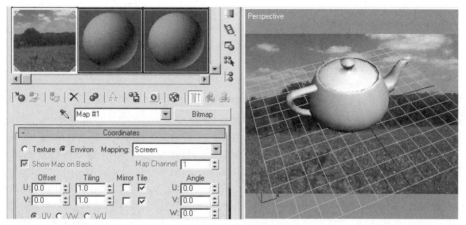

Figure 4-73: **Screen** environment map background

A **Spherical Environment** map places the map on the inside of an invisible sphere surrounding the scene. This is the same process as a default reflection map. If you open the Coordinates rollout for a reflection map, you'll see that Spherical Environment is the default mapping type. The difference between a Spherical reflection map and a Spherical background is that a reflection map is projected from the sphere onto objects in the scene, while a background is simply rendered behind all geometry in the scene.

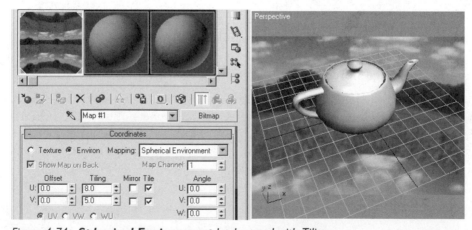

Figure 4-74: **Spherical Environment** background with Tiling

In contrast to Screen mapping, Spherical Environment maps will respect changes in the viewpoint. This is the most common choice for scenes with moving cameras. However, Spherical Environments have limitations. Most notably, spherical mapping requires very large bitmaps, about ten times the resolution of your final rendering, to avoid fuzziness. This is because the bitmap is stretched over an extremely large area, but the virtual camera can only see a small area at once. Changing the Tiling values of a Spherical Environment is one possible workaround, but photographs from the real world will exhibit very obvious seams when tiled across a Spherical Environment.

Also, you might see distortion near the top and bottom of a Spherical Environment, because this type of mapping results in pinching at the north and south poles of the sphere. There are two more types of environment mapping that may help allieviate this pinching.

The **Cylindrical Environment** option projects the map onto a large cylinder surrounding the scene. This can be helpful if the Spherical Environment map is introducing too much distortion to your background map, but this will only work if the camera is not pointing up or down at an extreme angle.

Shrink-wrap Environment projects the map onto a large sphere surrounding the scene, but does so differently than Spherical Environment mapping. Instead of the two poles of a spherical map, Shrink-wrap mapping has only one pole. A 2D map is "shrink-wrapped" around a sphere, and all four sides of a bitmap meet at the bottom of the sphere. This choice sometimes works better than Spherical Environment for reflection maps, because the pinching of the map can be hidden underneath the object. However, it is not appropriate for backgrounds.

Screen maps and Spherical Environment maps are the most commonly used. When using a map as a background for a still image, you will almost always want to use Screen mapping. When using a map as a background in a scene with an animated camera, you will probably use Spherical Environment.

Seeing the Background in a Viewport

To make a Environment Map show up in a viewport, activate the desired viewport and choose **Views > Viewport Background** from the Main Menu. In the **Viewport Background** dialog, check the **Use Environment Background** check box. You must also check the **Display Background** check box at the lower right of the dialog. Click **OK** to exit the dialog and cause the background to appear in the viewport.

MAPPING COORDINATES

When you use a 2D map as part of a material, you must tell 3ds Max how you want the map to be placed on the object. This is especially important for bitmaps. The mapping is oriented with *mapping coordinates*. Some of the types of mapping coordinates available in 3ds Max are shown in the following illustration.

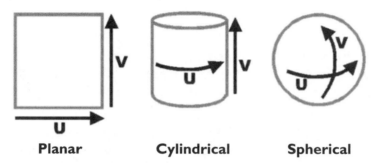

Planar   **Cylindrical**   **Spherical**

Figure 4-75: Mapping coordinate types

3D graphic applications use the letters **U**, **V**, and **W** to denote the three axes of mapping coordinates. This is to avoid confusion with the X, Y, and Z axes of Cartesian coordinate space. UVW space is relative to the surface of the object. It has no direct correlation to world space or to the local space of the object's Pivot Point.

A UVW map is like a sheet of rubber that you can paste onto an object. The U dimension is the width, and the V dimension is the height of the sheet. The W dimension is always perpendicular to U and V, so you can think of it as the distance away from the sheet.

Consider that a sheet of rubber is not always flat; you can stretch, bend, or even fold it. The same is true for UVW maps. If you bend a UVW map around so that two of its sides meet, you create a Cylindrical UVW map. Then, if you stretch and pinch the tops and bottoms of the sheet, the result will be a Spherical UVW map.

There are several ways mapping coordinates can be specified for an object:

- On a primitive, check the **Generate Mapping Coordinates** check box
- On a loft object, check the **Apply Mapping** check box
- On any kind of object, apply a **UVW Map** modifier

Generate Mapping Coordinates is automatically enabled for any primitive when a mapped material is applied to it. The type of mapping coordinates is determined by the type of object. For example, cylinders and capsules generate cylindrical mapping coordinates.

UVW MAP MODIFIER

The **UVW Map** modifier is very useful for setting up mapping coordinates exactly the way you want them. With the UVW Map modifier, you can specify the type of mapping to be applied to the object and control the placement of the map by transforming a gizmo.

The type of mapping you would want to use depends on the shape of the object and the type of effect you are trying to achieve. If you're not sure, try different types of mapping to see which looks best.

Each type of mapping available with this modifier can be altered through a gizmo. When a set of mapping coordinates is selected, a gizmo appears around the object to which the mapping coordinates are being applied. You can access the **Gizmo** sub-object level, and move, rotate, or scale the gizmo as needed.

DECALS

You can use mapping coordinates and tiling values to place a decal on an object without having it tile all over the object. On the bitmap's **Coordinates** rollout, turn off the **Tile** check box for both **U** and **V**. When rendered, the object will have a decal where the UVW Map gizmo is positioned, and the Diffuse Color everywhere else. Then you can also use the **Tiling** and **Offset** values to control the size and placement of the decal.

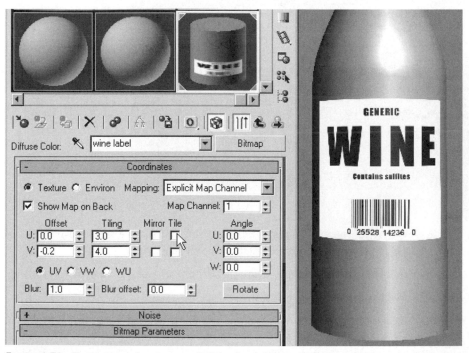

*Figure 4-76: To create a decal, turn off **Tile** check boxes, and adjust **Tiling** and **Offset** values*

EXERCISE 4.9: UVW Map Modifier

In this exercise, you'll experiment with the UVW Map modifier to get different effects on objects.

Default Mapping

1. Reset 3ds Max.

2. Activate the Perspective viewport. Click the **Box** button in the **Create > Standard Primitives** panel. In the **Creation Method** rollout, select **Cube**. Click and drag to create a cube with length, width, and height of about **30** units.

3. Also in the Perspective viewport, create a Sphere with a **Radius** of about **17** units. Move the sphere up in the global Z axis by about 17 units, so the bottom of the sphere is sitting roughly on the Home Grid.

4. Also in the Perspective view, create a Cylinder with a **Radius** of about **17** and a **Height** of about **30** units. Maximize the Perspective view.

   Your screen should look something like the following illustration.

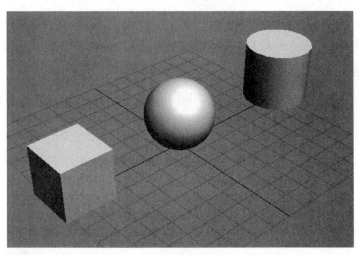

Figure 4-77: Cube, Sphere, Cylinder

5. Press the **<M>** key to open the Material Editor. In the first sample slot, assign the bitmap **Maps\Space\EarthMap.jpg** as a Diffuse Color map.

6. With the Material Editor rollout panel displaying the parameters for **EarthMap.jpg**, click the **Show Map in Viewport** button .

7. Select all three objects in the scene. In the Material Editor, click the **Assign Material to Selection** button .

The cube, sphere, and cylinder are mapped with default mapping coordinates, according to the type of primitive.

Figure 4-78: Mapped material applied to primitives

8.   In the Material Editor, open the Coordinates rollout for the Diffuse map. Click and drag the **U Tiling** spinner up, and observe the result in the viewports.

The textures on all three objects tile several times.

9.   Type a value of **4** into the **U Tiling** parameter.

The texture repeats four times across each face of the cube. It also repeats four times around the circumferance of the sphere and the cylinder.

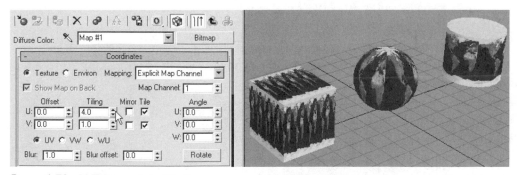

*Figure 4-79: **U Tiling** value of **4.0***

10. Enter a value of **4** into the **V Tiling** parameter and observe the results. Notice how the tiles are regular across the surface of the cube, but the texture is stretched on the sphere and cylinder. Just because the tiling values for U and V are equal does not mean that the texture will be the correct proportions.

Figure 4-80: **U** *and* **V Tiling** *values of* **4**

11. Reset the U and V Tiling values back to **1**.

Planar Mapping

1. Select the cube and apply a **UVW Map** modifier to it. This overrides the mapping coordinates applied at the level of the primitive Box object.

2. The default type of UVW mapping is a **Planar** map. Arc Rotate around the scene and observe the effect this has on the box. Note how the pixels stretch across some of the faces of the object.

Figure 4-81: Planar mapping causes the texture to stretch across some faces

3.   Enter the **Gizmo** sub-object level of the UVW Map modifier. Move the yellow and green planar mapping gizmo up in the global Z axis. Note that the texture does not change on the object.

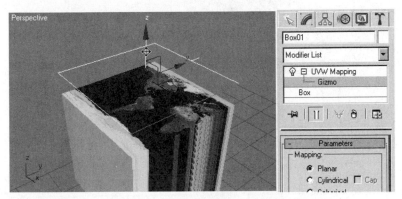

Figure 4-82: Move the planar mapping gizmo up in the global Z axis

4.   Move the mapping gizmo in the XY plane by selecting the XY plane of the Move Gizmo. The texture moves across the surface of the cube.

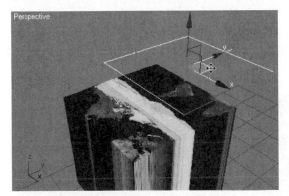

Figure 4-83: Move the planar mapping gizmo in the global XY plane

5.   Restore the mapping gizmo back to its original position. Scroll down to the **Alignment** section of the UVW Map modifier, and click the **Reset** button.

     The gizmo is returned to its default position.

6.   In the Alignment section, click the **X**, **Y,** and **Z** buttons. The mapping gizmo moves to align itself with the X, Y, or Z axes. Choose Y, so the map is placed right side up on one of the faces of the cube.

7.   Select the planar mapping gizmo and move it toward you in the Y axis. This will make it easier for you to manipulate the gizmo. Your screen should look like the following illustration.

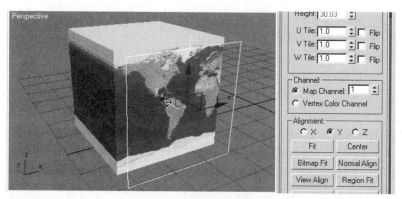

*Figure 4-84: Planar mapping gizmo **Reset**, aligned to Y axis, and moved forward*

8. Activate the Select and Scale tool on the Main Toolbar. Scale the planar mapping gizmo uniformly, to make it smaller.

    Note that this is very similar to changing the U and V Tiling values of the Material Editor. However, scaling the planar mapping gizmo only affects the cube. This is because you are manipulating the cube's UVW Map modifier, not the material applied to all three objects.

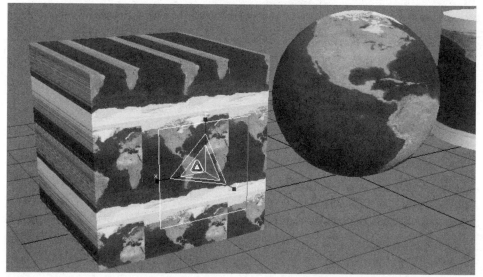

Figure 4-85: Scaling the planar mapping gizmo only affects the tiling of the cube's texture

9.   In the **Coordinates** rollout of the Material Editor, disable the **Tile** check boxes. The image becomes a decal on the surface of the cube.

The sphere and cylinder are not affected. This is because the **U Tiling** and **V Tiling** values of the map are at **1.0**, and because the sphere and cylinder are using default mapping coordinates. The map is repeating only once across the surface of the sphere and cylinder, so it makes no difference whether the Tile check boxes are on or off.

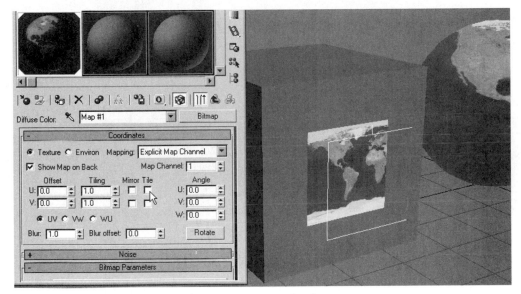

Figure 4-86: Turn off the Tile check boxes to create a decal

10.   Arc Rotate the scene to look at the back of the cube object.

The map is there, but is flipped from left to right. The planar map effectively penetrates through the entire object. Also note the seam on the back of the sphere and cylinder. The seam looks especially bad right now because tiling is disabled.

Spherical Mapping

1. Exit sub-object mode. Select the sphere. Apply a **UVW Map** modifier to it. Click the Y button to align the texture to the sphere. Move the planar mapping gizmo forward, just as you did with the cube object.

    Take special note of the texture stretching at the sides of the sphere. This is because the surface of the sphere is curved, and the planar map is a flat projection.

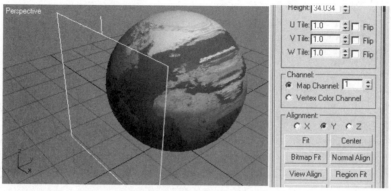

Figure 4-87: Severe texture stretching on the side of the sphere

2. On the UVW Map modifier panel for the sphere, change the mapping type to **Spherical**.

    The result is not good, because the mapping gizmo is still set up for planar mapping.

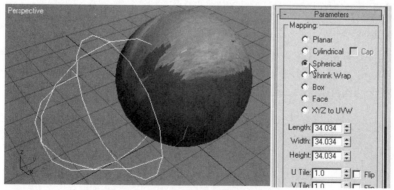

Figure 4-88: Misaligned spherical map gizmo

3. In the Alignment section, click the Reset button to restore the spherical mapping gizmo to its defaults. Click the Z button to align the gizmo with the world Z axis.

    The map is now rotated correctly, with the north pole at the top of the sphere.

4. It is difficult to see the mapping gizmo because it is exactly coincident with the object geometry. To make it easier to see, enter Gizmo sub-object mode and scale the gizmo up uniformly.

   This does not affect the mapping coordinates, because the gizmo is exactly centered on the center on the sphere. Spherical mapping is projected from the center of the gizmo.

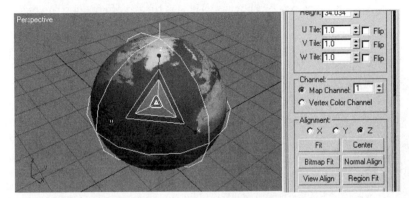

Figure 4-89: Scale the spherical mapping gizmo

5. Rotate the spherical mapping gizmo to hide any seams. While still in Gizmo sub-object mode, rotate the gizmo in the global Z axis. Stop when you see North America centered on the object from your present point of view in the Perspective view. Note that you are not rotating the sphere object; you are rotating the mapping coordinates.

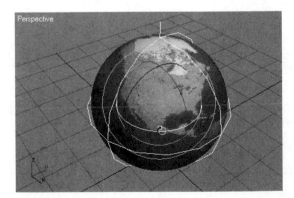

Figure 4-90: Rotate the spherical mapping gizmo in the Z axis

6. The `EarthMap.jpg` file does not include the entire surface of the earth; Asia is missing. When the map is tiled once around the circumferance of the sphere, the map looks stretched horizontally. Change the **U Tiling** value to **1.5** to restore the proportions of the map.

   If you Arc Rotate around the scene, you can clearly see that part of the map of the earth is missing. This is more obvious because Tiling is turned off, making the map a decal.

Figure 4-91: Map proportions restored, but the map is incomplete

Cylindrical Mapping

1. Exit sub-object mode. Select the cylinder. Apply a UVW Map modifier to the cylinder and observe the results.

   As always, the default mapping type is Planar. You can see severe texture stretching on the cylinder, just as with the sphere.

2. Change the mapping type to **Cylindrical**. Turn on the **Cap** check box.

   The Cap option applies planar coordinates to the top and bottom of the cylinder.

3. Enter sub-object Gizmo, and scale the gizmo uniformly, as you did with the sphere. Note that with cylindrical mapping, scaling does affect the mapping, even if the UVW map gizmo is centered on the cylinder.

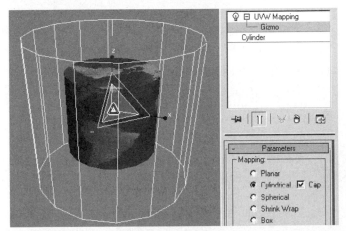

Figure 4-92: Uniformly scaling the cylindrical mapping gizmo alters mapping coordinates

4. Click the **Reset** button to restore the cylindrical mapping gizmo to its default scale.

5. From the Main Menu, pick **Customize > Preferences**. In the Preference Settings dialog, select the **Gizmos** tab. In the **Scale Gizmo** section, activate the **Uniform 2-Axis Scaling** option.

    This prepares the Scale Gizmo for the next step.

6.   Select the cylindrical mapping gizmo, and scale it non-uniformly in the XY axis. To do this, click and drag the XY plane of the Scale Gizmo.

The planar maps on the caps of the cylinder are affected. However, scaling in the XY plane has no effect on the sides of the cylinder, because it doesn't change how the mapping coordinates are projected from the gizmo to the object. (If the gizmo were not centered on the object, then scaling the gizmo *would* affect the mapping.)

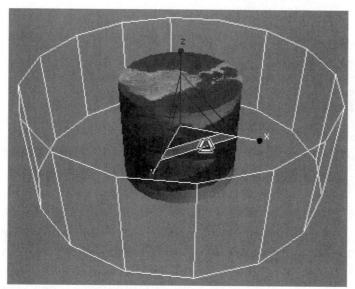

Figure 4-93: Scale the gizmo equally in X and Y axes

7.   Experiment with moving, rotating, and scaling the mapping gizmos on all three objects until you are familiar with how they work.

8.   Save the file as **UVWmapping.max** in your folder.

LOFT MAPPING

Mapping coordinates can be automatically generated by a Loft object. On a Loft object, the U axis is along the circumference of the loft shape, and the V axis is along the loft path.

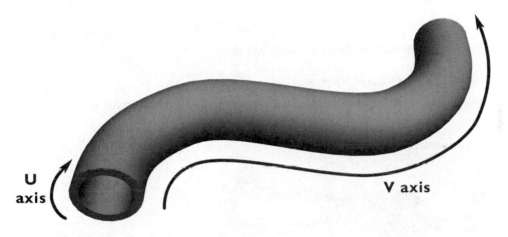

Figure 4-94: Loft mapping coordinates

To activate mapping, turn on **Apply Mapping** under the **Surface Parameters** rollout. This check box is automatically activated when you assign a mapped material to the Loft.

In the **Mapping** section of the Surface Parameters rollout, you have access to the **Length Repeat** and **Width Repeat** values. Length Repeat is the number of times the map repeats in the V axis, which is the length of the path. Width Repeat is the number of repeats in the U axis, which is along the circumference of the loft shape. Of course, you can also change similar parameters at the material level, using the Tiling values in the Coordinates rollout.

When **Normalize** is on, the map repeats are evenly distributed along the length of the object. When it is off, the distribution of vertices along the path determines how the mapping will lie along the length of the path. Where the path vertices are closer together, the map will be compressed. Where the path vertices are farther apart, the map will be expanded.

EXERCISE 4.10: Loft Mapping

In this exercise, you'll put a Checker material on a lofted object and experiment with mapping.

1. Load the file **Hose.max** from the disc.

2. Open the Material Editor. Select an unused sample slot.

3. Assign **Checker** as the **Diffuse Color** map type.

4. Click the **Show Map in Viewport** button 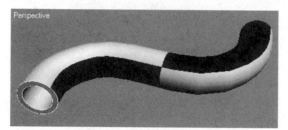.

5. Apply the material to the loft object. The Checker map appears on the object, tiled once.

Figure 4-95: Checker map on Loft object

6. Select the loft object and go to the Modify panel. Expand the **Surface Parameters** rollout.

    The **Apply Mapping** check box is active by default when a mapped material is applied to the object.

7. Increase the **Length Repeat** value and observe the effect on the object's mapping. Also experiment with the **Width Repeat** value.

8. Turn the **Normalize** check box off to see the effect.

    You'll see the most dramatic change if you have vertices very close to one another on the Loft path. Next, you'll add a vertex to the path object to demonstrate the concept.

9. Move the Loft object out of the way if it is in the same place as the loft path. Select the loft path object, which is a **Line** or Editable Spline. Enter **Vertex** sub-object mode. Click on the **Refine** tool, and click the path to add a vertex. Right-click to exit the Refine tool.

    The mapping on the Loft object is much more unevenly spaced.

10. In the Material Editor, go to the **Coordinates** rollout for the Checker map. Change the **U Offset** and **V Offset** values, and observe the results in the viewport.

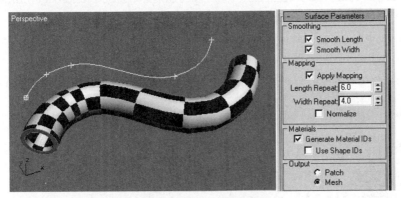

*Figure 4-96: Loft map with **Normalize** disabled*

11. Save the scene with the filename **HoseLoftMap.max** in your folder.

MATERIAL IDS

A **Material ID** is a number assigned to each face of an object. A face can only have one ID number, but many faces can share the same ID number. The ID number is needed to control certain features of the Material Editor. For example, Material IDs are necessary to assign more than one material to a single object.

Material IDs are assigned by default when the object is created. Many objects are automatically assigned more than one material ID from the start. For example, a box has six Material IDs, a different one on each side of the box.

A Material ID can be set manually. If the object is an Editable Mesh or Editable Poly, Material IDs can be assigned in the **Surface Properties** or **Polygon Properties** rollout. Enter Face or Polygon sub-object mode, select some polygons or faces, and open the Surface Properties or Polygon Properties rollout. Type a number into the field labeled **Set ID** to assign a material ID number to the currently selected polygons or faces.

In addition, you can apply a Material ID to selected faces or polygons by first selecting them with a selection modifier, then applying the **Material** modifier. This method works well if you don't wish to convert a parametric object, such as a Loft, to an Editable Mesh.

MULTI/SUB-OBJECT MATERIAL

As explained eariler, there are several different types of materials. The most commonly used material type other than Standard is the **Multi/Sub-Object** material.

The Multi/Sub-Object material is made up of two or more materials. Each sub-material is assigned to the faces on the object via material ID numbers.

To create a Multi/Sub-Object material, open the Material Editor and choose an unused slot. Click the Material Type button, which reads **Standard** by default. Choose Multi/Sub-Object from the Material/Map Browser.

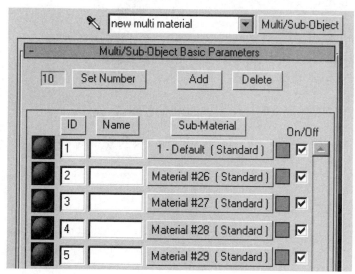

*Figure 4-97: **Multi/Sub-Object Basic Parameters** rollout*

Click the **Set Number** button to determine how many sub-materials are present in the current Multi/Sub-Object material. Each material is assigned to a material ID based on its number. By default, the first material in the list is assigned to material ID #1, the second to material ID #2, and so on.

The best way to work with Multi/Sub-Object in the Material Editor is by using instanced materials. Just drag and drop material samples onto the **Sub-Material** buttons, and choose **Instance** from the pop-up dialog. In this way, you can have all of the sub-materials and their parent Multi/Sub-Object material loaded in different slots, and visible at all times. You can easily move among different sub-materials, without the hassle and confusion of having to use the Go to Parent button or the Material/Map Navigator.

EXERCISE 4.11: Multi/Sub-Object Material

In this exercise, you'll use the **Multi/Sub-Object** material to make a beach ball.

Figure 4-98: Beach ball

Assign ID Numbers

1. Reset 3ds Max.

2. Create a **GeoSphere** with a **Segments** value of **6**. In the **Geodesic Base Type** section of the Command Panel, choose **Octa**.

    The Octa base type will make it easy to select polygons for material ID assignment.

3. Right-click the Geosphere and select **Convert To > Convert to Editable Poly**.

4. Activate **Polygon** sub-object mode. You can press the **<4>** key to do this. In the Top viewport, select the polygons in the upper-left quadrant of the object. Be careful not to miss any polygons or select any unwanted polygons.

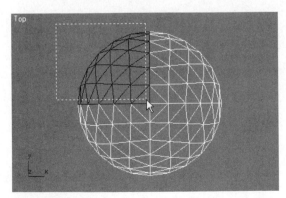

Figure 4-99: Select the polygons in the upper-left quadrant

5. In the **Named Selection Sets** pulldown on the Main Toolbar, enter the name **quadrant 1** for the current polygon selection. Press **<ENTER>** to create the new selection set. This is so you can reselect the polygons later if needed.

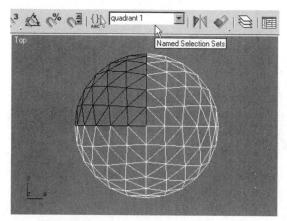

*Figure 4-100: Enter **quadrant 1** in the Named Selection Sets pulldown*

6. In the Modify panel for the Editable Poly, scroll down to the **Polygon Properties** rollout. In the **Material** section, enter the number 1 in the **Set ID** field. 3ds Max does not give you any feedback, but you have just assigned the selected polygons of **quadrant 1** to have material ID #1.

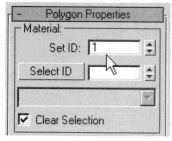

Figure 4-101: Assign material ID #1 to the selected polygons

7. Select the polygons in the upper-right quadrant of the Top viewport. In the Named Selection Sets pulldown, enter the name **quadrant 2**, and press **<ENTER>**.

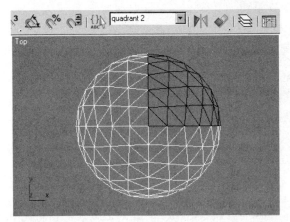

*Figure 4-102: Polygons saved as **quadrant 2** in **Named Selection Sets** pulldown*

8. On the Modify panel, look at the ID number displayed in the Polygon Properties rollout. It should read ID #2 by default. If it doesn't, then enter the number **2**.

9. Select the polygons in the lower-right quadrant. Save them as **quadrant 3** in the Named Selection Sets pulldown. Enter the number **3** as the material ID for the current selection.

10. Finally, select the polygons in the lower-left quadrant. Save them as **quadrant 4** in the Named Selection Sets pulldown. Enter the number **4** as the material ID for the current selection.

11. You now have four sections of the Editable Poly object. Each section has a named selection set, and each section is assigned a different material ID.

    To check your work, use the **Select ID** field on the Polygon Properties rollout. Type a number into the spinner field. When you click the **Select ID** button, all polygons with the material ID you entered are selected.

    Enter the number **1,** and then click the Select ID button. The polygons of **quadrant 1** should be selected. Repeat this process to select the polygons that use ID #2, then #3, then #4. See the following illustration.

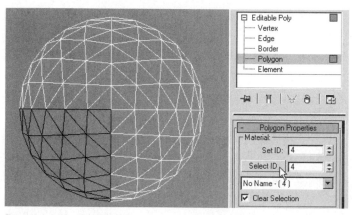

Figure 4-103: **Select ID**

WARNING: Don't change the ID numbers accidentally by entering a new number into the Set ID field. If you need to reassign ID numbers, make sure the correct polygons are selected first. Also, don't turn off the **Clear Selection** check box in the dialog, or else each time you choose a new ID number, those polygons will be *added* to the current selection.

12. Once you have checked your work, exit Polygon sub-object mode. Press the **<6>** key to exit Polygon sub-object mode.

NOTE: You may need to press the **<4>** key instead of the **<6>** key to exit sub-object mode, depending on the status of the **Keyboard Shortcut Override**

Toggle button. This mysterious button is found on the **Extras** toolbar, which is hidden by default. The state of this button determines which set of Hotkeys 3ds Max will use. That's right, there is more than one set of Hotkeys! If the Keyboard Shortcut Override button is on (the default), then when certain type of objects (such as Editable Poly) are selected, then 3ds Max uses a different set of Hotkeys. If the button is turned off, then the standard Hotkeys for the main interface are always used, no matter what type of object is currently selected.

To complicate matters, there is a bug in 3ds Max, which has persisted for years. The Keyboard Override Shortcut button is defaulted to being on, but when 3ds Max first starts, the button looks like it is turned off! If you don't want to use the specialized Hotkeys, and only want to use the standard Hotkeys for the main interface, then open the Extras toolbar by right-clicking any toolbar and choosing Extras. Then click the Keyboard Override Shortcut button a few times until you're certain it is off (not pushed in).

Create a Multi/Sub-Object Material

1. Open the Material Editor. Select the first sample slot. Click the **Material Type** button on the Material Editor toolbar. This button is not labeled, but it reads **Standard** by default.

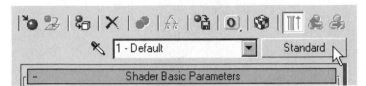

*Figure 4-104: Click the **Material Type** button, which reads **Standard***

2. The **Material/Map Browser** opens. Select **Multi/Sub-Object** by double-clicking it. When the **Replace Material** dialog comes up, you can choose to **Discard** the old material.

3. By default, a Multi/Sub-Object material has ten sub-materials within it. For the beach ball, we only need four. Click the **Set Number** button, and enter the number 4 into the pop-up dialog.

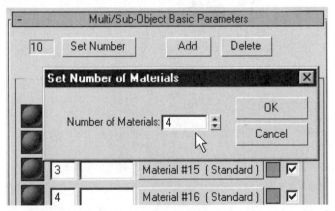

*Figure 4-105: Click the **Set Number** button and enter 4 into the dialog*

4. Name the current material **beach ball multi** by entering it into the pulldown on the Material Editor toolbar.

5. Select the second sample slot in the Material Editor. Change its Diffuse color to a saturated blue. Name this material **blue sub**.

    Do not change the Diffuse color of one of the sub-materials within the **beach ball multi** Multi/Sub-Object material.

6. Select the **beach ball multi** sample slot again. Click and drag the **blue sub** sample slot onto the first sub-material button on the rollout panel. See the following illustration.

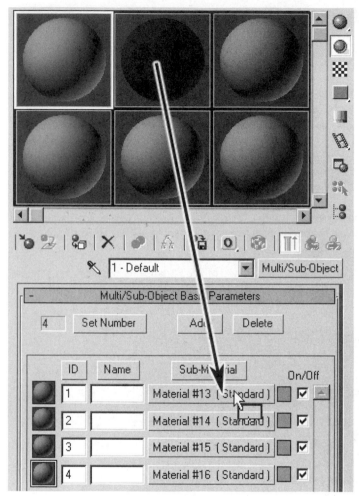

Figure 4-106: Click and drag the **blue sub** *sample slot onto the first sub-material button*

When the **Instance (Copy) Material** dialog comes up, choose **Instance**.

7.   Notice that the sample slot for **beach ball multi** now has blue patches on it. Assign the **beach ball multi** material to the ball in the scene.

One fourth of the ball is now blue. Any faces with ID #1 will receive the **blue sub** material.

Both of the sample slots you've worked with so far now have triangles in their corners, to indicate that they are hot, and present in the scene.

8. Activate the third sample slot. Give its Diffuse Color a bright yellow. Name the new material **yellow sub.**

9. Activate the **beach ball multi** material. Click and drag the **yellow sub** sample slot onto the second sub-material button. In the pop-up dialog, choose Instance.

    The second quadrant of the beach ball now has a yellow color.

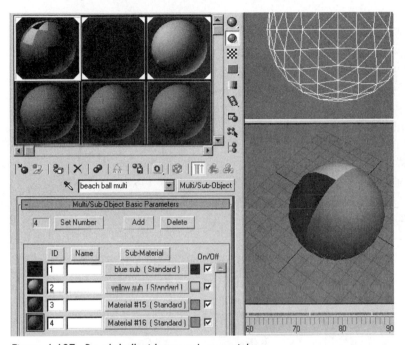

Figure 4-107: Beach ball with two sub-materials

10. In two more sample slots, create a red and a white material, and name them **red sub** and **white sub**. Assign them to the third and fourth sub-materials.

11. The beach ball is now complete. Note that you can select any of the sample slots devoted to the sub-materials, and change their parameters. Any edits done to the sub-materials will be seen on the beach ball, because the sub-materials are instanced.

12. Save the scene as **BeachBall.max** in your folder.

FACE SMOOTHING

As you recall from our discussion of polygonal mesh objects, they have no true curvature because the edges are all straight lines. As the number of edges increases, the approximation of curvature becomes more accurate. Eventually, the illusion of curvature becomes good enough to fool our eyes into believing that the object is actually curved.

However, you may have noticed that even a low-density mesh looks smooth when it is rendered. An object with a relatively low number of polygons can still give the illusion of smooth curvature, thanks to a technique known as *face smoothing*. A rendered image using this technique is called a *smooth shaded* rendering.

Take a look at the following image. On the left is a sphere with only 168 faces, shown in a wireframe rendering to illustrate the low level of detail. On the right is an identical sphere rendered with defalt Blinn shading. The shaded sphere looks very smooth in its interior area, and only at the boundary of the sphere can you see its flat faces and straight edges. The face smoothing rendering technique makes the object look round, except at the border between the object and its background. The object's profile is not affected by face smoothing, only by the level of polygonal detail.

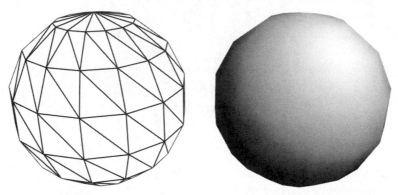

Figure 4-108: Identical spheres; the one on the right uses default face smoothing

Face smoothing works by creating a gradual transition between the colors of adjacent faces. The hard edges of mesh objects are smoothed over, so in a typical shaded rendering you don't see the individual faces.

NOTE: In the early days of computer graphics, there was no face smoothing. Shaded renderings all had a faceted look, and it was impossible to create a realistic rendering of a curved object. In 1971, a Ph.D. candidate at the University of Utah, named Henri Gouraud, developed the first smooth shaded algorithm. Since then, the concepts developed by Gouraud have been improved upon by other computer scientists such as Bui-Tuong Phong and Jim Blinn.

FACETED MATERIALS

In 3ds Max, it is easy to see the difference between objects with smooth shading and those without it. Simply turn on the **Faceted** check box in the **Shader Basic Parameters** rollout of the Material Editor. The Faceted option overrides any face smoothing on an object.

*Figure 4-109: Indentical spheres; the one on the left uses the **Faceted** option*

SMOOTHING GROUPS

Sometimes it is necessary to control the smoothing at a sub-object level instead of at the object level. You may need to create or eliminate creases in a rendered object. 3ds Max allows you to do this by manipulating the properties of polygons or faces within an Editable Mesh or Editable Poly object.

3ds Max organizes faces and polygons into *smoothing groups*. A smoothing group is similar to a material ID, in that each face has a number associated with it. When an object is rendered, 3ds Max smooths the edges among all faces in the same smoothing group.

If two adjacent faces are in the same smoothing group, then their shared edge will be smoothed over. If the two adjacent faces are in different smoothing groups, then the shared edge will not be smoothed over, and the rendering will show a sharp transition or crease between those two faces.

You can control which faces will be smoothed by activating Face or Polygon sub-object mode of an Editable Mesh or Editable Poly and opening the **Surface Properties** or **Polygon Properties** rollout. In the **Smoothing Groups** section of the rollout, you can manually or automatically assign smoothing groups to selected faces or polygons.

*Figure 4-110: **Smoothing Groups** section in the **Polygon Properties** rollout of Editable Poly*

When you select a face or polygon, the button for the smoothing group to which it belongs is highlighted on the rollout. To assign the selected face to a different smoothing group, click a different button. Multiple selected faces or polys can be assigned to a single smoothing group using the same method.

AUTO SMOOTH

The most common usage of the Smoothing Groups section of the rollout is to assign smoothing groups automatically. This is accomplished based on the angle between the normals of adjacent faces.

To assign smoothing groups automatically, select a collection of faces or polys. Usually you will select all of the faces in the object. Then enter a threshold value, in degrees, next to the **Auto Smooth** button. Click the Auto Smooth button, and the smoothing groups are reassigned. If the angle between two face normals is less than the threshold value, then the edge will be smoothed. If the angle between two face normals is greater than the threshold value, then the edge will not be smoothed.

In the following illustration, two identical spheres have been smoothed with the Auto Smooth button. All faces of the sphere on the left were auto smoothed with a threshold value of **20** degrees. All faces of the sphere on the right were auto smoothed with a value of **25** degrees. A value of **30** degrees or more will smooth all of the faces on this particular sphere.

*Figure 4-111: Identical spheres with different **Auto Smooth** threshold values*

Generally speaking, faces that join together at shallow, obtuse angles are good candidates for smoothing. An example of this is a sphere. The faces of a sphere meet one another at shallow angles of less than 30 degrees.

Faces which meet one another at sharp angles usually should not be smoothed. The sides of a box or cube meet at 90-degree angles. Those faces should not be smoothed, or the cube will not look like a cube anymore. The edges of the cube will not be distinct, and there will be strange problems with the lighting. In the following illustration, the faces of the cube on the right have been smoothed, resulting in a bizarre and unnatural rendering.

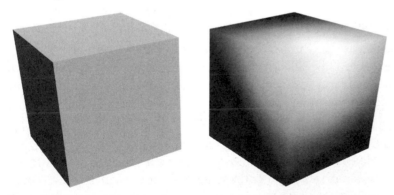

Figure 4-112: Identical cubes; the one on the right has smoothing applied to all faces

For objects other than Editable Mesh and Editable Poly, you can control smoothing with the **Smooth** modifier, which works the same as the Smoothing Group rollout. Usually, you will use the Smooth modifier to perform an Auto Smooth with a threshold value. It is also possible to use the Smooth modifier to manually assign a smoothing group to a Polygon or Face sub-object selection. In that case, you must make a selection somewhere lower in the Stack, probably with a selection modifier such as Mesh Select.

Finally, if two faces are in the same smoothing group, but are *not* adjacent to one another, then they will not be smoothed. In this way, 3ds Max can reuse a limited number of smoothing groups. For example, a default box has only three smoothing groups instead of six. Since faces on opposite sides of the box don't share an edge, it is OK for them to share a smoothing group.

MATERIAL ORGANIZATION

Efficient use of the **Material Editor** depends on your ability to organize your materials intelligently.

NAMING MATERIALS

As mentioned earlier, it is important that you name your materials as you go along. This will help you immensely when the scene becomes complex. You can also name maps if you like, but this is not as important as naming materials.

Remember that a material assigned in a scene must have a unique name.

COPYING MATERIALS

Because of the way material names work in 3ds Max, you can have several materials with the same name in the Material Editor, but only one at a time can actually be assigned in the scene.

You can drag and drop materials between sample slots to create working copies of material definitions. In this way, you can work with the material copy without changing the material that is being used in the scene. When you are happy with the changes you've made to the copy, you can use the **Put Material to Scene** button to assign the copy of the material to the scene. This will only work if the original material and the edited copy have the same name. If you change the material name, the Put Material to Scene button will be grayed out.

You may also use the **Make Material Copy** button to create a copy of a material in the selected sample slot. The copy appears in the same sample slot, replacing the original material. The only visible change to the sample slot is that the triangles at the corners of the sample slot disappear, indicating that the material is not assigned in the scene.

The original material is still present in the scene, and you can make changes to the copy without affecting the scene. When you are ready to reassign the copy, click the Put Material to Scene button.

MATERIAL SAMPLE DISPLAY

Changing the Sample Slot Layout

To change the number of sample slots visible at one time, right-click any slot to display a pop-up menu.

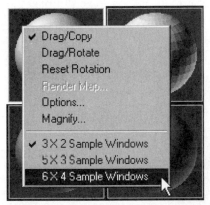

Figure 4-113: Right-click pop-up menu for changing display options

At the bottom of the menu are three options for sample slot display layouts. You may change the layout at any time. If the layout is set to **5 x 3**, then there will be 15 slots displayed at once instead of only six, but they will be much smaller. If the sample slot layout is set to **6 x 4**, all 24 of the available samples are displayed at once, in tiny slots. This means it is not possible to pan the display for more slots.

Changing the Sample Object

The sample slots are not limited to displaying spheres. Boxes and cylinders can also be displayed in a sample slot. To change the sample object being displayed, click and hold the **Sample Type** button at the upper-right corner of the **Material Editor**. Select the type of object desired from the flyout. The sample object changes only for the currently selected slot.

*Figure 4-114: **Sample Type** flyout*

Changing the sample slot to a cube or cylinder can be useful when the object to which the material will be applied resembles a cube or cylinder more than it does a sphere. In this way, you can get a better idea of what the material will look like in the rendered scene.

If you're using a cube or cylinder as a sample object, it's often useful to rotate the sample object. Use the middle mouse button to click and drag in a sample slot to rotate the object. To reset the rotation, use the right mouse button to access the pop-up menu.

Sample Windows

Occasionally you will want to see a larger sample of a material than the sample slots will allow. In that case, double-click a sample slot, and a new window will pop up. You can resize this window by dragging its sides or corners, to make the sample rendering as large as desired.

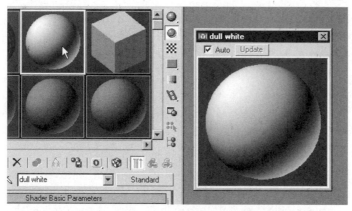

Figure 4-115: Double-click a sample slot to open a sample window

If the **Auto** check box is enabled in the sample window, then changes made in the Material Editor will be automatically updated in the sample window. If the size of the sample window is large, then this may take a long time to update. In that case, turn off the Auto check box, and use the **Update** button to re-render the sample window manually.

MATERIAL LIBRARIES

A *material library* is a collection of material definitions stored in a file. Saving a material to a material library makes it easy to use the material in another scene. It is not necessary to create a library to work with materials, but it can be very convenient. You can put together libraries for each type of work you do. For example, a material library can be made for architectural materials, or for favorite abstract materials.

A material library is saved in a file with the extension **.mat**. The default material library **3dsmax.mat** is loaded when you load 3ds Max. By default, when you browse for materials in the **Material/Map Browser**, you are looking in the **3dsmax.mat** file.

It's not a good idea to place your new, custom materials in the default **3dsmax.mat** library file. Instead, you should create a new library and place it in the same folder where you store your scene files, and any bitmaps used in the scene. That way, you can easily back up and restore all of the data needed for your project. You won't have to worry about losing your work if you have to reinstall 3ds Max, and your library won't be cluttered by a lot of materials you aren't using.

To create a new library, the easiest thing to do is to save the materials in the current scene or the materials currently loaded in Material Editor slots. Begin by selecting an empty sample slot. Click the **Get Material** button ![button] to open the **Material/Map Browser**. (In this case, you're not getting a material from the Material/Map Browser, but the Get Material button is the only way to open the Browser window.)

In the Material/Map Browser, look in the **Browse From** section. Choose **Mtl Editor**, **Scene**, or **Selected** under Browse From. All materials in the Editor, the scene, or the selected scene objects are displayed in the Browser window. Click **Save As**. Navigate to your project folder and save the new material library file.

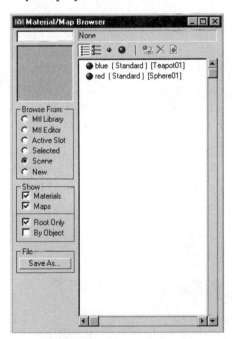

Figure 3-116: Save the materials in the current scene to a material library file

Once the new library has been saved, you'll need to manually open it in the Material/Map Browser to work with it. Choose **Mtl Library** in the Browse From section, then click the **Open** Button and navigate to your **.mat** file. When your new library is loaded, you'll see its filename listed in the window bar at the top of the Material/Map Browser.

Once you have a material library saved and opened, you can then go back to the Material Editor and add more materials to it as desired. Click the **Put to Library** button to save the material in the active sample slot into the currently open material library. You will see a pop-up dialog, in which you can change the name of the material if desired.

You can also clear the current library, save to a new file, then add materials to it. To clear the current library, click **Clear Material Library** on the Material/Map Browser toolbar. All materials in the Material/Map Browser are deleted. Then click **Save As**. Navigate to your project folder and save this new, empty library to a file. Now you can proceed to add materials to the library using the methods described earlier.

WARNING: Do not click **Save** immediately after clearing the library, or you will overwrite the active **.mat** file and destroy all of the materials in it!

The Material/Map Browser has a convenient **Merge** feature that allows you to combine material libraries. When you click Merge and choose a material library file, a dialog box opens. Here, you can select one or more materials to merge into the current library. When you save the library, all of the merged materials are included.

It is also possible to load materials from other **.max** scene files instead of material libraries. This can be very convenient, and you can even use this function to create material libraries by merging materials from several different scene files.

To load a material from an external scene file, click **Open** or **Merge** in the Material/Map Browser as if you were opening a **.mat** library file. When the **Open Material Library** dialog opens, simply choose **3ds Max** from the **Files of Type** pulldown at the bottom of the dialog.

END-OF-CHAPTER EXERCISES

EXERCISE 4.12: Table Scene

In this exercise, you'll create materials for the table scene and work with it further to make it more realistic.

1. Load the file **TableScene08Mirror.max** from the disc.

2. Add a light fixture to the scene, such as a lamp hanging from the center of the ceiling. This could be as simple as a sphere, or as complex as a chandelier. This might require you to place a ceiling in the scene.

    The light fixture shown in the following image was created with two modeling methods. The supports are circles lofted on curved paths. The lampshades were created by applying the **Lathe** modifier to half of a bell shape.

3. Create a material with high self-illumination and assign it to the light fixtures.

4. Create further materials as you like to improve the scene. For example, you could put wallpaper on the walls.

Figure 4-117: Table scene with materials

5. Save the scene with the filename **TableScene09.max** in your folder.

Hints

- The patterns in the **maps\Fabric** folder make excellent wallpaper.

- The patterns in the **Fabric** folder, when used as is, are usually too garish for a realistic scene. Soften the intensity of the map by setting the Amount for the Diffuse Color map to 60 or 70 and changing the Diffuse Color swatch on the Blinn Basic Parameters rollout to a light color.

- Increase the Tiling values for the map as much as you like. The patterns in the **Fabric** folder are designed so they don't have visible seams when tiled.

- For the picture on the wall, look through the images in the **Backgrounds** folder to find a suitable image.

- Name the materials as you go along. This will help you keep track of materials.

EXERCISE 4.13: Space Materials and Background

In this exercise, you'll put materials and a background in your space scene.

1. Load the scene **SpaceScene01.max** from your folder or from the disc.

2. To set up the background, open the Material Editor. Select an unused sample slot. Click **Get Material** and choose the **Noise** map from the **Material/Map Browser**.

3. Set the following options on the **Noise Parameters** rollout:

Fractal	enabled
Low	0.65
Size	0.1

   Leave all other settings at their default values.

   The **Low Noise Threshold** setting controls the amount of **Color #2** that appears in the scene. Increasing this value decreases the amount of white. Setting **Size** to **0.1** makes the white specks very tiny, like stars.

4. Change **Color #1** to a deep blue. Colors on backgrounds tend to appear lighter than expected, so make the color almost black.

   The material in the sample slot will not look as if there are any white specks present to form the stars. The specks are too small to be seen in the sample slot. You will have to set up the background and render the scene to see them.

5. Choose Rendering/Environment from the menu. Click and drag the **Noise** map to the **Environment Map** button. Choose **Instance** from the dialog.

6. Render the Perspective view to see the stars. They aren't very bright. Go to the **Output** rollout in the Material Editor. Increase the **RGB Level** to **2.0**. This causes the Noise map to be much brighter.

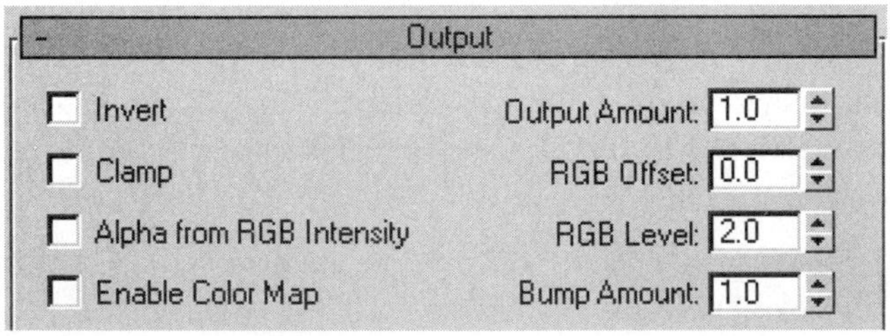

*Figure 4-118: Map **Output** rollout*

Re-render the scene. It may be necessary to readjust Color #1 to make it darker.

Figure 4-119: Starfield background

To improve the starfield, you'll add some large purple nebulas.

7. On the **Noise Parameters** rollout, click the button under **Maps** for **Color #1**. From the **Material/Map Browser**, select **Noise**.

There are now two Noise maps, one the child of the other.

8. Change the parameters to the following:

Fractal	enabled
Low	0.4
Levels	6.0

9. Change **Color #1** to a deep blue color. Change **Color #2** to a medium dark purple-pink.

10. Render the Perspective view to see the new background.

The background now has purple blotches that resemble distant nebulas.

11. Experiment with colors and settings to customize your starfield.

12. Create custom materials for the objects in the scene using the techniques you learned in this chapter.

Figure 4-120: Space scene with background and materials

13.   Save the scene with the file name **SpaceScene02.max** in your folder.

Hints

- The bitmaps in the **maps\Space** folder make good planet bitmaps.
- The bitmaps in the **maps\Ground** folder are good for making ground surfaces.
- Procedural maps such as Splat, Smoke and Noise make very good bump maps.
- Use submaps to make your materials more complex and interesting.

SUMMARY

Materials are set up in sample slots in the **Material Editor**. Materials have four main levels of definition: *material type*, *shading algorithm*, material *parameters*, and *maps*. Materials are organized in the scene by their names, so there can only be one material with a particular name in any given scene.

Most shading algorithms have three main color components: **Ambient**, **Diffuse**, and **Specular**. The Ambient component is the color of the object in shadow. Diffuse Color is the color of the object where it receives direct light. Specular color is the color of the highlights on an object.

A map is a pattern of colors or brightness values that can vary a material attribute across the surface of an object. Maps come in two varieties: bitmap files and *procedural maps*. Procedural maps are internally generated by algorithms within 3ds Max, while bitmaps are external 2D image files stored separately from the 3D scene.

The most common material attribute that a map can be applied to is the Diffuse Color. A diffuse map will produce a colored pattern on the object. Grayscale maps can also be applied to other attributes, such as **Bump**. A bump map creates the illusion of roughness on an object's surface. A map cannot be applied directly to an object; only a material can.

Maps can also be applied to the environment of a scene rather than to a material on an object. This is called an *environment map*. **Reflection** maps project images from the environment back onto the object to create reflections. Another type of environment map is a background for a scene.

Each face on an object is assigned a *material ID*. Material IDs can be used to assign multiple materials to one object with the **Multi/Sub-Object** material type.

The way a map lies on an object is determined by *mapping coordinates*. Bitmaps require mapping coordinates, or else the object will not render properly. Mapping coordinates are assigned to an object with the **UVW Map** modifier. Some objects, such as primitives and Lofts, can generate their own mapping coordinates.

Materials and maps can be dragged and dropped within the Material Editor to make copies and instances. This allows you to use the same map in more than one material and streamline your workflow when working with complex material hierarchies.

Face smoothing is a rendering technique that enhances the illusion of curvature on mesh objects. Mesh faces are organized into *smoothing groups*. If two adjacent faces are in the same smoothing group, the edge between the two faces will be smoothed.

Materials can be stored in external files called *material libraries*. Materials can be merged from other scenes, then saved in material library files.

REVIEW QUESTIONS

1.   What is a *map*? How does it differ from a *material*?

2.   If you want a map to define the main colors of a material, which map attribute do you assign it to?

3.   What is the relationship between tiling and the size of the map?

4.   What does a *bump map* do?

5.   How do you assign more than one material to an object?

6.   How does a *reflection map* work?

7.   What are *mapping coordinates*?

8.   How do you set up a bitmap to use as a background?

9.   How do you display more sample slots?

10. You assign a material to an object, and subsequently clear its sample slot in the Material Editor. Does the material remain on the object?

11. How many materials can you have in a scene? How many materials can you load into the Material Editor at one time?

12. What is *face smoothing*?

Chapter 5
Cameras and Lights

OBJECTIVES

In this chapter, you will learn about:

- Camera concepts such as *field of view* and *aspect ratio*
- Placing cameras and lights in a scene
- Working in camera and light viewports
- Different types of lights
- Basic lighting setups
- Casting shadows
- Varying light intensity with Falloff and Attenuation
- Using Advanced Effects to control how lights affect materials

ABOUT CAMERAS AND LIGHTS

In 3ds Max, cameras and lights are treated like objects in the scene. You can transform them just like geometric objects, but lights and cameras don't appear in rendered images.

When placed in a scene, lights affect the scene similar to the way a photographer's lights affect the subject being photographed. The intensity and color of lights can be adjusted individually.

A camera is used to set up a particular view. Once a camera is placed, one of the viewports can be replaced by the camera view. This view shows exactly what the camera sees. Doing this enables you to render the viewport, which means you can "shoot" the scene from exactly the angle you want.

Cameras can also be adjusted like real-world cameras. Virtual lenses can be changed, or even animated, for tight or wide-angle shots.

CAMERAS

In order to understand how cameras work in 3ds Max, it is necessary to understand a little about how perspective works.

PERSPECTIVE

Perspective is a way of representing a three-dimensional scene on a two-dimensional surface. The 3D scene is projected onto the 2D picture plane. One of the characteristics of perspective projections is that parallel lines appear to converge as they go into the distance.

Figure 5-1: Perspective drawing with converging lines

This is the way our eyes see the world, so a drawing done in this way looks realistic. Since cameras see the world in the same way our eyes do, photographs come with perspective built in.

Figure 5-2: Photographs show perspective

Perspective also comes into play in determining distances between objects. When a photograph contains objects that are both close and distant, you can usually tell roughly how far apart they are. After a lifetime of using your eyes, you're accustomed to how things should look when they're close together or far apart. The following image shows a wood sculpture among some rocks. Because you are accustomed to looking at similar environments, you can tell how far apart the objects are.

Figure 5-3: Objects near and far

Cameras in 3D computer graphics are designed to render images in perspective. The Perspective view and camera views use the principles of perspective to make the scene look realistic, as if it were a photograph or something seen with the naked eye.

CAMERAS IN 3DS MAX

A camera is created on the **Create** panel. Click the **Cameras** button ![icon].

There are two types of cameras, **Target** and **Free**. A Target is a look-at point. A Target camera is always pointed at its target. This makes it easy to point a camera at a particular object or place in the scene. Just move the target to where you want the camera to look.

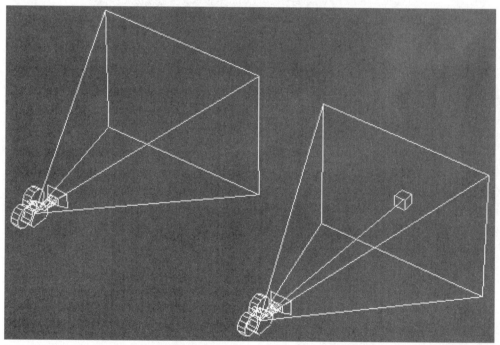

Figure 5-4: ***Free*** *camera and* ***Target*** *camera*

A Target camera is created by clicking and dragging in a viewport. The camera is placed at the first click point. When you drag and release the mouse, the target is placed. A visible line connects the camera with its target.

A Free camera does not have a target. To point the camera at a subject, move and rotate the camera with the standard transform tools.

To create a **Free** camera, click the screen. The camera is placed where you click. Like all other objects in 3ds Max, cameras are placed on the construction plane. When creating a Target camera, you should create it in the Top view or the Perspective view.

This will make the new camera sit horizontally in the scene, instead of looking straight up or down.

Free cameras should be created in the Front, Back, Left, or Right views. The camera will be aligned so that it faces along the Z axis of the current orthographic view, into the viewport. If you create a Free camera in the Top or Perspective view, it will face directly downward, which is usually not what you want.

Placing the Camera

For a Target camera, the target should be placed at the center of the scene, or at the object of interest. When you move the camera, it will still point in the correct direction.

To move both the camera and target at the same time, click the line between the camera and target. Both camera and target will be selected. Click **Select and Move**, and click and drag the line.

A Free camera must be rotated to point in the correct direction. Click **Select and Rotate** and rotate the camera as you would any other object. To pan the camera left to right, rotate it in the World Z axis. If you use the Local Y axis, you'll end up rolling the camera along its line of sight (local Z axis) relative to the world, which is usually not what you want. To tilt the camera up or down, rotate in its Local X axis.

Once you have placed a camera in the scene, you can change one of the viewports to display what the camera sees. This will enable you to render the camera view later on. You can use any viewport to display the camera view, but the Perspective view is the usual choice. Activate the viewport and right-click the viewport label. Choose the camera from the pop-up list. If you have only one camera in the scene, you can also press the <**C**> key on the keyboard. The viewport name changes to the camera name and displays what the camera sees.

Continue to move the camera and/or target until the camera view depicts the shot you wish to render.

Tips for Placing Cameras

- To look at the scene from an angle, move the camera rather than rotating objects.

- If a Target camera view doesn't show the part of the scene you want, move the target, not the camera itself.

- To move a camera forward and back, move it in its Z axis in the **Local** coordinate system.

- Don't place the camera too close to the object you wish to view.

EXERCISE 5.1: Attic Camera

In this tutorial, you'll place a camera in a scene.

1. Load the file **Attic.max** from the disc that comes with this book.

   This scene consists of several objects in an attic.

Figure 5-5: Attic scene

2. On the **Create** panel, click **Cameras** . Click **Target**. In the **Top** viewport, click and drag from the lower-right corner of the **Top** viewport to the center area of the attic, as shown the following figure.

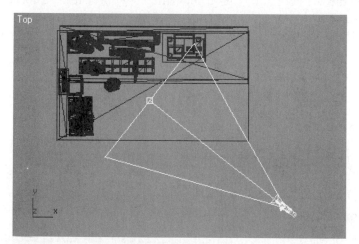

Figure 5-6: Camera created in Top viewport

3. Activate the **Perspective** view. Press the **<C>** key on the keyboard. The Perspective view changes to the **Camera01** view. The camera currently looks straight at the floor. Like geometric objects, cameras are created on the construction plane.

4. In the Camera01 view, right-click the viewport label and choose **Show Safe Frames** from the pop-up menu. The viewport is cropped to display the precise view of the camera. In any viewport, move the camera and target until the Camera01 view looks like the following illustration.

Figure 5-7: Attic scene

5. Save the scene in your folder with the filename `AtticWithCamera.max`.

PHOTOGRAPHY AND CAMERAS

The cameras in 3ds Max are designed to work similarly to cameras in real life. For this reason, it will help you to understand a little about how photographic lenses work.

Focal Length

The view you see through a camera is determined by its lens. Every lens has a specific *focal length*, which is the distance from the center of the lens glass to the film. Focal lengths are measured in millimeters. Focal length is sometimes called *lens length*. The most common focal lengths for cameras in real life are 35mm and 50mm. These lenses produce an image that is similar to what is seen with the naked eye.

Field of View

The length of the lens determines how wide an angle you will be able to view with the camera. The angle, called the *field of view*, is measured in degrees.

A *telephoto* lens has a longer lens length, such as 120mm or 150mm. These lenses can see detail very far away, but the camera can only see a small portion of what you can see with the naked eye. A shorter lens length, such as 28mm, can be used to take *wide-angle* pictures.

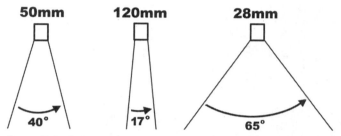

Figure 5-8: *Relationship between lens length and field of view*

The longer the focal length, the smaller the field of view. If you take a picture with a 150mm lens, you will only be able to see a small portion of the scene. A 15mm lens, however, can shoot a wide panorama.

Zoom Lenses

Most consumer cameras come equipped with a *zoom* lens. A zoom lens has a variable focal length, and therefore a variable field of view. This type of lens allows you to quickly zoom in tight for a telephoto shot, or zoom out to get a wide shot. Professional photographers often use lenses that have a fixed focal length, known as *prime* lenses, because they have superior optical image quality.

In 3D computer graphics, the virtual cameras work like zoom lenses: you can choose any focal length, but without the penalty of reduced optical quality.

Telephoto Lens Perspective Distortion

Telephoto lenses and wide-angle lenses tend to warp perspective in unexpected ways. This is not to say that they shouldn't be used. You just need to know what they do so you can purposefully go about getting the effect you want.

A telephoto lens tends to flatten out the sense of depth. Objects that are far apart appear to be close together. For example, suppose you're using a telephoto lens to photograph a friend. Right when you snap the picture, another person moves into the frame ten feet behind your friend. When the photo is developed, this person will look as if he is practically standing on your friend's shoes.

Wide-Angle Lens Perspective Distortion

A wide-angle lens makes objects that are close together appear to be far apart. A photo of your kitchen will make it look as large as the Taj Mahal. Photographers use this trick when taking photos of hotel rooms for advertising. When you arrive for your vacation, the room is somehow not quite as large as it looked in the brochure.

The disadvantage of a wide-angle lens is that it tends to distort the edges of the picture. This effect is not so noticeable with photos of landscapes and buildings, but it can cause pictures of people to look mighty strange.

Also, the virtual wide-angle lenses found in computer graphics programs work differently from real lenses. Real wide-angle lenses cause objects in the center of the frame to look bigger. This spherical distortion is commonly called the *fish-eye* effect. Computer graphic virtual lenses usually do the opposite: objects in the center of the frame appear smaller, and objects at the edges of the frame look bigger. This disparity can cause problems when compositing live action shots with computer graphic images.

Setting the Focal Length

Earlier you learned that cameras in real life commonly have a focal length of 35mm or 50mm. When you create a camera in 3ds Max, the default lens length is halfway between the two, around 43mm. The default field of view (FOV) is 45 degrees.

The focal length is set on the command panel with the **Lens** parameter. The field of view is set with the **FOV** parameter. When one of these parameters is changed, the other changes automatically.

Under the **Stock Lenses** section on the Modify panel is a series of preset lens lengths ranging from **15mm** to **200mm**. Click a button to enter the value for the **Lens** parameter.

These presets correspond to the standard prime lenses used by cinematographers when shooting 35mm motion picture film. The presets are included as a convenience for quick setting of the lens length. There is no difference between clicking a preset button or manually entering the same value in the Lens parameter field.

A radical change to the Lens value or FOV will often cause the camera to appear to be too close or too far from the subject. You can remedy the situation by *dollying* the camera. Dollying a camera means moving the camera toward or away from its subject. One way to dolly a camera is to move it in its local Z axis. You can also use the **Dolly Camera** viewport control, which is described later in this chapter.

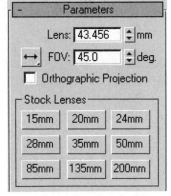

Figure 5-9: **Lens** *and* **Field of View** *controls on the Camera Modify panel*

EXERCISE 5.2: Long and Short Lenses

In this exercise, you'll work with camera lens lengths to see their effects on the view.

1. Load the file **Teapots&Cameras.max** from the disc that comes with this book.

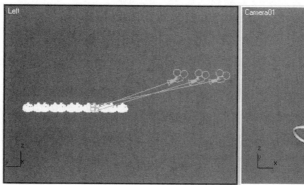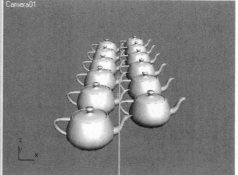

Figure 5-10: Teapots and cameras

This scene consists of several teapots in two rows. Three cameras have been set up to view the teapots. From left to right in the **Left** viewport, the cameras are named **Camera01**, **Camera02**, and **Camera03**. The view from **Camera01** is shown in the viewport at the lower right.

2. In order to see the effects of changing lens settings, you need to change the settings for the Home Grid. Right-click the **3D Snap** button on the Main Toolbar. The **Grid and Snap Settings** dialog opens. Choose the **Home Grid** tab, and disable the check box that reads **Inhibit Perspective View Grid Resize**.

The Home Grid extends to the edges of the camera viewports.

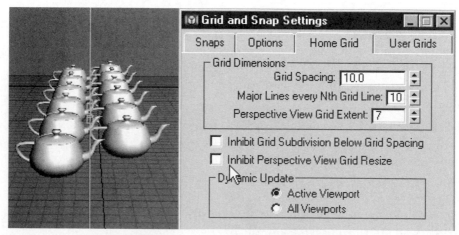

*Figure 5-11: Turn off **Inhibit Perspective View Grid Resize***

3. In the Left viewport, you can clearly see that the teapots are all the same size. In the **Left** viewport, select the object **Camera02**, which is the middle camera.

4. Go to the **Modify** panel. Under **Stock Lenses**, click the **15mm** button.

    The camera's field of view widens to about 100 degrees, as seen by the **FOV** parameter on the Modify panel. The small lens length has made for a wide field of view. In the Camera02 view, the rear teapots appear to be far from the camera, and all teapots appear to be farther away from one another.

5. At the top of the Modify panel, change the camera name to `Wide Angle`. The name of the viewport for this camera changes at the same time.

6. Activate the **Select and Move** tool on the Main Toolbar. Choose **Local** from the **Reference Coordinate System** pulldown. In the **Left** viewport, move the `Wide Angle` camera in its local Z axis, toward its target. Dolly the camera until the teapots at the front of the **Wide Angle** view appear to be about the same size as the teapots in the **Camera01** view.

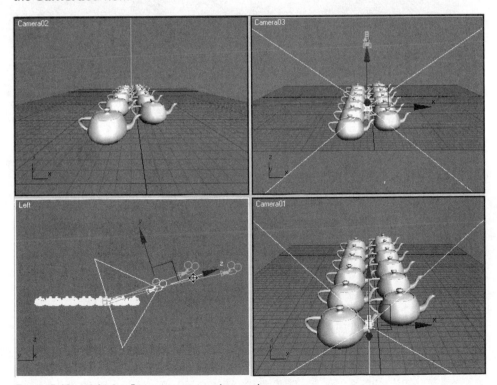

Figure 5-12: `Wide Angle` *camera moved toward teapots*

7. In the **Left** viewport, select **Camera03**, which is the camera farthest to the right.

8. On the **Modify** panel, click the **200mm** button. Change the camera name to `Telephoto Lens`.

The camera's field of view is now about 10 degrees. The teapots appear very large in the **Telephoto Lens** viewport.

9. In the **Left** viewport, move the camera away from the teapots until the teapots at the front of the Telephoto Lens viewport are about the same size as the teapots at the front of the Camera01 view.

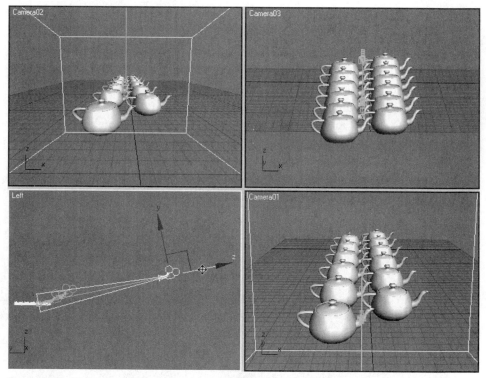

Figure 5-13: **Telephoto Lens** *camera moved away from teapots*

10. Look at the three camera views. In which view do the teapots at the back of the scene seem to be far away? In which view do they seem closest to the camera?

From this exercise, you can see that a wide-angle lens (a short focal length) makes objects appear to recede into the distance, while a telephoto lens (long focal length) makes objects appear to be sitting very close together.

Also note that the teapot near the front of the Wide Angle view is warped, with a slight oval shape. If you move this camera around to view the scene from different angles, you will find that objects near the edge of the view often appear to stretch and warp.

In 3ds Max, wide-angle lenses yield distortion, so that the objects at the edge of the frame look big, and objects in the center of the frame look small.

Figure 5-14: Serious teapot warpage

CAMERAS AND VIEWPORTS

As you gain experience with cameras, you will find that knowing all the ways to work with them in viewports will make your work easier and faster.

Camera Cone

Every camera has a cone that shows the direction of the camera and its field of view. The cone is visible when the camera is first created, and when it is selected. However, the cone disappears as soon as another object is selected.

When composing a scene, it can be helpful to have the camera cone displayed at all times. In this way, you can see the camera's field of view (FOV) while moving and placing other objects in the scene.

To display a camera's cone at all times, click the **Show Cone** check box on the **Parameters** rollout of the camera's Modify panel.

Hiding Cameras

If the camera is cluttering up your scene, you can hide it by going to the **Display** panel . On the **Hide by Category** rollout, check the **Cameras** check box. This will hide all cameras in the scene. You can later unhide all cameras by unchecking the check box.

Clipping Planes

When a camera gets too close to an object, it will not register parts of the objects closest to the camera. This is because the camera automatically clips off part of the object when it is a certain distance from the camera.

Clipping planes are useful if you have a lot of geometry in your scene, so you can hide many polygons and speed up interactivity. They are also indispensable for some modeling operations, so you can work on just part of your model.

*Figure 5-15: **Viewport Clipping***

You can control the clipping plane for any viewport, not just a camera view. Right-click the viewport label and choose **Viewport Clipping**. You will see a yellow line with two triangles on the right side of the viewport.

The triangle at the bottom sets the location of the *near* clipping plane. The near clipping plane hides all polygons that are between the near clipping plane and the viewplane. Move the triangle upward to move the clipping plane back into the scene, and clip the polygons of objects in the foreground.

The triangle near the top of the viewport sets the *far* clipping plane location. Move the triangle downward to move the far clipping plane toward the viewplane, and clip off polygons in the distance.

Viewport clipping planes do not affect how an image is rendered. If you need to clip the polygons in a rendered image, you can do so by activating **Clip Manually** in the **Clipping Planes** section of the camera's Modify panel. You'll see two red clipping planes intersecting the camera cone, and you can change the clipping distances with the **Near Clip** and **Far Clip** parameters.

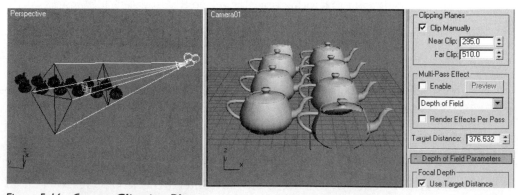

*Figure 5-16: Camera **Clipping Planes***

Aspect Ratio and Safe Frames

All film, video, and computer displays are rectangular. The rectangles come in various sizes and shapes. The most important characteristic of a screen is its *aspect ratio*, which is the proportion of screen width to screen height. Aspect ratio determines the shape of the screen, not its size.

Different visual media use different aspect ratios. A common aspect ratio for 35mm motion picture film is 1.85 to 1, meaning that the screen is 1.85 times as wide as it is tall. Widescreen 35mm film has an aspect ratio of 2.35 to 1.

Most video screens today use the aspect ratio of 1.33 to 1, which can also be expressed as 4 to 3. High definition television (HDTV) uses the aspect ratio of 1.78 to 1, or 16 to 9.

In 3ds Max, aspect ratio is determined by the resolution setting in the **Render Scene** dialog, not in the camera Modify panel. The default resolution is 640 x 480. If you divide 640 by 480, you get 1.33. So, the default aspect ratio of a rendered image in 3ds Max is 1.33, which is the same as a standard video screen.

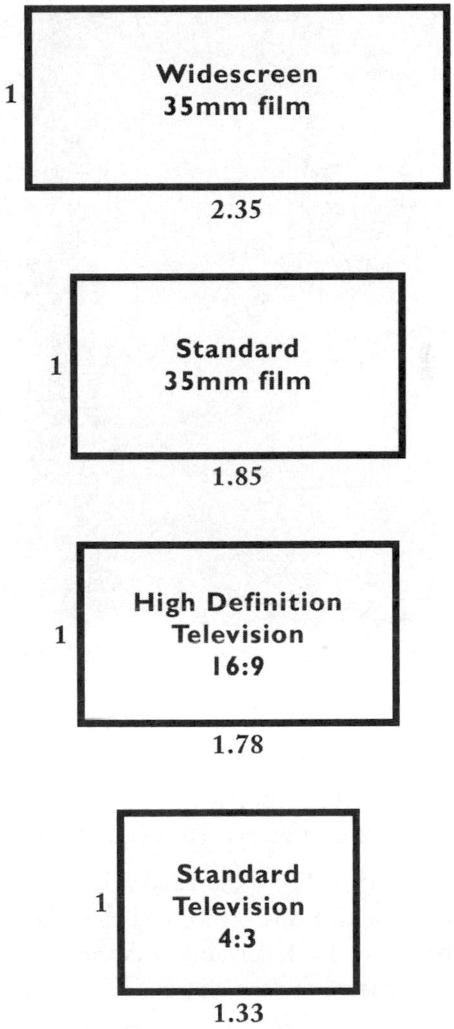

However, viewports can be any aspect ratio. They can even be taller than they are wide, which is almost never the case for a film or video screen. If the viewport is not the same

Figure 5-17: Common aspect ratios

aspect ratio as the rendered image, you can have big problems. You might compose the image very carefully in your viewport, only to find that the rendered image is different. Objects might be cropped in the rendered image, or objects might appear at the margins of the rendered image when you didn't want them to.

The solution to these problems is to always enable *safe frames* in a camera viewport. Right-click the camera viewport label, and choose **Show Safe Frame** from the pop-up menu. The viewport changes, and geometry that is outside the camera's field of view is cropped in the viewport. Now you can tell exactly what will be rendered, and what will not be rendered. You can now compose your shots with certainty, and you won't have any unpleasant framing surprises when you view your rendered images.

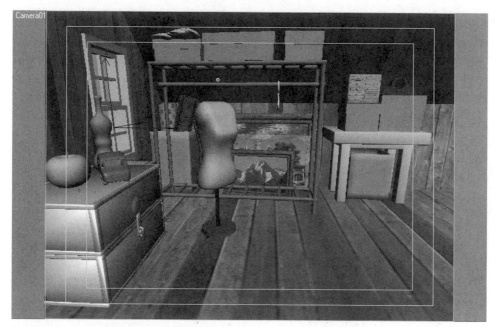

*Figure 5-18: Camera view with **Safe Frames***

There are three rectangles displayed when you enable safe frames. The outermost rectangle is the **Live Area**. Anything in the Live Area is what will be rendered.

The middle rectangle is the **Action Safe** area. This is the part of the image that probably won't be cropped when the image is projected or shown on a video monitor.

The edges of most film and video displays are cropped by five to ten percent, to eliminate unattractive dirt or noise at the edges of the frame. When you go to a movie theater, there are black curtains or fabric around the screen, and the image is projected slightly larger than the screen size. The edges of the image are projected onto the black fabric, and the resulting effect is that the image is slightly cropped. Standard video monitors use a similar technique, called *overscan*, whereby about ten percent of the recorded image is lost when the image is projected onto the inside of the video screen.

Keep all important objects within the Action Safe area to ensure that they will not be cropped.

Finally, the innermost rectangle is the **Title Safe** area. Keep all titles, logos, and important graphic elements within this area to ensure that they are not cropped, and that there is a margin. This works just like a printed page—you don't want to put type at the very edge of the page. You must leave a margin, so the type will be legible and the page will be aesthetically pleasing.

When designing a computer graphic image, safe frames are of paramount importance. However, if you are rendering an image for display *only* on the computer screen, you don't need to worry as much about the Action Safe area. Computer displays don't use the overscan process, because if they did, then important interface elements such as menus would be off the edge of the screen, and therefore invisible. Computer monitors are *underscan* displays, meaning that the image is projected slightly smaller than the available screen size, and there is a black border around the image at all times.

TIP: If you are an emerging computer artist and wish to send your work to companies, you will probably be sending it in the DVD or VHS format. Companies usually don't like it if you send them Web addresses or data discs. Therefore, you should render all of your work to video. That means you need to compose your shots with safe frames visible at all times.

Target Distance

You can change the distance of the camera to its target by moving either the camera or the target in the viewports. In addition, you can move a target toward or away from a camera with the **Target Distance** parameter on the camera's Modify panel.

This is most useful for Free cameras. Although a Free camera does not have a target look-at point, the target distance does affect the size of the camera cone. The blue camera cone always stops at the target. Changing the Target Distance of a Free camera helps you visualize what is in the camera's field of view when manipulating the camera in an orthographic or Perspective viewport.

Camera Viewport Controls

Sometimes it's easier to manipulate camera position and field of view controls from directly within a camera viewport. You can get more of a feeling of working with a real world camera, and it's often more intuitive.

When a camera view is activated, the **Viewport Control** icons will change. Instead of the controls for an orthographic or Perspective view, you now have access to specialized camera control icons.

 Dolly Camera moves the camera toward or away from its target, in the camera's local Z axis. If the camera is a free camera, dollying the camera moves it toward its viewing direction. If the camera is dollied past its target, the camera will flip over, rendering the scene upside-down.

 Dolly Target moves the target toward or away from the camera. This button is on a flyout from the Dolly Camera button. It is available only for Target Camera views.

Dolly Camera + Target moves the camera and target along the view direction. This button is on a flyout from the **Dolly Camera** button, and is available only for Target Camera views.

 **NOTE:** The Dolly icons look very similar. The difference is only one of color. The camera and/or target to be dollied is shown in red on the icon.

 Field-of-View changes the camera's focal length interactively. This tool works the same as it does in a Perspective view. In a camera view, the **Lens Length** and **FOV** parameters on the camera's Modify panel change automatically.

 Perspective alters a camera's position and field of view simultaneously, so the view in the viewport stays approximately the same. This tool is used to adjust the amount of perspective distortion in the image.

 Roll Camera rotates the camera around its local Z axis. This gives an effect as if the viewer tilted his or her head to the side. This is only very rarely used, for unusual shots that are designed to convey imbalance or disturbance. Normally the camera should not be rolled, and the top and bottom edges of the frame should be parallel to the horizon.

 Orbit Camera moves a camera around its target. The camera maintains a constant distance from its target as it orbits.

 Pan Camera rotates a camera around its local X axis and the world Y axis. This works the same as a real camera on a tripod. The Pan Camera button is on a flyout from the Orbit Camera button. In film and video terms, a pan is a horizontal camera sweep, achieved by rotating the camera on a tripod, so the image moves from side to side on the screen, while the camera stays in one location. Pan is short for *panorama*. A *tilt* is a vertical camera sweep, in which the tripod head is rotated so that the image moves up and down on the screen, while the camera remains in the same place.

 Truck Camera moves the camera and its target along the camera's local X and Y axes. (This is the same as the Pan button in orthographic or Perspective views, but the terminology of the buttons in those views is not technically accurate.) The Truck Camera control is used to achieve the effect of a moving camera shot. Side-to-side camera motion is known in film terms as a *tracking* shot or a *crab* shot, while up and down camera motion is a *pedestal* shot or a *crane* shot.

Walkthrough allows you to navigate the scene just as you would in a first person game. Select the Walkthrough button from the Truck Camera flyout to activate walkthrough mode. Then use the mouse and Hotkeys to move through the scene. Move the mouse to aim the camera and determine your direction of travel. The most commonly used Hotkeys are **<W>** or **<UPAR-ROW>** to move forward, **<E>** to move up, **<C>** to move down, and **<Q>** to accelerate. A complete list of Hotkeys is in the 3ds Max help file.

LIGHTS

When a light object is placed in a scene, it shines from the location of the light icon. Lights can be placed anywhere in a scene, such as inside a lampshade, on a ceiling, or above the model to simulate the sun.

HOW LIGHTS WORK

During the rendering process, 3ds Max determines which parts of which objects in the scene will be bright, and which will be dark. This is determined by the face angle.

Face Angle

In general, faces pointing toward the light receive the most illumination. The more a face is angled away from a light, the less illumination it receives.

To figure out how a face will be illuminated, 3ds Max uses face normals. If a face's normal is pointing right at the light, it will get the highest possible illumination from that light. If the face normal is pointing away from the light, the face will receive no illumination whatsoever.

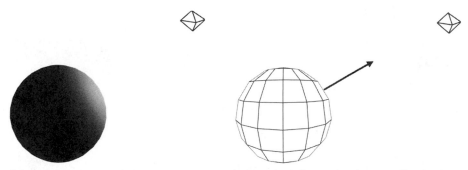

Figure 5-19: Faces with normals pointing directly at the light receive the most illumination

Shining Through Objects

By default, all lights shine "through" objects. This means that even though an object is placed behind another object, it will receive the same amount of light as the object in front of it.

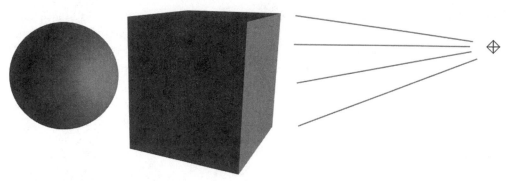

Figure 5-20: Light shines through objects

To prevent a light from shining through objects, you must enable shadow casting. Then the light will blocked from shining through an object, and will cast shadows instead.

Shadows

All light types can be made to cast shadows. The direction of the shadow is determined by the light's placement in the scene, while the darkness of the shadow depends on the intensity of the light in comparison to other lights in the scene.

In real life, shadows help to "ground" objects. You can tell how far above the ground an object is by its distance from its shadow.

In the following illustration, the only way you can tell whether the teapot is on the table or floating above the table is the position of the shadow. When the teapot is on the table, its shadow is in contact with the object. When it is floating above the table, its shadow is away from the object.

Figure 5-21: Shadow for object on a table and above the table

Leaving shadows out of scenes causes objects to look as if they're floating above the ground.

If the normals of an object are facing away from the light, those faces will not generate shadows. For example, you can put a light inside a sphere, and the light will illuminate other objects in the scene, whether you have shadows enabled or not.

LIGHT TYPES

There are several types of lights available in 3ds Max. These lights are designed to simulate a variety of real-world light sources from candlelight to moonlight. In this chapter we will cover the basic types of lights.

Default Lighting

When you start 3ds Max, there is a light already in the scene, so that you can see the contours of objects in shaded viewports. This is called *default lighting*.

Default lighting isn't visible as an icon in the scene. It is there so you can see shaded objects right away when you start your work. If the default lighting weren't there, objects in shaded viewports would appear to be black.

The default lighting is initially a single light that provides "over-the-shoulder" illumination in viewports. The default light is locked to the viewport, so that as you change the view, the light moves with the point of view. You can also choose a two-light default configuration by right-clicking a viewport label and choosing **Configure**. In the **Viewport Configuration** dialog, you will find an option for one or two default lights in the **Rendering Method** tab.

As soon as you place a light icon in a viewport, default lighting is turned off. This often means that initial placement of a light causes objects in shaded viewports to go black until you move the light to its correct position.

Sometimes the viewport approximation of your scene lighting will make it difficult to see your models. If this happens, you can activate the **Default Lighting** check box in the Viewport Configuration dialog. This causes default lighting to override whatever lights you have in the scene, but only in smooth shaded viewports. Rendered images always use any scene lights, ignoring the state of the Default Lighting check box.

Default lighting is fine for early stages of a project, when you're simply setting up objects. However, default lighting doesn't produce shadows, nor can it give you many of the other effects that you can produce with custom lights. Default lighting exists purely for viewport navigation, and not for artistic renderings.

Omni Lights

The simplest type of light is an *omnidirectional* or point light source. It sheds light on objects in all directions, and is useful for general illumination of a scene. 3ds Max calls this an **Omni** light. An Omni light appears in a scene as a small, yellow eight-sided object resembling a diamond (called an *octahedron*).

Omni lights are commonly used for *fill* light. Pho-

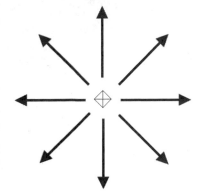

Figure 5-22: **Omni** *light*

tographers and filmmakers use fill lights to "fill in" dark areas on a studio set. That way, shadows are not too black, and the image will record well, and look good, on film or video.

Spot Lights

A **Spot** light shines in one direction, creating a cone of light that gets wider the farther it gets from the light source. Just like cameras, there are two types of Spot lights, **Target** and **Free**. A **Target Spot** uses a target look-at point to direct its light beam, while **Free Spot** is directed by rotating the light.

A Free Spot appears in a scene as a small yellow cone. When you select the light in a viewport, a larger blue cone also appears. The small cone indicates the light source, while the larger cone shows the area the light shines over.

A Target Spot also has two cones, plus a target, represented by a yellow cube.

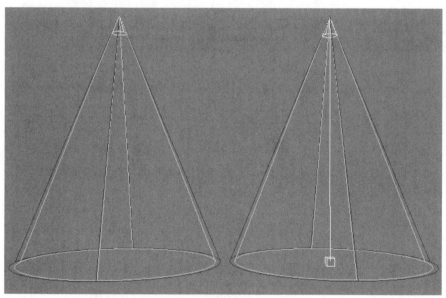

*Figure 5-23: **Free Spot** and **Target Spot** lights*

Free and Target Spots achieve identical results in renderings. The difference between the two is how the light is aimed in the scene, not the illumination effect.

By default, a Spot light shines infinitely in the direction of the cone. The cone gets larger as it moves farther from the light source. The cone icon stops at the target distance, but the light rays continue forever unless you enable Far Attenuation, which is discussed later in this chapter.

Direct Lights

A directional light, or **Direct** light, casts parallel rays. Instead of a cone of illumination, a Direct light projects a cylindrical beam. Unlike a Spot light, the rays of a Direct light do not diverge, and the lighted area does not become larger with distance.

A **Free Direct** light appears in a scene as a yellow arrow. The arrow indicates the light source, while the cylinder shows the area the light shines over. A **Target Direct** light also has a target to aim the beam.

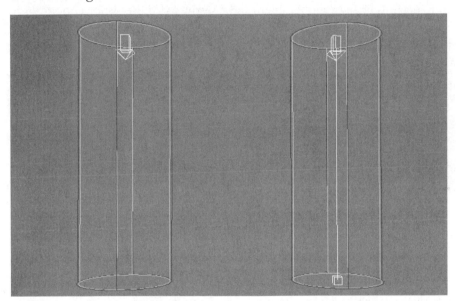

Figure 5-24: **Free Direct** and **Target Direct** lights

The main purpose of a Direct light is to simulate very distant light sources, such as the sun. The sun is so far away that its rays are effectively parallel by the time they reach the earth.

Except for the parallel rays of its light beam, a Direct light is identical to a Spot light.

Ambient Light

In life, ambient light is the overall light that pervades an entire scene. It is created when light rays bounce off of objects to illuminate other objects. In computer graphics, the term *ambient light* refers to a specific technique that attempts to approximate this effect in the simplest possible way: by merely brightening the rendered image.

In 3ds Max, the amount of ambient light is set with the **Rendering > Environment** menu option. The **Ambient** swatch in the **Environment** tab of the **Environment and Effects** dialog sets the color of ambient light in the scene.

By default, the Ambient light setting is set to zero. If your scene is too dark, you might be tempted to increase the Ambient light parameter to a high number. However, this is not usually a good idea.

The Ambient light parameter in 3ds Max simply raises the brightness, or *black level*, of an image. No color in a rendered image can be darker than the color defined in the Ambient color swatch. A high Ambient light setting will make shadows too bright, destroy any contrast, and make a scene look flat and featureless.

You will rarely need to change the Ambient light setting. If there is not enough light in your scene, work with the Omni, Spot, and Direct lights. Increased Ambient light should not be used as a substitute for good lighting. In fact, professional CG lighting artists almost never use Ambient lighting at all.

In the following illustration, the image on the left uses a single Spot light, with an Ambient light setting of **80**. The image on the right uses an Ambient light setting of zero. Six Omni lights placed strategically in the scene provide a much more realistic effect.

Figure 5-25: **Ambient** *lighting on left, multiple* **Omni** *lights on right*

Other Lights

Omni, Spot, and Direct lights are the most commonly used light types in 3D graphics. There are other types of lights in 3ds Max, such as **mr Area Omni** and **mr Area Spot** lights, which only work with the Mental Ray renderer. 3ds Max also offers **Photometric** lights, which are physically accurate lighting models based on distance and scale. These lights, and the techniques that utilize them, require a great deal more setup. This book devotes an entire chapter to this field: Chapter 12, *Advanced Lighting*.

PLACING LIGHTS

To create an Omni light or Free light, click the screen to place the light in the desired location. In orthographic viewports, Free lights are initially created so that they are aimed into the viewport, facing in the direction of the viewport's positive Z axis. In Perspective or Camera views, Free lights are created facing downward, in the negative Z axis of the world.

To place a Target light, click to set the light's location, then drag and release the mouse to set the target position.

Like all other objects, lights are created on the active construction plane. Most likely the light and/or target will have to be moved to get the effect you want.

Lights and targets can be selected like any other object. It can sometimes be difficult to click on a light or target in a complex scene. To help you do this, use the **Selection Filter** on the Main Toolbar. Choose **Lights** from the pulldown list. Now only lights and light targets can be selected in viewports. Remember to change the Selection Filter back to **All** when you have finished working with the lights.

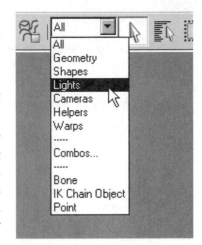

Figure 5-26: **Selection Filter**

Placing Free Lights

To aim a free light in the proper direction, rotate it in the viewports. Even though a free light doesn't have a target, it does have a **Target Distance**. This distance sets the length of the light cone or cylinder in viewports, and provides a virtual target for other light placement tools. These tools are covered later in this chapter.

Placing Target Lights

To move both a target light and its target at the same time, click the line between the camera and target. Both the light and target will be selected.

Light Cone

Direct and Spot lights have an icon that shows the area over which the light shines. The illumination area icon for a Direct light is cylindrical. For a spot light, it is shaped like a cone. In the 3ds Max interface, this is called a **Light Cone** for both light types.

By default, a Light Cone appears when the light is selected, but disappears when the light is not selected. Sometimes it can be helpful to see the Light Cone at all times. To do this, select the light and open the Modify panel. In the **Light Cone** section of the the **Spotlight** or **Directional Parameters** rollout, check the **Show Cone** check box.

Light Cones project a circle of light by default. However, you can also choose to project a rectangular cone. This comes in handy when using Projector Maps, which are discussed later in this chapter.

EXERCISE 5.3: Placing Lights

In this exercise, you'll place lights in a scene.

Omni Light

1. Load the file **AtticWithCamera.max** from your folder or from the disc.

2. Activate the **Camera01** view, and click the **Quick Render** button to render the view.

   Currently, only the default lighting is in the scene. This rendering gives you a frame of reference for further renderings after you have placed lights.

3. To place some lights in the scene, you first need to make some room in the **Top** viewport. Zoom out in the Top viewport.

4. On the **Create** panel, click the **Lights** button [icon]. Click **Omni**.

5. In the **Top** viewport, click at the lower right of the viewport to place an Omni light. The Omni light appears in the viewport as a small diamond.

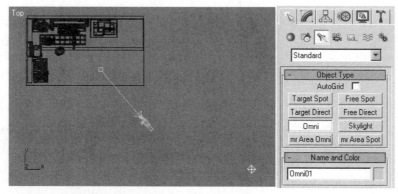

Figure 5-27: **Omni** *light created in Top viewport*

6. Render the **Camera01** view.

   In the rendering, you can see that the lighting in the scene hasn't changed very much, except the scene looks flatter, and the floor is a lot darker.

7. Click **Zoom Extents All** [icon].

8. Locate the light in the **Left** viewport. Note that the light is on the construction plane, near the floor of the attic.

9. In the Left viewport, move the light upward so it's about even with the highest part of the attic model.

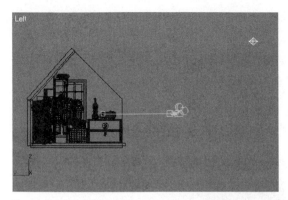

Figure 5-28: **Omni** *light moved up in* **Left** *viewport*

10. Render the **Camera01** view again.

    The floor now appears lighter. This is because the light has been moved so that it hits the faces of the floor object at a more direct angle.

Target Spot

1. Zoom out slightly in the Top viewport. On the Create panel, click **Target Spot**. In the Top viewport, click and drag from the right side of the viewport to the center of the scene to create a Target Spot light.

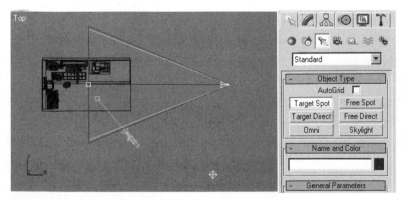

Figure 5-29: **Target Spot** *light created in Top viewport*

    The light source is placed where you first clicked, while the light's target is placed near the center of the scene.

2. Render the **Camera01** view.

    The Spot light is sitting on the construction plane and pointing straight at the far wall. For this reason, the far wall is affected strongly by the Spot light, while the other walls are not affected very much.

3.   In the **Front** viewport, move the Spot light source upward so the light points down at the scene.

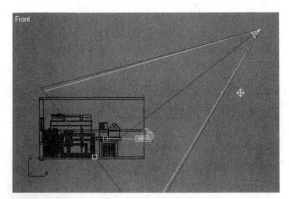

*Figure 5-30: **Spot** light moved up in Front viewport*

4.   Render the **Camera01** view.

The light still shines rather strongly on the far wall. We would like this light to shine mostly on the floor. To do this, we'll have to move the Spot light's target.

To move the target, you can pick it up and move it like any other object. If you try to do this, however, you will find that it is hard to pick the target in the scene. There are many other objects near the target, so clicking in that area is more likely to select one of the objects in the scene rather than selecting the target.

We'll get around this problem by using a *selection filter*. A selection filter allows some types of objects to be selected while others are prevented from being selected. Here, we'll use a selection filter to limit selection of objects to lights.

5.   On the **Main Toolbar**, locate the **Selection Filter** pulldown. This pulldown is currently set to **All**. Pull down the Selection Filter and choose the **Lights** option.

Setting the **Selection Filter** to **Lights** means that now you can select only lights on the screen.

6. In the Front viewport, click the Spot light target. Move the target downward until the light cone just touches the area where the far wall meets the floor.

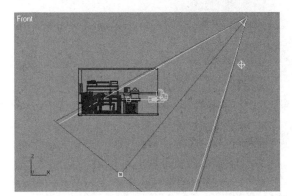

Figure 5-31: **Spot** *light target moved in Front viewport*

7. Render the **Camera01** view.

    The far wall is no longer so bright.

Direct Light

1. Pan the Top viewport over, so there is room to create a light on the left side of the viewport. On the Create panel, click **Target Direct**. Click and drag in the Top view to create a Direct light that goes through the window at an angle and to the opposite end of the room.

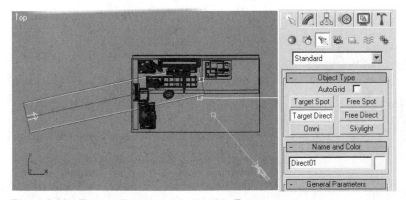

Figure 5-32: **Target Direct** *light placed in Top viewport*

2. In the **Front** viewport, move the light upward so it shines right through the window. Move the light target downward. The light should be positioned as shown in the following illustration.

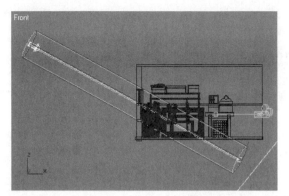

Figure 5-33: **Target Direct** *light and its target moved in Front viewport*

As you move the light, you can see the outlines of the direct light cone in the **Camera01** viewport.

3. Render the **Camera01** view. The scene now has lights in it. The lighting is far from perfect. You will learn to improve the lighting in the next exercise.

Figure 5-34: Rendered attic scene with lights

4. Save the scene in your folder, with the file name **AtticWithLights.max**.

LIGHT PARAMETERS

Each light is an object and has an object name that you can change. You can also adjust the light color and many other parameters. Just like geometric models, lights are created in the Create panel and modified in the Modify panel.

Color and Intensity

The color and intensity of a light is set by the color swatch in the **Intensity/Color/Attenuation** rollout. Click the swatch to open the 3ds Max Color Selector.

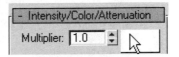

Figure 5-35: Light color swatch

When setting the color of a light, think about real-world lighting conditions. What is the color of the sun? It depends on the time of day. At noon, the sun is perfectly white, but the blue of the sky tends to cause the overall effect of bright sunlight to be slightly blue. At dusk, the color of light from the sun is a saturated orange. What is the color of a standard incandescent light bulb? It's slightly orange compared to noon sunlight. What is the color of light coming from a television screen? It is a fairly intense blue.

In general, most light sources are nearly white, with a slight blue or orange cast to them. Occasionally you will use very saturated lights, such as when simulating a stage lighting setup or a dance floor. However, most of the time your lights will have a high value and a low saturation, giving the color swatch a pastel appearance.

The color swatch is used to set the overall color of a light. In addition, each light has a **Multiplier** parameter. The value in the Multiplier field is multiplied by the color in the color swatch to determine the end result. With the default Multiplier value of 1.0, the color and intensity is solely controlled by the color swatch. If the Multiplier value is 2.0, the light will be twice as bright; if the Multiplier value is 0.5, the light will be half as bright.

The Multiplier can also be a negative number. A negative Multiplier value has the effect of removing light from the scene in the light's area of influence. This feature is handy for darkening an area of the scene that has received too much light.

Lighting Techniques

At first it may be challenging to get good results from your lighting setup. It is useful to study up on the the techniques of traditional studio lighting. Most photography or film books will have basic rules you can follow to achieve good results.

For example, the principle of *triangle lighting* is very widely known and employed throughout photography and filmmaking to light a single subject in a pleasing manner. The most significant and brightest light in a scene is called the *key* light. It is placed to one side of the camera. The key light creates the shading from dark to light on the subject, and also produces shadows.

To brighten up shadows and dark areas on the subject, a *fill* light is placed on the other side of the camera. The fill light is usually about one-third as bright as the key light. Finally, a *back* light is placed behind the subject, usually off to one side. The back light is used to enhance the edges of the subject, so that it stands out from the background.

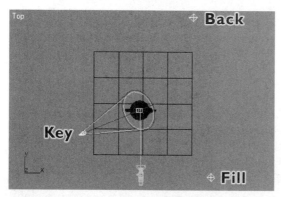

Figure 5-36: Top view of a traditional triangle lighting setup

Figure 5-37: Rendered image of the triangle lighting setup

Hotspot and Falloff

The Light Cone for a Spot light or Direct light is actually two cones in one. The inner light cone is called the **Hotspot**. In this area, the light shines at its full intensity. The outer light cone is the **Falloff** area. Outside this range, the light has an intensity of zero.

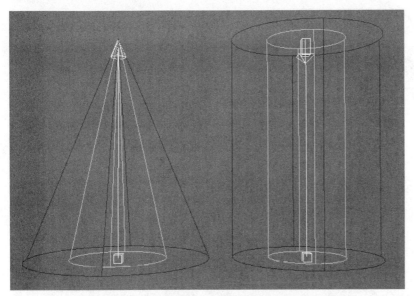

Figure 5-38: ***Hotspot*** *and* ***Falloff*** *angles of Spot and Direct lights*

Hotspot and Falloff are used in conjunction with one another. The **Hotspot/Beam** and **Falloff/Field** parameters on the Modify panel set the angles of these areas.

The difference between the Hotspot and the Falloff angles controls the softness of the light beam at its edge. The intensity of light "falls off" gradually between the Hotspot and Falloff. So, if the Hotspot and Falloff are nearly the same angle, the light beam will have a distinct, hard edge, creating a pool of light similar to a stage spotlight.

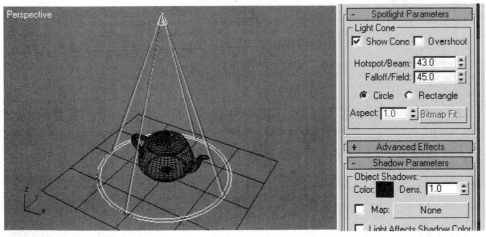

Figure 5-39: Default Spot light, with a Hotspot angle of **43** *degrees, Falloff angle of* **45** *degrees*

Figure 5-40: Rendered image showing the result of the default Hotspot and Falloff angles

If the Hotspot and Falloff angles are very different, then the transition between full intensity and zero intensity will be spread out over a larger area. The result is a much softer edge to the light.

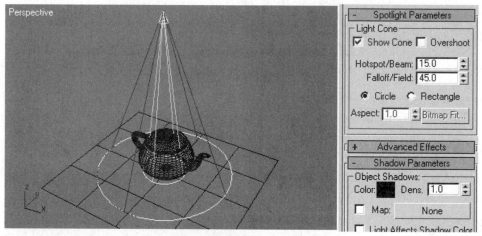

*Figure 5-41: Hotspot angle of **15** degrees, Falloff angle of **45** degrees*

*Figure 5-42: Rendered image showing a Hotspot angle of **15** degrees, Falloff angle of **45** degrees*

The Hotspot and Falloff angles have no effect on the sharpness or softness of shadows. Shadows are controlled separately.

Shadows

You allow a light to cast shadows by checking the **On** check box in the **Shadows** section of the **General Parameters** rollout. Many other shadow parameters can be found in the **Shadow Parameters** rollout. For now, we will use the default settings. Later in this chapter we will explore shadows in detail.

The darkness of shadows depends on several factors, including the intensity of the light and the amount of illumination coming from other lights.

A Spot or Direct light can only generate shadows within its Falloff radius. As you will learn later, this allows you to control the quality of shadows with great precision.

LIGHTS AND VIEWPORTS

Lights can be hidden like ordinary objects. This can be handy when you have finished working with lights for the time being and want to be able to click Zoom Extents without going to the extents of the lights, which are often far from the objects in the scene.

When a light icon is hidden, the light still affects the scene.

WARNING: Scaling lights can give very strange results. A light icon will not change shape if the light is non-uniformly scaled, but a Light Cone will change shape. However, since Omni lights don't have Light Cones, it is nearly impossible to tell if they are scaled. Don't scale lights until you become more familiar with 3ds Max.

Light Views

It is often very helpful to see what a light is aimed at, and exactly what geometry in the scene is within the light's area of illumination. You can do this by changing a viewport to a *light view*. Then you can see the scene from the point of view of the light, as if it were a camera. You also have camera-style controls for aiming the light and changing its Hotspot and Falloff parameters within a light viewport. This option is only available for Spot and Direct lights.

To change a viewport to a light view, activate a viewport and right-click its label. The viewport pop-up menu appears. Choose **Views**, and select the light's name from the list. The viewport changes to show you what the light "sees," and the viewport label displays the name of the light.

When a light view is active, the Viewport Control buttons at the bottom right of the screen change to a new set of controls. These buttons are very similar to the camera view controls, and they allow you to affect the light in a number of ways.

Light Controls

These viewport control buttons are available only when a light view is activated.

 Dolly Light moves the light toward or away from its target. A Free light doesn't have a target, but it does have a **Target Distance** parameter. This parameter sets the length of the light cone, and also sets the position of a virtual target. When used with a Free light, the Dolly Light button moves the light and its virtual target in the local Z axis.

 Dolly Target moves the target toward or away from the light. This button is a flyout from the **Dolly Light** button. It is available only for target light views.

 Dolly Light + Target moves the light and target along the light's direction. This button is a flyout from the **Dolly Light** button, and is available only for Target light views.

 Truck Light moves the light and its target (if any) in its local X and Y axes. This is the easiest way to control exactly where a light is shining.

 Light Hotspot adjusts the light's Hotspot angle interactively in a light viewport. As you move the cursor, the inner circle in the light view changes size, and the Hotspot/Beam parameter changes on the Modify panel.

 Light Falloff adjusts the light's Falloff angle interactively in the viewport. The Falloff angle is indicated in the viewport by the outer circle. This circle will remain the same while the Hotspot's inner circle appears to get larger and smaller. In fact, the Falloff angle is changing while the Hotspot remains the same size. A light viewport is always about the same size as the Falloff circle, and the viewport must retain its size on the screen. For this reason, the Hotspot will appear to change when it's really the Falloff that's changing.

 Roll Light rotates the light around its local Z axis. Unless you're using a rectangular light cone, this will have no effect on the lighting in the scene.

 Orbit Light moves a light around its target or virtual target. The target remains stationary.

 Pan Light rotates a light around its local X and Y axes. It is available on a flyout from the Orbit Light button. The light remains in the same place while the target or virtual target moves around the scene. Pan Light is often more intuitive for aiming Free lights than simply rotating them with the Rotate Gizmo.

EXERCISE 5.4: Intensity and Falloff

In this exercise, you'll adjust the intensity, falloff, and shadows of lights in the attic scene. You will adjust the light intensities to give you more contrast, and improve the overall look of the lights.

Adjusting Intensity

1. Load the file **AtticWithLights.max** from your folder or from the disc.

    This scene contains three lights and a camera.

2. Select the Direct light and note its Light Cone area. Then select the Target Spot and note its Light Cone.

3. Render the camera view. Right now, the three lights in the scene all have the same intensities and no shadows. This washes out the scene, giving you little contrast between object colors. There is also an unnatural pool of light in the center of the room.

4. Select the Target Spot light. Go to the Modify panel. On the **Intensity/Color/Attenuation** rollout, locate the **Multiplier** parameter. Change the Multiplier value to **0.8**. This reduces the intensity of the Spot light.

5. Select the Omni light. Change its Multiplier value to **0.4**. This reduces the intensity of the Omni light. The Camera view shows rough overall changes to lighting, but you can see exactly how the lighting has been affected only by rendering the scene.

6. Render the camera view. The scene now has more varying light and dark areas, but still has a pool of light in the center. This pool is caused by the cone of the direct light.

*Figure 5-43: Result after editing **Multiplier** values of Spot and Omni lights*

Sunlight Through the Window

1. Select the Direct light. Note that the size of the cone as it hits the floor is the same size as the pool of light. We want the light to shine only through the window, not through the entire wall. To cause this to happen, make the light a shadow-casting light.

2. Go to the Modify panel. On the **General Parameters** rollout, check the **On** check box in the **Shadows** section.

3. Render the camera view. The pool of light is gone, and has been replaced by a pattern of squares from the light shining through the window.

Figure 5-44: Direct light with shadow casting

4. To make sure that the Direct light is shining through the window correctly, you should look from its point of view in a light viewport. Right-click the Left viewport label. In the pop-up menu, choose **Views > Direct01**.

    It is difficult to see the scene in wireframe mode. Press the **<F3>** key to change the **Direct01** viewport to Smooth + Highlights mode.

5. Adjust the Direct light using the **Truck Light** tool in the viewport controls. You want the light to shine only into the window. The outer Falloff circle of the **Direct01** view should be positioned so that it encompasses the window, as shown in the following illustration.

Figure 5-45: Point of view from Direct01 light

Spot Light Parameters

If you look carefully at the area near the man's suit in the rendering, you can see that there is another pool of light in the same area, with the edge of the light cutting across the man's suit and/or surrounding objects. This is caused by the Spot light. The Spot light's Hotspot and Falloff angles are very close, giving the light a hard edge. The edge can be softened by increasing the difference between the Hotspot and Falloff angles.

1. Select the Spot light. On the **Spotlight Parameters** rollout, locate the **Falloff/Field** value. Increase the Falloff value to **55**.

    In the Front viewport, you can see that the Falloff cone has increased in size. This will cause the edge of the Spot light to become softer.

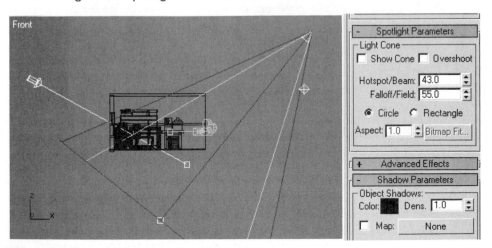

Figure 5-46: Spot light has larger Falloff cone

2. Render the camera view. The edge of the Light Cone on the suit is no longer visible.

Figure 5-47: Spot light edge no longer visible

Now you'll enable additional shadows to make the scene more realistic.

3. Select the Spot light. On the **General Parameters** rollout, check the **On** check box in the **Shadows** section.

4. Render the camera view. Shadows appear in the scene, most notably on the man's suit.

    If you were setting up the lights in this scene on your own, now would be the time when you would start experimenting by moving lights, increasing and decreasing Multiplier values, perhaps adding another light or two.

    To decrease the effect of the shadow on the man's suit, we'll decrease the intensities of the Spot light shining on it.

5. Select the Spot light. Decrease the Multiplier value to **0.5**.

6. Render the camera view. The lighting now resembles an attic in the late afternoon.

    There are a few ways the lights could be improved. These methods will be explored in the next tutorial.

7. Save the scene with the filename **AtticWithLights02.max** in your folder.

Figure 5-48: Scene with adjusted lights

WORKING WITH LIGHTS

On/Off

Lights can be turned on and off with the **On** check box on the General Parameters rollout, from the Quad Menu, or from the Light Lister, which is covered in the following section. Whe a light is turned off, the light object remains in the scene, but the light has no effect in viewports or the rendered image. The light can be turned back on at any time.

Converting Light Types

Any light can be converted to another type of light with the pulldown in the **Light Type** section of the **General Parameters** rollout. For Direct and Spot lights, you also have the option of converting between Target and Free lights by turning the **Targeted** check box on or off.

Converting a light to a different type will not change its name. You must remember to change the object name manually.

Exclude/Include

Lights can be set to include or exclude specific objects in the scene. If an object is excluded from a light, the light will not illuminate it. If an object is included, then all other objects are automatically excluded. To exclude or include objects, select the light and click **Exclude** on the **General Parameters** rollout. The **Exclude/Include** dialog appears.

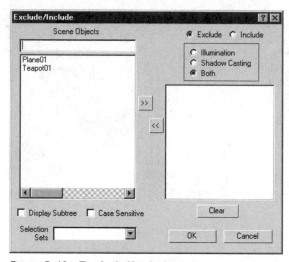

Figure 5-49: ***Exclude/Include*** *dialog*

Select the **Exclude** or **Include** option. Select objects to include or exclude, and click the right arrow button at the center of the dialog to move them to the list on the right. You can choose to exclude an object from Illumination, Shadow Casting, or Both. The effect of excluding an object from a light shows up only in a rendering, not in viewports.

NOTE: You can also control whether an object can cast or receive shadows by right-clicking the object and selecting **Properties** on the Quad Menu. In the Object Properties dialog you will find many options, including **Cast Shadows** and **Receive Shadows.** If these options are disabled in the Object Properties dialog, the object will not cast or receive shadows, regardless of any light's settings.

Light Lister

When working with numerous lights, the **Light Lister** panel comes in quite handy. It allows you to control the basic parameters of many lights all at once. It is found in the Main Menu, under **Tools > Light Lister**.

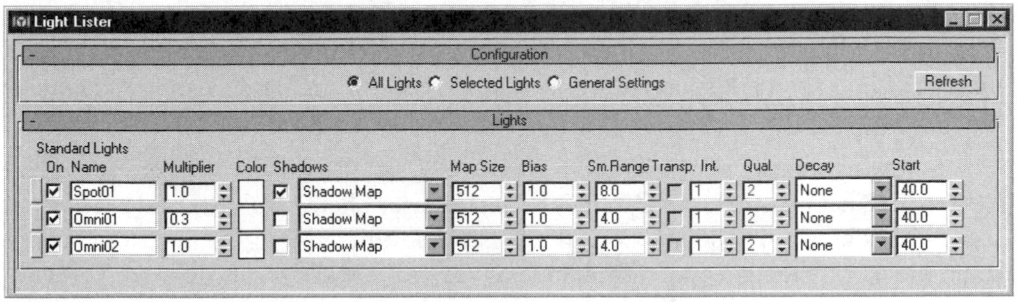

Figure 5-50: **Light Lister**

The Light Lister has many controls, arranged in rows. Each row displays the most commonly used parameters of a single light in the scene. You can change the On/Off status of a light, its object name, Multiplier value, color, shadow parameters, and so on. At the very far left of the panel are small unmarked buttons. Click one of these buttons to select a light in the scene. This makes it easier to tell which light you are controlling.

You can also use the Light Lister to control the parameters of several lights at once. This can save a lot of time. In the **Configuration** rollout, choose the **General Settings** button. In the **General Settings** rollout, you can affect the currently selected lights, or all lights at once, by choosing the appropriate button at the top of the rollout. Then, when you make changes to the parameters in the General Settings rollout, all selected lights, or all lights in the scene, will be updated with the values you have entered.

Bounced Light

In life, light bounces off surfaces to illuminate areas that don't receive light directly. An example can be found under your desk or table. Most likely, none of the light sources in the room you're sitting in right now shine directly on the floor under your desk. Yet when you look under the desk, there's sufficient light for you to be able to see what's down there. This is because light from light sources has bounced off the walls, the bottom of the desk, and the floor to eventually reach under the desk.

Although the Standard lights in 3ds Max simulate some aspects of real light sources such as shadows, they don't account for bounced light. Advanced lighting techniques such as *radiosity* do calculate the effects of bounced light. This is an example of a *global illumination* rendering technique. These techniques generally require a lot of setup, and in many cases manual control with Standard lights is quicker and easier. Radiosity is discussed in Chapter 12, *Advanced Lighting*.

In many scenes, you will have to account for bounced light by creating numerous Standard lights and placing them strategically to simulate the effect of light reflecting off of diffuse surfaces.

A good practice is to use one or two Spot or Direct lights for the key lights, or main illumination in your scene. Turn on shadows for one or both lights. Then use Omni lights to fill in the dark spots in the scene. These Omni lights will serve to simulate bounced light in the scene.

The following image uses 11 lights to simulate global illumination. There is one key light, which happens to be an Omni light in this case. There is a light inside the teapot to simulate illumination bouncing off the object. The other nine lights are placed outside the room to give the effect of light bouncing off of the walls.

Figure 5-51: Global illumination simulated with Omni lights

You will find the scene file that was created to render this image on the disc that came with this book. The file is called **OmniBounce.max**. This scene does not use any advanced lighting techniques, and it takes only a few seconds to render.

EXERCISE 5.5: Exclusion and Multipliers

In this exercise, you'll improve the attic lighting scene by working with the Multiplier parameter, and by excluding lights from geometry.

1. Load the file **AtticWithLights02.max** if it is not already on your screen.

2. Render the camera view. The lighting is pretty good, but could stand some improvement. The main problem is that the wall where the window resides is lighter than the far wall, which is unnatural when the main light is supposed to be coming through the window.

    This problem will be solved by adding another light to the scene that will exist exclusively to pull light from the window wall. This will be accomplished by giving the light a negative Multiplier value, and by including only the window wall in its effect.

3. In the Top viewport, create an Omni light directly across from the wall.

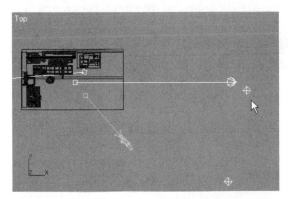

Figure 5-52: New Omni light in Top viewport

4. In the Front viewport, move the light upward so it's about the same height as the center of the wall.

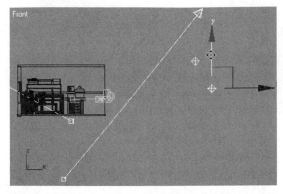

Figure 5-53: Omni light moved upward in Front viewport

5. On the Modify panel, change the **Multiplier** for the light to **-0.2**.

   When the Multiplier is a negative number, the light will remove illumination from the scene.

6. On the Modify panel, click the **Exclude** button. In the **Exclude/Include** dialog, select **Include** at the upper right. On the left side of the dialog, select the following objects:

   Pane divider bottom horizontal
   Pane divider bottom vertical
   Pane divider top horizontal
   Pane divider top vertical
   Wall window
   Window frame all
   Window frame bottom
   Window frame top

   Click the right arrow button to move these objects to the right side of the dialog.

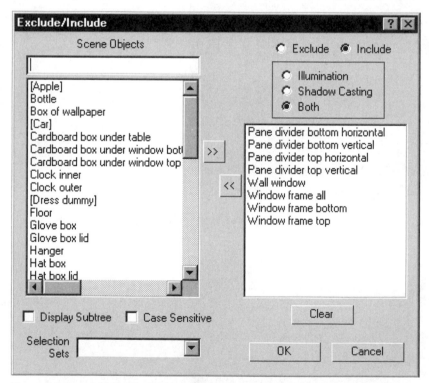

Figure 5-54: ***Exclude/Include*** *dialog with window wall objects included*

Click **OK** to exit the dialog and perform the inclusion operation. The window wall and all the window parts are now included from the Omni light with the negative Multiplier. The Omni light will only remove illumination from the window wall objects.

7.   Re-render the camera view.

Figure 5-55: Darker window wall

The window wall is now darker, without the rest of the scene becoming darker.

8.   There are currently no window glass objects in the window— it is just a blank space. Window glass objects were created for the original scene, but they are hidden.

On the **Display** panel, click **Unhide All**. The window glass objects appear in the scene.

9.   The window glass objects are named `Window glass bottom` and `Window glass top`. The objects have no material assigned to them. Open the Material Editor and find the material named **(watch) Glass**.

Rename the material to `Window Glass`.

Increase the **Reflection** amount to **50**.

In the Coordinates rollout for the `CHROMIC.JPG` reflection map, increase the **U Tiling** and **V Tiling** values to **5**.

10.   Assign the `Window Glass` material to the objects `Window glass bottom` and `Window glass top`.

11.   Re-render the camera view.

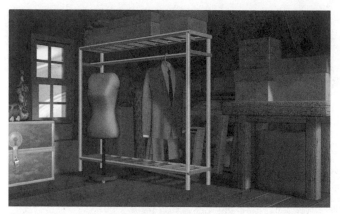

Figure 5-56: Reflective window glass

There are no longer dappled light squares on the floor. This is because the window glass is blocking the Direct light from coming through the window. Even though the **Window Glass** material has an Opacity of only **15%**, 3ds Max still treats the window glass objects as completely opaque objects, allowing no light to pass through. This is because the Direct light uses a type of shadow called a **Shadow Map**, which does not account for transparency. You'll learn about Shadow Maps later. Meanwhile, to solve this problem you can exclude the window glass from the Direct light.

12.   Select the Direct light. On the Modify panel, click the **Exclude** button. When the **Exclude/Include** dialog appears, select the objects **Window glass top** and **Window glass bottom,** and click the right arrow to move them to the right side of the dialog. Click OK to set the exclusion.

13.   Render the **Camera01** view. The dappled light reappears, because the light is no longer blocked by the window glass.

Figure 5-57: Glass no longer blocking Direct light

14. Next, we'd like to brighten up the sunlight coming through the window. This light's color swatch is already at its maximum, so the only way to brighten the light is to increase the **Multiplier** parameter.

Select the Direct light and increase its Multiplier to **1.5**.

15. Re-render the camera view. The sunlight on the floor is now a little brighter.

Figure 5-58: Brighter Direct light

16. Save the scene with the filename **AtticWithLights03.max** in your folder.

Attenuation

Attenuation is a method of varying the intensity of a light source over distance. In life, all light behaves this way. Light energy becomes dimmer as it spreads out away from a light source. So, the farther an object is from a light source, the less illumination it will receive. In computer graphics, attenuation is very useful for simulating realistic interior lighting.

Attenuation is set up by specifying a series of distances from the light source. These distances are called *ranges*. Attenuation ranges are displayed as spheres around an Omni light, and as discs in front of a Spot or Direct light.

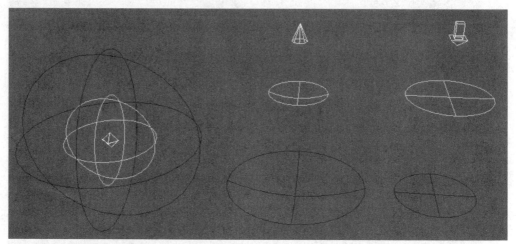

Figure 5-59: Attenuation ranges for Omni, Spot, and Direct lights

Each of these range icons corresponds to a parameter found on the **Intensity/Color/Attenuation** rollout. The attenuation range parameters indicate the distance from the light source in 3ds Max units.

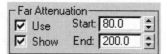

*Figure 5-60: **Far Attenuation** parameters*

There are two types of attenuation: **Near** and **Far**. Far Attenuation is much more useful. It controls the loss of brightness over distance. The light dims out between the **Start** range and the **End** range. The light is at full brightness until it reaches the Start range, then it gradually fades out to zero brightness. At the End range, the light has faded out to nothing.

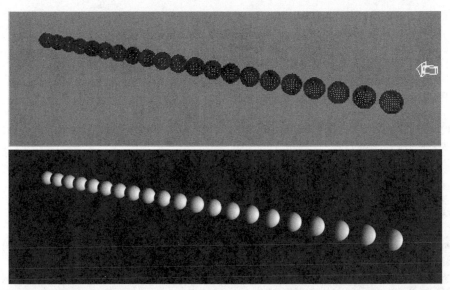

Figure 5-61: *Direct light with no Attenuation*

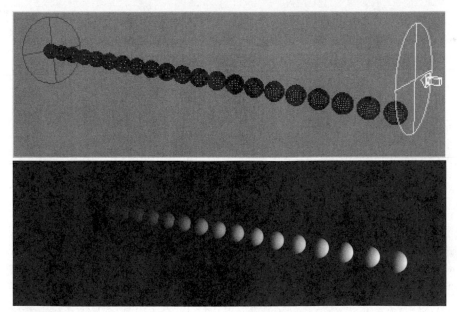

Figure 5-62: *Direct light with gradual* **Far Attenuation**

In the preceding images, the effect of Far Attenuation is clearly demonstrated. The first pair of images shows a Direct light with no attenuation at all. The second set of images shows the result of placing the Far Attenuation ranges at the extents of the geometry in the scene. The light fades out completely when it reaches the last sphere in the row.

To enable attenuation, click the **Use** check box. If you wish to see the range icons even when the object is not selected, activate the **Show** check box.

Near Attenuation is the opposite of Far Attenuation. It causes the light to fade out as it gets closer to the source. This is really only useful in two cases. First, Near Attenuation prevents an object like a lampshade from being overlit by a light inside it. Second, if you're using a negative Multiplier, you might want to remove more light from the scene as distance from the light source increases. Near and Far Attenuation can be used in combination with one another.

In review, attenuation has the effect of changing the brightness of a light source over distance, as indicated in the following:

Near Start range:	Zero intensity
Near End range:	Gradually increase to full intensity
Far Start range:	Full intensity
Far End range:	Gradually decrease to zero intensity

WARNING: Attenuation increases rendering time. Use it only when needed.

Decay

The **Decay** feature of lights attempts to create a more physically accurate simulation of how light dissipates over distance. However, in practice, the results of using the Decay feature are very hard to predict. You are better off using the Attenuation feature when working with Standard lights, because you have more control.

SHADOWS

Working with shadows is an art in itself. There are many things to consider if you want to achieve realistic renderings, and shadows are an important factor. Should the shadows in a scene be fuzzy or sharp? Should they be deep, dark shadows, or should they be lighter in color to simulate the effects of bounced light?

In addition, there are numerous technical considerations in working with shadows. It is extremely common to see jagged, aliased shadows in renderings made by beginners. If you understand the peculiarities of how shadows work in a 3D program, you are well on your way to making professional-looking renderings.

There are several types of shadows in 3ds Max. These types can be selected from the pulldown list in the **Shadows** section of the **General Parameters** rollout. This chapter will cover the most commonly used shadow types, **Ray Traced** and **Shadow Map**. Other shadow types are discussed in Chapter 12, *Advanced Lighting*.

In general, the focus of your work with shadows will be making them look the way you want, while keeping the rendering time as short as possible.

RAY TRACED SHADOWS

Ray Traced shadows work by tracing the paths of rays projected from the light source into the scene. In 3ds Max, standard Ray Traced shadows always have sharp edges. They are useful for renderings of outdoor scenes in bright sunlight, or for light coming in through a window. Ray Traced shadows respond accurately to the opacity of objects: a highly transparent material will allow light to pass through it, and won't generate a dark shadow.

Ray Traced shadows usually work well without much fuss. However, they are not appropriate for most scenes. Look at the shadows in the room around you. These shadows are most likely quite soft and fuzzy. If your goal is to create soft shadows, such as those seen in an indoor shot, or in an outdoor scene on a cloudy day, then standard Ray Traced shadows will not work. You will need to use one of the other shadow types, such as Shadow Maps.

In addition, Ray Traced shadows usually take longer to render than Shadow Maps. When rendering a long animated sequence, the render time for shadows is an important consideration. Experienced users make use of Shadow Maps as much as possible, and only use Ray Traced shadows when absolutely necessary.

SHADOW MAPS

Shadow Maps are used more commonly than Ray Traced shadows, because they are more versatile. They can achieve soft or hard edges, and are often faster to render than Ray Traced shadows. The main drawback of Shadow Maps is that they do not respect the **Opacity** parameter of object materials. Even a material with an Opacity of zero will still cast a Shadow Map.

Shadow Maps work by projecting a bitmap from the light onto the scene. Like any bitmap, a Shadow map has a resolution. In 3ds Max, the resolution of a Shadow Map is determined by the **Size** parameter.

Size

Size is the single most important parameter of a Shadow Map. In general, the larger the Size is, the sharper the shadow will be.

Size sets the resolution of the bitmap used to calculate the shadow, in pixels. For example, the default Size of 512 creates a Shadow Map with a resolution of 512x512 pixels.

If the Size is too small for the map to make a decent approximation of the shadow, the shadow won't look very good. Symptoms of the Size being too small are:

- The shadow edges appear very blocky (not smooth).
- The shadow appears as a series of rough lines across an object.
- The shadow does not appear at all.

In the following renderings, the **Size** was set to **64** for the rendering on the left, then was increased to **1024** for the rendering on the right. The pixels of the low-resolution Shadow Map are extremely obvious and unpleasant.

*Figure 5-63: Shadow with Size of **64**, then with Size of **1024***

The only way to know if the Size is large enough is to render the scene and look at the shadow edges. You can then try larger Size values and see if the shadow looks different or better. The Size parameter works best if the number you enter is a power of two, such as **64, 128, 256, 512, 1024**, and so on. For your final rendering, set the Size parameter as low as you possibly can while still retaining a good-looking shadow.

How Big Should the Size Go?

A Size of **1024** should be sufficient for most of your purposes. A Size of **4096** is about as high as you should go. Any higher than this will increase your rendering time dramatically.

Increasing the Size parameter can greatly increase your rendering time by using up all the available memory. The amount of memory needed for a Shadow Map can be found by multiplying the Size by itself, then multiplying by four. For example, a Size of **4096** would require 4096x4096x4 bytes, or 64 MB RAM.

Depending on the amount of RAM you have and the number of shadow-casting lights you use, you can easily exceed the amount of available memory, which will cause 3ds Max to start using the hard disk to store data during rendering. This situation immediately doubles your rendering time. For this reason, you should exercise care when increasing the Size parameter.

If you find that a Size above **4096** is required to get a decent shadow, then you should adjust the scene to decrease the Size of the map needed.

Optimizing the Scene for Shadow Maps

The Size of the Shadow Map needed to create a clean shadow is determined by three primary factors:

- The distance between the light and the shadow-casting objects. The farther the light is from the objects, the larger the Size will have to be.

- The size of the light's Falloff area. The area covered by a Shadow Map is determined by the Falloff area. A large Falloff angle means a larger Size will be needed to make a clean shadow. A smaller Falloff angle will allow you to reduce the Size parameter.

- The resolution of the rendered image. The shadows in an image rendered at 320x240 might look fine, but when the image is rendered at 640x480, a larger Size might be needed.

To keep the Size low, use these guidelines to change the scene.

- Move the light source closer to the objects.

- Decrease the Falloff angle of the light.

- If you are using an Omni light to cast shadows, change it to a Spot or Direct light. The Size of a Shadow Map cast by an Omni light must be about six times the Size of a Spot or Direct light in order to get comparable shadows.

It is worth your time to find the lowest Size value you can use for your scene while still getting the shadow to look the way you want it to. An hour or two spent experimenting with the Size value will pay you back with a much lower overall rendering time for an animated sequence.

Overshoot

Spot and Direct lights have a handy feature called **Overshoot**, located in the **Spotlight Parameters** or **Directional Parameters** rollout. Overshoot makes a Spot or Direct light illuminate an entire scene, in a similar manner to an Omni light.

You might wonder why such a feature would be necessary. In fact, Overshoot is extremely useful. First, it allows a Direct light to shine parallel rays throughout the scene, without having to make the Falloff radius the size of the whole scene. This is perfect for sunlight.

In addition, Overshoot lets you optimize your Shadow Maps. You can reduce the Falloff of a Spot or Direct light so that the Falloff radius only encompasses the objects that need to cast shadows. Normally this will create a pool of light, because geometry outside the Falloff will receive no illumination. However, with Overshoot turned on, you can also illuminate the rest of the scene from the same light. You get the best of both worlds: the Shadow Map size can remain small without the limitation of a visible pool of light.

EXERCISE 5.6: Shadow Map Size

In this exercise, you'll practice working with the Size parameter for Shadow Maps.

Setting the Scene

1. Reset 3ds Max.

2. In the Top viewport, create a Plane with a **Length** of **200** and a **Width** of **300**.

3. In the Top viewport, create a Cylinder with a **Radius** of **20** and a **Height** of **65**. Move the Cylinder so it sits at the center of the plane.

4. Place a camera in the scene to view the scene from the front. Change the Perspective view to the camera view. Your camera view should look similar to the following image.

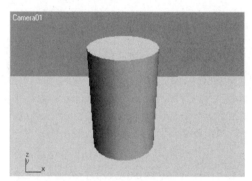

Figure 5-64: Camera view of scene

5. Assign a light gray or white material to both objects.

6. Zoom out of the Top viewport. Place an **Omni** light at the lower-left corner of the **Top** viewport. Place a **Target Spot** light at the lower right of the Top viewport, with its target at the center of the Cylinder. Move the lights upward in the Front or Left viewport so they shine down at the scene at about a 45-degree angle, as shown in the following illustration.

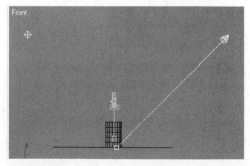

Figure 5-65: Light and camera placement as seen in the Front view

7. Select the Omni light. In the **Intensity/Color/Attenuation** rollout, change the light's **Multiplier** value to **0.3**.

Shadow Map Size

1. Select the Spot light. In the **Shadows** section of the **General Parameters** rollout, check the **On** check box to enable shadow casting.

2. On the **Shadow Map Params** rollout, set the **Size** parameter to **64**.

3. Render the camera view with a resolution of 640 x 480.

   The shadow is so chunky that you can see the square blocks of the Shadow Map around the base of the cylinder.

*Figure 5-66: Shadow Map with **Size** of 64*

4. Increase the Shadow Map **Size** to **256**. Render the camera view.

   The shadow is smoother.

5. Increase the Size to **512**, and render the camera view. Then increase it to **1024** and render the camera view again. Each time you increase the **Size** parameter, the shadow appears sharper.

6. Set the Size back to **512**.

Spot Light Distance

Next you'll test the effect of increasing the distance from the spotlight to the shadow-casting object.

1. Right-click the **Front** viewport label. In the pop-up menu, choose **Views > Spot01.**

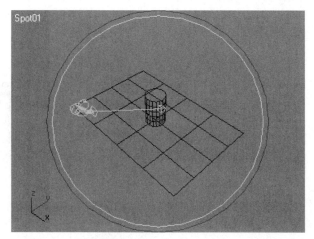

Figure 5-67: **Spot01** *viewport*

2. Use the **Dolly Light** [⚓] viewport control to move the Spot light away from the scene. Watch the number near the top of the **General Parameters** rollout. This is the **Target Distance**. Dolly the light back until the Target Distance is about **3000** units.

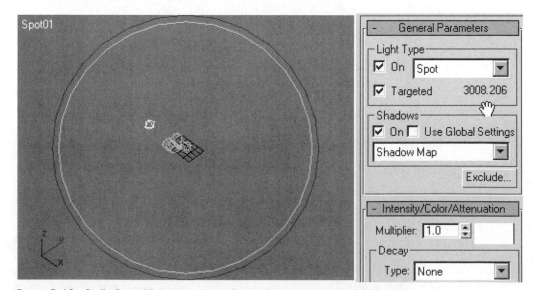

Figure 5-68: Dolly **Spot01** *back until the Target Distance is about* **3000** *units*

3.   Render the camera view. The shadow is blocky again. This is because the light is very far from the objects in the scene.

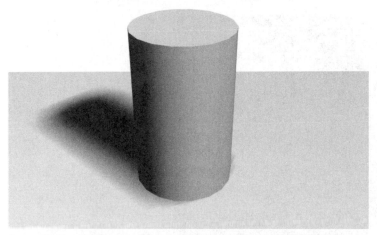

*Figure 5-69: Shadow Map Size of **512**; Spot light dollied back*

4.   Change the Shadow Map Size to **4096**. Render the camera view. The scene takes a little longer to render, and the shadow is sharper.

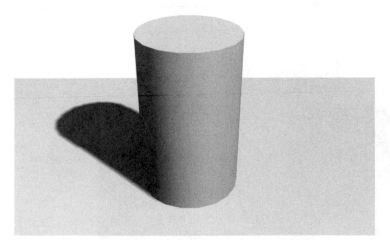

*Figure 5-70: Shadow Map Size of **4096**; Spot light dollied back*

5.   Change the Shadow Map Size to **32**. Render the camera view. The Shadow Map Size is so small that no shadow appears at all, except for perhaps a few dark squares.

6.   Change the Shadow Map Size to **1024**. Dolly the Spot light forward so that the Falloff radius is just larger than the cylinder. Use the **Truck Light** viewport control 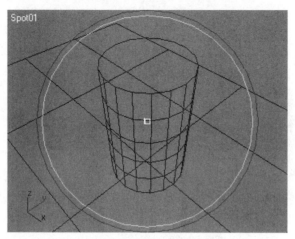 to center the cylinder within the Falloff radius.

Figure 5-71: Dolly and Truck the Spot light to center it on the Cylinder

7.   Render the camera view. The shadow edge is the best quality seen so far, but there is an obvious pool of light. Also, shading on the cylinder is different, and the shape of the shadow is distorted. This is because the Spot light is too close to the cylinder.

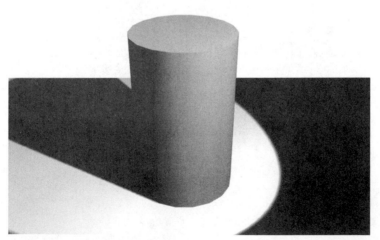

Figure 5-72: Spot light moved too close to the Cylinder

Falloff and Overshoot

1. Dolly the Spot light back so that it is at approximately the same height as the Omni light. Use the **Light Falloff** ⊚ and **Light Hotspot** ⊚ viewport controls to make the Falloff and Hotspot circles slightly larger than the Cylinder.

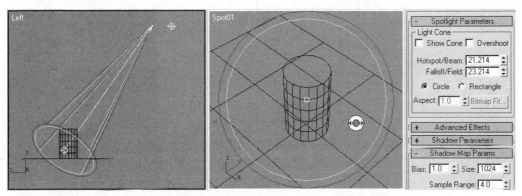

*Figure 5-73: Using the **Light Falloff** viewport control*

2. Render the camera view. The shadow has a nice, clean edge, but the pool of light is still present.

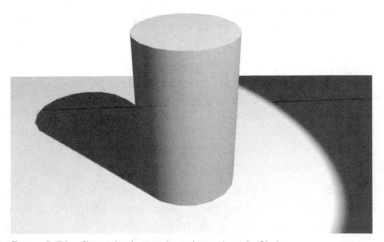

Figure 5-74: Clean shadow with undesired pool of light

3. In the Spotlight Parameters rollout, turn on the **Overshoot** check box. Render the camera view. The entire scene is lit, but the shadow hasn't changed.

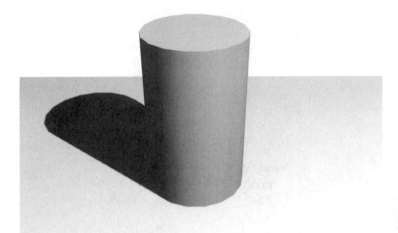

Figure 5-75: Overshoot

4. You may notice a little bit of blockiness on the Cylinder. This is the Shadow Map appearing where it shouldn't. To fix this, look in the **Shadow Map Params** rollout. Change the **Bias** parameter to **2** and re-render the scene.

5. Save the file as **ShadowMapOvershoot.max** in your folder.

Sample Range

Sample Range is essentially a blur factor for Shadow Maps. Higher values cause the shadow map to be sampled more times, resulting in a softer edge. Lower values produce crisper shadows, but if the Size parameter is low, then a low Sample Range will result in a jagged, aliased edge. The default Sample Range is **4**, which is a good balance, and a good starting point for customizing your Shadow Map edges.

A high Sample Range value can make your render times much longer, especially if there are many lights in the scene. Although the Sample Range parameter goes up to a maximum value of **50**, a value of **10** or higher will increase render times noticeably. As with the Size parameter, you should use the lowest Sample Range value you can while still retaining the desired look of the shadows.

When customizing Shadow Maps, work with the Size and Sample Range parameters to make the shadows appear as sharp or fuzzy as you like. For soft, diffuse shadows, use a low Size, such as **256**, and a high Sample Range, such as **10**. For sharp, crisp shadows like those created by Ray Traced shadows, use a high Size, such as **2048**, and a low Sample Range, such as **2** or **4**.

EXERCISE 5.7: Sample Range

1. Load the file **ShadowMapOvershoot.max** from your folder or from the disc.

    This is a simple scene containing a Plane and a Cylinder, which you made in the previous exercise.

2. Select the Spot light and go to the Modify panel. On the **Shadow Map Params** roll-out, set the spotlight's **Size** parameter to **512**. Render the camera view.

3. Set the **Sample Range** to **1, 2, 4, 8**, and **12**. Render the scene after each change. Observe the changes to the shadow.

    Higher Sample Range values make the shadow fuzzier and also make it smoother.

4. Set the Sample Range parameter to **1**. Set the Size parameter to **64**. Render the camera view.

5. An interesting blocky shadow with somewhat sharp edges is generated.

6. Set the Size parameter to **2048**. Render the camera view.

7. The shadow is sharp and smooth.

8. Experiment with different combinations of the Size and Sample Range parameters until you feel comfortable working with them.

9. Save the scene as **SampleRange.max** in your folder.

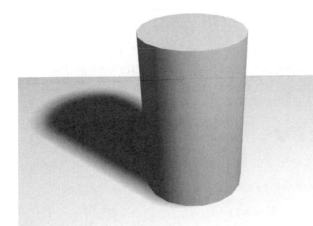

Figure 5-76: Shadow Map Size of 256; Sample Range of 15

Shadow Bias

Sometimes shadows appear to be shifted slightly toward or away from the object casting the shadow. This problem is only noticeable when the object casting the shadow is sitting against the object receiving the shadow, as with a vase sitting on a table. Even worse, an object can sometimes cast a shadow on itself incorrectly. This was seen on the cylinder at the end of Exercise 5.6.

The most obvious problems occur with thin objects, such as the legs of a chair on a floor. If the legs of the chair are sitting right on the floor, the shadows should touch the legs. If they don't, then the **Bias** parameter needs to be changed.

Lower Bias values pull the shadow closer to the object casting the shadow, while higher values push the shadows away from the object. The default value for the Bias is 1, which is usually correct.

The only time you should change the Bias is if there is a problem. It is difficult to predict when that might happen. You won't have to change the Bias very often, but it's important to know about it for those times when you need it.

The **Ray Bias** parameter is available for Ray Traced shadows. It is found on the **Ray Traced Shadow Params** rollout, and it works the same as the Bias parameter does with Shadow Maps.

In a rendered animation, the edges of Shadow Maps sometimes appear to flicker due to recalculation of the Bias on each frame. When the **Absolute Map Bias** check box is enabled, calculations of the map are stabilized throughout an animation. Check this check box only if shadows flicker in a rendered animation.

Note that there can be other causes of flickering shadows. A low Shadow Map Size can also be the cause of this problem. Adjust the Size parameter first. If that doesn't fix the flickering at the edges of shadows, then try Absolute Map Bias.

EXERCISE 5.8 Shadow Bias

In this exercise, you'll work with the Bias parameter.

1. Load the file **ShadowMapOvershoot.max** from your folder or from the disc.

2. Select the Cylinder, and change its **Radius** to 2.

3. Select the Spot light. Change the parameters in the **Shadow Map Params** rollout to the following values:

Size	512
Sample Range	4
Bias	1

4. Render the camera view.

The Cylinder's shadow touches the Cylinder, as it should.

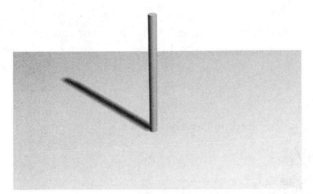

Figure 5-77: Shadow touching object

5. Change the **Bias** parameter to **5**. Render the camera view.

    The Cylinder's shadow is pushed away from the object, making the Cylinder look as if it's floating above the box.

Figure 5-78: Shadow with Bias of 5

6. Change the Cylinder's **Radius** to **20**. Render the camera view.

    Because the object is thicker, the shifted shadow is not noticeable.

7. Save the scene as **ShadowBias.max** in your folder.

GLOBAL SHADOW SETTINGS

To use the same shadow parameters for two or more lights, you can check the **Use Global Settings** check box.

When you check this check box and there are no other lights in the scene with Use Global Settings checked, nothing happens immediately. However, if you select another light and check Use Global Settings, then the selected light's settings change to match the other light for which Use Global Settings is checked.

The Use Global Settings check box affects the shadow type and the values in the **Shadow Map Params** or **Ray Traced Shadow Params** rollouts.

WHERE'S THE SHADOW?

If shadows don't show up in the scene when you expect them to, there could be many reasons. Use this checklist to get the shadows to appear.

- Make sure **On** is checked in the Shadows section of the General Parameters rollout.

- If other lights in the scene are too bright, they'll wash out the shadow. Try turning off one or more lights and see if the shadow appears.

- Shadows don't appear on a material with medium to high Self-Illumination. Lower the Self-Illumination value to make shadows appear.

- If the light is nearly parallel to the surface that is supposed to receive shadows, the shadows will not appear. Move the light to a higher angle.

- If the light cone is very large and you are using Shadow Maps, the Size parameter on the Shadow Map Params rollout must also be very large. Try doubling or quadrupling the Size parameter. If that doesn't work, try Ray Traced shadows.

LIGHTING EFFECTS

There are a few ways you can customize your lights to make special effects. Some of these effects increase rendering time significantly, and so should be used only when you need them.

PROJECTOR MAP

A **Projector Map** is a bitmap or other map projected by the light onto the scene. This feature is useful for simulating the light from a movie projector. It can also be used to create the illusion of shadows cast from objects that are not visible to the camera, such as trees overhead. This takes less time to render than putting actual shadow-casting objects in the scene, and it also saves you the trouble of modeling these objects.

Figure 5-79: Rendered scene without and with a projector, and the bitmap used for the projector

To use a map as a projector, go to the Material Editor and select an unused sample slot. Click **Get Material** and choose a map from the Material/Map Browser. Set up the map as desired.

To assign the map as a projector, look in the **Advanced Effects** rollout of the light. Click and drag the map from the Material Editor slot to the button in the **Projector Map** section of the Advanced Effects rollout.

After you click and drag the map to the button, a pop-up dialog appears. Choose **Instance**, and click OK. In this way, you can change the map in the Material Editor to update the projector map.

The effects of Projector Maps are visible only in renderings, not in viewports. By default, a Projector Map is visible in a rendering only where the light strikes an object. Just like shadows, Projector Maps can only be seen within the Falloff radius of a Spot or Direct light.

AFFECT SURFACES

In real life, lights have different effects on the surfaces of objects. Some lights are focused, others are more diffuse. Diffuse light means that the light is more scattered. The 3ds Max lighting parameters we have covered so far can go a long way toward making convincing renderings, but there are still more tools at your disposal. One of the most important is the **Affect Surfaces** section of the **Advanced Effect**s rollout.

Material Component Illumination

In real-world studio photography and cinematography, the fill light is usually very diffuse. It is usually undesirable for the fill light to generate a highlight or hotspot on the surface of the subject. To scatter the illumination coming from the fill light, a large piece of translucent plastic or cloth is placed in front of the fill light. This is called a diffusion gel in lighting terminology. The result is that the fill light illuminates the subject, filling in the shadows and dark areas of the surface, without generating a distracting specular highlight.

To achieve the effect of a very diffused light source, you can disable the **Specular** check box in the Advanced Effects rollout for the light. The light will only affect the Ambient and Diffuse material components of any object it illuminates.

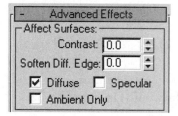

Figure 5-80: **Advanced Effects** *rollout with* **Specular** *illumination disabled*

In the following illustration, there are two lights. The key light is a Spot light on the right side of the camera, and the fill light is an Omni light on the left side of the camera. In the first image, the Specular material component has been enabled for both lights, and there are two distinct series of highlights on the object. In the second image, the Specular component of the Omni fill light is disabled. The second image shows a classic example of a highly diffused fill light that generates no specular highlights.

Figure 5-81: Omni fill light to the left of the camera, with and without Specular highlights

You can also use the check boxes in this section of the Advanced Effects rollout to help control the placement of highlights on the surfaces of objects. To do this, turn off the Diffuse check box and turn on the Specular check box. Then the light will *only* generate specular highlights, and you can move the light to where you need it in order to place the highlight. You can then use another light to illuminate the Diffuse component.

Contrast

Real-world lights and cameras have interesting properties that happen naturally, like contrast. For example, a tightly focused searchlight will produce extremely bright illumination where it strikes an object directly, but it won't contribute much to the Ambient component of the object. The difference between the Diffuse and Ambient areas of the object will be very distinct.

This is an example of high contrast. The contrast of an image is determined by many factors, including the focus of lights, the type of film being used, and the atmosphere. Space scenes are usually very contrasty, because there is no air in space to scatter the light.

The contrast level of a scene can be controlled by lights. Simply increase the **Contrast** parameter in the Advanced Effects rollout of the light. In the two images that follow, the intensity of the light is the same, but the Contrast parameter has been changed.

*Figure 5- 82: Key light with a Contrast of **0**, and with a Contrast of **80***

Soften Diffuse Edge

For more precise control over the focus of a light, you can use the **Soften Diffuse Edge** parameter. It controls the sharpness of the transition between the Diffuse and Ambient components of an object. This provides more realism than you can achieve by simply disabling the Specular highlights of a light.

In the following illustration, the Soften Diffuse Edge parameter is used to simulate a very soft key light, such as sunlight on a cloudy day. The first image uses the default Soften Diffuse Edge value of zero. In the second image, Soften Diffuse Edge has been increased to its maximum value of 100. Note the lack of a distinct transition between the Diffuse and Ambient areas of the object.

Figure 5-83: Soften Diffuse Edge value of 0, then increased to 100

ADVANCED LIGHTING AND RENDERING

There are several techniques you can employ to even further enhance the realism of your scenes, even to the point of *photorealism*. Photorealism is a word used to describe a CG scene that is so convincing that it can hardly be distinguished from a photograph. These techniques will be covered in later chapters. Just to give you a taste of what's ahead, the advanced lighting and rendering techniques fall into these main categories:

Atmospheres: Light passing through air, such as visible shafts of light and distance haze. Atmospheres can also be used to create clouds and fire.

Render Effects: Approximation of the effects of a real camera lens, such as flares, glows, streaks, and de-focusing.

Ray Tracing: Tracking simulated light rays as they move through a scene. Ray tracing produces the most convincing reflections and refractions. *Refraction* is the effect of light bending as it passes through a transparent object.

Photometric Lights: Accurate physical simulation of lighting intensity, based on measurements of real lights in real environments.

Light Tracer: Advanced lighting technique that yields realistic renderings of diffuse surfaces with little setup time. Great for scenes with a lot of ambient light, like outdoor sunny days.

Radiosity: Physics-based simulation of how light bounces around in a scene. Excellent for realistic renderings of interiors.

Atmospheres and Render Effects are discussed in Chapter 8, *Special Effects*. Ray tracing is implemented as a material in 3ds Max, so it is covered in Chapter 11, *Advanced Materials*. Photometric Lights, Light Tracer, and Radiosity are the focus of Chapter 12, *Advanced Lighting*.

END-OF-CHAPTER EXERCISES

EXERCISE 5.9: Table Scene

In this exercise, you'll set up a camera and lighting for the table scene.

1. Load the file **TableScene09.max** from the disc.

    This is the latest version of the table scene you have been working on throughout this book.

2. Zoom out in the Top viewport. Place a target camera to view the scene from an angle.

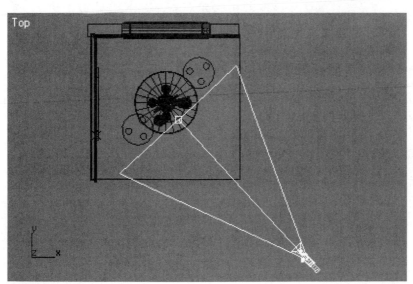

Figure 5-84: Camera placed in table scene

3. Activate the **Perspective** view and press the **<C>** key. The viewport changes to the **Camera01** view.

4. Turn on **Show Safe Frame** in the **Camera01** viewport right-click menu.

5. Adjust the camera until the Camera01 view looks similar to the following illustration.

Figure 5-85: Rendering of the Camera view of table scene

6. In the Top viewport, create a **Free Spot** light centered on the light fixture on the ceiling. In the Front viewport, move the light up so it originates just below the light fixture, as shown in the following illustration.

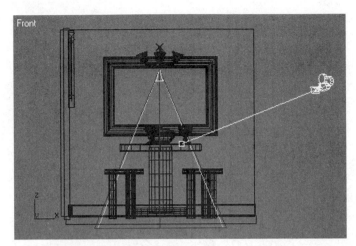

Figure 5-86: Spot light from chandelier

7. Set the light's **Multiplier** to **0.7**, and enable shadow casting by checking the **On** check box. Adjust the light's **Falloff** parameter so it is about twice as wide as the **Hotspot** angle, as shown in the following image.

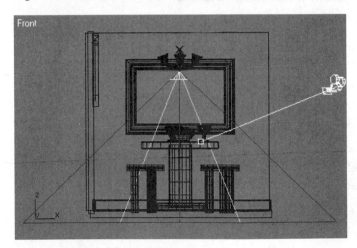

*Figure 5-87: Adjust the **Falloff** angle of the Spot light*

8. Exclude the light fixture from the light source so the light won't be obstructed by the light fixture. To do this, click **Exclude** on the **General Parameters** rollout to access the **Exclude/Include** dialog. Select all of the **LampLoft** and **Light Bulb** light fixture objects from the list on the left. Click the right arrow button to move them to the right side of the dialog. Click OK to set the exclusion.

9. In the Top viewport, place two Omni lights on either side of the camera, as shown in the following illustration. Set the Multiplier value for each light to **0.4**.

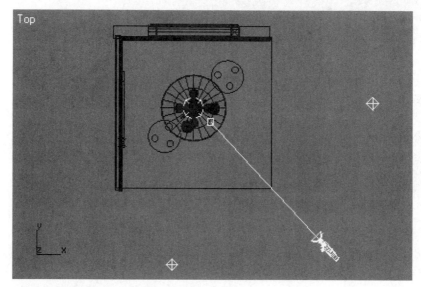

Figure 5-88: Omni lights placed around camera

10. In the Front viewport, move the lights upward so they're between the table and the ceiling area.

11. Create a Direct light coming through the window at an angle to simulate sunlight. Set the Direct light's Multiplier value to **0.8**, and enable shadows. Select the Omni light facing the wall with the sunlit window, and change its Multiplier value to **0.2**.

12. Render the Camera01 view.

13. Adjust materials and other elements as necessary. Experiment with the light intensities and positions until you are satisfied with the scene.

14. Save the scene with the filename **TableScene10.max** in your folder.

Hints

- It can sometimes be hard to select the camera's target without selecting other objects in the scene. You can get around this problem in a number of ways. One is to select the target line between the camera and target, which selects both camera and target together. You can then move both at once.

- To move both the camera and target at the same time, you can use the **Truck Camera** tool  in the camera view.

- At any time, you can change the **Selection Filter** on the Main Toolbar to **Cameras** or **Lights**. This will allow you to select only cameras or lights in the scene, and not geometric models.

Figure 5-89: Rendered scene

EXERCISE 5.10: Space Scene

In this exercise, you'll set up a camera and a light for the space scene you made earlier.

1. Load the file **SpaceScene02.max** from your folder or from the disc.

2. Right-click the Perspective view to activate it. Then choose **Views > Create Camera from View** from the Main Toolbar.

   A Target Camera is created, and both the camera and its target move to match the Perspective view.

3. In the Perspective viewport, press the **<C>** key to activate the **Camera01** viewport. The viewport label changes to indicate that we are looking through the Camera01 virtual lens instead of the Perspective viewport.

   In space, there is no atmosphere to scatter the light. This results in very deep shadows and very sharp contrast. We will create this effect with a single Direct light, designed to reproduce the light from a nearby sun.

4. In the **Top** viewport, place a Direct light in the scene as shown in the following illustration.

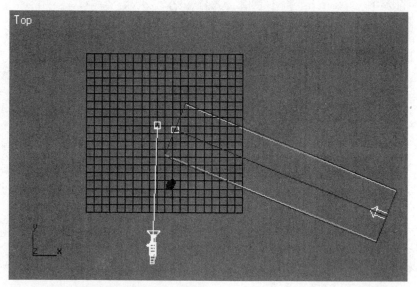

Figure 5-90: Direct light in space scene

5. In the **Front** viewport, move the light upward so it shines down at the terrain at an angle.

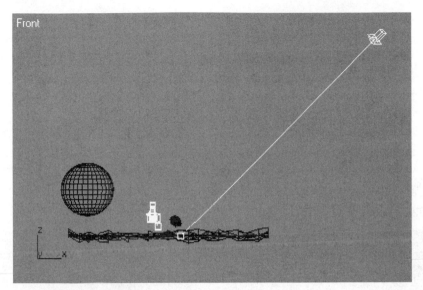

Figure 5-91: Direct light at an angle

6. Render the camera view. There is a single pool of light in the center of the terrain.

7. Select the Left viewport. Right-click the viewport label and choose **Views > Direct01**.

    The viewport changes to show the Direct light's point of view.

8. Increase the **Falloff** and **Hotspot** of the Direct light to encompass the terrain, but not the planet in the background, as shown in the following figure.

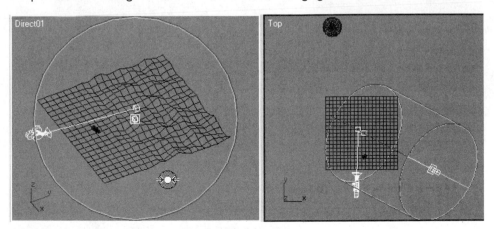

*Figure 5-92: Increase **Falloff** and **Hotspot** of Direct light*

9. Turn on shadow casting for the Direct light. Change the shadow type to **Ray Traced Shadows**. Render the camera view.

    The planet in the background is not receiving any light.

10. Activate **Overshoot** in the **Directional Parameters** rollout. Re-render the camera view. Now the planet in the background is illuminated.

11. In the **Advanced Effects** rollout, increase the **Contrast** to **50**, to reproduce the harsh lighting of outer space. Notice how the edge of the illuminated area of the planet in the background becomes sharper and more distinct.

12. Place a single **Omni** light below the surface of the terrain, near the falling asteriod. Use the **Exclude** list to Include only the asteroid. This will give the effect of light bouncing off of the terrain and illuminating the asteroid from below. Turn the **Multiplier** of the Omni light down to **0.4**. Render the scene.

13. Continue to experiment with lights and materials until you get the look you desire.

Figure 5-93: Space scene

14. Render the scene and save it as **SpaceScene03.tga** in your folder. Save the scene with the filename **SpaceScene03.max** in your folder.

SUMMARY

Cameras and lights are placed and moved like any other object in the scene. A viewport can be changed to show what a camera or light sees. The viewport controls at the lower right of the screen can be used to control the view.

Each camera has a *focal length* and a *field of view* associated with it. These two parameters work together to produce different perspective effects. As the focal length increases, the field of view decreases, and perspective tends to flatten out. This is called a *telephoto* shot. As the focal length decreases, the field of view increases, and perspective becomes exaggerated, for a *wide-angle* shot.

All images, including camera views, have an aspect ratio, which is the proportion of image width to height. To be certain that the aspect ratio you see in a viewport is the same as the rendered image, activate **Show Safe Frame** from the camera viewport pop-up menu.

The primary types of standard lights in 3ds Max are **Omni, Spot** and **Direct**. Omni lights shine in all directions and are good for *fill* light, which fills in dark areas of a scene. Spot and Direct lights shine in a single direction and are good for primary or *key* illumination.

The rays of light coming from a Spot light spread out in a cone-shaped area of illumination. Direct lights cast parallel rays of light, in a cylinder-shaped area. This area of illumination is called the **Light Cone**. Spot and Direct lights have **Hotspot** and **Falloff** parameters, which control how hard the edge of the Light Cone is.

The color, intensity, and shadows of each light can be set individually. Lights can be made to ignore objects with the **Exclude** list.

All lights shine through objects unless shadow casting is enabled. The two primary types of shadows are **Shadow Maps** and **Ray Traced** shadows. Shadow maps use a bitmap projected onto the scene, so it is necessary to control the resolution and blurring of the shadow map with the **Size** and **Sample Range** parameters. Standard **Ray Traced** shadows always create hard-edged shadows. A Spot or Direct light can only create shadows within its Falloff area.

Light intensity can be made to fade off with distance, using the **Attenuation** parameters. A map can be used to project an image through a light. This is called a **Projector Map**, and it is handy to simulate shadows cast by offscreen objects such as trees and clouds overhead.

A light can be made to only affect certain components of a material, such as the Diffuse Color component. The overall "look" of a light can be altered with the **Contrast** and **Soften Diffuse Edge** parameters.

The effects of light parameters can only really be seen when a viewport is rendered.

REVIEW QUESTIONS

1. What is *perspective*?

2. What is the relationship between the camera's *focal length* and its *field of view*?

3. True or false: A long lens length makes faraway objects appear to be closer to the camera than a short lens length.

4. Why are shadows important in a rendering?

5. What is the difference between a **Free Spot** light and a **Target Spot** light? How do you decide which to use?

6. What is the difference between a **Spot** light and a **Direct** light?

7. Why is it better to use a Spot or Direct light as a shadow-casting light, rather than an **Omni** light?

8. Why do you usually have to move a light after placing it?

9. What parameters are used to set the intensity of the light?

10. True or false: Adding more lights increases rendering time.

11. What is **Attenuation**?

12. What are some of the reasons a shadow might not appear in a scene?

13. What are the two main types of shadows? How are they different?

14. What kinds of lights allow you to see from the light's point of view in a viewport? What would you use this type of viewport for?

15. What is an *aspect ratio*? How do you set up a viewport to display the same aspect ratio as the rendered image?

16. What is the purpose of a light's **Falloff** parameter?

17. What are some of the ways you achieve the effect of an extremely soft light source?

18. What is the simplest way to simulate bounced light?

Chapter 6
Keyframe Animation

OBJECTIVES

In this chapter, you will learn about:

- Using the Track Bar and Auto Key mode
- Motion panel parameters
- Editing object position with Trajectories
- Track View editor windows
- Editing keyframes with the Dope Sheet
- Editing function curves with the Curve Editor
- Looping and repeating animation
- Linking objects to one another in an animation hierarchy

ABOUT ANIMATION

In Chapter 2, you learned to create simple animation using the **Auto Key** button. You created *keyframes* (or simply *keys*) of object transforms and modifier parameters, and 3ds Max filled in the in-between frames for you, creating smooth motion. In this chapter, you will learn some of the many ways to work with keyframes to customize and fine-tune your animation.

TRACK BAR

At the bottom of the screen are several controls for quickly creating and editing keyframes. The **Track Bar** shows the keyframes for selected objects and allows you to perform simple editing tasks.

To animate an object, turn on the **Auto Key** button in the Status Bar area at the bottom of the screen. The Auto Key button, the Time Slider, and the active viewport are highlighted in red. Then move the **Time Slider** to a frame in time, and then change an object's transforms or parameters. A keyframe is created automatically.

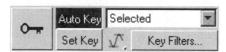

Figure 6-1: **Auto Key** button

Keyframes for currently selected objects appear as small rectangles in the Track Bar. You can select keyframes by clicking on them, or by dragging a selection rectangle around them. Selected keyframes are highlighted in white. You can drag the keyframes in the Track Bar to move them to a different frame in time.

Figure 6-2: **Track Bar** with keyframes at frame **0** and frame **20**, **Time Slider** at frame **14**

To delete keyframes from the Track Bar, simply select them and press the **<DELETE>** key on the computer keyboard.

You can easily move to the next or previous keyframe by activating **Key Mode** in the Time Controls toolbox. Then you can use the **Prev Key** and **Next Key** buttons to skip through keyframes for the selected object.

Figure 6-3: Using **Key Mode** to skip to the **Next Key**

MOTION PANEL

One way of editing animation with with the **Motion** panel. To open the Motion panel, click the tab that looks like a wheel ⚙ at the top of the Command panel.

The Motion panel has two primary subpanels: **Parameters** and **Trajectories**. The default mode is Parameters. In the Parameters subpanel, you can see and change the values of keyframes with spinners and buttons. With the Trajectories subpanel, you can control position keyframes in the viewports.

PARAMETERS

Keyframes for position, rotation, and scale transforms can be edited in the Parameters subpanel. When an object is selected in the viewport, you will see several rollouts, such as **PRS Parameters**. If no object is selected, the Parameters subpanel will be nearly blank.

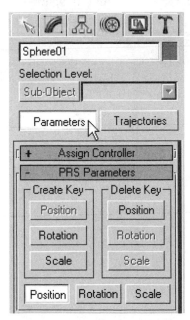

Figure 6-4: **Motion** *panel in* **Parameters** *mode*

The PRS Parameters rollout works in conjunction with other rollouts in the Motion panel to create, delete, and edit keyframes. On the Motion panel, you can only work with the keyframe data for one transform at a time, and one axis at a time. Click the **Position** button to edit position keyframes for the currently selected object. Click the Rotation or Scale button to edit the keyframes for those transforms.

The PRS Parameters rollout is another way of creating and deleting keys. To create a keyframe, move the Time Slider to the desired frame. In the **Create Key** section of the PRS Parameters rollout, click a button to create a position, rotation, or scale keyframe at the current time. A new keyframe is created based on the current value of the transform for the selected object. That way, you can create a keyframe without using the Transform Gizmos.

To delete a keyframe with the Motion panel, simply move the Time Slider to the keyframe you wish to delete, and click a button in the **Delete Key** section of the PRS Parameters rollout.

When you create a transform keyframe, 3ds Max records the X, Y, and Z transform values for the object. In the Motion panel, you can see and edit position and rotation X, Y, and Z values separately.

For example, to access the X, Y, or Z transform values for position transforms, click the Position button at the bottom of the PRS Parameters rollout. Directly below the PRS Parameters rollout, you'll see the **Position XYZ** parameters rollout. Here, you can select which axis you want to work with by clicking the X, Y, or Z button.

If the currently selected object has position keyframes, you can see and edit the position values in the **Key Info (Basic)** rollout.

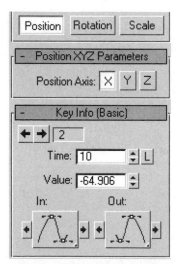

Figure 6-5: **Key Info (Basic)** *rollout*

If the Time Slider is located at a keyframe, you can edit the **Time** and **Value** of the keyframe with the Key Info (Basic) spinners or by typing in values. If the Time Slider is not at a keyframe, the parameters are grayed out. You can view the current values, but you can't edit them. That is because the values in between keyframes are automatically interpolated by 3ds Max.

At the top of the Key Info (Basic) rollout is the number of the current keyframe. This is not the number of the animation frame in time, but the number of the keyframe in sequence. In the previous illustration, keyframe #2 is located at frame 10 on the animation timeline.

Next to the keyframe number are two arrows that allow you to skip forward and backward to the next or previous keyframes. This is similar to Key Mode in the Time Controls toolbox, except that the buttons on the Motion panel only work for the current transform, and the current axis. Key Mode skips to any keyframes on the Time Slider, and the Time Slider displays all keyframes by default.

At the bottom of the Key Info (Basic) rollout are two very large buttons that display images of Bezier curves. These buttons allow you to select different methods of keyframe interpolation. You will learn about this later in this chapter, when we discuss the Curve Editor.

EXERCISE 6.1: Motion Panel

In this exercise, you'll gain practice using the basic features of the Motion panel.

1. Reset 3ds Max.

2. Maximize the **Perspective** viewport. Create a **Box** approximately **20** units long, **30** units wide, and **10** units in height. Place it on the left side of the screen, at approximate coordinates **(-50, 0, 0)**.

3. With the box selected, open the **Motion** panel 🔘. In the **Create Key** section of the **PRS Parameters** rollout, click the **Position** button. A position keyframe is created at frame **0** on the Track Bar.

4. Move the **Time Slider** to frame 50. On the Motion panel, click the Position button in the Create Key section of the PRS Parameters rollout. Another keyframe is created at frame **50**.

5. Play back the animation. The box does not move, because both keyframes are identical.

6. In the **Key Info (Basic)** rollout, use the arrow keys to go to keyframe #2, at frame **50** on the timeline.

7. In the **Position XYZ Parameters** rollout, make sure that the X axis is currently selected. In the Key Info (Basic) rollout, increase the **Value** of the keyframe to approximately **50**.

8. Play back the animation. The box slides from the left side of the screen to the right side.

9. Save the file as **MotionPanel.max** in your folder.

TRAJECTORIES

3ds Max has a wonderful feature called **Trajectories** that allows you to visualize the position transforms of an animated object. A *trajectory* is the path that an object describes as it moves through space.

All 3D programs have *path animation*, which allows you to attach an object to a spline, and move the object along the length of the spline. 3ds Max does have path animation, in the form of the Path Constraint. Even better, 3ds Max has Trajectories, which shows you the motion curve of any moving object, not just those that are constrained to a spline curve.

To use Trajectories, select an animated object and open the Motion panel. Click the Trajectories button. The path of the selected object is displayed as a red curve with white boxes and dots. The white boxes represent position keyframes. The white dots, called *ticks*, represent interpolated frames.

In the following example, there are three keyframes. The location of the object at each of the in-between frames is indicated by white ticks. There is one tick for each frame in the animation that is not a keyframe.

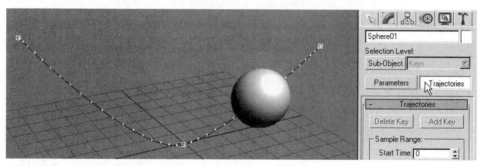

Figure 6-6: **Motion Panel** *in* **Trajectories** *mode*

Trajectories are extremely useful for several reasons. First, they allow you to see the speed of an object as it moves. The distance between frame ticks shows you how fast the object is moving. The farther the ticks are from one another, the faster the object is moving.

Each tick is one frame, and the amount of time between frames does not change during an animation. The Trajectory of a fast-moving object will have frame ticks farther apart than a slow-moving object, because the fast-moving object travels a farther distance between frames.

Naturally, Trajectories also display changes in velocity for a single object. As an object slows down, the ticks become closer together. In the following illustration, the object moves more slowly at the bottom of its arc.

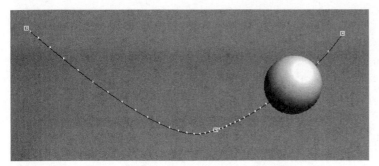

Figure 6-7: Trajectory showing an object changing speed

Another important feature of Trajectories is the ability to edit keyframes directly in the viewport. This only works for position keyframes applied to the object. In Trajectories mode, click the button labeled **Sub-Object**. Now you can select a keyframe in the viewport and move it in space with the Move Gizmo.

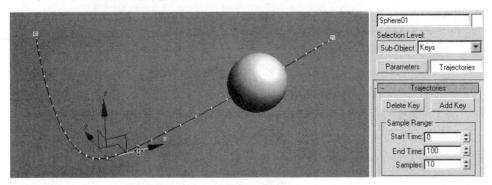

*Figure 6-8: Moving a **Key** sub-object of a Trajectory*

In Sub-Object mode, you can also delete or create keyframes. To delete a key, select it in the viewport, then click the **Delete Key** button on the Motion panel (or press the **<DELETE>** key on the keyboard). To create a keyframe, click the **Add Key** button on the Motion panel. Then click anywhere on the Trajectory in the viewport to create a new keyframe.

One of the great advantages of Trajectories is that the motion path of an object is always shown, even if the object is not directly animated. Quite often an object will be controlled by other objects, such as in a hierarchical linkage. (You'll learn about animation hierarchies at the end of this chapter.) 3ds Max Trajectories always show you the resulting path of the object, even if very specialized or complicated animation techniques are used. This comes in very handy when creating loops such as walk cycles, because the Trajectory will clearly show you any glitches in the cycle.

TRACK VIEWS

The **Track View** editor window is used for controlling and adjusting animation, and for setting up various aspects of the scene, such as sound. There are two specialized forms of Track View, each designed for different jobs. They have different toolbars and different ways of displaying animation data. The first Track View we'll discuss is called the **Dope Sheet**.

DOPE SHEET

In traditional hand-drawn animation, the animators use a written page called a *dope sheet*. This is a list of the actions and spoken word dialog for each frame of the animation. 3D graphic programs use the term "dope sheet" to refer to a keyframe editor.

The 3ds Max Dope Sheet looks similar to the Track Bar, because it lets you move keyframes around in time. Any of the operations you can do in the Track Bar or the default Motion panel can be done in the Dope Sheet. However, the Dope Sheet is much more than that. The Track Bar and Motion panel are fine for simple scenes, but if you have a scene with more than just a few objects, you'll need to work with the Dope Sheet. There are also many operations that can *only* be done in the Dope Sheet.

To open the Dope Sheet, choose **Graph Editors > Track View - Dope Sheet** from the Main Menu.

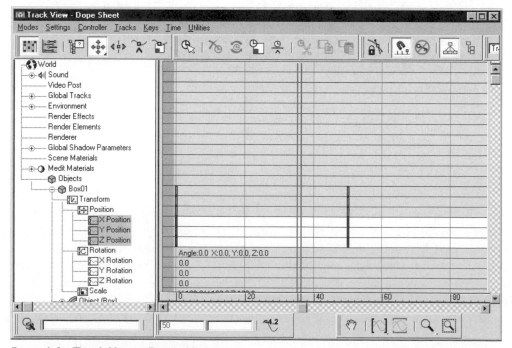

Figure 6-9: ***Track View - Dope Sheet***

Controller Window

On the left side of a Track View window, such as the Dope Sheet, is the **Controller Window**. Here you will see a hierarchical listing of all available animation *tracks*.

A track is a single animation channel, or animatable parameter, such as **X Position** or **Bend Angle**. For example, if you move an object with the **Auto Key** button on, you create keys on the X, Y, and Z Position tracks for that object.

All animatable scene tracks, including object transforms, modifier parameters, and materials, are listed in the Track View. The tracks are organized in a hierarchy. To navigate through the scene hierarchy in the Controller Window, click the plus sign next to an item. Any subordinate objects or tracks will expand and become available for editing.

A Track View will automatically scroll to display, expand, and select the tracks of the object currently selected in the viewport. If you wish, you may disable this behavior from the **Settings** menu at the top of the Track View window. The **Manual Navigation** menu item will disable all of the Track View's automatic navigation options.

When there are many objects in the scene, the Track View can become cluttered. You can use the **Filters** button on the Track View toolbar to limit the displayed tracks to just those that you want to see. For example, you can filter the Track View so that only animated tracks are displayed.

NOTE: In 3ds Max, a **Controller** is a program module that creates animation. For example, position transforms use the **Position XYZ** Controller by default. You will learn more about Controllers later.

Key Window

On the right side of a Track View you will find the **Key Window**. This is where you edit animation keyframes.

You can control the Key Window display with the **Navigation** toolbar at the lower right corner of the Track View window. The **Pan, Zoom,** and **Zoom Region** tools work very much the same as they do in the viewports. There are also a few other navigation tools that come in handy when the Track View is in Curve Editor mode. We will discuss them in the Curve Editor section later in this chapter.

The Dope Sheet Key Window displays a grid of rows and columns. The rows correspond to animation tracks, and the columns indicate frames in time. As you zoom into the Dope Sheet, the grid becomes visible. At the bottom of the Key Window is the **Time Ruler**, a horizontal bar that measures the frames of the animation. The Time Ruler may be moved up and down if desired.

The currently active frame of animation in the scene is indicated within Track View by a vertical bar called the **Track View Time Slider**. As you move the Time Slider in the main 3ds Max interface, the Track View Time Slider moves also. In addition, you can drag the Track View Time Slider to scroll through your animation.

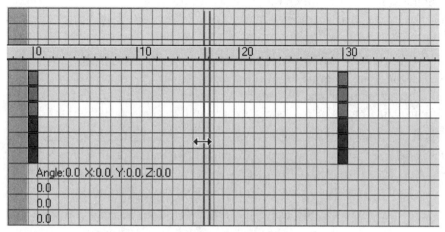

Figure 6-10: **Time Ruler** *and* **Track View Time Slider**

Edit Keys

A Track View can represent animation keys with lines, boxes, or curves. The Dope Sheet displays keys as boxes by default. This is called **Edit Keys** mode, and it is active whenever the Edit Keys button ▨ is pressed on the **Keys** toolbar at the top left of the Dope Sheet.

In Edit Keys mode, keyframes are displayed as boxes, just like in the Track Bar at the bottom of the 3ds Max viewport area. By default, Position keys are red, Rotation keys are green, Scale keys are blue, and object and modifier parameter keys are gray.

To select keyframes and move them in time, use the **Move Keys** tool ✛, which is activated by default. To select a keyframe, click a key box. The key turns white when selected. The key time (the frame on which the key sits) is displayed at the bottom of the Track View window. To select multiple keys, hold down the <**CTRL**> key while clicking keys. You can also select multiple keys by drawing a bounding box around them.

Notice that when you select a single key, more than one key box is highlighted. The Dope Sheet displays keyframes for individual animation tracks, and by default it also displays keyframes for other levels of the Controller Window hierarchy, such as objects and transforms. This means you can select and edit keys within individual tracks, and you can also select and edit all transform keys for a single object at once.

For example, when you click a key in the **Y Rotation** track, you will see that keyframe highlighted in white. In addition, key boxes in the **Rotation** track, the **Transform** track, and the track for the object itself, are all highlighted. This is because Y Rotation is a type of rotation, and it's a type of transform, and it's a track within a particular object.

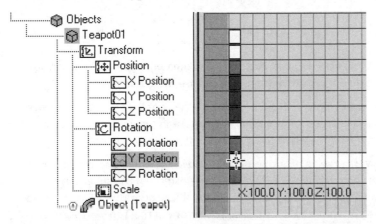

*Figure 6-11: Selecting a keyframe in the **Y Rotation** track also selects keys in the **Rotation**, **Transform**, and **Teapot01** tracks*

This process also works in the reverse manner: if you click on a key box in the Rotation track, the keys at that frame in all three Rotation tracks will be selected. If you click on a key box in the Transform track, the keys at that frame in the Position, Rotation, and Scale tracks will be selected. Clicking the key box in the object track selects all keys for that object at that frame. 3ds Max calls this **Modify Subtree**, and it is activated from a button 🔱 on the **Dope Sheet** toolbar, at the upper right of the window. Modify Subtree is on by default when you open the Dope Sheet.

When Move Keys is on, selected keys can be moved in time by clicking and dragging. To copy one or more selected keys, hold down the **<SHIFT>** key while moving the keys to another frame.

To see the values associated with a key box, right-click the key. A **Key Info** dialog appears that displays values for the key. This is the same information that is displayed on the Motion panel in Parameters mode.

To add keys, click **Add Keys** 〰 and click in the Key Window. A key is created at that frame. The value of the new key is derived from the existing interpolated value.

To delete keys, select the keys and press the **<DELETE>** key on the keyboard.

EXERCISE 6.2: Basic Keyframing

In this exercise, you will create a simple bouncing ball animation. You will control its path with Trajectories and edit the keyframes in the Dope Sheet.

Setting the Scene

1. Reset 3ds Max.

2. Create a **Plane** in the **Top** viewport. Make it approximately **200 x 200** units in size. Center it on the world coordinate system. This will be the ground plane.

3. Switch to the Perspective viewport and click **Zoom Extents**. This zooms the viewport out to make all scene geometry visible. Arc Rotate in the Perspective view so that the Plane is lined up with the viewport, as shown in the following illustration.

4. Create a **Sphere** with a radius of about **10** units. Move the Sphere so that it hovers over the left side of the ground plane. The absolute position of the sphere in world coordinates should be approximately **(-100, 0, 50)**. Make sure the ball is correctly positioned over the ground plane by checking your work in the Top viewport.

Figure 6-12: Sphere and Plane positioned in Perspective viewport

Keyframing the Ball's Movement

1. With the **Time Slider** positioned at frame **0**, and the sphere selected, open the

   Motion panel ⊛ . In the **PRS Parameters** rollout, locate the **Create Key** section and click the button labeled **Position**. Notice the small, red rectangle on the far left of the **Track Bar**, just below the Time Slider. This is the position key you just created.

2.　Click the **Auto Key** button; it turns red. In the Time Controls area at the bottom right of the screen, locate the **Current Time** field. Type **50** in the field. The Time Slider moves to frame **50**.

*Figure 6-13:　Type **50** in the **Current Time** field to move to frame **50***

3.　Switch to the **Front** viewport. Move the sphere so it hovers over the center of the ground plane by selecting the XY plane of the sphere's Transform Gizmo. The sphere should just barely touch the plane, as shown in the following image. Since the Auto Key button is still on, another keyframe is created at frame **50**.

If you wish, you can use the Align tool to position the sphere at frame **50**.

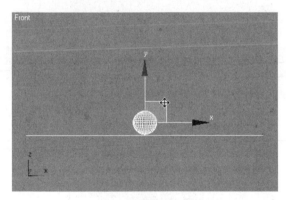

*Figure 6-14:　Position the sphere so it touches the plane at frame **50***

4.　Rewind the animation to frame **0** and play it. The sphere should fly from its initial position on the upper left to land on the ground plane at frame **50**.

5.　Fast forward the animation to frame **100** and move the sphere to the upper right of the Front viewport, automatically creating another key at frame **100**. Playing the animation shows that the sphere floats across the screen in an arc. To make it bounce, we must add more keyframes.

6.　Go to frame **15** by typing it in the Current Time field. Move the ball up in the world Z axis, a little higher than it was at frame **0**. Repeat this process to create another key at frame **85**. Play the animation again; you should have more of a bouncing motion now.

7.　Turn off the Auto Key button.

Editing the Ball's Trajectory

1. With the sphere still selected, open the **Motion** panel and click **Trajectories**. Now you can see the path of the ball.

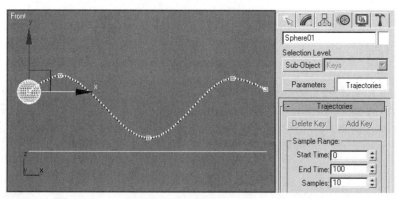

*Figure 6-15: Viewing the **Trajectory** of the ball*

2. In the Motion panel, click **Sub-Object**. Under **Trajectories**, click **Add Key**. Position your cursor over the sphere's Trajectory. Your cursor turns to a cross. Add two more keyframes close to the bounce point, one on either side of the impact at frame 50.

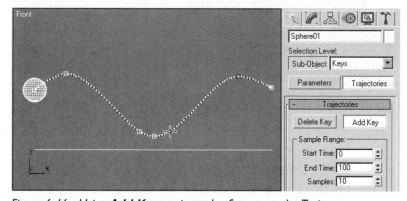

*Figure 6-16: Using **Add Keys** to insert keyframes on the Trajectory*

3. Turn Add Key off. Use the **Select and Move** tool to select and adjust keyframes. Click on a key box on the Trajectory and drag to move the keyframe in space. To make sure the ball bounces in a straight line, take care to only move the keys in the global XZ plane, never in the Y axis.

    This is easiest if you only use the Front viewport. If you're in the View coordinate system, simply move the keyframe in the XY axes.

    If you do choose to edit your trajectory in the Perspective view, remember to use the XZ plane of the Move Gizmo.

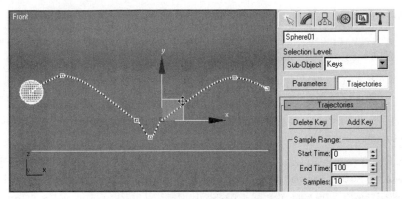

Figure 6-17: Use the Move Gizmo to move keyframes on the Trajectory

4. Play the animation. Adjust the keys more to get a better bouncing motion. The keys can be moved through space by using the Transform Gizmo, or moved in time by dragging the keyframe icons in the Track Bar. Tweak the animation until it looks more convincing, adding additional keyframes where necessary.

Remember that the small white ticks on the Trajectory indicate the speed of motion. Try to position keyframes in space and time so the ball keeps a constant velocity and doesn't speed up or slow down.

You will find that it is difficult to achieve a good, solid bounce with this technique. Don't worry too much about that right now. We will fix this problem in the next exercise.

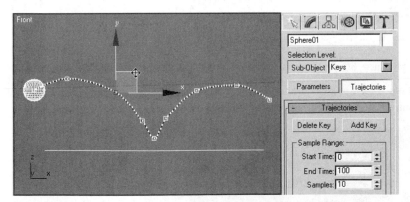

Figure 6-18: Add and move keyframes as needed to achieve a better bounce

When you are reasonably happy with the animation, turn off Sub-object mode.

Putting Spin on the Ball

1. Next we will make the ball spin as it bounces. To see the effect of the ball rotating, we need to place a map on the sphere.

    Make certain that Auto Key is turned off. Open the **Material Editor**. Place a **Checker** map in the **Diffuse Color** map channel. Adjust the **Tiling** parameters, and assign this new material to the ball. Remember to turn on **Show Map in Viewport** 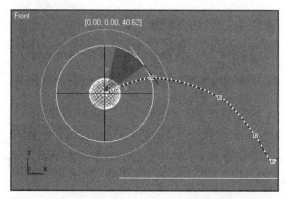, or you won't see the checker pattern on the ball.

2. Turn Auto Key on again. Rewind the animation to frame **0**. Zoom in on the ball in the Front viewport. Rotate the ball about **40** degrees in the world Y axis, as shown in the following illustration. The easiest way is to rotate the ball in the Z axis of the Front viewport.

*Figure 6-19: At frame zero, rotate the ball **40** degrees in the world Y axis*

3. Advance to frame **50**. Rotate the ball about **-80** degrees in the world Y axis.

    It might help to use shaded mode in the Front view, so you can see the checker pattern on the ball.

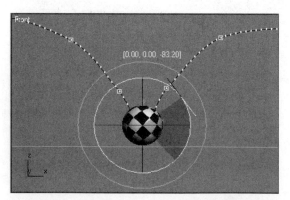

*Figure 6-20: At frame **50**, rotate the ball **-80** degrees in the world Y axis*

4. Turn off Auto Key. Play the animation in the Perspective viewport. Arc Rotate around the scene to check your work.

    The spin doesn't look right. The ball should keep spinning after it bounces. You'll fix this in the Dope Sheet.

5. Choose **Graph Editors > Track View - Dope Sheet** from the Main Menu. Look in the Rotation tracks. 3ds Max has created keyframes for X, Y, and Z rotation, even though the ball is only rotating in the global Y axis.

    Click the **Move Keys** button . Select all four of the **X Rotation** and **Z Rotation** keyframes by holding down the **<CTRL>** key and clicking the keyframe boxes.

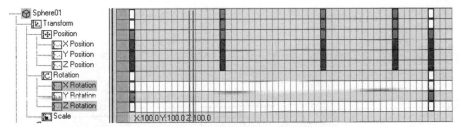

*Figure 6-21: Select all four **X Rotation** and **Z Rotation** keyframes*

    When all four keyframes are selected, press the **<DELETE>** key to remove them.

6. Click the **Y Rotation** keyframe at frame **50**, and drag it to the right until you reach frame **100**. The current frame is displayed at the bottom of the Track View.

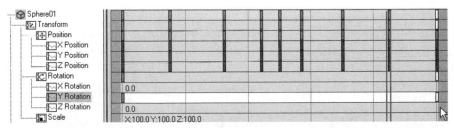

*Figure 6-22: Click and drag the **Y Rotation** keyframe from frame **50** to frame **100***

7. Play back the animation. The ball continues to rotate after it bounces.

8. Adjust the animation further if you wish. Save the scene as **KeyframeAnimation.max** in your folder. We will use it for the next exercise.

Edit Ranges

Sometimes you will want to shift entire chunks of keyframes in time, or scale the keyframes so the action happens faster or slower. It's possible to do this by selecting multiple keyframes by dragging a box around them, but this is not optimal. If you have many keyframes or you need to work with many objects at once, then dragging keyframes is not practical. It is tedious, and sometimes the interactivity within the Dope Sheet can be very slow if you have many keyframes selected.

The solution is found in a Dope Sheet mode called **Edit Ranges** , which lets you edit an entire track at once. Instead of individual keyframes, animation in tracks is displayed as range bars that you can move. Click and drag the middle of the range bar to move it. Click and drag the ends of a range bar to scale the animation, making the action speed up or slow down.

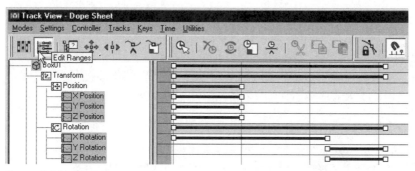

Figure 6-23: **Edit Ranges** *mode*

The range bars appear at all levels of the Track View hierarchy. For example, you can edit the range bar for a single track, such as X Position, or you can edit the range bar for the entire object.

Modify Subtree

As previously mentioned, the **Modify Subtree** button allows keyframe editing on animation tracks to affect all sub-tracks. For example, moving keyframes on the **Transform** track will move keyframes in all Position, Rotation, and Scale tracks.

If Modify Subtree is turned off when the Dope Sheet is in **Edit Keys** mode, the result is a kind of hybrid between Edit Keys mode and Edit Ranges mode. You can edit individual keys in sub-tracks such as X Rotation, but you can also edit range bars for high-level tracks such as Transform.

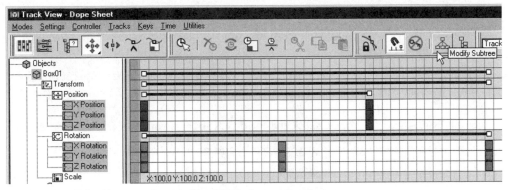

Figure 6-24: **Edit Keys** *mode with* **Modify Subtree** *disabled*

EXERCISE 6.3: Range Bars

In this exercise you will tighten up the bouncing ball animation by editing range bars.

1. Open the file **KeyframeAnimation.max** from your folder or from the disc. Play the animation. The ball is moving far too slowly. It doesn't look natural.

2. Select the ball in the viewport. Then choose **Graph Editors > Track View - Dope Sheet** from the Main Menu.

3. In the Dope Sheet, activate Edit Ranges mode by clicking the button near the top left of the window.

4. Locate the **Sphere01** track. Click the box at the far right of the **Sphere01** range bar and drag it to the left. Release the mouse button when you reach frame **40**.

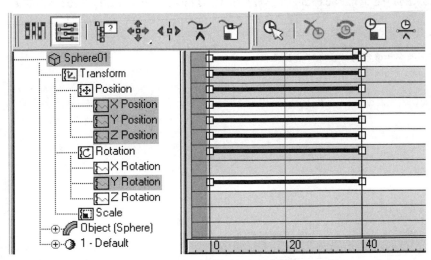

Figure 6-25: Click and drag the **Sphere01** *range bar to frame* **40**

5. Play the animation. The ball now moves much more naturally.

   To get practice working in Edit Keys mode, we will do the same operation again with a slightly different method.

6. Hold the **<CTRL>** key down and press the **<Z>** key to undo the last operation. The range bars are restored to their original state, and the ball moves very slowly again.

7. Enter **Edit Keys** mode. Click the **Modify Subtree** button to turn it off.

   Now you can see range bars for some tracks and keyframe boxes for others.

8. Click and drag the right edge of the **Sphere01** range bar from frame **100** to frame **40**.

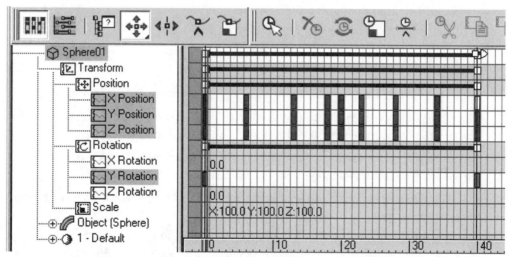

*Figure 6-26: Edit the **Sphere01** range bar in **Edit Keys** mode, with **Modify Subtree** turned off*

9. Play the animation in the viewports. The ball moves more naturally once again.

10. Save the scene as **RangeBars.max** in your folder.

CURVE EDITOR

The other type of Track View is called the **Curve Editor**. It allows you to see and edit the interpolation between keyframes. Keyframes are represented by boxes, and the values in between keyframes are drawn as graphs called *function curves*, or *fcurves* for short.

You can have the Curve Editor and the Dope Sheet open at the same time. However, it is easier to simply switch back and forth between the two modes. Select **Curve Editor** or **Dope Sheet** from the **Modes** menu, and the Track View will switch to the other mode.

Once you have set your basic keyframes with Auto Key, the Track Bar, and the Dope Sheet, the Curve Editor is where you refine the animation.

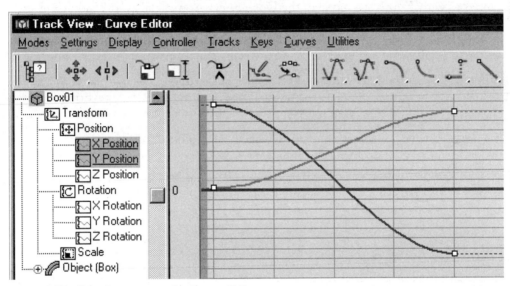

*Figure 6-27: Function curves in the **Curve Editor***

By default, the Curve Editor only displays function curves for animation tracks that are highlighted in the Controller Window. To see the function curve for a track, click on the track label in the Controller Window. To see multiple function curves, hold down the **<CTRL>** key and click additional track labels.

A function curve displays animation data for a single track in two dimensions. The horizontal dimension represents time, just as it does in the Dope Sheet. Frame numbers increase from left to right. The vertical dimension of a function curve represents the value of the track. Higher values are higher on the graph; lower values are lower on the graph.

Since the vertical dimension measures the value of the track, the rate of change is indicated by the steepness of the curve. Flat lines show no change in value over time, indicating that the track's parameter value is constant and unchanging. Steep curves indicate rapid-speed animation.

The function curves for transforms are red, green, and blue. These colors correspond to the X, Y, and Z axes, just as they do for the Transform Gizmos.

The keyframes on the curves are represented as boxes. If a keyframe is selected, it is highlighted in white, and its track label is underlined in the Controller Window. The time and value of a single keyframe is displayed at the bottom of the Curve Editor.

Keyframes can be selected, moved, and copied with the **Move Keys** button, in the same manner as with **Edit Keys** mode in the Dope Sheet. You can also add and delete keys, just as you can in the Dope Sheet.

Tangent Types

Although you set the exact time and value of each keyframe, 3ds Max interpolates the track values between each key. Controlling keyframe interpolation by editing function curves is a very important skill.

The most basic way to change the interpolation is to choose a different curve type. To change the interpolation type, select the keyframe or keyframes, and click a button on the **Key Tangents** toolbar at the top of the Curve Editor. There are seven different types of animation curves available in 3ds Max, giving you a wide variety of choices for how in-between values are interpolated.

*Figure 6-28: **Key Tangents** toolbar*

 Auto tangents produce Bezier function curves. The vector handles of the curves are controlled by 3ds Max automatically. This is the default interpolation type. Auto tangents are a good starting point, but usually you'll need to adjust the interpolation with Custom tangents.

 Custom tangents also create Bezier function curves, but the Bezier vector handles can be edited manually. This interpolation type is the most commonly used, because it gives you the most freedom to customize the shape of the curve.

Fast tangents cause interpolated values to change more quickly as the curve approaches the keyframe. The closer a frame is to a keyframe, the more rapidly the value of the curve will change. As a result, motion is accelerated near the keyframe.

Slow tangents cause interpolated values to change more slowly as the curve approaches the keyframe. The Slow tangent type will create an automatic *ease in* or *ease out* effect. Easing in or out refers to natural animal motion. The movement of living organisms tends to accelerate out of a resting position, and decelerate into a resting position.

Step tangents essentially have *no* interpolation. Instead of creating new values in between keyframes, a Step tangent simply holds the last value until the next keyframe is reached. The curve instantly jumps to the new value when the next keyframe is reached. As a result, the curve has a stair-step appearance. Step keys are sometimes called *hold* keys in computer animation.

Step or hold keys are very useful when you need a value to remain absolutely constant and unchanging over time. Step keys are necessary if you want a value to change instantly. For example, if you wanted to animate the Multiplier value of a light to create a strobe effect, you'd need to use a Step curve.

Linear tangents create straight-line segments between keyframes. Linear motion has a constant velocity, and does not accelerate or decelerate. This is often useful for mechanical motion, or for objects that do not speed up or slow down over time.

Smooth tangents create curved segments between keyframes, but you cannot edit the curvature. This is the same as a Smooth spline vertex.

Smooth animation curves are not generally as useful as Custom curves. The Smooth tangent type tends to create a phenomenon known as *overshooting*, in which interpolated values exceed the keyframed values. The result is a curve that behaves differently from what you might expect. For that reason, Auto tangents are the default interpolation type in 3ds Max 5.

You can mix tangent types within a function curve in the same way that you can mix vertex types in a spline. More often, however, you will simply use the Custom tangent type, and edit the Bezier vector handles to change the shape of the curve.

By default, new keyframes are created with the Auto tangent type. If you wish to use a different type of interpolation when you create a new keyframe, choose a tangent type from the flyout labeled **Default In/Out Tangents for New Keys** , found in the Animation Controls section near the bottom right of the main 3ds Max interface.

In and Out Curves

Each keyframe has an *incoming* curve before the key, and an *outgoing* curve after the key. When you click on a button on the Key Tangents toolbar, by default you change the tangent type on both sides of the keyframe. However, if you need to change the curve on just one side of the key, you can select a flyout button from one of the Key Tangents buttons.

Each Key Tangent button has three buttons on a flyout. A standard button changes the tangency on both sides of the keyframe. A button with a black arrow on the left side of the button will change the In tangent type. A button with a white arrow on the right side of the button will change the Out tangent type.

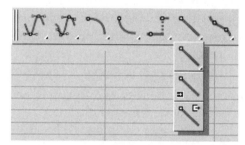

Figure 6-29: **Key Tangents** *flyout showing* **In** *and* **Out** *options*

To see more information about a key, right-click it. The **Key Info (Basic)** dialog appears, allowing you to see and edit the keyframe time and value. At the bottom of the Key Info dialog are two large buttons, labeled **In** and **Out**. These buttons give you another way to change the key tangent type, in addition to the Key Tangents toolbar. Click the In or Out button and choose a tangent type from the flyout.

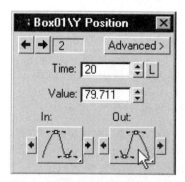

Figure 6-30: **Key Info (Basic)** *dialog*

Bezier Function Curve Editing

When you create transform keys, 3ds Max uses Auto tangent interpolation by default. If you open the Curve Editor and select keyframes, you'll see light blue Bezier vector handles. These are Auto tangents. If you click and drag one of the vector handles with the Move Keys tool, the vector handle will automatically change to black, indicating that the tangent has been converted to Custom.

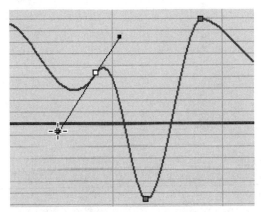

*Figure 6-31: Edit a **Custom** tangent with the Move Keys tool*

Initially, the Bezier vector handles on either side of the key are locked. If you move one handle, the other one moves also. If you need to control the two handles separately, you can unlock them by holding down the **<SHIFT>** key and moving a handle.

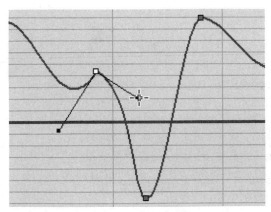

*Figure 6-32: Hold down the **<SHIFT>** key and drag to unlock vector handles*

If you need to lock the vector handles, right-click the key to access the Key Info dialog. Click the **Advanced** button. The Key Info dialog now shows you the values for the vector handles. Click the padlock icon to lock the vector handles.

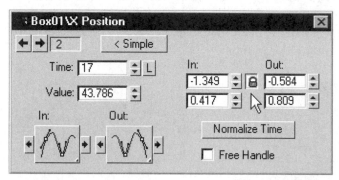

Figure 6-33: **Lock Handles** button in the **Key Info (Advanced)** dialog

If the handles are locked, they will maintain the same angle with respect to one another. So, if you move one handle, the other will also move, but they are not necessarily in a straight line. In many animation situations, you will want the vector handles to be colinear, in a straight line. This produces a smooth movement through the keyframe.

If you have unlocked the handles, you'll have to take extra steps to restore the vector handles to colinear and locked. Convert the keyframe to Auto. This resets the vector handles to colinear. Then right-click the keyframe to open the Key Info (Advanced) dialog. Lock the vector handles with the padlock button. Now the handles are colinear and locked, and will behave the same way as they did before you unlocked them.

EXERCISE 6.4: Function Curves

In this exercise you will use function curves to fine-tune the bouncing ball animation, making it more realistic.

Cleaning Up the Function Curves

1. Open the file **RangeBars.max** from your folder or from the disc.

2. Play the animation. Right now the ball stops moving at frame **40**, but there are **100** frames in the active time segment of the animation.

3. Click the **Time Configuration** button 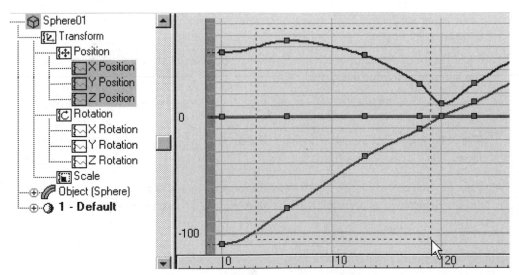 in the Time Controls toolbox. In the Animation section, change the End Time value to **40**. Click OK to exit the dialog.

    Now the animation runs from frame **0** to frame **40**.

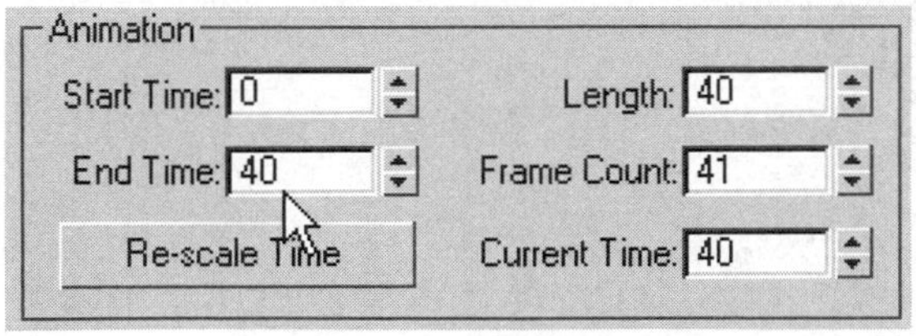

*Figure 6-34: Set **End Time** to **40** in the **Time Configuration** dialog*

4. Select the ball and open the Curve Editor. From the Main Menu, select **Graph Editors > Track View - Curve Editor**.

    In general, fewer keyframes result in smoother motion. The many keyframes we created for the Position and Rotation tracks of the ball are making it move unnaturally.

5. Hold down the **<CTRL>** key and click the Y Rotation track to deselect it. With the **X Position**, **Y Position**, and **Z Position** tracks highlighted, draw a selection box around the keyframes between frame **0** and frame **20**, as shown in the following illustration.

    Make sure you don't select the keyframes at frame **0** or **20**. Use the **<ALT>** key to deselect keyframes if necessary.

*Figure 6-35: Select keyframes between **0** and **20***

6. Press the **<DELETE>** key on the keyboard to remove the selected keyframes.

7. Select the keyframes between frame **20** and frame **40**. Press the **<DELETE>** key to remove them. The Curve Editor should look similar to the following illustration.

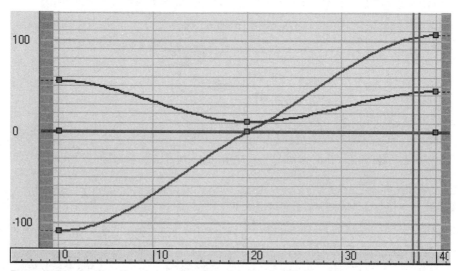

Figure 6-36: Position function curves after deleting keyframes

Converting Tangent Types

The goal of this exercise is to create a natural-looking bouncing ball. If the ball obeys the laws of physics, it will not speed up or slow down by itself. So, the speed of the ball's movement in the world X axis should be constant. To achieve this, we will use the **Linear** tangent type.

1. Highlight the **X Position** track by clicking its label in the Controller Window. Select the keframe at frame **20** and delete it.

2. Select the keyframes at frame **0** and **40**. Click the **Linear** button on the **Key Tangents** toolbar at the top of the Curve Editor.

    The X Position function curve is now a straight line, indicating a constant speed of movement.

NOTE: In the real world, the ball would slow down very slightly after the bounce. For the purposes of this animation exercise, that loss of inertia is not significant.

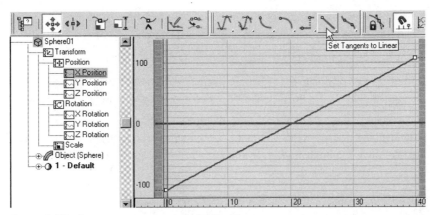

Figure 6-37: **Linear** *tangents on the X Position curve*

The rotation of the ball will obey the same physical laws of Newtonian mechanics. It will continue to spin in the same direction at a constant rate.

3. Select the **Y Rotation** track in the Controller Window. Select the keyframes and convert them to Linear tangents. The Rotation track is now also a straight line.

Editing Bezier Handles

Play the animation. The movement of the ball in the X axis, and the spin on the ball, look more natural. However, the ball is not bouncing, but gliding through the scene.

1. In the Curve Editor, select the **Z Position** track in the Controller Window. Click the vector handle of the keyframe at frame **0**. Drag the vector handle up and to the right to edit the function curve, so it looks similar to the following illustration.

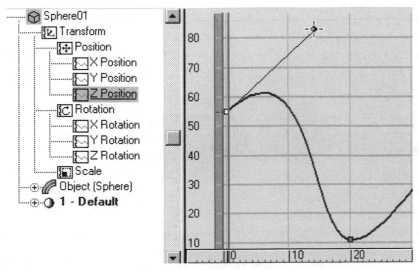

Figure 6-38: Edit the vector handle of the first keyframe

2. Select the keyframe at frame **20**. Hold down the **<SHIFT>** key and click one of its vector handles. Drag the vector handle upward. Release the <SHIFT> key.

   The vector handles are now unlocked.

3. Click and drag the opposite vector handle of the keyframe at frame **20**. There is no need to use the **<SHIFT>** key; the handles now move independently of one another.

4. Edit the vector handles of all keyframes until you have a function curve that resembles the following illustration. Play the animation, then adjust the function curves. You may also need to move the keyframes at frame zero and **40** to get the result you want.

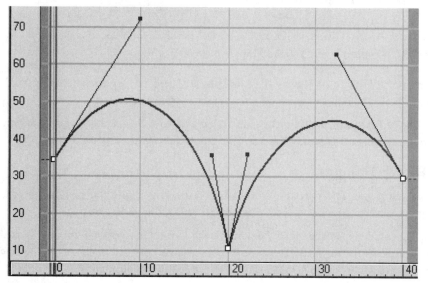

Figure 6-39: **Z Position** *function curve*

5. It may be helpful to view the animation in **Trajectories** mode. However, you cannot edit vector handles of keyframes in the viewports. Continue to adjust the Z Position function curve in the Curve Editor until you are happy with the animation.

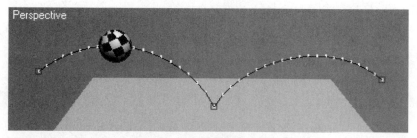

Figure 6-40: Completed animation in **Trajectories** *mode*

6. Save the animation as **FunctionCurves.max** in your folder.

ADDITIONAL TRACKS

You can add certain types of animation tracks to the Track View for specific purposes. The two most important additional tracks are **Visibility** and **Sound**.

Visibility

A **Visibility** track is used to make an object disappear or reappear in renderings. To create a Visibility track, select the object name in the Controller Window. In the Track View Menu, select **Tracks > Visibility Track > Add**. A new track, named Visibility, appears under the object name in the Controller Window.

To animate the visibility of an object, activate **Add Keys** and click in the Visibility track for the object. A Visibility keyframe value of zero means the object is hidden, and a keyframe value of one means the object is visible. Values between zero and one will cause the object to render as partially transparent.

You can make objects fade in or out of visibility. Simply edit the function curve of the Visibility track to create a gradual change in the Visibility value.

Sound

A sound file can be imported into your scene to help with synchronization to recorded audio, such as character voices.

To add a sound file, look for the **Sound** track near the top of the Track View Controller Window. Select the Sound track label, right-click and choose **Properties**. The **Sound Options** dialog appears. Click the **Choose Sound** button and browse your hard drive for a sound file to accompany the animation.

If you expand the Sound track by clicking the plus sign in the Controller Window, a waveform appears in the Key Window. The peaks of the waveform represent high sound volume. You can use the waveform display to help you align animation keys to specific parts of the sound file.

To see the sound waveform while working in the viewports, right-click the Track Bar, and choose **Configure > Show Sound Track**.

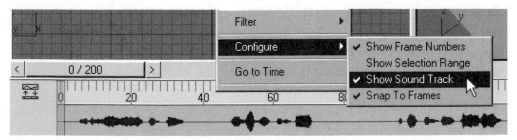

*Figure 6-41: Right-click the **Track Bar** to display the Sound Track*

3ds Max audio support is extremely basic. You can only import a few types of files, such as standard Microsoft .WAV files and .AVI files. If your audio is in another format, such as Macintosh .AIF, you will need to convert it to .WAV with an audio program before loading it into 3ds Max.

The sound file cannot be edited in any way in 3ds Max. The only thing you can do is move the sound so that it starts at a different time. To do this, enter Edit Ranges mode in the Dope Sheet, and drag the sound's Range Bar. In addition, you can only hear the sound during playback in 3ds Max if the **Real Time** option is checked in the **Time Configuration** dialog 🔳 .

WARNING: If you know that you will be synchronizing animation with sound, especially character animation with *lip sync*, then you must have your audio ready **before** you attempt to create any animation! Lip sync is the movement of a character's lips and jaw to match an existing soundtrack. If lip sync is required for a project, the audio is always recorded first. It is nearly impossible for an actor or voiceover artist to match the lip movements of an existing character.

You must at least have a basic *scratch track* for animation timing purposes. A scratch track is a reference audio track that is accurate in time, but might not be fully mixed or sweetened. (*Mixing* is combining multiple sound sources, and *sweetening* is adding the finishing touches to a mixed soundtrack.)

OUT-OF-RANGE ANIMATION

You may have noticed that function curves extend before and after the frames that you have keyed. The function curves in these time ranges are indicated by dashed lines.

To determine how an object should be animated before and after the keyframed range of time, 3ds Max *extrapolates* the animation of an object. Extrapolation is very similar to interpolation, but is based only upon the keyframes at the beginning and end of the curve.

Just as with interpolation tangent types, there are several different types of extrapolation techniques. 3ds Max calls these by the awkward name of **Parameter Curve Out-of-Range Types**. To select the extrapolation type, open the Curve Editor, select one or more tracks, and click the Parameter Curve Out-of-Range Types button 🔳 on the Curve Editor toolbar. A dialog pops up allowing you to choose an extrapolation type for the currently selected tracks.

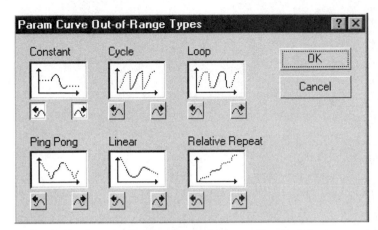

Figure 6-42: ***Param Curves Out-of-Range Types*** *dialog*

Choose the type of cycle you would like the track to have.

Constant is the default. This causes the curve to hold the values of the keys at the edges of the keyframed range.

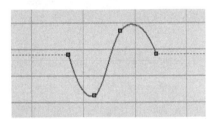

Figure 6-43: ***Constant*** *Out-of-Range Type*

Cycle causes the animation to jump to the value of the first key immediately after the last key is reached.

Loop is similar to cycle, except it attempts to make a smooth transition between the last key and first key. Loop works best when the last key matches the first key.

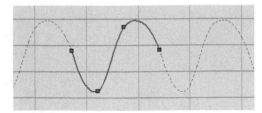

Figure 6-44: ***Loop*** *Out-of-Range Type*

Ping Pong plays the animation forward, then backward, then forward.

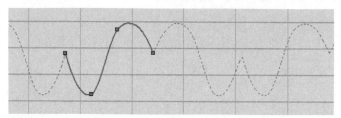

*Figure 6-45: **Ping-Pong** Out-of-Range Type*

Linear continues the curve in a straight line based upon the tangent type and vector handles of the edge key.

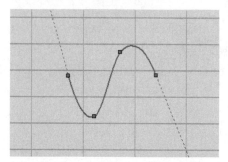

*Figure 6-46: **Linear** Out-of-Range Type*

Relative Repeat repeats the animation and offsets it. If the first and last key are different, the result will be that the extrapolated curve is offset more and more with each cycle. This is extremely useful for walk cycles. You can make a character walk across the scene by animating one cycle of movement and using Relative Repeat. (If you used Cycle or Loop, the character would pop back to its starting location at the beginning of each cycle.)

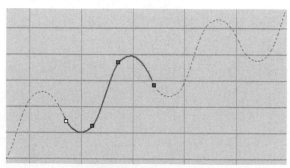

*Figure 6-47: **Relative Repeat** Out-of-Range Type*

After changing the Parameter Curve Out-of-Range Type, it is usually necessary to adjust the function curves in **Track View** for smooth movement.

EXERCISE 6.5: Looping

In this tutorial, you'll make a ball bounce repeatedly.

1. Load the file **Bounce.max** from the disc that comes with this book.

   Play the animation. The ball bounces once.

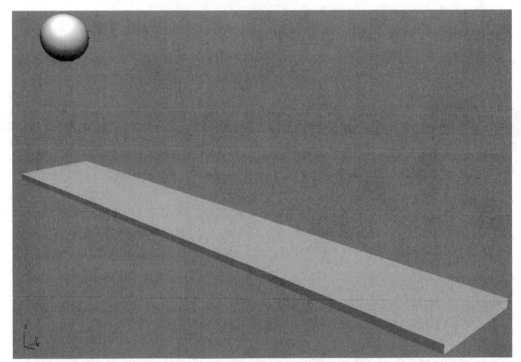

Figure 6-48: Bouncing ball

2. Open the Curve Editor. Right-click **Filters** ▣ and select **Animated Tracks Only** from the pop-up menu. The **Z Position** and **Scale** tracks for the ball are displayed in the Controller Window.

3. The **Z Position** and **Scale** tracks should be highlighted. If they aren't, use the **<CTRL>** key and click the track labels to select them both.

4. Click the **Parameter Curve Out-of-Range Types** button ▣ . In the dialog, click the **Loop** button, and click OK to exit the dialog.

5. Play the animation. The ball now bounces up and down throughout the animation.

6. Turn on the **Auto Key** button. Open the **Motion** panel and activate **Trajectories** mode.

7. Fast-forward to the end of the animation. Move the ball in the world X axis until it is at the end of the plank.

    Play the animation. The ball eases in and out of its positions at each end of the plank. That's not what we want.

8. In the Curve Editor, select the **X Position** track. Convert both keyframes to **Linear** tangents. The function curve is now a straight line.

    Play the animation. The ball bounces several times and travels the length of the plank.

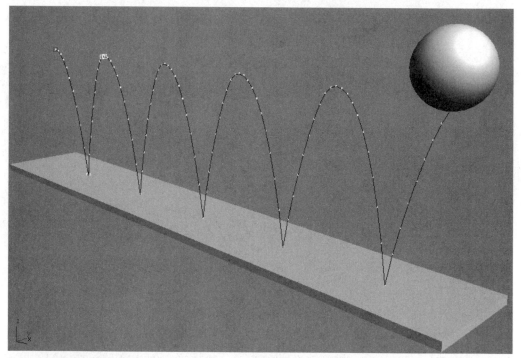

Figure 6-49: Looped bounce

9. Save the file as **BounceLoop.max** in your folder.

LINKING

One of the most fundamental concepts of computer animation is *linking*. It goes by many names. The terms *linking*, *parenting*, *animation hierarchy*, and *forward kinematics* all mean the same thing: several objects connected to one another. As one object in the chain moves, rotates, or scales, objects that are linked to it are also transformed.

In an animation hierarchy, one object is always the *child* or subordinate of another. The leading object is called the *parent*. Child objects *inherit* the transforms of their parent. The child always follows the parent, but the parent does not follow the child.

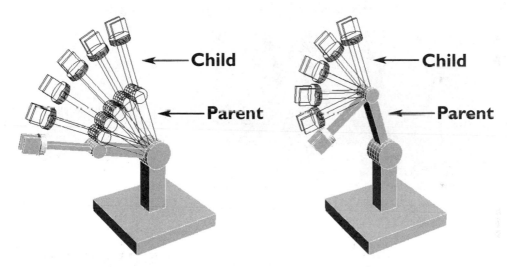

Figure 6-50: Child follows parent; parent does not follow child

A series of linked objects is often called a *chain*. However, an animation hierarchy is really more like a tree than a chain. A parent object may have many children, but each child may only have one parent. Child objects may have their own children, and the tree can get quite deep for complex hierarchies such as characters.

In keeping with the tree metaphor, objects in an animation hierarchy are sometimes called *nodes*. In biology, a node is a place where a new branch sprouts.

There is always one node in the hierarchy that is not linked to any other object. This object is called the *root*. The root is not the child of any object in the scene.

To link an object to a parent, click **Select and Link** ⬚ on the Main Toolbar, and click and drag from the object to another object, which will be the new parent. When the link cursor ▦ appears, release the mouse button. The second object (the parent) flashes white momentarily to show that it has been successfully linked.

EXERCISE 6.6: Linking

In this exercise, you'll practice linking a simple chain of objects.

1. Reset 3ds Max.

2. In the Top viewport, create a box with the following dimensions.

Length	20
Width	20
Height	50

3. In the Front viewport, create two copies of the box above the original.

Figure 6-51: Three boxes in Front viewport

4. Next, you'll link the boxes together. In the Front viewport, select the topmost box.

    On the Main Toolbar, click **Select and Link**. Click and drag from the topmost box to the box just below it until the link cursor appears. Release the cursor. The second box flashes white to show that the selected object has been linked to it.

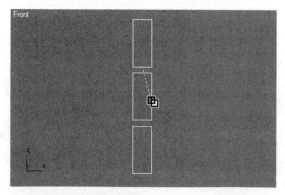

Figure 6-52: Linking boxes

When linking, drag the cursor over the box's wireframe, not its interior.

5. Select the second box. Link it to the box below it in the same manner.

    You now have a linked chain of three objects. The bottom-most box in the **Front** viewport is the root node in the chain.

6. Select the root parent box. Click **Select and Move**, and move the box around.

    The two child objects move with the root parent.

7. Select the middle box. Click **Select and Rotate**. In the Front viewport, rotate the box in the Z axis.

    The topmost box rotates with the second box, but the bottom box does not.

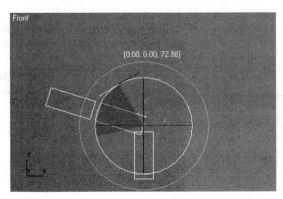

Figure 6-53: Rotate the middle box; the top box also rotates

8. In the Front viewport, rotate the top box.

    The top box is the last child in the chain, so it does not affect any other objects.

9. Save the scene with the filename **LinkedBoxes.max** in your folder.

CHECKING THE LINK STRUCTURE

Once the objects are linked together, it's a good idea to check the link structure before attempting to animate with it. There are several ways to do this, each described in the following section.

Double-Click

The easiest way to check the link structure is to double-click an object in the viewport. The object will be selected, and any children of the object will also be selected. Any *descendants* (children of children) will be selected. The object's parent and any other *ancestors* (parents of parents) will not be selected. Double-clicking the root node will select the entire hierarchy.

This technique doesn't give you any detailed information about the structure of the hierarchy, but it does provide a quick way of checking to see which objects are descendants of the selected object.

Select Objects Dialog

A more thorough way to check the link structure is with the **Select Objects** dialog. Click **Select Object** on the Main Toolbar, then click **Select by Name**. The **Select Objects** dialog appears. In the Select Objects dialog, check the **Display Subtree** check box. The object names are indented according to their links.

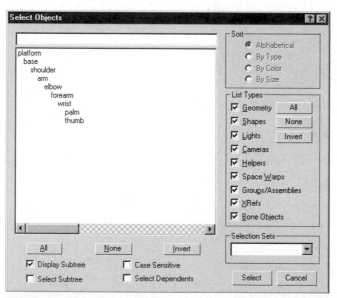

*Figure 6-54: Indented linked objects in the **Select Objects** dialog*

WARNING: Be sure to click **Select Object** (or another selection tool) before clicking Select by Name. If the **Select and Link** button is still on when you click Select by Name, the **Select Parent** dialog will appear. 3ds Max expects you to select a parent object. The Select Parent and Select Objects dialogs look very similar.

The root object is all the way over to the left. Objects linked to the root are indented slightly to the right. Objects farther down the chain appear farther to the right.

Here, the importance of naming objects intelligently is very clear. If you don't rename objects with meaningful names, then you won't be able to tell much by looking at a list of objects.

The **Selection Floater** is identical to the Select Objects dialog, so it can also be used to view the link structure. To access this floater, choose **Tools > Selection Floater** from the Main Menu. Check the **Display Subtree** check box to see the child objects indented.

Schematic View

The **Schematic View** window shows a visual representation of linked objects. It also shows a partial data flow of each object, such as modifiers and materials.

To open the Schematic View window, click the **Schematic View** button on the Main Toolbar. You can also choose **Graph Editors > New Schematic View** from the Main Menu. You'll see objects represented by boxes, called *nodes*, connected by lines.

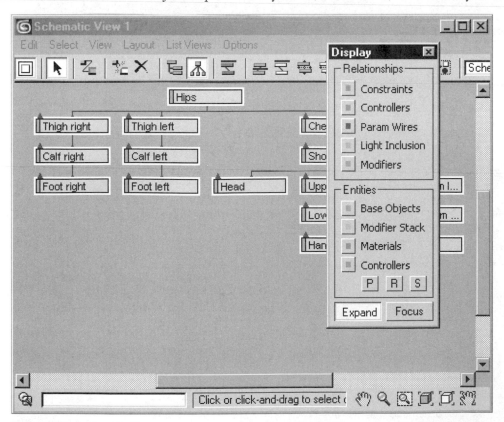

Figure 6-55: Schematic View

You can move nodes around in the Schematic View to arrange them as you choose. However, 3ds Max will rearrange the display automatically. To prevent 3ds Max from rearranging the Schematic View, disable the **Always Arrange** button 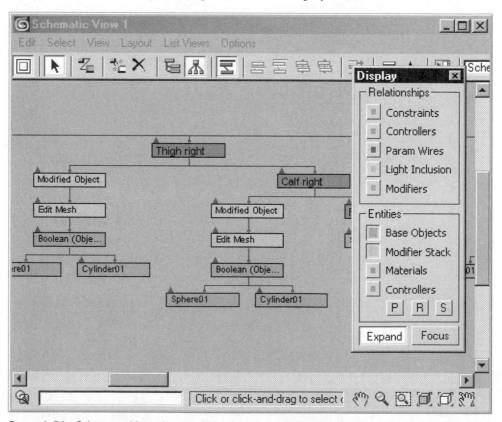 on the Schematic View toolbar.

To expand or collapse the display of connected nodes, select a node and click one of the red arrows, or use the **Expand Selected** or **Collapse Selected** buttons on the toolbar.

You can also link and unlink objects within the Schematic View. Use the **Connect** and **Unlink Selection** buttons on the Schematic View toolbar. Unfortunately, Schematic View will always rearrange the display of nodes when you link or unlink objects, even if you have Auto-Arrange Graph Nodes turned off.

Schematic View can be used to visualize and edit other relationships, not just hierarchical links. The Schematic View **Display** floater lets you turn other node types on or off. For example, you can display an object's Modifier Stack, materials, or animation Controllers by activating those options on the Display floater.

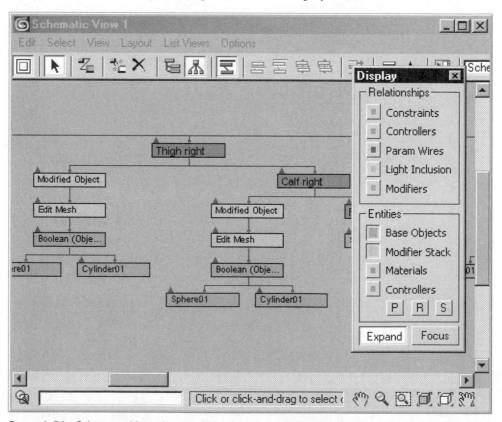

Figure 6-56: Schematic View showing Base Objects and Modifier Stacks

Once you have the Schematic View set up the way you like it, save the layout by typing a name into the text entry field in the Schematic View toolbar. Hit the **<ENTER>** key to save the layout. You can load the layout later by choosing it from the list in the Main Menu. Click **Graph Editors > Saved Schematic Views,** and choose your saved layout from the list.

LINKING AND PIVOT POINTS

Objects always transform relative to their Pivot Points. When linked objects are rotated, they rotate around their Pivot Points. When working with a linked chain, Pivot Points must be placed properly *before any animation is created.*

To move an object's Pivot Point, select it and open the **Hierarchy** panel . The **Pivot** button at the top of the panel should be active. Click **Affect Pivot Only**. The object's Pivot Point is displayed in the viewports as a large red, green, and blue axis tripod.

Use **Select and Move** to move the Pivot Point. You may also click on other objects and move their Pivot Points.

When you have finished moving Pivot Points, turn off the **Affect Pivot Only** button before continuing with your work.

LINKING AND SCALING LIMITATIONS

Beginners with 3ds Max often experience a strange effect related to scaling and linking. When an object is scaled with the Select and Scale transform, and then later linked, there will be problems. Any rotation performed on the object will behave erratically. You will see this as a very bizarre skewing of child objects.

The rule to follow is that linked objects must have Scale transform values of 100,100,100. If you really need to resize an object before linking it, *don't use the standard Select and Scale tool.* Instead, do one of the following:

- Use object parameters such as **Height**
- Use the **XForm** modifier to scale the object
- Scale all of the faces within the object at the sub-object level

TIP: If you do find yourself in a desperate situation with a scaled object, you can use a tool found in the **Adjust Transform** rollout of the Hierarchy panel. In the **Reset** section of the rollout is a button labeled **Scale**. If you click the Scale button, the scale of the selected object will be reset to **100,100,100**, while the shape of the object is retained.

COMBINING SCENES AND OBJECTS

MERGING SCENES

When working with scenes in 3ds Max, you will often find that you wish to merge objects from one scene to another. For example, you or another artist may model an environment in one scene, a character in another, and a prop in yet another. All of these scene elements must be merged together.

To do so, open the target scene. This is the scene you wish to merge objects into. From the Main Menu, choose **File > Merge**. A file selector appears. Choose a **.MAX** file in order to merge objects into the current scene.

After the scene is selected, a list of objects appears. Every object in the selected scene is displayed, including cameras, lights, space warps, and hidden and frozen objects. Select the objects you want to bring into the scene.

If the names of any of the incoming objects have the same names as objects in the scene, the **Duplicate Name** dialog appears.

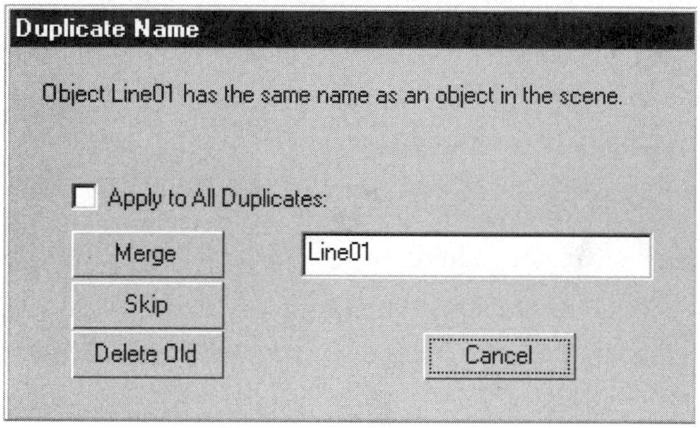

Figure 6-57: ***Duplicate Name*** *dialog*

Here you can choose to **Merge** the object anyway, which will result in two objects in the scene with the same names. You can also choose to **Skip** the object and not merge it at all. The **Delete Old** option deletes the object in the current scene.

Another option is to enter a new object name in the entry area. The object will be merged into the scene with the new name.

Right here is a very good argument for naming objects intelligently. If all your object names are the default names, it will be very hard for you to figure out what you're merging.

Object Scale

Sometimes you will merge an object into a scene only to find that it is much too large or too small for the scene. If you know in advance that you will be merging objects, you can prevent this problem by preparing ahead of time. Model all of your objects to the size they would actually be in the real world. In this way, all your merged objects will already be to scale.

To do this, you'll need to tell 3ds Max what a Unit should represent. By default, 3ds Max displays Generic Units, which are not related to any real world measurements. You could decide, for example, that each Unit represents one centimeter. So, before you even begin your project, go to the **Customize > Units Setup** and change the **Display Unit Scale** in the dialog.

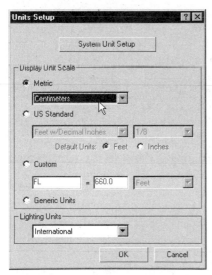

Figure 6-58: Units Setup dialog

When you change the Display Unit Scale, 3ds Max displays sizes and distances in the units you specify. The absolute size of your scene does not change. So, you can merge scenes even if their Display Unit Scale is not the same. Let's say you modeled a conference table in Meters. Then you modeled a telephone in another scene, set to Inches. As long as you built the objects so they matched real world measurements, you could easily merge the two scenes and there would be no problem.

WARNING: The button at the top of the Units Setup dialog is labeled **System Unit Setup.** This button launches a dialog that *does* change the absolute size of your scene. This should be left alone unless you are modeling a scene that is incredibly huge or incredibly tiny. Do not change the System Unit Setup! Leave it at the default of **1 Inch.** Changing the System Unit Setup can even corrupt your scene.

Fixing Problems with Scale

If you do experience problems with merged objects being the wrong size, there are several things you can do. Your first instinct may be to simply Scale the offending objects so they match your scene, but this is generally a bad idea. Remember that scaling can cause problems with linked objects in an animation.

If you do use the standard Scale transform, remember to use the **Reset Scale** command on the **Hierarchy** panel. This restores the Scale values to **100%** without changing the size of the object. Once the Reset Scale command is complete, you can link and animate the objects normally.

Instead of using the standard Scale transform, you can apply an XForm modifier. This is a better idea, but it unnecessarily complicates the scene by adding modifiers.

There is a better solution. On the **Utilities** panel you will find a button labeled **More**. Click the More button to access a dialog that lists miscellaneous **Utilities**. Select **Rescale World Units** from the Utilities dialog, and click OK.

A new rollout appears on the Utilities panel. Click the **Rescale** button, and the **Rescale World Units** dialog appears. Here you can change the size of the entire scene, or of selected objects, without altering the Scale transform values or adding unnecessary XForm modifiers. The scale of the scene or selected objects is multiplied by the value in the **Scale Factor** spinner field. For example, a value of **10** makes the objects ten times their original size, and a value of **0.1** makes the objects one-tenth of their original size.

WARNING: Be aware that complex setups, such as character animation rigs, generally will not scale well at all. If you build a character rig to the wrong size and then attempt to scale it later to fit an existing scene, you will most likely have many problems. Avoid these problems by constructing all of your scene elements to real-world scale.

GROUPS

In 3ds Max, you have the option of putting objects into Groups. A **Group** is a collection of objects that have been glued together, and can be unglued at any time. A Group has a separate name that appears on selection lists. The individual objects inside a Group do not appear on selection lists unless the Group is opened.

To group objects, select them, then choose **Group > Group** from the Main Menu. You will be prompted for a name for the new Group.

Any time you click any one of the grouped objects, the entire Group is selected. The Group moves, rotates, and scales as one unit. Any modifiers or animation applied to the Group are applied to all objects together. In selection lists, the Group name appears in brackets to indicate that it is a Group and not an object.

You can ungroup the objects at any time by selecting the Group and choosing **Group > Ungroup** from the Main Menu. This destroys any animation applied to the Group.

If you want to temporarily work with an object inside the Group, you can select the group and choose **Group > Open** from the Main Menu. Individual objects in the Group appear on selection lists. Choose **Group > Close** when you are finished.

Generally speaking, Groups are not good for animation. If you want to animate several objects together, it is better to create a Point Helper, and parent the objects to the Helper. However, grouping is handy for managing a collection of objects that are not animated, such as a bunch of furniture in a room. Grouping the objects would allow you to position all the furniture by moving one Group.

Grouping can also help when merging scenes. Before saving a scene that you know you will merge with another scene later on, group all the relevant objects together with a meaningful name. It will be very easy to pick the Group from the selection list when you merge the objects later on.

END-OF-CHAPTER EXERCISES

EXERCISE 6.7: Hovering Spaceship

In this exercise, you'll add motion to the space scene.

1. Load the file **SpaceScene03.max** from your folder or from the disc.

2. Hide the asteroid. Select it, then right-click and choose **Hide Selection** from the Quad Menu.

3. Using any modeling method, create a spaceship or flying saucer to appear in the scene.

4. Animate the spaceship to have it fly quickly into the scene, stop in midair for a few moments, then move offscreen quickly. Use the techniques you learned in this chapter to keep the spaceship from drifting as it hovers.

5. Save the scene with the filename **SpaceScene04.max** in your folder.

EXERCISE 6.8: Linked Character

In this exercise, you'll link a simple character together and change Pivot Points as necessary. Then you'll animate the character.

1. Load the file **Guy.max** from the disc that comes with this book. This scene consists of a very simple character made from spheres and cylinders.

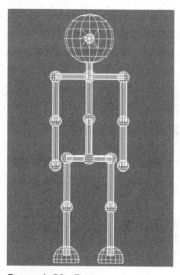

Figure 6-59: Guy

2. Select and move a few of the objects. Hold down **<CTRL>** and press **<Z>** to undo each movement, so the model is restored to the way it was originally.

    By moving the objects around, you can see that none of them are linked together.

3. Click **Select by Name** and look at the list of objects. The objects have been named for body parts, such as **Upper arm right**.

    Next, you'll link the body parts together.

4. Maximize the Front viewport. Select the left foot. Click **Select and Link** on the Main Toolbar. Click and drag from the foot to the calf. When the link cursor appears, release the mouse button.

5. Select the left calf. Click and drag to the thigh to link it to the thigh.

6. Select the right foot, and link the right leg parts together in the same way.

7. Link the arms together in the same way, always starting from the outer extremities and working inward. For example, you should link the hand to the lower arm, and the lower arm to the upper arm.

8. Link both upper arms to the shoulder object. Link both thighs to the hip object.

9. Link the head to the shoulder object.

10. Link the shoulder object to the chest, and link the chest to the hip object.

The hip object is the only object that is not linked to any other object. This makes the hip object the root parent.

Next, you'll check the link structure.

11. Click **Select Object**, then click **Select by Name** 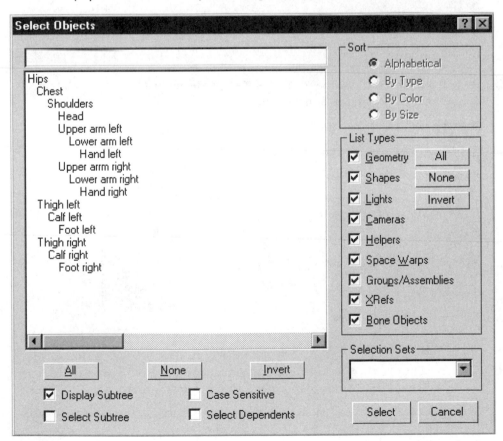. In the **Select Objects** dialog, check the **Display Subtree** check box.

12. Look carefully at the indented child objects, and ensure the object Hips is the only one directly against the left side of the dialog display. If there are other objects at the left, these objects have not been properly linked. Try again to link the objects correctly until the display on the **Select Objects** dialog looks right. See the following illustration.

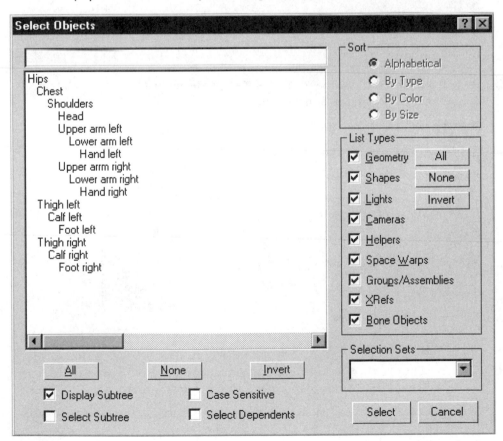

Figure 6-60: Select Objects dialog

13. Save the scene with the filename **GuyLinked.max** in your folder.

14. Click **Maximize Viewport Toggle** to return to the four-viewport display.

15. In the Left viewport, rotate one of the feet. The foot rotates from its center, which is incorrect. Hold down **<CTRL>** and press **<Z>** to undo the rotation.

The foot's Pivot Point is in the wrong place.

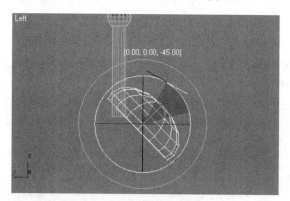

Figure 6-61: Foot rotates from center

16. Go to the **Hierarchy** panel 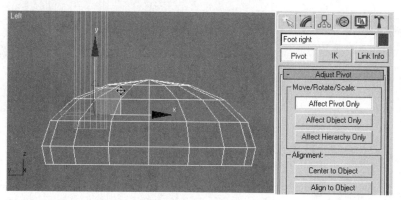. Click **Affect Pivot Only**. Move the foot's Pivot Point to the area where the leg meets the foot.

Figure 6-62: Pivot Point moved

17. Move the other foot's Pivot Point in the same way.

18. Check the Pivot Points of all other objects by rotating them. Hold down **<CTRL>** and press **<Z>** after each rotation to return the object to its original rotation.

The Pivot Point for the head needs to be moved to the base of the neck. The rest of the Pivot Points are already in the correct positions.

19. Save the scene again.

Next, you'll animate your linked guy to make him do a dance.

20. Turn on the **Auto Key** button. Go to a frame other than **0**, such as **20**. Rotate the character's arms so they raise in the air.

21. Go to frame **35**. Raise one of the character's legs.

22. Continue to animate the character in any way you like.

    Animate the character on frames that are multiples of **5**, such as **15, 20, 25**, and so on. This will make it easier to return to existing keyframes later on if changes are needed.

23. Save the scene with the file name **GuyDancing.max** in your folder.

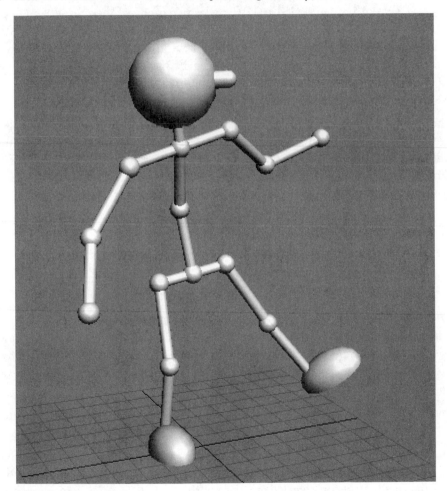

Figure 6-63: Guy Dancing

SUMMARY

Animation is accomplished with *keyframes* or *keys*. Keyframes can be created for any transform or parameter if the **Auto Key** button is on. Keys for selected objects can be viewed, moved, and copied using the **Track Bar** in the Status Bar area.

Keyframe data can be edited in the **Motion** panel, in **Parameters** mode. **Trajectories** mode lets you see and edit the path of an animated object's position.

Keyframes are stored in *tracks*, which are channels of animation data. To access individual tracks, open a **Track View** window. Track View has two modes: **Dope Sheet** and **Curve Editor**. Each is designed for a different type of animation editing. The Dope Sheet is primarily used for dealing with the location of keyframes in time. Two of the main modes within the Dope Sheet are **Edit Keys** and **Edit Ranges**. Edit Keys gives you precise control over each keyframe. Edit Ranges allows easy changes to entire tracks, such as shifting or scaling keys.

When animation is created, 3ds Max *interpolates* the values between keyframes. Keyframe interpolation can be edited using the Curve Editor. In the Curve Editor, animation data is displayed as graphs, where the height of the graph indicates the keyframe value. For each keyframe, you can select one of seven different interpolation types for the incoming and outgoing curve. 3ds Max calls these **Tangent Types**.

Most function curve editing is done with a single type of keyframe interpolation: **Custom Tangents**. Custom Tangents keyframes are Bezier curves with editable vector handles. The keyframes or handles may be moved to change the shape of the curve.

The Track View lets you add specialty tracks. Each object may have an animatable **Visibility** track. You can also add a soundtrack to your animation within Track View.

3ds Max must determine what values an animation track has before and after the keyframes. You can choose how an *extrapolated* animation curve will behave by choosing one of the **Parameter Curve Out-of-Range Types**. In this way, you can make an animation track repeat in an endless loop, or other options.

Linking or *parenting* attaches objects together. This creates an *animation hierarchy*. A *child* object is linked to a *parent*. A child always inherits the transforms of its parent, but the parent does not inherit the transforms of the child.

A hierarchy of linked objects is also called a *chain* or a *tree*. A parent may have many children, but a child may have only one parent. Children may have children of their own, and the tree may become very deep and complex.

Objects within an animation hierarchy are called *nodes*. The parent object that is not linked to any other object is called the *root node*.

Before animating, you should always check the location of Pivot Points. They must be located in the correct place for the animation to work properly.

REVIEW QUESTIONS

1. Locate the **Track Bar** on your screen. What does it display?

2. What can you do with the **Motion** panel?

3. What kind of animation data can you edit with **Trajectories**?

4. What is **Track View** for?

5. What are the two modes of Track View?

6. How do you make Track View display only animated tracks?

7. How do you copy keys in Track View?

8. What is a *function curve*?

9. How do you edit function curves?

10. What are the different types of *keyframe interpolation* in 3ds Max?

11. What is the most commonly used type of keyframe interpolation?

12. How do you make an animation loop?

13. What is *linking*?

14. True or false: In a standard *animation hierarchy*, parent objects inherit the transforms of their children.

15. What is a *root* object?

16. Name two ways to check the link structure after linking objects together.

17. What should you do if you need to scale an object that will be part of an animation hierarchy?

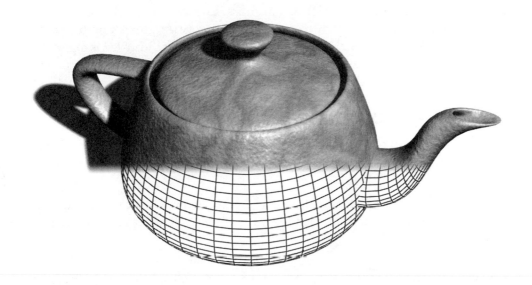

Chapter 7
Rendering

OBJECTIVES

In this chapter, you will learn about:

- Options in the Render Scene dialog
- File types and image sequences
- Production and Draft rendering
- ActiveShade for interactive rendering
- Preview movies
- Rendering a portion of a viewport
- The Rendered Frame Window
- Antialiasing
- Motion blur

ABOUT RENDERING

So far, you have seen that rendering an image is as easy as clicking a button in 3ds Max. However, there are a number of ways you can improve and fine-tune the rendering process to get the best possible renderings from your scenes. 3ds Max is a professional production tool. For this reason, there are many choices for the rendering process. In this chapter, you'll learn about the most important and often used rendering features.

PREPARING TO RENDER

A rendering can be initialized in a number of ways with 3ds Max. First, click in the viewport you wish to render. Then you can begin the rendering process in any of the following ways:

- Choose **Rendering > Render** from the Main Menu. The **Render Scene** dialog appears. Set parameters and click the button labeled **Render** at the bottom of the dialog.

- Click the **Render Scene** button on the Main Toolbar, or press **<F10>,** to launch the Render Scene dialog.

- Click the **Quick Render** button on the Main Toolbar. The current viewport renders with the last settings entered in the Render Scene dialog.

- The **Quick Render (ActiveShade)** button is available on the flyout that includes the Quick Render button. This launches a new ActiveShade window, which allows you to see an interactive preview of how lights and materials will look when rendered. You can also load ActiveShade into a viewport. ActiveShade is discussed later in this chapter.

RENDER SCENE DIALOG

The **Render Scene** dialog sets rendering parameters. Here, you can control nearly every aspect of rendering.

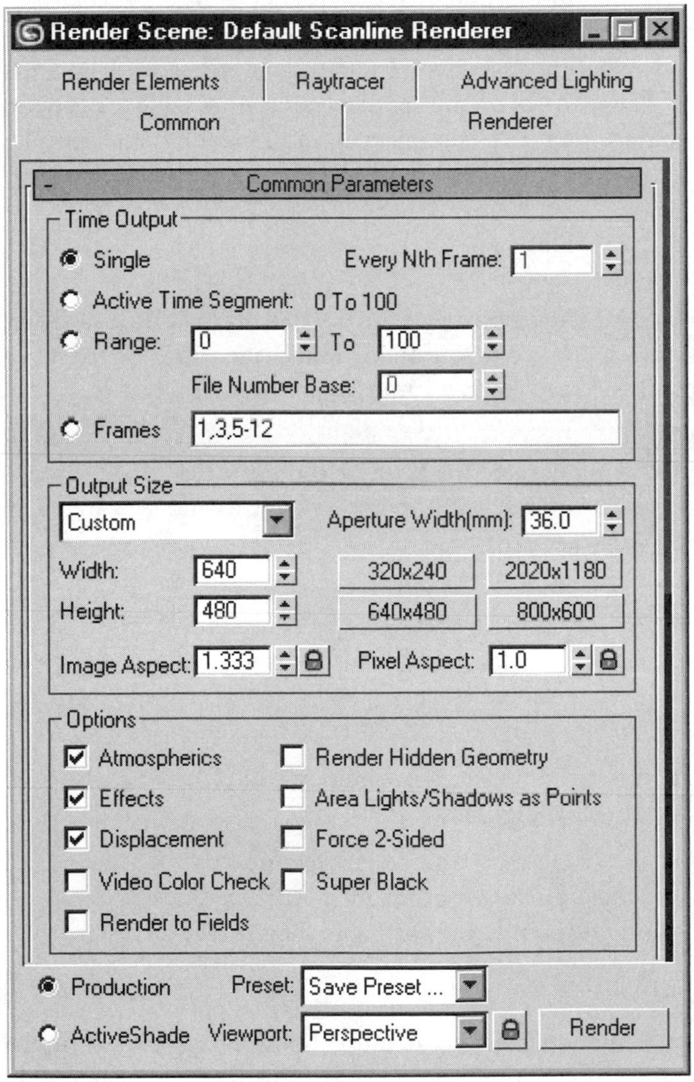

Figure 7-1: **Render Scene** *dialog*

The **Common** tab of the dialog is where you define the most basic aspects of the rendering, such as frame range, resolution, and file type. You can also choose a different rendering software module from the **Assign Renderer** rollout. 3ds Max ships with two rendering modules: the **Default Scanline Renderer** and the **mental ray** renderer.

The other tabs in the Render Scene dialog are more specialized. The **Renderer** tab holds parameters that are specific to the particular rendering software currently assigned. So, if you assign **mental ray** as your current renderer, you'll see the mental ray parameters in the Renderer tab.

The **Raytracer** and **Advanced Lighting** tabs control the options for advanced shading techniques, which are discussed in Chapters 11 and 12. The **Render Elements** tab allows you to split out different parts of the rendered image to separate files, to allow more control during compositing. For example, you can render out the specular highlights separately from the Diffuse Color component of the scene, so the highlights can be adjusted later.

COMMON TAB

The **Time Output** section of the Common tab controls the number and range of frames to be rendered. To render just a single frame, choose the **Single** option. To render a sequence of frames, choose the **Active Time Segment**, or enter a **Range** of frame numbers to render.

You can skip frames by entering a number into the **Every Nth Frame** field. For example, entering the number **10** in this field will tell 3ds Max to render every tenth frame, and skip the rest.

The **Output Size** section of the dialog determines the resolution of the rendered image or animation. You can enter any resolution you like for the **Width** and **Height** of the image. When practicing with 3ds Max, use 320x240 or 640x480. In a professional situation, the image resolution is determined by such factors as a client's needs or the native resolution of a video editing system.

There are many preset resolutions available from the pulldown list. You may also customize the preset buttons by right-clicking them.

The **Options** section allows you to enable or disable certain features of the renderer. The **Atmospherics** check box, for example, refers to fog and other Atmospheric effects, which you might want to turn off during test renderings to make the process go faster.

FILE OUTPUT

To save the rendered image or animation to a file, you must choose a file type and enter a filename in the **Render Output** section of the Common tab in the Render Scene dialog. To set a filename, click the **Files** button. The **Render Output File** dialog appears.

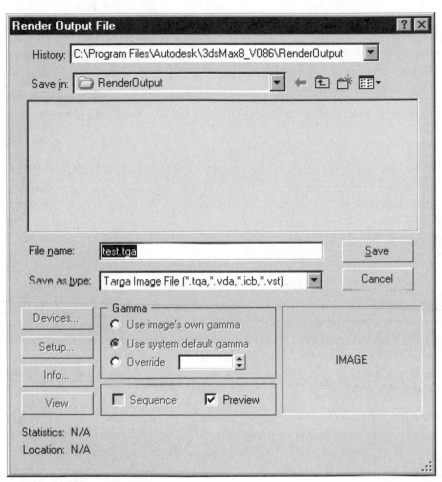

*Figure 7-2: **Render Output File** dialog*

On the **Save as type** pulldown, choose a file type. These are all of the file types to which 3ds Max can render.

Animation formats such as Microsoft **.AVI** and Apple QuickTime **.MOV** are generally used for test renderings. For final production renders, you should always render to a still image format such as **.TGA**. The result will be a series of sequentially numbered files. For example, if you render an animation and set the filename as **CUBE.TGA,** the animation will be rendered as a series of files named **CUBE0000.TGA, CUBE0001.TGA, CUBE0002.TGA,** and so on. This is called an *image sequence*.

After you have chosen a file type and entered a filename, click **Save**. A **Setup** dialog may appear. Each Setup dialog is different, and provides options especially for the selected file type. When you click **OK** and return to the **Render Scene** dialog, the selected folder and filename appear next to the **Files** button.

The **Save File** check box is also automatically checked when a filename is selected. If you later decide to render to the screen only, you can uncheck the **Save File** check box.

File Types

Big animation jobs are *never* rendered directly to movie files such as **.AVI** or QuickTime. Rendering to a movie file is a dangerous practice. If the rendering fails due to human or machine error, low disk space, or any other reason, you will probably lose the entire rendering. This is not much fun if you've been rendering for hours or days and have to start over again, and it could mean losing the job.

However, if you render to a still image format and something bad happens, you will only lose one frame, instead of hundreds or thousands of frames.

Also, **.AVI** and QuickTime movies tend to take up a huge amount of disk space. One minute of animation rendered to uncompressed video resolution will be nearly two gigabytes in size. This presents real problems for file management and backup. For example, a file of that size won't fit on a CD. Again, rendering to an image sequence is the answer. You can break the sequence up into groups of frames, storing several hundred of them on a CD.

For most animation jobs, you'll choose the **.TGA** format, because it is the most widely supported file type for video. If you need to store an alpha channel for compositing, save as a 32-bit **.TGA**. If you won't be compositing the animation later, save as a 24-bit **.TGA**, which will save a lot of disk space.

Compression

Another reason **.AVI** and QuickTime files are undesirable is that these formats usually use *lossy compression*. This is a technique for discarding information to save disk space and speed up transmission of images over the Internet and other networks such as digital cable and satellite systems. There are many examples of lossy compression algorithms, such as MPEG, DivX, Sorenson, Indeo, Microsoft Video, DV, and so on.

You should never, ever render a final production animation to a lossy compression format. Lossy formats should only be used for test renders. Data is discarded during the rendering process, you are left with a poor-quality image, and you can never restore the missing data. Your only choice would be to re-render the project from the beginning.

Instead, always render to an uncompressed format. If you need to compress the animation to put it on the Internet or distribute it on a CD or DVD, you should render to a **.TGA** sequence, then use a program such as After Effects, Premiere, Combustion, or Cleaner to compress the movie. That way, you can try out lots of different compression settings until you find the right combination that works for your project.

TIP: If you don't have a compositing or compression application, you can use the **Video Post** module within 3ds Max to convert an image sequence to a compressed movie. Video Post is discussed in Chapter 8, *Special Effects.*

The **.TGA** format can be compressed or uncompressed. However, the compression used in **.TGA** files is *lossless*, which means no information is discarded. The file size is merely compacted. It is safe to render to a compressed **.TGA** sequence, and in fact this is the default in 3ds Max.

TEST RENDERS

RENDERING PRESETS

3ds Max allows you to keep multiple complete rendering setups available at all times. They are stored in presets. These presets are saved to disk in a special format with the extension **.RPS**.

Numerous tests are required before the final rendering is started. Full production renders, with fancy options such as Raytracing and Advanced Lighting, take a long time to render. Usually there isn't enough time for full renderings during the animation process, so artists must render lower quality "draft" renders for their tests. The settings used for draft rendering greatly differ from those for final production rendering. If you had to reset all the parameters manually when you were ready to do your production rendering, you could easily miss one or two, leading to a lot of wasted rendering time.

Default render presets are included as a convenience for animators working under deadlines in production environments. You can quickly choose a preset by selecting one from the **Preset** pulldown list at the bottom of the Render Scene dialog. Another dialog, labeled **Select Preset Categories**, appears. Here you can opt to load some or all of the parameters stored in the preset you have chosen. Usually you'll leave all of the categories highlighted, and click **Load**. See the following illustration.

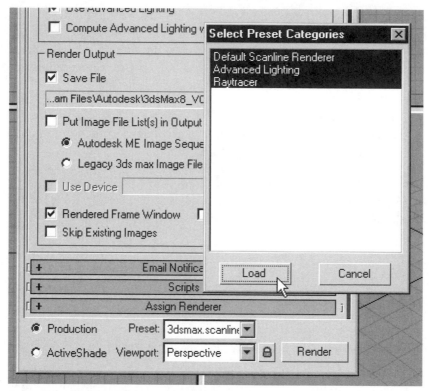

Figure 7-3: **Select Preset Categories** *dialog*

To use your own customized render presets, choose the **Load Preset** or **Save Preset** items from the bottom of the Preset pulldown list. By default, **.RPS** files are stored in the **renderpresets** directory of your 3ds Max program folder.

In addition, the **Render Shortcuts** toolbar gives you access to the same rendering presets. This toolbar is hidden by default; to show it, right-click an empty area of another toolbar and choose Render Shortcuts from the pop-up menu.

The Render Shortcuts toolbar also gives you the ability to store quick presets. Once you have a render setup you like, simply hold down **<SHIFT>** and click one of the Render Shortcuts buttons, labeled A, B, or C. The current settings from the Render Scene dialog are stored in a **.RPS** file in the **renderpresets** directory, and the settings can be recalled at any time by clicking the button. These files, **A.RPS**, **B.RPS**, and **C.RPS**, are not saved with the 3ds Max scene.

Figure 7-4: **Render Shortcuts** *toolbar*

ACTIVESHADE

ActiveShade provides a nearly real-time update to changes in materials and lighting. This allows you to see the results of lighting and material edits almost instantly, without having to wait for a complete render of the scene. It can be a real time-saver.

ActiveShade works by rendering in a two-step proces. First, the ActiveShade renderer calculates scene geometry and camera parameters. Second, materials and lighting are applied to the scene. ActiveShade holds the scene geometry and camera data in memory, so it doesn't have to recalculate it. Then, when you change lights or materials, the renderer only has to calculate those changes, which greatly speeds up the rendering.

To use ActiveShade, choose **Quick Render (ActiveShade)** from the flyout on the far right of the Main Toolbar, or select **Rendering > ActiveShade Floater** from the Main Menu. This launches a new window which renders the current viewport with the ActiveShade renderer. You can also load ActiveShade into the current viewport by selecting **Rendering > ActiveShade Viewport** from the Main Menu, or by right-clicking on the viewport label and choosing **Views > ActiveShade**. Only one ActiveShade viewport or window may be displayed at a time.

ActiveShade may take a moment to calculate the scene geometry. During this time, the window will be black. You can see a progress indicator as red and green lines at the top and the right edge of the window. If an object is selected before opening ActiveShade, only that object will be rendered.

When ActiveShade has completed its first rendering, you may then make changes to the lighting and materials in the scene. These changes will be updated almost immediately in the ActiveShade window.

However, ActiveShade won't respond to changes made to objects or cameras. For example, if you move an object, ActiveShade will not change. To re-render the ActiveShade window, right-click in the window and choose **Initialize** from the Quad Menu.

ActiveShade render settings are completely separate from those of the current production renderer. At the bottom of the Render Scene dialog, you can choose the **ActiveShade** radio button instead of the default, which is **Production**. Now, any changes you make to the various rendering parameters will only affect ActiveShade, and not the production renderer. In addition, ActiveShade works with rendering presets.

In some cases you can use ActiveShade to test your renderings. However, the only way to be certain how a scene will look is to use the production renderer.

ANIMATION PREVIEWS

A *preview* is a limited rendering that can be made very quickly to help you see how your animation is coming along. In a complex scene, the real-time viewports cannot render the scene at an acceptable frame rate, making it impossible to evaluate the animation. If your scene is too "heavy" to render in real time, you can render a preview to a file, and play that file back from your hard disk.

Previews can be rendered much more quickly than draft renderings using the standard 3ds Max renderer. If you are using Direct3D or OpenGL hardware acceleration, the preview renderer will take advantage of this, further speeding up the preview rendering process. The 3ds Max production renderer is a software renderer, which means it does not use 3D acceleration.

A preview can only render what you see in a viewport. Previews don't render shadows, reflections or atmospheric effects, and only maps that display in shaded viewports can be seen in the preview. However, previews have the advantage of being able to include nonrendering elements such as lights and cameras. This enables you to see how they interact with the scene.

Rendering a Preview

To make a preview, choose **Animation > Make Preview** from the Main Menu. In the **Make Preview** dialog, choose the range of frames, the objects to appear in the preview, and various other options. The default file format is .AVI, but you can choose another format if you wish.

Click **Create** to create the preview. If you haven't yet chosen a *codec* (compression/ decompression algorithm), the **Video Compression** dialog will pop up.

A progress bar at the bottom of the 3ds Max user interface shows you how much of the preview has been rendered. If you render to an `.AVI` file, Windows Media Player automatically opens and plays the file.

By default, the preview is saved in a file called `previews\_scene.avi`. The next time you create a preview, this file will be overwritten. To save a preview from being overwritten, choose **Animation > Rename Preview** from the Main Menu.

RENDERING A PORTION OF A VIEWPORT

Instead of rendering the entire viewport each time, you can specify a portion of the viewport to be rendered. This is accomplished with the **Render Type** pulldown on the Main Toolbar.

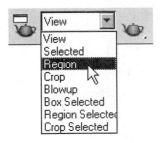

Figure 7-5: **Render Type** *pulldown*

The following options are available on the **Render Type** menu:

View renders the entire viewport. This is the default setting.

Selected renders selected objects only. The previously rendered image remains in the image display, and selected objects are rendered over it. If two or more objects are selected, the area between selected objects renders as the background.

Region renders a specified region. When you click Render in the Render Scene dialog, or click the Quick Render button, a rectangular selection region appears in the current viewport.

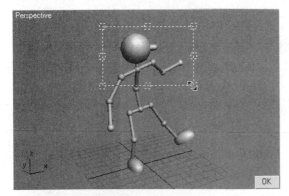

Figure 7-6: Render Region

Move the handles of the selection region to surround the region you wish to render. Hold down the **<CTRL>** key when moving handles to preserve the current aspect ratio. Click and drag the center of the selection region to move it. Click the **OK** button at the lower right of the viewport to start the rendering. The previously rendered image remains in the image display, and the selected region is rendered over it.

The Selected and Region render types can save rendering time by rendering only objects or regions of the screen that have changed since the last render. There are other Render Types, but Selected and Region are the most commonly useful.

EXERCISE 7.1: Rendering a Portion of a Viewport

In this tutorial, you'll practice rendering a portion of a viewport.

1.   Load the file **AtticWithCamera.max** from your folder or from the disc.

     This is a scene of an attic with a camera placed in it. The lower-right viewport shows the **Camera01** view.

2.   Activate the Camera01 view.

3.   Click the **Render Scene** button on the Main Toolbar. The **Render Scene** dialog appears. The following settings should be visible in the dialog by default:

Time Output:	Single
Output Size:	640x480
Render Output:	**Save File** unchecked

4.   Click **Render** to render the scene.

Figure 7-7: Rendered attic scene

Next you'll change one of the materials in the scene.

5. Open the **Material Editor**. Select the first sample slot.

6. On the Material Editor toolbar, click **Pick Material from Object** , then click on the dress dummy to put its purple color in the sample slot.

7. On the **Blinn Basic Parameters** rollout, change the **Diffuse** color from purple to yellow.

   Next, you'll render only that portion of the scene that has changed.

8. From the **Render Type** pulldown, choose **Region**.

9. Click the **Quick Render (Production)** button on the **Main Toolbar.** A dotted rectangular selection region appears in the Camera01 viewport.

10. Click and drag the handles in the rectangular selection region so they surround the yellow dress dummy.

Figure 7-8: Selection region around dress dummy

11. When the region is placed, click the **OK** button at the bottom right of the viewport.

    The rendering is very quick because only the selected region is rendering.

VIEWING RENDERED IMAGES

In order to learn how to get the best possible renderings from 3ds Max, you must first understand the components that make up the rendering process.

RENDERED FRAME WINDOW

When an image is rendered, it is displayed in a separate window called the **Rendered Frame Window**, or RFW for short. A *frame buffer* is a piece of hardware that displays video images on a screen. The Rendered Frame Window is a software component of 3ds Max that displays images in a separate window.

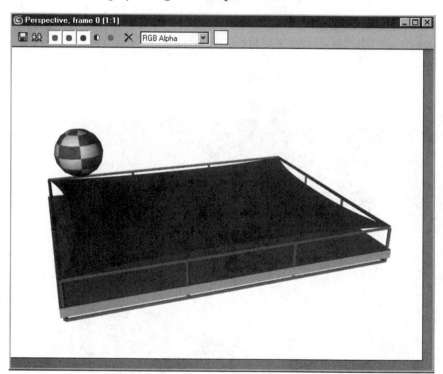

Figure 7-9: **Rendered Frame Window**

The buttons on the RFW toolbar change how an image is displayed but don't affect how the image is saved on disk.

 Save Bitmap stores the image to disk.

 Clone Rendered Frame Window makes a copy of the RFW window. This is very helpful for comparing changes to the rendering. Click the Clone button and leave the cloned copy of the RFW open. When you re-render the scene, the original RFW window is overwritten, and you can compare it to the cloned copy.

Display Alpha Channel shows the alpha channel of the current rendering. Refer to Chapter 1 for a discussion of alpha channels.

 Monochrome displays the rendering as a grayscale image. The saturation of all pixels in the image is reduced to zero, giving a black-and-white image.

 Clear erases the contents of the RFW.

For information on the rendered image, right-click and drag across the RFW. The cursor turns to an eyedropper, and a pop-up window appears with information about the image. As you move the cursor across the image, the data for the pixel under the cursor is displayed in the pop-up window. When you release the mouse, the last color accessed appears in a color swatch at the upper right of the RFW. You can drag this color elsewhere; for example, to a Material Editor color. You can also click the color swatch on the RFW toolbar to access the Color Selector and see the color's values. Any changes to the Color Selector are not recorded in the image.

The camera and frame number are displayed in the window bar at the very top of the RFW. In addition, the *zoom ratio* is also displayed there. If the rendered image is too high resolution to fit on your screen, 3ds Max will automatically shrink it to fit. If the image is displayed at half of its actual height, the zoom ratio is 1:2. If the image is displayed at double its actual height, the zoom ratio is 2:1.

If the zoom ratio is not 1:1, then you will see jagged edges in the RFW. Don't worry, your image is not really rendered that way; it has merely been resized to something other than its actual resolution. You can zoom into the RFW by holding down the **<CTRL>** key and clicking on the image. To zoom out, hold down **<CTRL>** and right-click in the RFW. If you have a mouse with a wheel, you can use the wheel to zoom in and out of the RFW. To pan the image in the RFW, click and drag with the middle mouse button.

The viewing options in the RFW window are designed to help you analyze your renderings. Although you can change the RFW options to view the image in many different ways, you cannot save those changes.

What the RFW Does

The RFW serves as a viewing window for your rendered images. Since the RFW is in its own window, you can copy the RFW and keep various versions of your rendering as you work.

By default, when you render a scene, the RFW appears, and the rendered image appears in it line by line as the scene is rendered.

In the **Render Scene** dialog, the **Rendered Frame Window** check box is on by default. If you uncheck this option, the RFW will not appear when the scene is rendered. This may speed up your rendering slightly, but you won't be able to catch any mistakes.

When rendering a single frame that you want to save as an image file, you have the choice of using the Render Scene dialog to enter a filename before rendering, or rendering the image without a filename and clicking the **Save Bitmap** button in the RFW to save the image. Animation, on the other hand, can only be saved by entering a filename through the Render Scene dialog.

Show Last Rendering

There will be times when you need to see the results of the most recent rendering after you have closed the Rendered Frame Window. You could re-render the scene from scratch, but this might take a long time. Instead, you can simply view whatever is left over in the RFW. Choose **Rendering > Show Last Rendering** from the Main Menu.

When you exit 3ds Max, the RFW is flushed.

View Image File

You can also use the Rendered Frame Window to view images stored on disk. Choose **File > View Image File** from the Main Menu. The **View File** dialog opens. Browse your system for an image file. Remember that the dialog box will only show files that match the format specified in the **Files of type** pulldown, so it's best to choose **All Files (*.*)**. That way you can see all files, regardless of their type.

When you click **Open** in the View File dialog, a Video Frame Buffer window appears, displaying the image. You can view any image file supported by 3ds Max. It doesn't have to be a file rendered in 3ds Max; it could be a photograph.

The View Image File command will also let you view movies, such as **.AVI** or Quick-Time files. In that case, the movie player application, such as Windows Media Player or Apple QuickTime Player will play the file, not the RFW.

The Rendered Frame Window is invoked in other places in 3ds Max. For example, if you click **View Image** within the Material Editor, a RFW window opens, displaying the bitmap used as a map within your material.

RAM PLAYER

The **RAM Player** loads an animation file or a sequence of frames into memory and plays them back at a selected frame rate. This allows you to view a series of sequential images as an animation.

The RAM Player is superior to standard playback programs such as the Windows Media Player because it plays the animation from RAM, not from the hard disk. If an **.AVI** or **.MOV** file is very large, it might not play back smoothly from disk. When an animation file is played with the RAM Player, the playback is always very smooth. The only disadvantage of RAM Player is that you might not have enough RAM in your computer to play back a very long sequence, or a very high-resolution sequence.

To access the RAM Player, choose **Rendering > RAM Player** from the Main Menu. When the RAM Player window opens, load an animation file or animated sequence.

Click the **Open Channel A** button ⌸ next to **Channel A** on the RAM Player toolbar. The **Open File** dialog appears. Choose the animation file, or choose the first file in a numbered sequence. If you wish to load a sequence, click the **Sequence** check box.

If you have chosen an image sequence, and clicked the Sequence button, the **Image File List Control** dialog will appear. Here, you can choose a frame range and other options. When you click **OK**, 3ds Max saves an **.IFL** file in the same folder as your image sequence. This is an **Image File List**, which is simply a text file that lists the names of image files. The **.IFL** file is employed to load images into the RAM Player, as well as other modules in 3ds Max.

Once you have chosen the file(s) to load into RAM Player, the **RAM Player Configuration** dialog appears.

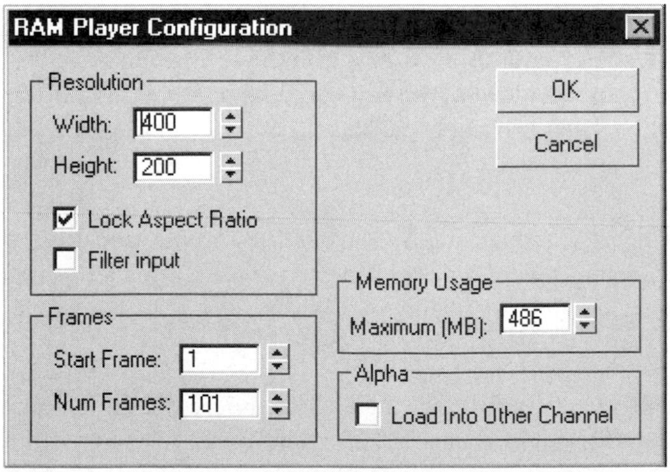

Figure 7-10: ***RAM Player Configuration*** *dialog*

Sequentially numbered files are loaded until the end of the sequence is reached, or until your computer runs out of RAM. The **Loading File** message box appears, showing the status of the animated sequence being loaded.

Each frame of the animation is displayed as it is loaded. It might take the **RAM Player** a few minutes to fully load the animation.

Figure 7-11: **RAM Player** *with animation loaded*

To play the animation, click the **Playback Forward** button. You can change the playback speed with the pulldown list at the top right of the RAM Player window. The default speed is 30 frames per second.

To analyze the animation more thoroughly, step through the animation frame-by-frame with the **Previous Frame** and Next Frame buttons on the RAM Player toolbar, or with the arrow keys on the computer keyboard. You can also right-click and drag the mouse to scroll forward or backward through the animation.

Comparing Sequences

RAM Player can also load two sequences that can be played back simultaneously. This allows you to compare two animated sequences right on the screen. To load a second sequence, click the **Open Channel B** button next to **Channel B** on the RAM Player toolbar.

The display area is split in half vertically to show the two channels. At the top and bottom of the display area are small triangles that you can slide from left to right to show more of one channel and less of another. You can also click the **Horizontal/Vertical Split Screen** button to toggle between a horizontal and vertical split.

RENDERING OPTIONS

ANTIALIASING

Aliasing is the phenomenon of harsh, jagged edges in bitmap images. *Antialiasing* refers to an algorithm designed to smooth the transition between between dark and light pixels.

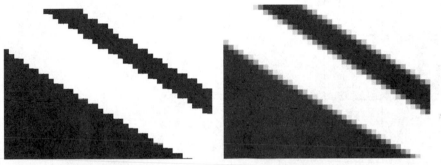

Figure 7-12: Aliased line on the left, antialiased line on the right

3ds Max gives a choice of several filters (methods) for antialiasing. To choose a filter for the Default Scanline Renderer, open the Render Scene dialog and select the **Renderer** tab. In the **Antialiasing** section, select an antialiasing algorithm from the **Filter** pulldown.

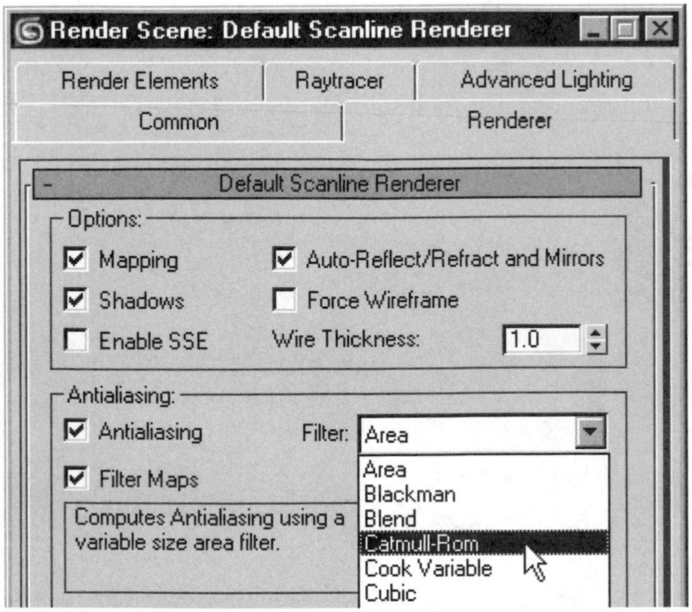

Figure 7-13: **Antialiasing Filter** *pulldown*

As each filter is selected, a short description of the filter appears in the box below the pulldown list. For some filters you can enter a custom **Filter Size**. The **Filter Size** is not the size of the rendered pixels; it's the size of the area that is looked at to compute antialiasing.

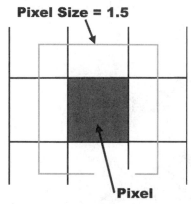

Figure 7-14: Filter Size

Filter Size is expressed in pixels. For example, a **Filter Size** of 1.5 causes the render to look at a radius of 1.5 pixels around each pixel when calculating antialiasing. The higher the Filter Size value is, the more blur will be applied to the image.

MOTION BLUR

Motion blur is a rendering effect that blurs fast-moving objects. When objects move quickly past the naked eye or a film camera, they appear blurred. Motion blur can be used to make fast-moving objects appear to be moving more smoothly and realistically.

Figure 7-15: Same object without and with motion blur

Types of Motion Blur

There are four types of motion blur available in 3ds Max: **Object, Image, Scene, and Camera** motion blur.

Image motion blur is the simplest to use and renders the fastest. Rather than engaging in a lot of calculation, Image motion blur simply applies a smearing effect to objects in motion.

Object motion blur takes a picture of the object at short intervals (fractions of a frame) before and after the current frame, then layers the pictures together. The result is more physically accurate than Image motion blur.

Some very nice effects can be made with Object motion blur, but it takes a lot longer to render. Object motion blur also requires careful setup and testing of parameters. Image motion blur, on the other hand, has fewer parameters and is easier to get working.

Scene motion blur works similarly to Object motion blur in that it takes pictures of the scene at many short intervals and layers them together. The difference is that Scene motion blur operates on the entire scene, not just one object. Use Scene motion blur when you want an exaggerated effect of motion in your rendering.

Camera motion blur is the most accurate type of motion blur, but is by far the slowest. Similar to Scene motion blur, Camera motion blur also works by rendering the scene multiple times. However, Camera motion blur looks a lot better. Unfortunately, this comes at the price of extremely long rendering times.

Setting up Motion Blur

Both Image and Object motion blur are enabled in the **Object Properties** dialog. To access the Properties for an object, select the object and right-click in the viewport. Choose **Properties** from the Quad Menu. You can also select an object and choose **Edit > Object Properties** from the Main Menu.

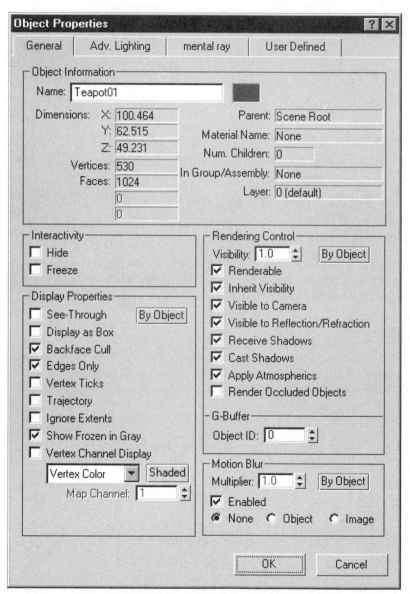

Figure 7-16: **Object Properties** *dialog*

At the bottom right of the Object Properties dialog, choose **Object** or **Image** motion blur. For Image motion blur, you can change the **Multiplier** parameter to increase or decrease the amount of motion blur.

To set the same motion blur properties for several objects at once, select the objects before accessing the Object Properties dialog. The settings you put on the Object Properties dialog will go into effect for all selected objects.

There are further settings for Object and Image motion blur on the Renderer tab of the Render Scene dialog. In general, these settings tell the renderer how many pictures to take, and at what intervals.

Scene motion blur is set up in the **Environment and Effects** dialog, or in **Video Post**, a compositing module within 3ds Max. You will learn about these features in Chapter 8, *Special Effects*.

Camera motion blur is defined in the camera's Modify panel. In the **Parameters** rollout you will find a section called **Multi-Pass Effect**. Select **Motion Blur** from the pulldown list, and click the check box to **Enable** motion blur. Then you can adjust the options in the **Motion Blur Parameters** rollout.

EXERCISE 7.2: Rendering with Motion Blur

In this tutorial, you'll practice setting up and using motion blur in a scene.

Figure 7-17: Spaceship with motion blur

1.   Load the file **SpaceScene04.max** from your folder or from the disc.

      The space scene contains a spaceship that flies onscreen, stays for a moment, then flies offscreen.

2. Select the spaceship. Right-click in the viewport, and choose **Properties** from the pop-up Quad Menu. The **Object Properties** dialog appears.

3. In the **Motion Blur** section of the dialog, choose the **Image** option. Change the **Multiplier** parameter to **4.0**. Click **OK** to close the dialog.

4. Click the **Render Scene** button. In the Render Scene dialog, change the **Output Size** to render a frame with a **Width** of **400** and a **Height** of **200** pixels. Render a still image to test the motion blur effect.

   As you render the still frame, notice that at least two passes are made, one for the image and one or more for calculating the motion blur.

   Entering new pixel values into the Output Size changed the aspect ratio of the rendering. You may need to go back to the Camera viewport and adjust the camera position so that the scene is framed properly. Safe Frames should be enabled. If they are not, right-click the camera viewport label and choose **Show Safe Frame.**

5. Render the camera view to a sequence of .**TGA** frames. This may take a while, perhaps a half hour or more. This is because there are a lot of polygons in the scene, and the light uses Ray Traced shadows.

6. When the animation has finished rendering, play it. Choose **Rendering > RAM Player** from the Main Menu. In the RAM Player, click **Open Channel A** and load the sequence. In the next two dialogs, click **OK** to accept the default values.

   Press **Play Forward** to view the animation. Note the blur on the ship as it moves.

7. If the motion blur is too fuzzy or not fuzzy enough, adjust the motion blur **Multiplier** parameter and re-render the scene. Higher **Multiplier** values make a fuzzier motion blur.

8. Save the scene with the filename **SpaceScene05Blur.max** in your folder.

SUMMARY

Although basic rendering is very easy in 3ds Max, there are a number of tools you can use to customize your renderings.

Movie files such as **.AVI** and QuickTime should only be used to test render an animation. A final production rendering is always in the form of a numbered *image sequence*, which may then be compressed and output to a movie file later.

Render Presets can be used to set up completely different rendering setups. For example, you can use one preset for test renderings and another for final renderings. This saves a lot of time and prevents the possibility of forgetting to change one of many important render settings.

ActiveShade allows you to make changes to lights and materials and to see an interactive preview of the final rendering.

A **Preview** can show you what your animation will look like if it won't play back smoothly in the viewports. A Preview can only render what you see in a viewport.

Part of a viewport can be rendered using the settings on the **Render Type** pulldown.

The **Rendered Frame Window** displays the output of the 3ds Max renderer and lets you view still images on disk. The **RAM Player** allows you to view and analyze rendered sequences more accurately than Windows Media Player or QuickTime, because playback from system memory is faster than playback from hard disk.

Antialiasing is a type of rendering algorithm designed to remove jagged edges. 3ds Max offers several different antialiasing filters.

Motion blur can be used to blur a fast-moving object, making it look more realistic.

REVIEW QUESTIONS

1. What work problem does using Render Presets help avoid?

2. How does a **Preview** differ from a full render? Name as many ways as you can.

3. How do you make a Preview?

4. How is *motion blur* enabled?

5. What are the options on the **Render Type** pulldown on the main toolbar used for?

6. What does **RFW** stand for? What is it used for?

7. Why is the **RAM Player** better than Windows Media Player?

8. Why should you output your final rendering to an image sequence?

Chapter 8
Special Effects

OBJECTIVES

In this chapter, you will learn about:

- Using Space Warps to deform objects
- Standard Particle Systems
- Particle Flow
- Using Space Warps and Particle Systems together
- Simple compositing with Video Post
- Render Effects for glows and lens effects
- Atmospheres such as Fog and Volume Lights

ABOUT SPECIAL EFFECTS

Computerized special effects are used when the real thing would be too costly, too dangerous, or just plain impossible. In years gone by, the only way to make an explosion scene for a film or television program was to actually blow something up. In the current age of computer animation tools, these special effects can be created on a computer. Not only are computers less dangerous to use than explosives, but they can also be used to create effects that can't be made with real-world tools.

3ds Max comes with several built-in special effects tools. Most are easy to use in their simplest forms. To make a special effect look realistic takes practice. This chapter is designed to get you started with special effects so you can experiment with them on your own.

SPACE WARPS

In 3ds Max, a **Space Warp** is a non-rendering object that affects other objects by distorting space. For an object to receive a Space Warp effect, it must be *bound* to the Space Warp icon. Multiple objects may be bound to a single Space Warp, and a single object may be bound to multiple Space Warps.

3DS MAX DATA FLOW

In Chapter 3, *Modeling*, you learned that each object in 3ds Max has an internal data flow, or order of operations. In a typical object, the data flow is processed in this order: object parameters, then modifiers, then object transforms, and finally object properties such as materials.

If an object is bound to a Space Warp, an additional step is made in the data flow. The effect of the Space Warp is calculated *after* the object's position, rotation, and scale transforms. The Space Warp is dependent on object transforms. Often this means that the effect of the Space Warp is localized. As a bound object moves closer to a Space Warp icon, the effect is more pronounced.

The term "Space Warp" is quite literal. The effect works by distorting the world coordinate system in the region affected by the Space Warp. Whereas most modifiers operate in the local space of the object, Space Warps work in world space. For that reason, a Space Warp is sometimes referred to as a **World-Space Modifier**.

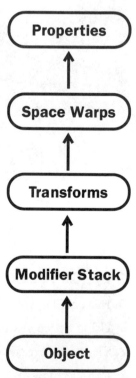

*Figure 8-1: Data flow of an object that is bound to a **Space Warp***

TYPES OF SPACE WARPS

There are several different types of Space Warps. Some of them affect geometric objects, primarily by deforming the objects. This is the type of Space Warp we will explore first.

Other Space Warps work with **Particle Systems**. A *particle system* is a collection of many animated objects, such as rain droplets or a school of fish. You will learn about Particle Systems later in this chapter.

Some Space Warps are designed to work within a **Dynamics** simulation. Dynamics systems allow objects to interact based on the rules of physics, just as they would in the real world. For example, a Dynamics simulation could automatically animate the interaction of the balls on a billiard table. 3ds Max uses the **reactor** system of dynamics.

There are six main categories of Space Warps. They are all found on the **Create** panel. Click the Space Warps button ![button] to access them. Then, select one of the Space Warp categories from the pulldown list.

Forces are simulations of energy forces such as gravity and wind.

Deflectors allow objects and particles to behave as solid objects, bouncing off one another.

Geometric/Deformable and **Modifier-Based** Space Warps change the shape of geometric objects. Most of them, such as Ripple and Bend, are the same as the similarly named modifiers, except that the deformation effect depends upon the distance from the object to the Space Warp icon.

Particles & Dynamics and **reactor** Space Warps are specialized to work in certain circumstances, such as with a simulation of a crowd of objects or characters.

Figure 8-2: **Space Warps** *on the* **Create** *panel*

SPACE WARP BINDING

To use a Space Warp to deform a geometric object, follow these steps.

1. Create a geometric object with a fairly large number of segments. If the object doesn't have enough segments, you won't see much (or any) deformation.

2. On the Create panel, click the **Space Warps** button ≋.

3. From the pulldown list, choose **Geometric/Deformable** or **Modifier-Based**.

4. Click a Space Warp button, such as **Wave** or **Stretch**. Click and drag in a viewport to create the Space Warp icon.

5. On the Main Toolbar, click the **Bind to Space Warp** button. Click the object you wish to bind to the Space Warp, and drag from the object to the Space Warp.

   Your cursor will change to a Space Warp cursor when you position it over a Space Warp icon in the viewport.

Once an object is bound to a Space Warp, a Space Warp binding item appears at the top of the object's Modifier Stack. For example, if a box is bound to a **Wave** Space Warp, then the item **Wave Binding** appears in the box's Modifier Stack.

This is a little confusing, because by now you should be accustomed to the idea that object transforms are always processed after the entire Modifier Stack. The appearance of the Space Warp binding in the Modifier Stack may mislead you to think that the Space Warp is not affected by object transforms, but this is not the case. Remember, the order of operations is as follows: object parameters, modifiers, transforms, space warps, and finally object properties.

The only indication that a Modifier Stack entry is in fact a Space Warp is the **WSM** in parentheses. WSM stands for World-Space Modifier. You can't drag a Space Warp Binding below a standard modifier in the Stack, because World-Space Modifiers are always calculated after the object transforms, whereas standard Object-Space Modifiers are calculated before the object transforms.

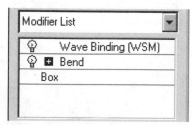

*Figure 8-3: Modifier Stack showing a **Bend** modifier and a **Wave** Space Warp*

To unbind an object from a Space Warp, select the binding entry in the Modifier Stack and click the **Remove Modifier from the Stack** button 🗑 .

SPACE WARP ANIMATION

Animating the effect of a Space Warp can be accomplished in two main ways. First, you can animate the transforms of an object or a Space Warp icon, so that geometry transforms relative to the Space Warp effect.

Also, the parameters of a Space Warp can be animated directly. Select the Space Warp icon, and its parameters will be displayed on the Modify panel. Turn on **Auto Key** mode, and change the parameters to create keyframes.

Many Space Warps will have a **Phase** parameter. You can animate this to vary the effect of the Space Warp– for example, to create rippling waves. Small changes to the Phase parameter, such as a change from zero to one over 100 frames, will yield slow and subtle effects.

EXERCISE 8.1: Wave Space Warp

In this exercise, you will gain basic experience using a Wave Space Warp to deform a box as it travels through space.

1. In the Top viewport, create a Box with the following parameters:

Length	80
Width	140
Height	10
Length Segs	4
Width Segs	7
Height Segs	1

2. Click the **Space Warps** button 〰 on the Create panel. Choose **Geometric/ Deformable** from the pulldown list. Click the **Wave** button to begin creating a Wave Space Warp.

3. In the Top viewport, click to place the center of the Wave Space Warp. Hold the mouse button down, and drag to determine the **Wave Length**. Watch the Modify panel. When the Wave Length parameter reads approximately **60** units, release the mouse button. Then drag the mouse again to define the **Amplitude** of the Wave. The amplitude of a wave is its height.

    When the Amplitude parameter reaches approximately **20** units, click the mouse button to complete the creation of the Space Warp icon.

4. Position the Box and the Wave Space Warp as shown in the following illustration.

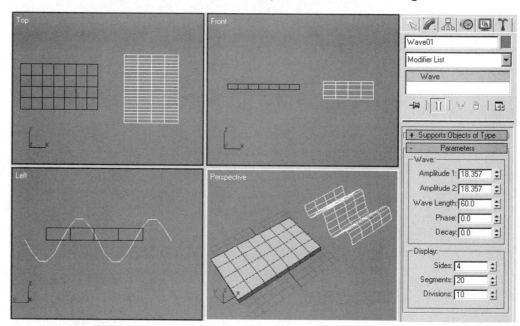

*Figure 8-4: **Box** and **Wave Space Warp** positioned in viewports*

5. On the Main Toolbar, click the Bind to Space Warp button. Click the Box in the viewport, and drag to the Wave icon. Release the mouse button when you see the Space Warp cursor.

The Box is now deformed by the Wave.

6. Maximize the Perspective viewport. Move the Box in the global X axis, toward the Wave Space Warp icon. Nothing happens. This is because the movement of the box is at right angles to the Wave deformation.

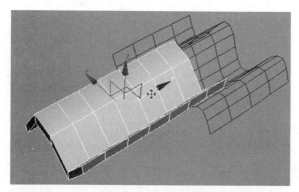

Figure 8-5: Move the Box in the global X axis

7. Press the **<A>** key to enable **Angle Snap**. Rotate the Space Warp icon **90** degrees around the global Z axis. Notice how the Box changes shape as you rotate the Space Warp.

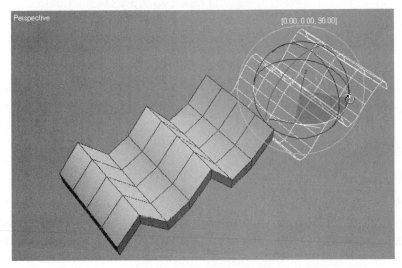

*Figure 8-6: Rotate the Space Warp icon exactly **90** degrees*

8. Select the Box. Increase the number of **Width** segments to **20**. Reduce the number of Length segments to **1**. Move the Box in the global X axis.

    The Box deforms as it moves toward or away from the Space Warp.

9. Select the Wave Space Warp and increase its **Decay** parameter to **0.02**. Move the Box in the X axis, toward and away from the Space Warp icon.

    When the Box is far away from the Space Warp, it is not deformed.

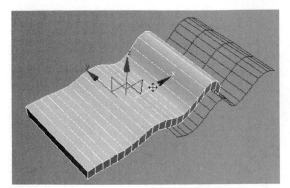

*Figure 8-7: Wave Space Warp with **Decay***

10. Experiment with the Wave parameters, such as Wavelength and Amplitude. Save your file as **WaveSpaceWarp.max** in your folder.

EXERCISE 8.2: Conform Space Warp (The Blob)

There's something in the bathtub, and it's coming this way! In this tutorial, you'll learn to use the **Conform** Space Warp to make an object ooze over another object.

1. Load the file **Bathtub.max** from the disc that comes with this book. Alternately, you can load the file **BathtubSlime.max** that you created earlier.

Figure 8-8: Bathtub scene

2. On the **Create** panel, click **Space Warps** [≋]. Select **Geometric/Deformable** from the pulldown list. Click the button labeled **Conform**. In the **Top** viewport, click and drag at the upper-left corner of the viewport to create the Conform Space Warp.

   The size and location of the icon is irrelevant when using the **Conform** Space Warp. All that matters is what object is chosen on the Command Panel.

3. On the Command Panel, in the **Wrap To Object** section, click the **Pick Object** button, then click the bathtub.

   After clicking the bathtub, the Wrap To Object is listed as **Bathub** on the Command Panel.

4. Choose the Select Object tool from the Main Toolbar. Select the oozing object, which is called **Bacteria**.

5. In the **Front** viewport, move the **Bacteria** object up above the bathtub.

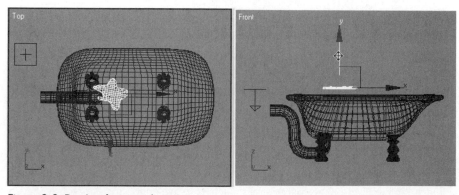

Figure 8-9: **Bacteria** *moved*

6. On the Main Toolbar, click **Bind to Space Warp** . Click and drag from the **Bacteria** to the **Conform** Space Warp. Binding the **Bacteria** to the Space Warp causes it to move in the direction of the arrow on the Space Warp icon, conforming to the object picked by the Space Warp, which is the bathtub. The result is that the **Bacteria** is stuck to the bottom of the bathtub.

7. In the **Top** viewport, move the **Bacteria** object around the bathtub. The **Bacteria** object moves slowly, but it always stays stuck to the surface of the tub.

Figure 8-10: **Bacteria** *sticks to bottom of bathtub*

8. Save the scene with the filename **BathtubConform.max** in your folder.

TIP: Note how the **Bacteria** is perfectly flattened against the surface of the bathtub. To keep some thickness to the slime, select the **Bacteria** object, temporarily disable the **Conform** binding using the light bulb icon in the Modifier Stack, and add a **Mesh Select** modifier. In the Front viewport, select the bottom vertices of the **Bacteria** object, and enable the **Use Soft Selection** option. Re-enable the Conform binding. Then, when you enable the **Use Selected Vertices** check box in the Conform Space Warp's Modify panel, the different vertices of the **Bacteria** object are more or less affected by the Conform depending on their Soft Selection. Adjust the **Falloff** value in the Mesh Select modifier to change the thickness of the **Bacteria** object. Enable **Show End Result** to see the results in real time. See the file **BathtubConformSoftSelect.max** on the disc.

PARTICLE SYSTEMS

Particle Systems can be used to create many types of effects, such as snow and spray. These effects can also be used for water fountains, particles from an explosion, and anything else where many objects need to move randomly or in a group.

CREATING A PARTICLE SYSTEM

To access the particle systems, go to the **Create** panel, click **Geometry** , and choose **Particle Systems** from the pulldown menu. A list of seven particle systems appears.

Figure 8-11: **Particle Systems** *on the Create panel*

Every particle system has an *emitter*. This is the icon from which particles are generated. **Spray** and **Snow** are the two simplest types of particle systems. Click Spray or Snow, then click and drag in the Top viewport to create the emitter. Move the Time Slider to see the particles fall. The particles always travel in a direction perpendicular to the emitter. Since the emitter is pointing down, the particles fall to the ground.

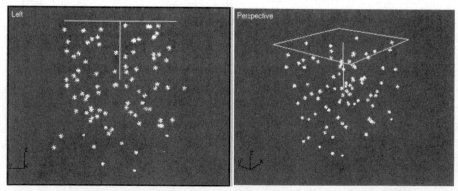

Figure 8-12: Particles falling from **Snow** *emitter in Left and Perspective viewports*

Particles are said to be *born* at a specific frame, *live* for a specified number of frames, and then *die*. These terms are used in setting the length of time a particle appears in the scene. For example, the **Life** parameter sets the number of frames any given particle will live after it is born.

Speed sets the speed of particles in units per frame. Increase **Life** and **Speed** to make the particles cover more area in the scene.

If the particles don't appear in a rendering, increase the **Drop Size** or **Flake Size** until the particles are visible.

Viewport Particles

When rendered, particles may appear as many types of objects. To save time in redrawing the screen, particles in viewports can be displayed as **Dots** or **Ticks**. Dots are small specks, while ticks are small crosses. The way particles are displayed in viewports has nothing to do with how the particles look when rendered. The viewport display tells you how fast the particles are moving and where they are rather than how they will look when rendered.

The number of particles can also be fewer for the viewport than for rendering. This also speeds up screen redraw time.

Rendered Particles

For the **Spray** particle system, two types of particles can be rendered. **Tetrahedron** uses a four-sided pyramid. When **Speed** is increased, the tetrahedrons become long and thin. The **Facing** selection uses rectangles that always face the Perspective or camera view.

EXERCISE 8.3: Rainy Day

In this tutorial, you'll make rain for a rainy-day scene.

1. Load the file **OvercastDay.max** from the disc that comes with this book.

    This scene contains just one object, a box, to which the **Noise** modifier has been applied to make a bumpy ground plane.

2. Render the camera view.

    The scene shows a cloudy background over a grassy plain. The background is an image of a lake on a cloudy day, but because of the way the ground plane is positioned, the lake isn't visible.

WARNING: If you can't see the maps in the scene, make sure your bitmap paths are configured correctly. If necessary, add the **/maps** path from the disc that comes with this book. An explanation of how to use the **Customize > Configure Paths** feature can be found in Chapter 4, *Materials*.

Figure 8-13: Grassy plain on overcast day

3. On the Create panel, click Geometry , and choose **Particle Systems** from the pulldown menu.

4. Click **Spray**. In the Top viewport, click and drag to create a Spray particle system a little bit larger than the box in the scene.

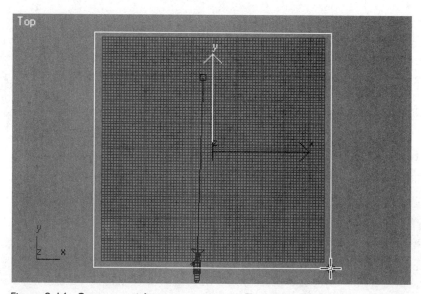

Figure 8-14: **Spray** *particle system emitter in Top viewport*

5. In the Front viewport, move the emitter up high enough so its plane cannot be seen in the camera view.

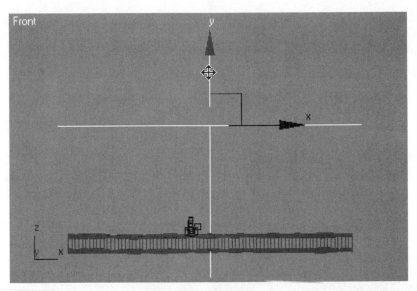

Figure 8-15: Emitter moved up in Front viewport

6. Pull the Time Slider back and forth. As you do so, rain particles emit from the emitter and fall to the ground.

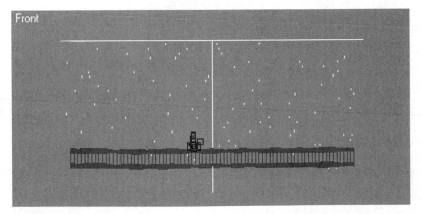

Figure 8-16: Rain falling from emitter

7. Open the Material Editor. Select the first material, called **Droplets**, This is a white, somewhat self-illuminated material. Apply this material to the Spray emitter.

8. Render the camera view at frame **100**.

    A few paltry raindrops appear in the scene. This is not very convincing!

On the Modify panel, in the **Particles** section, the first two parameters are **Viewport Count** and **Render Count**. Viewport Count determines the number of particles that will appear in the viewport, while Render Count is the number in the rendering.

9. Change the **Render Count** parameter to **20000**. Change the **Speed** parameter to **5.0**.

10. Render the Camera01 view.

There are now many more raindrops in the scene. However, some of the rain isn't reaching the ground. To make the rain fall farther, you'll increase the **Life** parameter.

11. Under the **Timing** section, change the **Life** parameter to **90** and re-render the scene.

Figure 8-17: Many raindrops

Next, you'll apply motion blur to the raindrops to make them look more like real rain.

12. With the Spray emitter selected, right-click in the viewport and choose **Properties** from the Quad Menu. In the **Object Properties** dialog, under the **Motion Blur** section, select the **Image** option. The value of the **Multiplier** parameter should be **1.0**.

13. Render the Camera01 view. Each raindrop is now blurred.

Figure 8-18: Blurred raindrops

Rain often falls at an angle. To simulate this effect, you'll rotate the **Spray** emitter.

14. In the **Front** viewport, rotate the emitter by about **-10** degrees in the viewport's Z axis.

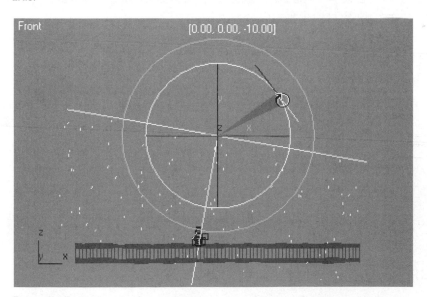

Figure 8-19: Rotate the Spray emitter

15.   Render the Camera01 view to see the angled raindrops.

Figure 8-20: Rotated blurry raindrops

Now for one last touch. At frame **0**, you will find that there is no rain at this frame, and that the rain starts falling right after frame **0**. To make continuous rain from the start, you can change the **Start** parameter.

16.   Under the **Timing** section, change the **Start** parameter to **-90**.

This causes the rain to start falling at frame **-90**. By the time the animation reaches frame **0**, the rain is in full swing.

17.   Save the scene with the filename **RainyDay.max**.

18.   Render the camera view to an image sequence called **Rain.tga**. View the image sequence with the **RAM Player**.

To see a rendered version of this animation, view the file **Rain.avi** on the disc that comes with this book.

SUPER SPRAY

Super Spray emits a stream of particles from a single point. The size of the Super Spray icon has no effect on the size of particles or of the spray. Create a Super Spray particle system by clicking the Super Spray button, then clicking in a viewport to create the emitter. Move the Time Slider to see the particles shoot away from the icon. Leave the Time Slider at a higher frame number such as **100** so you can see the particles as you adjust parameters.

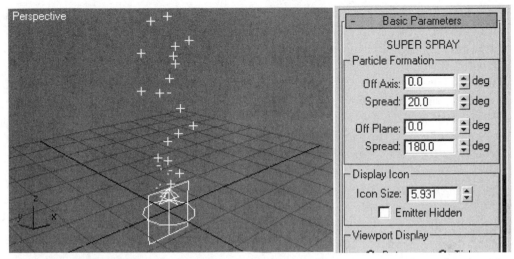

Figure 8-21: **Super Spray**

Super Spray is the most versatile of standard particle systems in 3ds Max. It has many parameters. Here we will go over the ones you will use most. The parameters in Super Spray are also found in other particle systems.

Spread

The **Spread** parameters determine the angle of particle spray. The first Spread parameter, just under the **Off Axis** parameter, spreads the particles out along the emitter's local X axis. The Spread parameter under the **Off Plane** parameter spreads the particles out along the emitter's Y axis. Larger Spread values make the particles shoot off into a wider spray. If both Spread parameters are set to **180** degrees, the emitter will shoot particles out evenly in all directions.

The Off Plane Spread parameter won't work unless you have defined an Off Axis Spread amount as well.

Particle Timing

The timing of particle motion is set on the **Particle Generation** rollout. Under the **Particle Quantity** section, you can set the birth rate of particles. You have the option of using the **Use Rate** parameter to set the number of particles born on each frame, or using the **Use Total** parameter to set the total number of particles born between the Emit Start and Emit Stop frames. **Percentage of Particles**, found on the **Basic Parameters** rollout, sets the percentage of the total number of particles that will appear in viewports, as opposed to renderings, which always display all particles.

Emit Start sets the frame at which particles begin to emit, while **Emit Stop** sets the frame at which they stop. **Display Until** determines the last frame at which particles will display. By default, Emit Start is **0** and Emit Stop is **30**. Ordinarily, you will want to change Emit Start to a negative number, to cause the particles to start emitting before frame **0**. That way, they'll already be in full motion by frame **0**. Change Emit Stop and Display Until to the last frame of the animation to cause the particles to continue emitting right up until the end of the animation. If you change the length of the animation, be sure to set Emit Stop and Display Until to the last frame if you want to see particles throughout the entire animation.

Variations

Super Spray has a number of **Variation** parameters, each of which randomizes the parameter above it by a specified percentage. Change the Variation parameters under **Speed**, **Life**, and **Size** to **10** or **20** to make a natural-looking variation. The effects of changing these parameters can really only be seen in a rendered animation, and not in a viewport.

Particle Types

Under the **Particle Type** rollout are several options for setting the shape of particles. Changing the particle type changes the appearance of particles in the rendering, but not in viewports. The only way to see what the particles will really look like is to render the scene.

Choosing **Standard Particles** enables the choices under the **Standard Particles** section of the rollout. These particle shapes render fastest of all the selections.

Instanced Geometry uses an object from the scene as a particle. This feature can be used to cause teapots to rain from the sky, or a school of swimming fish to be randomly generated on command. However, every face of every particle counts as a face in the scene during rendering. If a complex object is used as a particle, render times can skyrocket.

MetaParticles are particles that appear to stick together, like water droplets or gooey candy. Although it might seem like a great idea to use MetaParticles for water droplets, it is usually not necessary to get a good effect. In addition, MetaParticles take much longer to render than Standard Particles.

TIP: The effect of MetaParticles can be better achieved using the **BlobMesh** Compound Object. This object treats each particle, or each vertex of a mesh object, as a blob. It is more versatile and calculates more quckly than MetaParticles.

Whenever possible, use Standard Particles for your particle systems. They serve most purposes, and render quite quickly. This is important when you consider that most times when you use particle systems, you will be rendering an animated sequence. On a long animated sequence, the difference between render times for Standard Particles and other particle types can sometimes be measured in days.

Particle Size

The particle size is controlled by the **Size** parameter on the **Particle Generation** rollout. Common values for the Size parameter are between **2** and **20**.

The **Grow For** parameter causes particles to grow from nothing to full size over a period of frames, just after the particles are born. For example, when **Grow For** is **10**, particles will grow from nothing to full size over the first ten frames of their lives. **Fade For** does the opposite, causing particles to reduce in size over the last frames of their lives.

Seed

As with other 3ds Max features, such as the Noise modifier, the **Seed** parameter is the root number for the randomness of the effect. Changing the Seed is useful when you have two or more particle systems in the scene, and want them to have all the same parameters such as Life, Speed, and Type, but don't want the particle systems to look identical. Changing the Seed parameter changes the random arrangement of particles without changing anything else about them.

EXERCISE 8.4: Super Spray

In this tutorial, you'll set up a **Super Spray** particle system to simulate water coming out of a spigot.

1. Load the file **Spigot.max** from the disc that comes with this book.

    This file contains a tube to represent a spigot from which the Super Spray will flow. The purpose of the spigot is so you can place the Super Spray correctly for the scene that comes later.

2. On the Create panel, click Geometry. From the pulldown menu, choose **Particle Systems**.

3. Click the button labeled **Super Spray**. In the Top viewport, click and drag at the center of the spigot to make the Super Spray emitter icon.

    The size of the Super Spray icon does not affect the size of the spray. However, it makes sense to make the Super Spray icon about the same size as the spigot, or perhaps a little larger.

4. Move the Super Spray icon upward so the arrow is just above the top of the spigot.

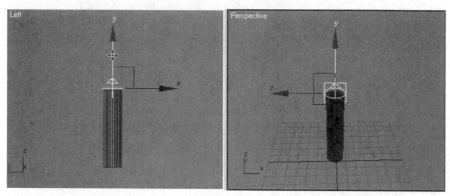

Figure 8-22: **Super Spray** *emitter icon moved*

5. Pull the time slider to frame **30** to see the default Super Spray motion. The particles fly up out of the emitter.

6. Go to the Modify panel. In the **Particle Formation** section of the **Basic Parameters** rollout, change **Spread** under **Off Axis** to **12**, and the **Spread** parameter under **Off Plane** to **90**. This spreads out the spray.

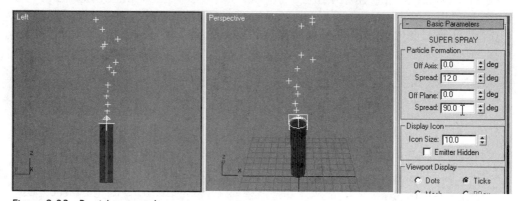

Figure 8-23: Particles spread out

7. On the **Particle Generation** rollout, change the following parameters. Leave the remaining parameters at their default values.

Speed	12
Variation (Speed)	20%
Emit Start	-100
Emit Stop	100
Display Until	100
Life	60
Variation (Life)	5

8.   Play the animation. The particles fly upward, fanning out as they go.

9.   On the **Particle Generation** rollout, under the **Particle Size** section, change the following parameters:

Size	2
Variation (Size)	10%
Grow For	0
Fade For	0

10. On the **Particle Type** rollout, under the **Standard Particles** section, choose the **Constant** option.

The Constant particle type makes small round disks, which will look like water particles when rendered.

11. Open the Material Editor. Select the first sample slot. This is a material that has already been set up. It is a shiny, self-illuminated white material called `Water particles`. Assign this material to the Super Spray.

12. Arrange the Perspective viewport so you can see some of the particles.

Next, you'll use motion blur to make the particles look more like water shooting from a spigot.

13. Select the Super Spray object. Right-click the Super Spray object, and choose **Properties** from the Quad Menu. On the **Object Properties** dialog, under the **Motion Blur** section, choose the **Image** option. Change the **Multiplier** to **5.0**.

14. Render the Perspective view.

Figure 8-24: Water blurred with motion blur

The particles are blurred, making them look more like water drops.

15. Save the scene with the filename **SuperSpray01.max** in your folder.

PARTICLE MAPS

In the Material Editor are two maps created especially for particles, **Particle Age** and **Particle MBlur**.

Particle Age assigns different colors to particles depending on their ages. This map can be used as an opacity map to make older particles fade away as they die. Particle Age can also be used as a diffuse map to cause particles to change color as they age. This is good for fire and sparks.

Particle MBlur is used to blur fast-moving particles. This map works only with certain types of particles under very specific circumstances. If you want particles to blur as they move, Image motion blur is just as effective, and is much easier to set up.

EXERCISE 8.5: Particle Maps

In this exercise you'll apply a new material to the Super Spray from the last exercise, to make it look like smoke.

1. Load the file **SuperSpray01.max** from your folder, or from the disc. This file shows a spigot with particles rising out of it.

2. Open the Material Editor. Select an unused sample slot. Name the material **SmokeParticleMaterial**. On the **Blinn Basic Parameters** rollout, change the **Diffuse** color to a light gray. Change the **Specular Level** and **Glossiness** parameters to **0**.

3. On the **Maps** rollout, assign a **Gradient** map to the **Opacity** parameter. The Gradient map will make the particle appear solid at the center, gradually becoming transparent at the edges.

4. On the **Gradient Parameters** rollout, change **Gradient Type** to **Radial**.

5. Click the **Background** button 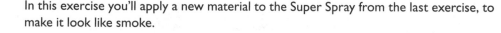 on the Material Editor.

    This displays a background in the sample slot, allowing you to better see the effects of the **Opacity** map. The sample slot sphere is opaque on the inside and gradually becomes transparent toward the outside.

6. Click **Go to Parent** 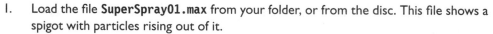. On the Blinn Basic Parameters rollout, check the **Face Map** check box. This will map the gradient onto each facing particle.

7. Uncheck the **Self-Illumination Color** check box, and increase the value to **100**.

8. Apply the material to the particles.

9. Render the Perspective view.

    The particles are tiny and blurred. We wish to get the effect of smoke. First you'll turn off the motion blur that was applied to the particles previously.

10. Select the particle system. Right-click in a viewport and choose **Properties** from the Quad Menu. Under the **Motion Blur** section, uncheck the **Enabled** check box.

11. On the Modify panel, change the **Size** on the **Particle Generation** rollout to **20**.

12. On the **Particle Type** rollout, under the **Standard Particles** section, choose the **Facing** particle type. This type of particle is a square that always faces the camera or Perspective view.

13. Zoom out the Perspective view so you can see most of the particles.

14. Render the scene again.

    The particles now face the Perspective view, are large enough to see, and are not blurred.

Figure 8-25: Smoke particles

15. On the Maps rollout of the Material Editor, assign a **Particle Age** map to the **Diffuse Color** parameter.

    The **Particle Age** map changes the color of each particle depending on its age. A particle is 0% old when it is emitted (born), then 100% old when it is about to disappear (die). The default colors will work fine.

    To see the particles better, we will change the background color to white.

16. Choose **Rendering > Environment** from the Main Menu. In the **Background** section at the top of the Environment and Effects dialog, click the swatch labeled **Color**. In the **Color Selector**, change the background color to white, and close the Color Selector.

17. Render the **Perspective** view. If the smoke globs are too small or too large, change the **Size** parameter for the **Super Spray**.

18. Enable motion blur again, adjusting the **Multiplier** parameter as necessary for the desired smoke effect.

*Figure 8-26: Smoke with **Particle Age** diffuse map and **Image** motion blur*

19. Render the animation in the Perspective view to an image sequence. In the rendered animation, the smoke billows up and away.

20. Save the scene as **Smoke.max** in your folder.

PARTICLE SYSTEMS AND SPACE WARPS

The category of Space Warps called **Forces** can be used to control particle systems. For example, the **Gravity** Space Warp can force particles to change direction, as with water spraying from a fountain.

EXERCISE 8.6: Fountain

In this exercise, you'll use the **Gravity** Space Warp to affect the **Super Spray** you made earlier, creating a fountain effect.

1. Load the file **Fountain.max** from the disc that comes with this book.

This file contains a fountain scene. There is water in the lower part of the fountain, but nothing coming out of the spigot.

2. Merge in the Super Spray scene you made earlier. To do this, choose **File > Merge** from the Main Menu. Select the **Super Spray01.max** file you created earlier. The Merge dialog appears. Choose the **Super Spray01** object from the list, and click **OK**.

   The Super Spray object appears in the scene at the center of the spigot.

   Next you'll place a Gravity Space Warp in the scene.

3. On the Create panel, click **Space Warps**. The pulldown menu reads **Forces** by default.

4. Click the button labeled **Gravity**. In the Top viewport, click and drag anywhere in the viewport to create a Gravity Space Warp of any size.

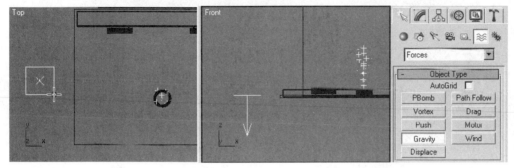

*Figure 8-27: **Gravity** Space Warp placed in scene*

   The Gravity Space Warp works in the direction of the Space Warp arrow. It works infinitely in all directions, meaning the Gravity Space Warp can be anywhere in the scene. It will work the same way regardless of whether it is above, below, or next to the object it is bound to.

5. Click the **Bind to Space Warp** button . In the Front viewport, click and drag from the Super Spray emitter to the Gravity Space Warp icon.

6. Pull the time slider to view the effect.

   The particles now fall back down after shooting up due to the effect of the Gravity Space Warp. However, the particles are not going up high enough. Select the Super Spray01 object, and increase the Speed parameter to 20.

7.   Render the camera view.

Figure 8-28: Fountain with water

8.   Adjust the Size, Speed, Quantity, Motion Blur, material, or any other parameters of the Super Spray to get the result you wish. Try a **Particle Type** of **Sphere** with a transparent material.

9.   Save the scene with the filename **FountainSpray.max** in your folder.

10.   Render the scene to an image sequence and view it with the RAM player.

To see a rendered version of this animation, view the file **FountainSpray.avi** on the disc that comes with this book.

PARTICLE FLOW

So far in this book we've looked at the so-called "standard" particles in 3ds Max. These can go a long way toward getting good effects, but if you need more control, you can also use the **Particle Flow** object. Particle Flow is a very advanced system for creating and controlling particles using an event-based flowchart diagram. It is extremely powerful and easy to use. All you need is a basic understanding of how the system works, and you can create nearly any particle effect you can imagine.

Particle Flow objects are created from the Create panel, just like standard particle systems. Click the button labeled PF Source, and then click in the viewport to create a Particle Flow emitter. Play the animation to see the default particles. Then go to the Modify panel, and with the PF Source emitter still selected, click the button that says **Particle View**. Now you can begin editing the flowchart to control the generation and movement of your particles.

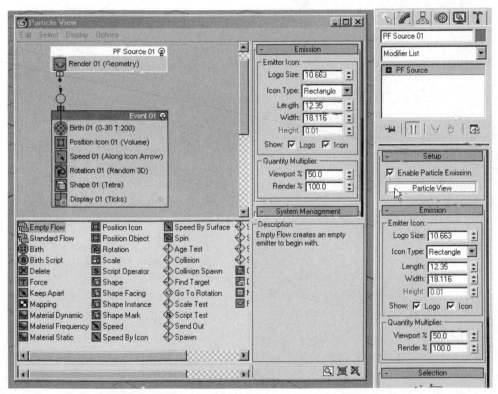

Figure 8-29: **Particle View** *window*

Particle View is similar to the Graph Editors in that there's only one Particle View window, but you can view multiple Particle Flow systems within the window. This window is divided into four panes. At the upper left is the **Event Display**, which shows you a graphic representation of the connections between events. To the right of the Event Display is the **Parameters** panel, where you can control the parameters for individually selected events. This panel looks very similar to the Modify panel in the main 3ds Max interface.

At the bottom left of Particle View is the **Depot**, where you will find all of the various Particle Flow actions from which to choose. When you select one of these actions, you'll see a **Description** appear at the bottom right of the Particle Flow window.

The general procedure for Particle Flow is to create and connect various *events* that include *operators* and *tests*. An event is a rectangular holder for one or more operators or tests. Events are connected to one another via their inputs and outputs that project outside the event rectangles. Outputs are shown as blue dots, and inputs are circles. Click and drag from an input to an output to connect them.

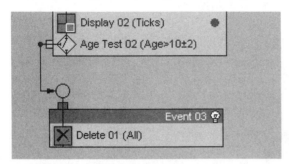

Figure 8-30: Two connected events

Operators do things like determine what type of particles are created, how fast they move, if they are affected by Space Warps, and so on. Tests check to see if certain conditions are met, and if so, permit other operators to perform functions. For example, an Age Test could cause a particle to change color, die, or spawn more particles.

You can drag operators or tests from the Depot into an empty area of the Event Display to create new events. Or, if you drag an operator or test onto an existing event, it will be added to that event. It's easy to rearrange operators and tests to move them within an existing event, or to another event.

If you click on the text label of any event, operator, or test, its parameters are displayed in the Parameters panel on the upper right of the Particle View window. Most of these parameters are animatable, just as they would be if they were on the Modify panel of the main 3ds Max interface.

Events can be disabled by clicking the light bulb icon at the top left of the event rectangle. Operators and tests within events can be disabled by clicking their icons at the left side of the event rectangle. Disabled events, operators, and tests are grayed out.

The top-level rectangle of a Particle Flow system is special. It is represented in the Event Display pane as a rectangle with the name of the PF Source object you see in the viewports. This is called a *global event*. Any operators placed in the global event will affect all events that are downstream. For example, if you wish to assign a material to all particles in a system, place a material operator in the global event. To prevent problems, don't use similar operators in the global event and in the individual events that follow it.

You don't need to bind a Particle Flow to a Force such as Gravity or Wind. Create the Space Warp as you normally would, but use a Force operator within Particle View.

Exercise 8.7: Magic Wand (Particle Flow)

In this exercise, you will learn the basics of working with Particle Flow.

1. Launch or reset 3ds Max.

2. In the Time Configuration [⊞] dialog, change the animation **End Time** to **300** frames.

3. Create a **Cylinder** with a **Height** of **80** units and a **Radius** of **0.5**. Rename it **Wand**.

4. Add a **Volume Select** modifier to the Cylinder. In the Parameters rollout, choose Vertex as the Stack Selection Level. Enter **Gizmo** sub-object mode, and move and scale the Volume Select gizmo to select the vertices at the top of the wand. Exit sub-object mode by clicking the word Gizmo in the Modifier Stack.

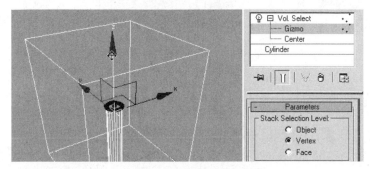

Figure 8-31: Move and scale the Volume Select Gizmo

5. On the Create panel, choose **Particle Systems** from the pulldown list. Click the **PF Source** button, then click and drag anywhere in the Perspective view to create the Particle Flow emitter. Play the animation; particles emit from the PF Source object.

6. On the Modify panel, click the button labeled **Particle View**. In the Particle View window, find the **Event 01** rectangle. Click the text label of the **Birth 01** operator. In the Parameters panel to the right, change the **Emit Stop** value to **300**.

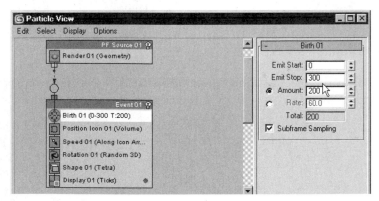

*Figure 8-32: Change **Emit Stop** to 300*

Play the animation. Particles emit from the PF Source icon for the full ten seconds of the active time segment.

7. Next we will make the particles emit from the selected vertices of the wand. In the Event 01 box in the Particle View, click the text label of the operator labeled `Position Icon 01 (Volume)`. Press the **<DELETE>** key to remove this operator.

Then look in the **Depot** area at the bottom of the Particle View window. You'll see an operator labeled **Position Object**. Click and drag a Position Object operator into the Event Display pane of the Particle View window. Place your cursor directly below the `Birth 01` operator within **Event 01**. When you see a blue line appear below the `Birth 01` operator, release the mouse button to add the new Position Object operator.

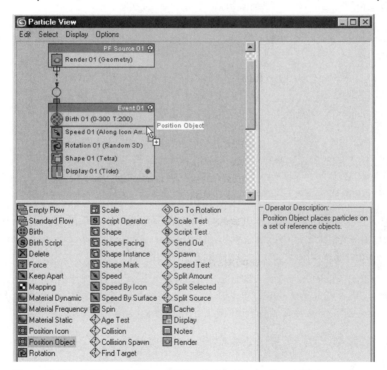

*Figure 8-33: Add a **Position Object** operator to Event 01*

8. Select the text label of the new `Position Object 01` operator. In the Parameters panel, look for the Emitter Objects section. Click the **Add** button, then click the `Wand` in the viewport. (You can also use the By List button.)

Look for the Location section of the Parameter panel. There is a pulldown list that currently reads Surface. Choose **Selected Vertices** from the pulldown. Play the animation in the viewport. Now the particles emit from the vertices at the tip of the wand. However, they are moving downward through the wand, and too fast.

9. Click the text label of the **Speed 01** operator. Change the **Direction** pulldown to read **Random 3D**. Change the **Speed** to 20.

    Play the animation. Now the particles are shooting out in all directions from the tip of the wand. However, they continue moving into the distance forever and never die.

10. Go to the Depot and find the **Age Test** action. Drag an Age Test to the very bottom of the Event 01 rectangle. When the new Age Test appears inside the **Event 01** rectangle, select the **Age Test 01** text label. In the Parameters panel, change the **Test Value** to 10.

11. Drag a **Spawn** test from the Depot into an empty area of the Event Display pane. 3ds Max automatically creates a new rectangle called **Event 02**, which includes the new **Spawn 01** action and a **Display 02** action.

12. Click and drag from the blue dot at the left of **Age Test 01** to the circle at the top of **Event 02**. Now you have connected the output of **Age Test 01** to the input of **Event 02**.

13. Click the text label of **Spawn 01** to access its parameters. Activate the checkbox labeled **Delete Parent**.

    Now, when a particle reaches an age of ten frames, it passes the Age Test. Then the Spawn action causes a new particle to be born. Play the animation. Click in the viewport to deselect the PF Source emitter. Notice that there are two different colors of particles on the screen.

    Back in the **Spawn 01** parameters, change the **# Offspring** value to **5** and the **Variation %** value to **30**. Now more particles are generated.

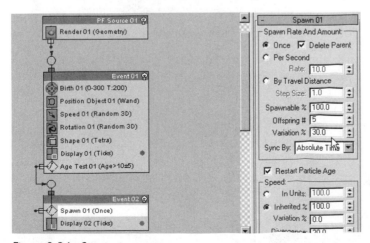

Figure 8-34: Spawn parameters

14. Add another Age Test action, this time at the bottom of **Event 02**. Then drag a **Delete** operator to an empty area of the Event Display pane. A new rectangle labeled **Event 03** is created with the **Delete 01** operator inside it. Connect the output of **Age Test 02** to the input of **Event 03**. In the **Age Test 02** parameters, set the **Test Value** to 10.

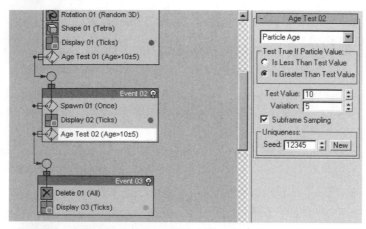

Figure 8-35: Delete operator after second Age Test

Play the animation. The spawned particles no longer live forever. Now they die after ten frames, plus or minus five frames, as defined by the **Variation** parameter.

15. To make the effect more interesting, let's modify the speed of the spawned particles. Add a **Speed** operator to **Event 02**, directly below **Spawn 01**. Change the **Speed** value to **10**, and set the **Direction** to **Random 3D**. Now the initial particles and the spawned particles are moving at different rates, giving a sparkly effect.

16. Experiment with different parameters until you get the look you want. Save the scene as **magic_wand.max** in your folder.

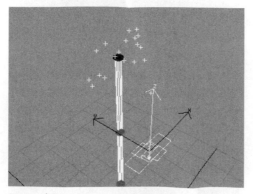

Figure 8-36: Magic wand sparkles

VIDEO POST

Video Post is a basic compositing program built into 3ds Max. A full-blown compositing application such as discreet Combustion or Adobe After Effects will give you a lot more power, but if you don't have one of those programs, Video Post can do simple jobs.

Video Post can be used to composite images or apply effects to an entire animation. It can also be used to convert an image sequence to a compressed movie file, such as an **.AVI** or QuickTime movie. This will allow you to try out different compression settings without having to re-render your entire scene.

WARNING: Remember that you should never render your finished scene directly to a compressed movie unless it is merely a test. It's better to render to an uncompressed format and apply compression later.

To open the Video Post window, choose **Rendering > Video Post** from the Main Menu.

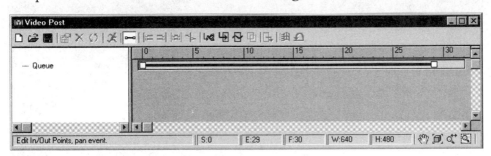

Figure 8-37: **Video Post** *window*

Video Post works with a queue, or list, of actions. The queue is found on the left side of the window. Items in the queue are processed from the top down.

Video Post actions, called *events*, are added to the queue with the buttons on the Video Post toolbar. The right side of the Video Post window sets the range of frames for each event.

VIDEO POST EVENTS

To add the current scene to the queue, click the **Add Scene Event** button on the Video Post toolbar. When you execute the Video Post sequence, this Scene Event invokes the 3ds Max renderer, so you may perform rendering and compositing in one step. Be sure to use a Camera view when you add a Scene event to Video Post; some effects don't work with Perspective views.

To add a still image or an image sequence to the queue, click the **Add Image Input Event** button . You could use this image as a background for compositing, apply an effect, or simply convert a sequence to a compressed movie file.

If you wish to layer one image over another, you will need to add a Video Post **Filter** event to the queue. First, add the two events you wish to layer, such as an image event and a scene event. The image you want to be layered on top must have an alpha channel, and it must be *lower* in the queue. Then select both events in the queue, using the **<CTRL>** key.

When both events are highlighted, click the **Add Image Layer Event** button 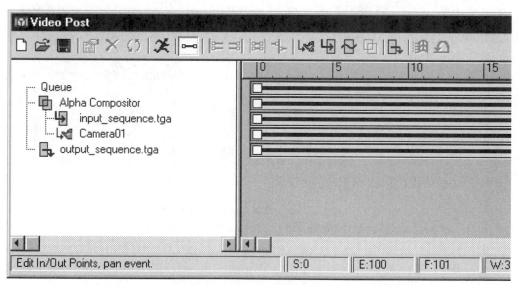. In the pop-up dialog, select **Alpha Compositor** from the **Layer Plug-In** pulldown. When you click OK, the Alpha Compositor filter appears in the queue, with the two other events appearing as subordinate events.

In order to save what the **Video Post** queue is doing, you must add an output event to the queue by clicking **Add Image Output Event**. To ensure the entire queue is output, make sure that no events on the queue are selected before clicking this button.

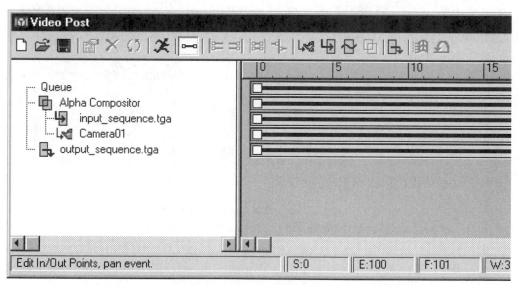

Figure 8-38: ***Video Post*** *queue set up to render a Camera view over an existing image sequence*

When the queue has been all set up, click **Execute Sequence** to run the sequence and render the final image. You can perform test renders at any time by clicking Execute, but the final image will not be saved. Be sure to add an Image Output event before performing the final render.

TIP: For the best image quality, all of your Input and Output events should have the same resolution. When you click Execute Sequence, be sure to choose the same rendering resolution as your Input image. Otherwise, the Input image will be scaled up or down, which degrades the picture quality.

RENDER EFFECTS

In the real world, human eyes and camera lenses produce interesting light effects such as glows, streaks, and defocusing. Although you can use Video Post to add these effects to your renderings, it's easier to use the **Render Effects** module.

Render Effects applies 2D special effects to your images immediately after they are rendered. Lens flares, glows, and other effects are easily applied to lights and objects, making renderings more realistic or more fantastic. To use these effects, select **Rendering > Effects** from the Main Menu. The **Environment and Effects** dialog opens up with the **Effects** tab active. The **Effects** rollout shows a list of all Render Effects you have chosen. Click the **Add** button to insert an effect into the queue. The **Add Effect** dialog pops up, and you can select from a number of effects plug-ins.

LENS EFFECTS

The most commonly used plug-in is **Lens Effects**, which creates a variety of effects, such as lens flares, glows and starbursts. If you add Lens Effects to the queue, you'll see the **Lens Effects Parameters** rollout directly below. Here you must choose one or more elements (sub-effects), such as **Glow**, **Ring**, or **Streak**. Select the name of the element you wish to use from the left side of the dialog, and click the right arrow button to send it to the right side of the dialog. This activates that element.

Scroll down in the Environment and Effects dialog to access the **Lens Effects Globals** rollout. The parameters in this rollout will control all elements within the current Lens Effects plug-in. So, if

Figure 8-39: **Effects**

Figure 8-40: **Lens Effects** applied to a light

you have a Glow and a Ring element active, the parameters in Lens Effects Globals will control both of them.

When you select an active element in the Lens Effects Parameters rollout, the parameters for that particular element appear in a rollout at the very bottom of the dialog. For example, if you highlight a Glow element, the **Glow Element** rollout is available at the bottom of the Render Effects dialog.

Render Effects have many parameters. For example Lens Effects have a **Size** and an **Intensity**. For the most part, these parameters are fairly self-explanatory. However, the procedure for applying a Lens Effect to objects in your scene is not obvious.

Lens Effects can be applied to a light, a material, or an object. Lights are the most straightforward. To apply a Lens Effect to a light, simply click the **Pick Light** button on the Lens Effects Globals rollout, then click on a light in your scene. Adjust the parameters, such as **Size** and **Radial Color**, to customize the effect. The Size of the effect is measured as a percentage of the image height. Radial Color gives you two colors to work with, one at the center of the effect and one at the edge of the effect.

Render the image to see the Lens Effect applied to the light. If you activate the **Interactive** option in the Effects rollout, then the effects will be updated and re-rendered automatically whenever you change a parameter.

Assigning a Lens Effect to an object or a material is a little more complicated. It involves the use of ID numbers, similar to the ID numbers used for Multi/Sub-Object materials. First we will discuss assigning a Lens Effect to an object.

Graphics Buffer Channels

When a Lens Effect is applied to an object, it is actually triggered by the pixels of a rendered image. The 3ds Max **Graphics Buffer** is a way for rendered pixels to be marked for special uses, such as triggering a Lens Effect. The Graphics Buffer lets you tell the Lens Effect which pixels in your scene should be considered for inclusion in the effect.

Each object in a 3ds Max scene has an ID number that is referred to as the **G-Buffer Object ID**. If this ID number is the same for the object and the Lens Effect element, then the rendered pixels of that object will potentially trigger the effect.

To assign the G-Buffer channel for an object, select the object and right-click in the viewport to access the Quad Menu. Select **Properties** from the Quad Menu. In the **Object Properties** dialog, look near the lower right for the **G-Buffer Object ID** spinner. Choose a number, such as channel **#1**. This is now the G-Buffer Object ID number for the selected object. Close the Object Properties dialog.

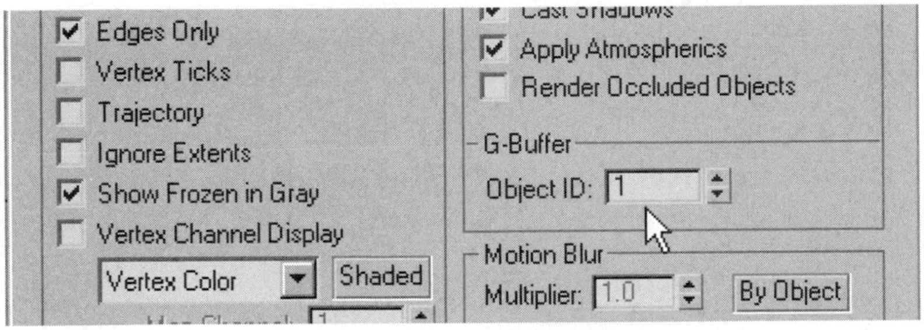

Figure 8-41: **G-Buffer Object ID** *in the* **Object Properties** *dialog*

Now you must assign the same number to the Lens Effects Element. In the element rollout, such as Glow Element, click the **Options** tab. In the **Image Sources** section, check the **Object ID** box, and enter the same channel number in the spinner. If you chose channel **#1** in the Object Properties dialog, then choose channel **#1** here also. The numbers must match for the effect to work.

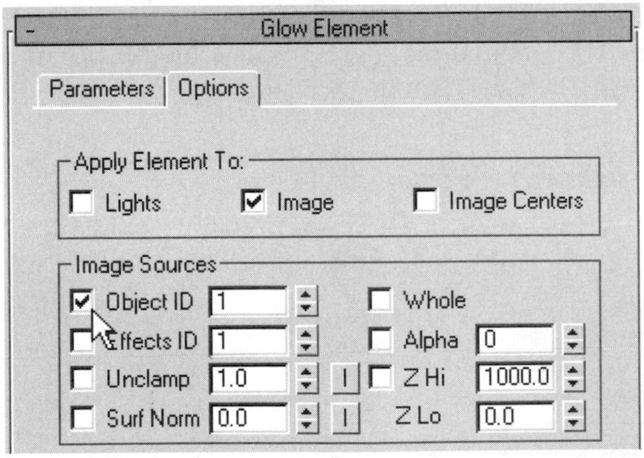

Figure 8-42: **Object ID** *channel in the* **Environment and Effects** *dialog*

Now, when you render the image, a Lens Effect will be applied to any objects that share the same G-Buffer Object ID number you picked in the Element rollout. You can apply the effect to multiple objects by assigning them the same ID number in the Object Properties dialog. Just select the objects in the viewport, right-click, and access the Object Properties for all selected objects at once.

Material Effects Channels

A similar principle applies to materials. A material can be assigned an ID number, so that any object that uses that material will potentially trigger an effect.

Material Effects Channels are assigned from the Material Editor toolbar. Select a sample slot, then choose a channel number other than zero from the flyout, as shown in the following.

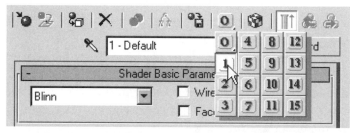

*Figure 8-43: **Material Effects Channel** flyout on the Material Editor toolbar*

Then you must check the **Effects ID** option in the Element rollout of the Environment and Effects dialog, and choose the same channel number that you chose in the Material Editor.

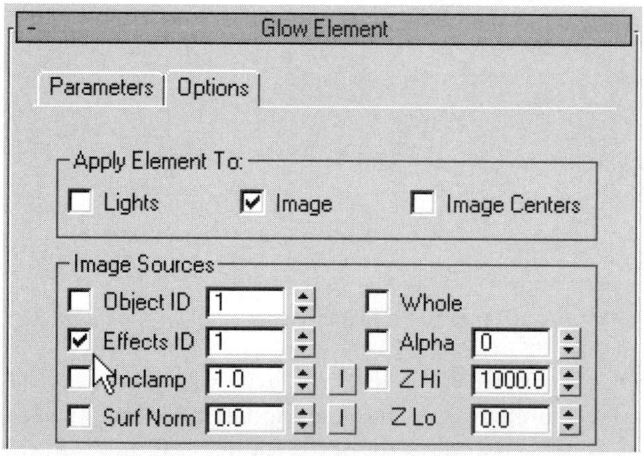

*Figure 8-44: **Effects ID** channel in the Environment and Effects dialog*

There are only sixteen Material Effects Channels, but there is no limit to the number of G-Buffer Object IDs.

EXERCISE 8.8: Lens Effects Glow

In this exercise, you will use Lens Effects to add glow to a neon sign.

1. Open the file **NeonSign.max** from the disc that comes with this book. The file consists of three Editable Spline objects that are made renderable.

2. Create three different materials, one for each object. Make a bright red material for the word **EAT**, an orange material for the word **AT**, and a yellow material for the word **JOE'S**. In all three materials, increase the **Self-Illumination** parameter to **100%**.

3. Render the camera view. The objects are bright, but not glowing.

4. Select all three objects in the viewport. Right-click to access the Quad Menu, and select **Properties**.

5. In the **Object Properties** dialog, change the **G-Buffer Object ID** to **1**. This assigns ID #1 to all three objects.

6. Choose **Render > Effects** from the Main Menu to open the Environment and Effects dialog. In the **Effects** rollout, click the **Add** button. Choose **Lens Effects** from the **Add Effect** pop-up dialog.

7. In the **Lens Effects Parameters** rollout, click the word **Glow** on the left side of the dialog. Then click the right-facing arrow to activate the Glow element.

8. Scroll down to the **Glow Element** rollout. Click the **Options** tab. In the **Image Sources** section, click the **Object ID** check box to activate it. The number in the spinner should already be ID #1.

9. Render the camera view again. The letters glow with a scattered white light.

*Figure 8-45: **Lens Effects Glow** with default parameters*

The Glow effect is too large. We will now reduce the size of the Glow.

10. In the Environment and Effects dialog, activate the **Interactive** option on the Effects rollout. Now, any changes to the effect will be updated and re-rendered automatically.

11. On the **Lens Effects Globals** rollout, **Parameters** tab, reduce the **Size** to **2**. The Glow effect is now much smaller, more like a neon sign.

*Figure 8-46: Lens Effects Globals **Size** of **2***

12. Select the **Parameters** tab in the Glow Element rollout. Click the white color swatch on the left, in the **Radial Color** section. The **Color Selector** dialog appears. Choose a very bright green color. The rendered image updates immediately, showing the letters glowing bright green.

13. Instead of a green glow, we will cause the objects to glow in the same colors as the materials you designed. On the Glow Element Parameters tab, increase the **Use Source Color** spinner to **100**. This causes Lens Effects to choose the object's rendered pixel colors as the basis for the Glow color.

14. To get a more realistic neon glow, reduce the **Intensity** to **85** in the Glow Element parameters. Reduce the **Size** to **5** in the Glow Element parameters.

Figure 8-47: More realistic neon glow

15. Render and save a final image in **.TGA** format.

16. Save the scene as **LensEffectsGlow.max** in your folder.

ATMOSPHERES

Render Effects are 2D effects applied to the image during rendering, so you can't see them in viewports. **Atmospheres** are similar, except they are true 3D effects. They can be used to create fog, clouds, explosions, and lighting effects. Atmospheres are truly *volumetric*, meaning they exist in three dimensions. So, for example, you can create Atmosphere clouds and animate a camera flying through them.

Atmospheres are added in the **Environment and Effects** dialog, in the **Evironment** tab. Choose **Rendering > Environment** from the Main Menu, or use the hotkey, which is **<8>**. In the Environment tab of the dialog, scroll down to the **Atmosphere** rollout. Click the **Add** button, and choose an effect from the **Add Atmospheric Effect** dialog that pops up. After you click OK, the Atmosphere effect is added to the queue in the Environment and Effects dialog.

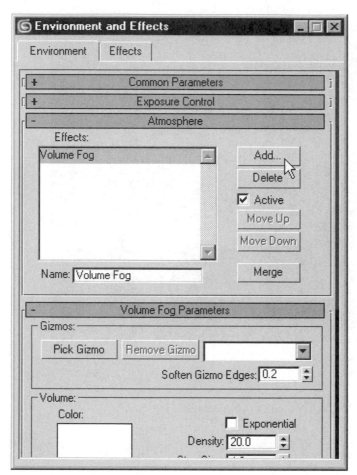

Figure 8-48: **Atmosphere** *rollout in the* **Rendering > Environment** *dialog*

FOG

The simplest type of Atmosphere is the **Fog** effect. Fog is extremely useful for adding realism, especially to outdoor scenes. You can use Fog to create the effect of mist in the air, or to add a touch of haze to a sunny scene. Even on a clear day, particles of dust in the air cause objects in the distance to look less contrasty and slightly blue. This is called *atmospheric perspective*.

The Fog Atmosphere has basic parameters such as its **Color**. In addition, you can choose either **Standard** or **Layered** Fog. Standard Fog uses the **Environment Ranges** parameters in the camera's Modify panel to help determine how much fog or haze to create at certain distances from the camera. Using the controls on the Environment and Effects dialog and the controls in the camera's Modify panel, you can fine-tune Standard Fog to achieve anything from a thick, cloudy fog to a light atmospheric haze.

EXERCISE 8.9: Misty Mountains

In this exercise, you will apply Standard Fog to a landscape to enhance its realism.

1. Open the file **Mountains.max** from the disc that comes with this book. This scene consists of a Plane with a Noise modifier added to create a mountainous terrain.

2. Render the camera view.

Figure 8-49: Mountains with no fog

3. Choose **Rendering > Environment** from the Main Menu. In the Environment and Effects dialog, scroll down to the **Atmosphere** rollout. Click the **Add** button, and choose **Fog** from the pop-up dialog.

4. Re-render the camera view.

Figure 8-50: **Standard Fog** *with default parameters*

5. In the Left viewport, select the camera. Open the **Modify** panel, and scroll down to the **Environment Ranges** section of the **Parameters** rollout. Click the check box labeled **Show**. Now you can see the camera ranges that determine the extents of the Fog effect. Use the spinner to change the **Far Range** to about **400** units.

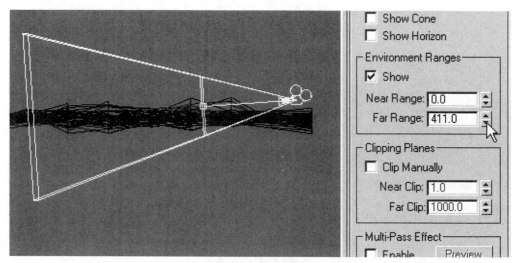

*Figure 8-51: Adjust the camera's **Far Range** parameter to approximately **400** units*

6. Re-render the camera view.

*Figure 8-52: Changing the camera's **Far Range** parameter results in a very thick fog*

7. Save the scene as **MountainsFog.max** in your folder.

TIP: You can apply a map to the Fog, using the Environment Color Map or the Environment Opacity Map. However, this only works for still images, or animations with static camera shots. For moving cameras, use Volume Fog.

VOLUME FOG

Standard Fog produces an effect throughout the scene based on camera distance, while Layered Fog works based on height in the world Z axis. A more complex form of fog can be created using the **Volume Fog** effect. This is excellent for creating clouds or patchy fog.

To add a Volume Fog effect, click the **Add** button in the **Atmospheres** rollout of the **Environment and Effects** dialog, and choose **Volume Fog** from the pop-up dialog. Then the Volume Fog rollout appears in the Environment and Effects dialog.

Volume Fog requires a **Helper** object to contain the atmospheric effect. This type of helper gizmo is called an **Atmospheric Apparatus**. You must create an Atmospheric Apparatus gizmo, and assign the Volume Fog to that gizmo.

To create an Atmospheric Apparatus, go to the Create panel and click the Helpers button . Choose **Atmospheric Apparatus** from the pulldown list. You can choose from three different primitive gizmos: **BoxGizmo**, **CylGizmo**, or **SphereGizmo**. (For clouds, you'll probably want to use a SphereGizmo.) Click one of the buttons, and click and drag in the viewport to create the gizmo.

*Figure 8-53: Creating an **Atmospheric Apparatus** Helper object*

In the Environment and Effects dialog, click the **Pick Gizmo** button on the Volume Fog Parameters rollout. Click the Atmospheric Apparatus gizmo you created in the viewport. Render the scene to see the results. Then go back to the Environment and Effects dialog and adjust the Volume Fog parameters, such as **Density** and **Noise** parameters.

You can transform the Atmospheric Apparatus any way you like, using the standard Move, Rotate, and Scale tools. Non-uniform scaling is a good way to stretch the Volume Fog to get more wispy clouds. However, modifiers will not affect an Atmospheric Apparatus.

It is possible to have multiple Atmospheric Apparatus gizmos in the scene, all sharing the same Environment settings. If you want the gizmos to look different, select a gizmo and choose a different **Seed** value on the Modify panel. If you wish to have totally different Environment settings in each gizmo, then you'll need to add multiple Volume Fog effects to the Atmosphere queue on the Environment dialog, and assign each Volume Fog to a different gizmo.

EXERCISE 8.10: Sunny Day

In this exercise, you will change the weather in the mountainous scene to make it look like a sunny day with a few scattered clouds.

Change the Fog to Atmospheric Haze

1.   Open the file **MountainsFog.max** from your folder or from the disc.

2.   Select the Left viewport and click **Zoom Extents** 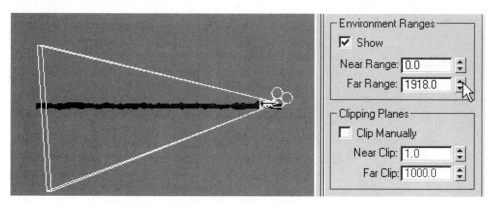. Select the camera and open the Modify panel. In the **Environment Ranges** section, increase the **Far Range** to approximately **1900** units, so the Far Range stops just short of the edge of the scene geometry.

*Figure 8-54: Increase the **Far Range** parameter to about **1900** units*

3.   Render the camera view. The fog is much more subtle.

*Figure 8-55: Subtle fog with adjusted camera **Far Range** parameter*

4. Select **Rendering > Environment** from the Main Menu. In the Environment and Effects dialog, select the **Fog** effect in the **Atmosphere** queue. Click the color swatch in the **Fog Parameters** rollout. In the Color Selector dialog, choose a light blue color. Use the following values:

Hue	170
Saturation	25
Value	255

Re-render the camera view. The fog now has a light blue color.

5. Now we will create a better sky by placing a map into the background. To see this, turn off the **Fog Background** option in the Fog Parameters rollout.

Render the camera view again. There is no fog applied to the background.

*Figure 8-56: Subtle blue haze with **Fog Background** option disabled*

Create a Gradient Ramp Sky

1. Open the **Material Editor**. Select an empty sample slot and click the **Get Material** button ![icon]. In the **Material/Map Browser**, double-click the **Gradient Ramp** map to load it into the sample slot. Close the Material/Map browser.

    In the Material Editor, change the name of the Gradient Ramp map to **SkyMap**. Look in the **Coordinates** rollout for SkyMap. Change the Coordinate type to **Environ**, and the Mapping type to **Screen**.

    Change the **W Angle** parameter to **-90** degrees, so the map is black at the top of the sample slot, and white at the bottom.

    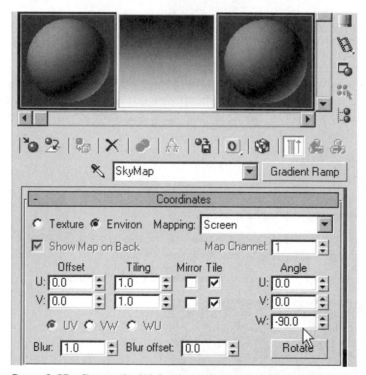

    *Figure 8-57: Change the **W Angle** parameter to **-90** degrees*

2. Scroll down to the **Gradient Ramp Parameters** rollout. Here you see the controls for the color gradients. Each of the arrows is a color flag that you can move or edit to control the map. Right-click the color flag to the far left of the gradient display, and select **Edit Properties** from the pop-up menu.

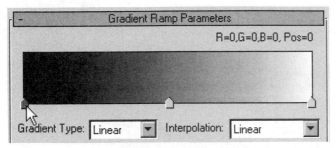

Figure 8-58: Right-click the far left color flag and choose **Edit Properties**

The **Flag Properties** dialog appears. Click the color swatch to open the **Color Selector** dialog.

Figure 8-59: Click the color swatch in the **Flag Properties** *dialog*

In the Color Selector, change the color flag from black to a medium blue. Use the following values:

Hue	170
Saturation	120
Value	255

Leave the Color Selector and Flag Properties dialogs open.

3.   Click the color flag in the center of the Gradient Ramp. In the Color Selector dialog, choose the following values to create a lighter blue:

Hue	170
Saturation	25
Value	255

This is the same color you used for the Fog Atmosphere.

4.   Drag the **SkyMap** sample slot in the Material Editor to the **Environment Map** slot at the top of the Environment and Effects dialog, and choose **Instance** from the pop-up dialog. This assigns the Gradient Ramp as the background image.

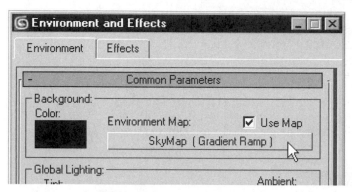

Figure 8-60: **SkyMap** *assigned to* **Environment Map** *slot*

Render the camera view. The background is now a blue gradient.

Figure 8-61: Blue gradient in the background

Make Clouds with Volume Fog

1.   In the Environment and Effects dialog, click the **Add** button on the Atmosphere roll-out. In the pop-up dialog, choose **Volume Fog** and click OK.

2.   Select the Top viewport and click **Zoom Extents**. Go to the Create panel and click the **Helpers** button. Choose **Atmospheric Apparatus** from the pulldown list. Finally, click the **SphereGizmo** button and click and drag in the Top viewport to create the gizmo.

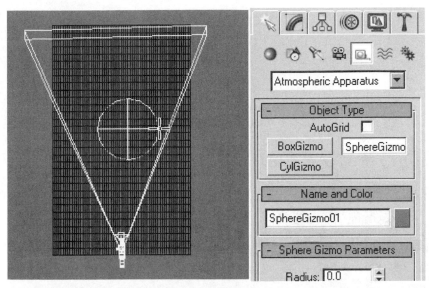

Figure 8-62: Create a **SphereGizmo Atmospheric Apparatus**

3.   Select the SphereGizmo and use the Move, Rotate, and Non-uniform Scale tools to place it in the scene so it is where a cloud should be. Check your work in the camera viewport.

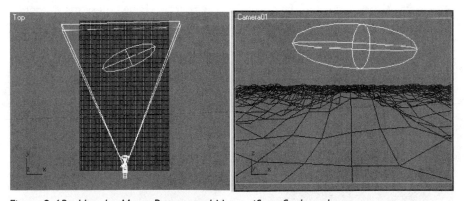

Figure 8-63: Use the Move, Rotate, and Non-uniform Scale tools

4.   In the Environment and Effects dialog, click the Volume Fog entry in the Atmosphere queue. In the **Volume Fog Parameters** rollout, click the **Pick Gizmo** button. Then click the SphereGizmo object in the viewport to assign the Volume Fog effect to the gizmo.

Render the camera view with the Volume Fog effect.

Figure 8-64: **Volume Fog**

5.   Although the Volume Fog correctly appears in the rendering, it is not very convincing. In the Volume Fog Parameters rollout, change the following parameters:

Soften Gizmo Edges   1
Exponential   checked

And in the **Noise** section at the bottom of the rollout, change these parameters:

Type   Fractal
Size   **50**

Render the camera view to see the results.

Figure 8-65: **Fractal Noise** *Volume Fog*

Finishing Touches

1. The Standard Fog isn't quite right. In the Environment and Effects dialog, choose the **Fog** effect from the Atmospheres queue. In the Standard section, enable the **Exponential** check box and re-render the camera view. The scene is now extremely foggy.

2. In the Environment and Effects dialog, reduce the **Far** % parameter to **70%**, and re-render.

3. To add the effect of shadows cast by clouds, create a **Noise** Projector Map for the Directional light in the scene. Go to the Material Editor and click Get Material. Select **Noise** in the Material/Map Browser. On the **Noise Parameters** rollout, change the **Type** to **Fractal**. Change the **High Noise Threshold** to **0.5**, and the **Low Noise Threshold** to **0.1**.

4. Unhide the light by going to the **Display** panel and uncheck the **Lights** option in the **Hide by Category** rollout.

    Select the light in the viewport, then go to the Modify panel. Increase the **Falloff/Field** parameter to around **300** units, so the projector map will be large enough in the scene.

    Drag the Noise map from the Material Editor slot to the **Projector Map** button on the Modify panel, and choose **Instance** from the pop-up dialog. Re-render the scene.

5. Continue to experiment with the scene, adding additional cloud gizmos and adjusting parameters. Continue to make render tests after every change you make. When you are happy with the results, render a single **.TGA** file and save it to your folder.

6. Save the scene as **MountainsSunny.max** in your folder.

*Figure 8-66: Completed rendering with multiple clouds and **Projector Map***

VOLUME LIGHTS

Another type of Atmosphere is **Volume Light**. This produces the effect of a visible light beam, as if the light were shining through fog or dust.

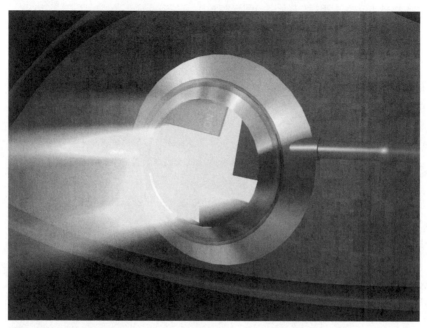

Figure 8-67: Volume Light

A volume light can be created from any kind of light. Volume Lights are set up in the same way as other Atmospheres. They tend to work best when the light is attenuated. Attenuation is the loss of light intensity over distance. This prevents the volume light effect from taking over the entire scene and washing it out. Enable **Far Attenuation** in the light's Modify panel, and adjust the ranges with the spinners.

In addition, there are **Attenuation** controls in the **Volume Light Parameters** rollout of the Environment and Effects dialog. These let you control the intensity of the Atmosphere effect separately from the illumination properties of the light itself. If you reduce the **Attenuation End %** value on the Environment and Effects dialog, the Volume Light effect will fade out before the light's Far Attenuation End Range is reached.

You can use a **Projector Map** with a Volume Light to project the image through the light. When a 2D map is projected through a volume light, the map projection is the same all along the light. When a 3D map is used, the projection changes in three dimensions all through the light.

WARNING: Volume Lights take a long time to render, so don't use them just because you can. Use them only when they are really needed.

END-OF-CHAPTER EXERCISE

EXERCISE 8.11: Oil Tank Explosion

In this exercise, you will apply what you have learned about Particle Systems, Space Warps, and Atmospheres to make an oil tank explode.

You will also learn how to use a few other tools, including:

- **PArray**, a Particle System based on a mesh object
- **PBomb**, a Force Space Warp that sends particles flying as if from an explosion
- **Deflector**, a Space Warp that simulates particles hitting a surface
- **Fire Effect**, an Atmosphere that renders fire and explosions

Create Object Fragments

1.   Open the file **OilTank.max** from the disc that comes with this book.

2.   On the Create panel, select **Particle Systems** from the pulldown list. Click the button labeled **PArray** to create a Particle Array system. Then click and drag anywhere in the viewport to create the PArray. The size and location of the PArray icon does not matter. Right-click to exit PArray creation mode.

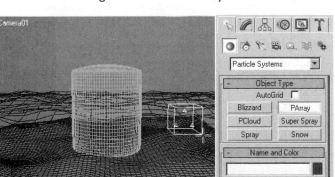

*Figure 8-68: Create the **PArray** icon*

PArray is a Particle System that uses a mesh object as its emitter. In this case we will use the PArray to create fragments of the oil tank.

3.   Select the PArray icon in the viewport, and open the Modify panel. In the **Basic Parameters** rollout, click the **Pick Object** button in the **Object-Based Emitter** section. Then click the **OilTank** object in the viewport.

The oil tank is now the emitter for the PArray particles.

4. Play the animation. Particles fly out from the oil tank. Position the Time Slider at frame **10**.

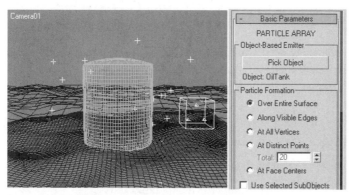

Figure 8-69: The **OilTank** *object is now a particle emitter.*

5. Right now the particles are simply Standard Triangles. Scroll the Basic Parameters roll-out down to the **Viewport Display** section, and select **Mesh**. The viewport updates, displaying the particles as triangles instead of ticks.

6. Open the **Particle Type** rollout. In the **Particle Types** section, choose **Object Fragments**. The viewport updates to display thousands of triangular faces.

7. Scroll the Particle Type rollout down to the **Object Fragment Controls** section. Choose **Number of Chunks**. The viewport updates to display larger chunks of the exploded oil tank.

8. Select the original **OilTank** object (not the object fragments) and hide it. Then play the animation. The PArray creates the effect of an exploding oil tank.

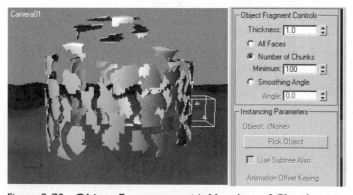

Figure 8-70: **Object Fragments** *with* **Number of Chunks** *option*

TIP: If the viewport playback is too slow, you may render a preview. On a very slow computer, you may need to change the Viewport Display option back to **Ticks**.

Scatter the Fragments with PBomb

This explosion is not very convincing, because the object fragments all move together, at the same speed and direction. To add realism, we will use the PBomb Space Warp.

1. With the object fragments selected, scroll the Modify panel up to the **Particle Generation** rollout. Change the **Speed** parameter to **0**. This makes the fragments stay in one place. Change the **Life** parameter to **61**. This allows the particles to exist throughout the length of the animation.

2. Switch to the Top viewport. On the Create panel, click **Space Warps**. Click the **PBomb** button, and click and drag to create a PBomb icon in the center of the oil tank fragments.

   The size of the PBomb icon does not matter, but the location of the icon determines how the Space Warp will scatter particles.

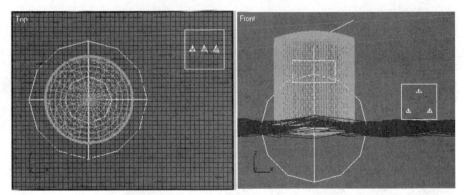

*Figure 8-71: Placement of the **PBomb** icon*

3. Bind the PArray icon to the PBomb Space Warp. Click **Bind to Space Warp**. Then click the PArray icon and drag to the PBomb icon. Play the animation.

*Figure 8-72: **PArray** Particle System bound to **PBomb** Space Warp*

4. The explosion is now more realistic, but it starts too late. Select the PBomb in the viewports, and open the Modify panel. Change the **Start Time** parameter to **0**.

   The explosion now begins at frame **0**.

Make the Fragments Fall

1. Go back to the Create panel. In the Space Warps creation panel, click the **Gravity** button. Click and drag in the Top viewport to create the Gravity Force.

   The size and location of the icon does not affect the simulation. The only transform that affects Gravity is rotation. Make sure that the icon arrow is pointing down.

2. Click the **Bind to Space Warp** tool. Click the PArray icon and drag to the Gravity icon. Play the animation.

3. The Gravity effect is a little too strong. Reduce the Gravity **Strength** parameter to **0.5** and play the animation again.

4. The object fragments are falling through the ground. To make the fragments stop at the ground plane, create a **Deflector** Space Warp. On the Space Warp Create panel, choose **Deflectors** from the pulldown list. Then click the **Deflector** button.

   Click and drag in the Top viewport to create the Deflector icon. Make the icon slightly larger than the high-density portion of the ground plane, as shown below.

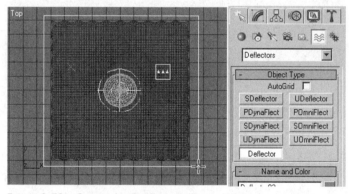

*Figure 8-73: Create the **Deflector** Space Warp*

The location and size of the Deflector icon is critical. The icon is a surface that particles can react with. If necessary, change the **Length** and **Width** of the Deflector icon in the Modify panel, or move the icon with the Select and Move tool.

5. Use the Bind to Space Warp tool to bind the PArray particle system to the Deflector Space Warp.

   Play the animation. The object fragments now bounce off of the Deflector icon.

6. Select the Deflector icon and open the Modify panel. Reduce the **Bounce** parameter of the Deflector to **0**. The fragments do not bounce, but slide across the Deflector.

7. Increase the **Friction** parameter of the Deflector to **100%** and play the animation. Now the object fragments are staying put.

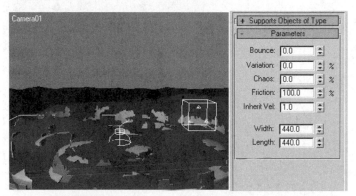

*Figure 8-74: Deflector **Bounce** of **0**, and **Friction** of **100%***

8. The object fragments are still moving too mechanically. Select the PArray and open the Modify panel. Select the PArray object in the Modifier Stack. Open the **Rotation and Collision** rollout. Change the **Spin Time** parameter to **150**, and play the animation.

    The object fragments spin throughout the animation, even after they have landed. To make the fragments stop spinning, we will animate the Spin Time parameter.

    Spin Time works by rotating the particles 360 degrees over the amount of time specified in the parameter field. So, if the Spin Time is **150**, it will take **150** frames to completely rotate the particles once.

9. Turn on the **Auto Key** button. Move the Time Slider to frame **10**. Change the Spin Time parameter to **0**. Turn off Auto Key, and play the animation.

    Because of the unusual way that Spin Time works, as the Spin Time value is reduced, the fragments spin faster. As Spin Time value reaches **0**, the fragments spin faster and faster, then suddenly stop rotating. This looks very unnatural.

    The solution is to use **Step** keyframe tangents, which holds the parameter value at **150** until frame **10**, then immediately jumps to a value of **0**. The fragments will spin at a constant rate, then stop.

10. With the PArray selected, open the **Curve Editor**. In the Controller Window, look for a track called **Object (PArray)**. Open that track and look for the **Spin Time** track. Select the Spin Time track. The animation curve is displayed in the Key Window.

11. Select both Spin Time keyframes by drawing a selection box around them. Then click **Set Tangents to Step** on the Curve Editor toolbar.

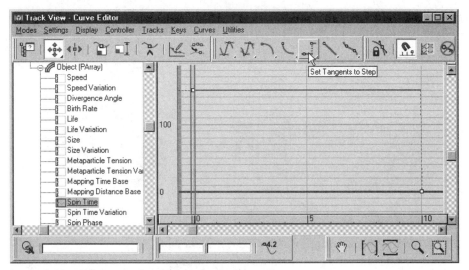

*Figure 8-75: Change the **Spin Time** keyframes to **Step** tangents*

The Spin Time parameter jumps to **0** at frame **10**. Close the Curve Editor and play the animation. The fragments rotate more smoothly.

Render the Fire

1. All the animation needs now is fire. Choose **Rendering > Environment** from the Main Menu. In the **Atmosphere** rollout, click **Add** and choose the **Fire Effect** from the pop-up dialog.

2. Open the Create panel and choose **Helpers**. Select **Atmospheric Apparatus** from the pulldown list, and click the **SphereGizmo** button. Create a SphereGizmo in the Top viewport, centered on the oil tank, and about the same size as the Deflector.

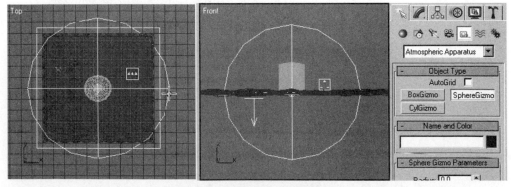

*Figure 8-76: Create a large **SphereGizmo** Atmospheric Apparatus centered on the oil tank*

3. Use the Scale tool to increase the scale of the SphereGizmo slightly in the world Z axis.

4. In the Environment and Effects dialog, select the Fire Effect in the Atmosphere queue. Click the **Pick Gizmo** button, and click the SphereGizmo in the viewport.

5. Scroll the Environment and Effects dialog down to the bottom of the Fire Effect Parameters rollout. Click the **Explosion** check box. Then click the button labeled **Setup Explosion**. In the pop-up dialog, enter **40** for the **End Time**.

6. Render a few test frames at 320 x 240. It appears that the SphereGizmo isn't large enough. Increase its **Radius** parameter to about **320**.

7. Experiment with different settings, such as the Fire Effect and PBomb parameters. When you are happy with the result, render a **.TGA** sequence at 640 x 480, and view it in the RAM Player. Save the scene as **OilTankExplosion.max** in your folder.

To see a finished animation, view the file **OilTankExplosion.avi** on the disc.

Figure 8-77: Completed explosion

Hints

-   To start the explosion later, change the PBomb **Start Time** parameter. However, the cracks in the PArray can be seen. You can render the original OilTank object before the explosion, then replace it with the PArray object at the first frame of the explosion. To do this, animate the **Visibility** tracks of the PArray object and the original OilTank.

-   To create the effect of a crater, add a **Ripple** modifier to the Ground Plane object, and animate its parameters.

SUMMARY

A number of tools are available in 3ds Max for creating special effects.

Space Warps can be used to deform objects. The difference between a Space Warp and a standard object modifier is that Space Warps operate in world space and are affected by object transforms.

A **Particle System** is a single object that produces many particles, such as snow or rain. Certain types of Space Warps, such as **Forces** and **Deflectors**, can influence the movement of particles. This allows particle systems to imitate wind and gravity effects.

Particle Flow is an advanced tool for creating event-driven particle effects. It gives you more power and freedom than the standard 3ds Max particle systems.

Video Post is a simple compositing program built into 3ds Max. It can be used to render layered images or to convert image sequences to compressed movie files.

Render Effects are an easy way to create 2D glows and other lens effects. Render Effects rely on the **Graphics Buffer** to apply effects to rendered pixels. You can choose to assign Render Effects using Object ID numbers or Material ID numbers.

Atmospheres are 3D rendering effects. They can produce fog, clouds, or fire. Some Atmospheres require an **Atmospheric Apparatus** Helper object to contain the effect.

Volume Lights are Atmospheres that render light beams.

REVIEW QUESTIONS

1. What does a **Space Warp** do?

2, How is a Space Warp associated with an object?

3. What are **Particle Systems**?

4. What is the difference in the effect when you blow something up with **PBomb** versus exploding it with **PArray**?

5. What is **Video Post** used for?

6. What is a **Volume Light**?

7. What is the difference between **Forces** and **Deflectors**?

8. How do you assign **Render Effects** to objects?

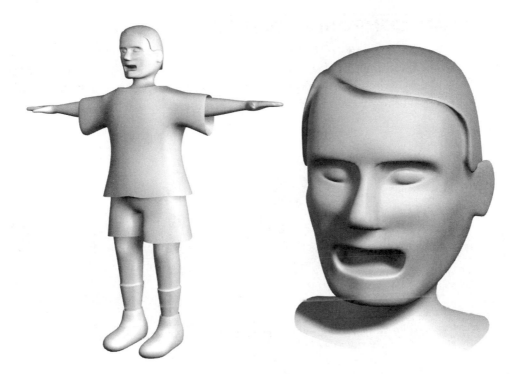

Chapter 9
Advanced Modeling

OBJECTIVES

In this chapter, you will learn about:

- Using Boolean compound objects
- Organic modeling with Bezier Patches
- Using splines to create Patches with Surface tools
- Subdivision surface modeling with Editable Poly and MeshSmooth
- NURBS basics
- Sculpting with Paint Deformation
- Merging objects and scenes

BOOLEAN COMPOUND OBJECTS

Boolean operations in computer graphics work with two overlapping objects. The two objects are called *operands*. Boolean operations can be used to combine objects, or to "carve" one object with another object. When a Boolean operation is performed, the result is a *union* (addition), *subtraction*, or *intersection* (overlapping volume).

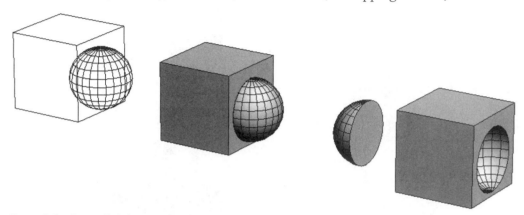

Figure 9-1: Original objects, and Boolean union, intersection, and subtraction

NOTE: The term "Boolean" is named after George Boole, a nineteenth-century English mathematician who developed symbolic logic, which applied mathematical theory to logic. Symbolic logic is one of the fundamental building blocks of all computing; without it, computers could not exist.

USING BOOLEANS

To perform a Boolean operation:

- Create two objects. Position them so they overlap.
- Select one object. On the Create panel, click **Geometry**, and choose **Compound Objects** from the pulldown menu.
- Click the button labeled **Boolean.**
- On the **Pick Boolean** rollout, click **Pick Operand B**. Click the other object in the viewport.
- In the **Operation** section of the **Parameters** rollout, choose the operation.

BOOLEAN RULES AND TIPS

Sometimes a Boolean operation will cause both operands to disappear or react in other strange ways. Before performing the operation, make sure there are no holes or open edges on either operand! A hole in your geometry results in an illegal operand, which is guaranteed to make the operation fail. You can use the **STL Check** modifier to verify the geometry of the operands. Boolean operations also work best on objects with similar levels of detail.

Because Boolean operations can sometimes fail, it's a good idea to save your scene before performing a Boolean operation. Also, if you select the **Instance** option before choosing Operand B, you can make changes to the original objects, and these changes are reflected in the instanced operands in the final Boolean. The **Move** creation option actually deletes the original objects.

If you are performing a Boolean operation with several objects, you must perform the operation with two objects at a time. The best approach is to convert all of your objects to Editable Mesh or Editable Poly, and use the **Attach** command until you have only two objects left. Then use those two objects as the Boolean operands.

WARNING: Don't use a Boolean Compound Object as an Operand in another Boolean! If you end up with a Modifier Stack of Booleans stacked on top of Booleans, 3ds Max will have a difficult time calculating the result, and it may even crash. Collapse the Stack before applying another Boolean operation.

Booleans are great for architectural or hard surface modeling. Don't use Booleans for objects that you plan to deform, such as characters. The Boolean algorithm often results in long, thin faces, which deform very poorly. For this reason, you should never use Booleans for organic character modeling.

Boolean operands retain their materials after the operation. This means that you can use Booleans to create material setups that might otherwise be difficult to produce.

EXERCISE 9.1: Multicolored Bead

In this exercise, you'll create a multicolored bead from a sphere and a cylinder.

1. Reset 3ds Max.

2. Press the **<S>** key to enable 3D Snap. In the Top viewport, create a sphere centered on the world origin.

3. In the **Top** viewport, create a cylinder with a Radius roughly one-third of the sphere's Radius. Make certain the cylinder is precisely centered on the world origin.

4. Adjust the **Height** and position of the cylinder so it passes through the center of the sphere. Make sure the cylinder is considerably taller than the sphere, and passes completely through it.

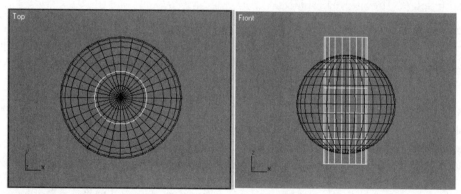

Figure 9-2: Cylinder through sphere

5. Open the Material Editor. Create a bright red material, and assign it to the sphere. Create a black material, and assign it to the cylinder.

6. To prepare for the Boolean operation, save the scene with the filename **Bead01.max** in your folder.

7. Select the sphere. In the **Create** panel, in the **Geometry** category, choose **Compound Objects** from the pulldown list. Click the button labeled **Boolean.**

8. Make sure the operation is set to **Subtraction (A-B)**.

9. Click **Pick Operand B**, and click on the cylinder.

10. The **Material Attach Options** dialog appears. Click **OK** to accept the default option, which is **Match Material IDs to Material**.

    3ds Max automatically assigns material IDs to the faces of each operand, creates a Multi/Sub-Object material, and assigns the new material to the new Boolean Compound Object.

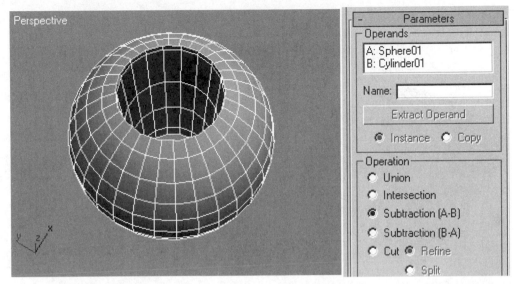

Figure 9-3: Multicolored bead

11. Save the scene as **Bead01.max** in your folder.

PATCH MODELING

Standard polygonal mesh modeling is excellent for simple objects, but it fails to combine the sophistication and ease of use needed for effective modeling of complex curves. For example, it's really hard to model a human head to a high level of detail using Editable Mesh alone. To meet this demand, 3ds Max offers **Bezier Patch** modeling.

A Bezier patch is a surface derived from Bezier splines. The splines are joined at their end vertices, and the surface is interpolated based upon the positions of the vertices and their vector handles.

In 3ds Max, you can create Bezier patch models in several ways:

- Use the **Patch Grids** pulldown on the Create panel
- Select **Patch Output** within a modifier such as Lathe
- Convert an existing object to an **Editable Patch**
- Use the **Surface** modifer on an Editable Spline object

The Surface modifier is an entire technique in itself, and it is covered in the next section of this chapter.

There are two types of **Patch Grid** objects: **Tri Patch** and **Quad Patch**. The Tri Patch is based on triangles, each with three control vertices and three splines. The Quad Patch is based on rectangles with four CVs and four splines.

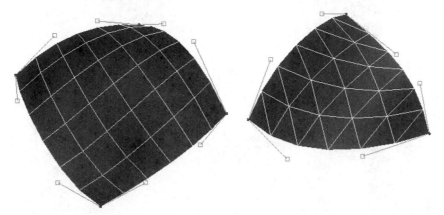

Figure 9-4: **Quad Patch** *and* **Tri Patch**

When you create a Quad Patch Grid from the Create panel, you can define the number of **Length Segments** and **Width Segments**. This determines the number of control vertices in the object. More CVs gives you greater control, but you can have too much of a good thing. Fewer CVs can actually be better in many cases, because you'll get smoother curves without incurring the penalty of reduced interactive performance.

One of the stranger things about 3ds Max is that in order to edit the control vertices of a Patch Grid object, you must first convert it to an Editable Patch. Select the Patch Grid, then right-click and choose **Convert To > Convert to Editable Patch** from the Quad Menu.

You can also convert any mesh object into an Editable Patch. It's usually most useful to start from a Box or other primitive with relatively few segments.

There are five sub-object types within Editable Patch:

Vertex: control vertex that can be moved to control curvature

Handle: Bezier vector, which can be selected independently of vertex

Edge: Bezier spline curve connecting two CVs

Patch: Bezier surface bounded by either three or four splines

Element: collection of Patch sub-objects, resulting from an **Attach** command

In addition, there are two types of control vertices: **Corner** and **Coplanar**. A Corner vertex has freely adjustable vector handles, while the vectors of a Coplanar vertex always remain locked to a plane. When you move one vector handle of a Coplanar vertex, the other handles move so that they all lie on the same plane.

Just as with Bezier splines, you convert between vertex types with the right-click Quad Menu. In addition, you can use the **Lock Handles** check box on the **Selection** rollout of the Modify panel. This lets you move vertex handles as a group, without being constrained to a plane.

The Modify panel of Editable Patch has several commands that work very similarly to those in Editable Mesh, such as Attach, Normals, Material IDs, and even Extrude and Bevel. The big difference is that the level of detail of a Patch object can be changed quickly and easily using the **Steps** parameters in the **Surface** section of the **Geometry** rollout. Just as you can change the Interpolation settings for a shape or spline, you can alter the smoothness of a Patch object with the Steps parameters. You can even have different settings for the viewport and the renderer, so your renderings have very smooth curves without slowing down viewport interactivity.

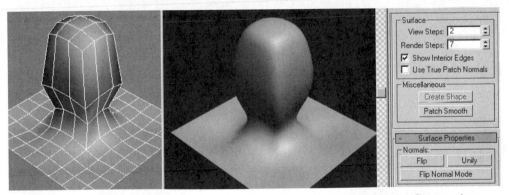

*Figure 9-5: An extruded Editable Patch with different **View Steps** and **Render Steps** values*

WARNING: Beware of converting high-density meshes to Editable Patches! Every polygon or face in the original mesh is converted to a patch with either three or four Bezier curves. This can slow your computer to a crawl, or even make it crash, as 3ds Max attempts to calculate the many Bezier curves and surfaces.

By default, when you convert an Editable Mesh or Poly to an Editable Patch, the Steps parameter is set to 0. This is a safeguard to prevent your computer from crashing if you inadvertently convert a high-density mesh object to a Patch.

SURFACE TOOLS

An even more powerful and versatile technique based on the same principles is often used instead of Editable Patch. This technique employs two modifiers, **CrossSection** and **Surface**, which are collectively known as **Surface Tools**.

Surface Modifier

The **Surface** modifier automatically creates Bezier patches from a network of splines within a single Shape, usually an Editable Spline object. Instead of creating a Patch object directly, you create a *spline cage* and let the Surface modifier create the patches for you. The curves within the Editable Spline object must be arranged in such a way that they *always* form four-sided or three-sided sections.

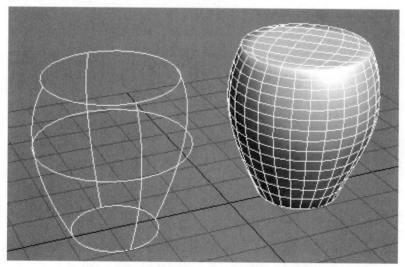

*Figure 9-6: Editable Spline before and after applying the **Surface** modifier*

CrossSection Modifier

Optionally, the Surface modifier can be used with the **CrossSection** modifier to create a patch surface from a collection of splines. CrossSection reduces the amount of work you need to do, because it automatically draws Bezier curves connecting several splines.

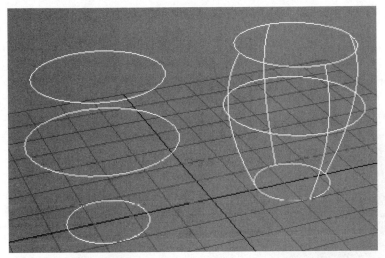

*Figure 9-7: Editable Spline before and after applying the **CrossSection** modifier*

In order for **CrossSection** to work, the splines must be all part of the same object. Each spline must have the same number of vertices, and the first vertices of each spline should be aligned.

CrossSection connects splines to one another in the order in which they were attached. You'll need to use the **Attach** command to bring several shapes into a single Editable Spline object. Make sure to attach the shapes in the correct order, or else your Cross Section splines will be distorted.

For more complex models, such as the mask you'll create in the following tutorial, you'll need to analyze the curvature of the object and create the crossing splines yourself.

EXERCISE 9.2: Building a Mask with the Surface Modifier

In this exercise, you'll make a mask using the Surface modifier. When doing this exercise, don't be discouraged if your first attempt doesn't come out very well. It takes some practice to use the Surface modifier, but the rewards are great. If necessary, practice making the mask three or four times until you are fully familiar with the way the Surface modifier works.

In order to use the Surface modifier, all splines must be part of the same object. To accomplish this, turn off the **Start New Shape** check box when creating lines, or use the **Attach** button to attach individual lines into one shape. Best of all, you can use the **Create Line** command on the Editable Spline Modify panel to build new spline subobjects within the current object.

Figure 9-8: Mask

Create Basic Contours

1. Reset 3ds Max.

2. In the Front viewport, use the **Line** tool to draw basic contours for half of the mask, as shown in the following illustration. Make all the lines part of the same object. To prevent problems later, make certain that the straight side of the mask is perfectly straight, and touching the world Z axis. The mask will be reflected along the world X axis to create the other half.

3. Press the **<S>** key to activate 3D Snap. Right-click **3D Snap Toggle** to display the **Grid and Snap Settings** dialog. Check the **Vertex** and **Grid Lines** options, and uncheck any other options. In Vertex sub-object mode, move the center vertices so they snap to the Z axis.

4. Save the scene with the filename **mask01_basic_contours.max** in your folder.

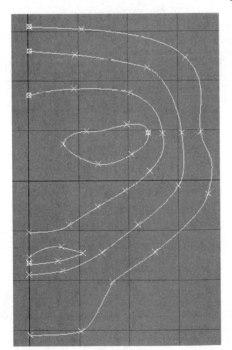

Figure 9-9: Basic contours of half of a mask

The key to building a spline cage is that it's constructed of numerous individual splines that cross each other and overlap. Wherever two splines overlap or touch, there must be two vertices, one on each spline. The crossing vertices don't have to be precisely in the same place, but it helps.

You need to create two lines at the center of the mask. These are the profile lines. For every vertex at the end of one of the contour lines, there must be a corresponding vertex on the profile line. To aid in this process, you'll use 3D Snap once again.

Add Connecting Lines

1. Right-click 3D Snap Toggle to display the Grid and Snap Settings dialog. Check the **Vertex** option and uncheck any other options.

2. Make sure you're still in **Vertex** sub-object mode. On the **Geometry** rollout, click the button labeled **Create Line**. In the Front viewport, click the vertex at the top center of the forehead. Then click each vertex down the profile until you reach the upper lip. Right-click to finish creating the line.

   Then click the vertex at the center of the lower lip, and click the other two vertices at the center of the chin. Right-click to finish creating the second profile line.

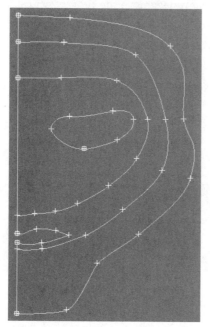

Figure 9-10: Contours and profile lines

3. Next, create additional contour lines to give the mask detail. Use the Create Line tool in conjunction with Vertex Snap to add a nose and lips.

   Remember that at every place where a spline touches another spline, you must have a vertex on each spline. Use the **Refine** tool to add vertices wherever two splines touch or cross. If the new vertices don't appear where you want them to, simply move them.

 TIP: You can also use the **CrossInsert** tool. If two splines are overlapping, you can click the CrossInsert button on the Geometry rollout. Then click at the place where the two splines cross. A new vertex is inserted on each spline.

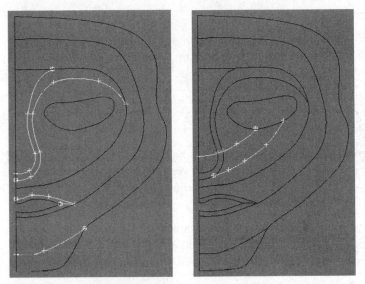

Figure 9-11: Additional contour lines added

4. Next you will create the remainder of the connecting lines to form the mask. Before continuing, take a good look at the contour lines in the Front viewport, and work out roughly where the connecting lines should fall. The following illustration shows a layout that will work for this mask. The connecting lines are shown in white.

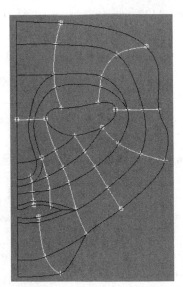

Figure 9-12: Connecting lines added

5. Save the scene as **mask02_connecting_lines.max** in your folder.

Apply the Surface Modifier

The **Surface** modifier will create surfaces only on areas bounded by three or four lines, and with each vertex connected to others by a line. The connecting lines are designed to form these areas.

1. On the **Modify** panel, go to the **Vertex** sub-object level. Click **Refine**. Add vertices to splines where necessary so connecting lines can be formed. Delete any vertices that will not be used for connecting lines.

2. Turn on 3D Snap Toggle, and click **Create Line** on the **Geometry** rollout. Draw the connecting lines, taking care to click existing vertices as you do so. If you find that some vertices are missing, click **Refine** and add them, then click Create Line again and resume drawing lines.

    Next, you will find out how well you have done connecting vertices and creating three- and four-sided areas by applying the Surface modifier and seeing if the surface appears on the entire mask.

3. Apply the **Surface** modifier to the mask.

    If nothing seems to happen, it is probably because the face normals are pointing the wrong way.

4. On the **Parameters** rollout, check the **Flip Normals** check box if necessary.

    Parts of the mask appear, but other parts may be missing. This is normal. Even experienced users often have to fiddle with a model before the Surface modifier will work properly on it.

5. Go down the Modifier Stack to the level of the Line or Editable Spline object. Turn on

    Show End Result ⏸ . Now you can see the surface and the spline cage at the same time.

    The most common problem that occurs with this type of model is that some vertices did not quite snap correctly.

6. Move one or more vertices and snap them into place. When you correct a problem, the surface patches will appear.

    Other common problems with the Surface modifier are areas with more than four vertices. Examples are the eye and mouth areas, which have more than four vertices. In this case, we want these areas to remain open.

7. Further correct the original mask by deleting vertices, drawing connecting lines, and so on, until surface patches appear across the entire mask.

    It can take some work, and you might think at times that the Surface modifier is not working correctly. It takes practice to get used to the intricacies of modeling with Surface. If you persevere, you will eventually get the surface to appear in all the right places.

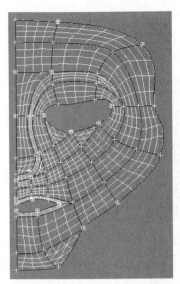

*Figure 9-13: Mask with spline cage and **Surface** modifier*

8. Save the scene with the filename **mask03_surface.max** in your folder.

Mirror the Mask

Now that you have half of a flat mask with a surface, it's time to mirror the mask. To make it easiest to do so, move the Pivot Point of the mask to the world origin.

1. Exit sub-object mode. Go to the **Hierarchy** panel and click **Affect Pivot Only**. Select the Move tool. Type **0,0,0** into the **Transform Type-In** fields near the bottom of the screen.

The Pivot Point moves to the world origin. Turn off Affect Pivot Only.

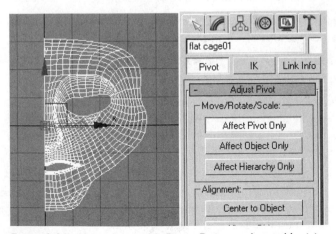

*Figure 9-14: Move the mask's **Pivot Point** to the world origin*

2. Select the Line or Editable Spline object in the Modifier Stack. Add a **Mirror** modifier. In the **Options** section of the **Parameters** rollout, enable the **Copy** check box.

The mask is mirrored. If necessary, adjust the **Offset** parameter.

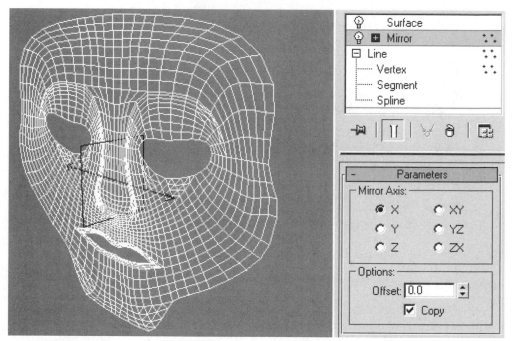

Figure 9-15: Perspective view of mirrored mask

Create Depth

Your final task is to bring this flat mask into the third dimension.

1. Enter sub-object Vertex. Make sure Show End Result is still on. Press **<F3>** to enter **Smooth+Highlights** viewport rendering mode.

2. Click the **Select and Move** button. Press the **<X>** key to hide the Transform Gizmo. Press **<F6>** to activate the global Y axis.

3. On the Modify panel, look for the **Area Selection** check box and activate it. This allows you to select several vertices at the same time with a single mouse click. The spinner next to Area Selection is the selection threshold. Vertices must be closer together than this value in order to be selected with a single click.

4. Maximize the Perspective view. Select the vertex at the tip of the mask's nose. Look on the Modify panel at the bottom of the Selection rollout. It should read **2 Vertices Selected**. If it says only one vertex is selected, increase the Area Selection threshold to **1.0**, and then click the vertex at the tip of the nose again.

5. Move the two vertices at the tip of the nose forward in the global Y axis.

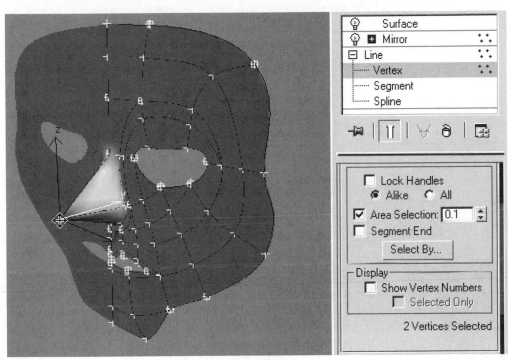

Figure 9-16: Move the two vertices forward in the Y axis

6. Continue moving the vertices forward to create a 3D mask. Convert vertices to Smooth or Bezier as necessary.

7. If necessary, change, add, or remove CVs and splines. Work in the Front viewport and the Perspective viewport.

8. Save the scene with the filename **mask04.max** in your folder.

Hints

*   When moving vertices, don't use the Move Gizmo! Make sure it is turned off with the **<X>** hotkey, not with the **Views > Show Transform Gizmo** menu item. Leave the Views menu item enabled, and use the **<F5>** through **<F8>** hotkeys to select the axis restrictions. That way, you can see what axis is selected, without the Move Gizmo getting in your way.

*   It might be easier to see what you're doing if you temporarily disable the Surface modifier.

*   If viewport performance is bogging down, you can disable the Mirror modifier, or reduce the number of Surface **Steps** while you're working.

*   In some cases, the Surface modifier may create duplicate patches, resulting in rendering errors. The **Remove Interior Patches** option in the Surface modifier will fix this.

- Smooth vertices are usually easier to manipulate than Bezier vertices. To keep your work simple, try to use Smooth vertices wherever possible, especially for areas on the interior of the model.

- If you try to enter Segment or Spline sub-object mode, your Mirror modifier stops working. This is because only the selected Segment or Spline is being passed up to the stack, so only the selected sub-object is mirrored. You can get around this by inserting a SplineSelect modifier below the Mirror modifier. This clears out any selection lower in the Stack.

- To soften up any creases, add an **Edit Patch** modifier to the top of the Stack, and increase the **Relax Value** parameter. Increase the **Render Steps** parameter for a smoother surface.

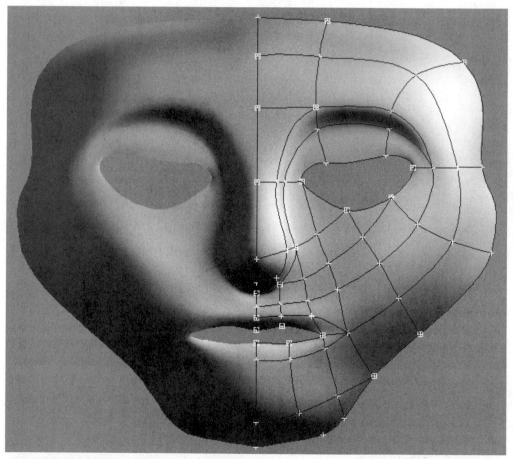

Figure 9-17: Completed mask

NURBS

NURBS stands for *Non-Uniform Rational Basis Splines*. This is a certain type of spline mathematics that uses weighted control points. The meaning of the term NURBS is not as important as understanding what NURBS does. It's a way of modeling that creates smooth curves and surfaces. NURBS modeling is useful for organic objects, draped cloth, and other smooth or complex surfaces that can be difficult to create with other modeling tools.

NURBS CONCEPTS

NURBS is similar to Bezier Patches and the Surface modifier, since you use control points to influence the curvature of a surface. However, NURBS is more powerful, and also more complicated. As with all models in 3ds Max, a NURBS object is converted to polygons when rendered.

Figure 9-18: A comfy NURBS sofa

NURBS is characterized by several concepts that are found only in NURBS modeling. For example, you can extract a curve from a surface, use a curve to trim part of a surface, or blend surfaces together seamlessly.

WHEN TO USE NURBS

Many users find that they need NURBS only occasionally, and avoid it the rest of the time. Other users find NURBS fascinating, and use it every chance they can. NURBS is not an easy tool to master, but it can be a pleasure to play with. If you'd rather stick to the more traditional modeling tools most of the time, that's fine too. Just make sure you get the basics so you'll be able to use NURBS when you need it.

CREATING A NURBS OBJECT

NURBS Surfaces can be created directly from the Create panel. Just select **Geometry**, and choose **NURBS Surfaces** from the pulldown menu. You can create NURBS patch grids just like Bezier Quad Patches. The Create panel lets you specify how many control points you wish to start out with.

Likewise, you can create NURBS Curves directly by selecting **Shapes** from the Create panel, and choosing **NURBS Curves** from the pulldown menu.

There are two ways to edit NURBS objects: **Points** and **CVs**. **Point Curves** and **Point Surfaces** use edit points that always lie on the curve or surface, just like standard splines and Bezier patches. **CV Curves** and **CV Surfaces** use control vertices that lie off the curve or surface. NURBS Points might seem to be easier to work with at first, but in fact, NURBS CVs are much more effective.

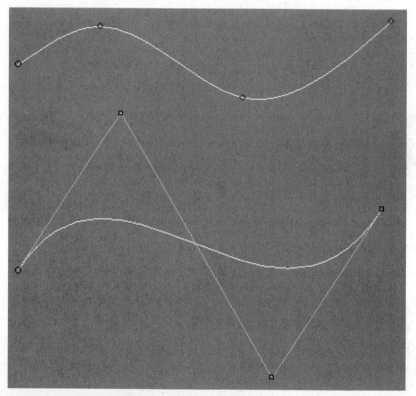

Figure 9-19: NURBS **Point Curve** and **CV Curve**

You can convert from Points to CVs and vice versa, but this is not optimal. The curvature always changes when you convert from one editing type to another, although you can minimize the damage by tweaking the settings in the conversion dialog. This is a limitation of how NURBS are implemented in 3ds Max.

You can also convert 2D or 3D primitives, Editable Splines, and Patches to NURBS. To do this, select the object and right-click to access the Quad Menu. Choose **Convert To > Convert to NURBS**. Be aware that the conversion process will always alter the form of the object to some degree. You can't convert Editable Mesh or Editable Poly objects to NURBS.

When a NURBS object is selected, the Modify panel has many options for editing the object. The object type listed in the Modifier Stack may read NURBS Curve, even if the object contains surfaces.

The options under the **Create Points**, **Create Curves** and **Create Surfaces** rollouts let you build new NURBS sub-objects from scratch, or by combining existing sub-objects or pieces of sub-objects. For example, you can create a surface across two or more curves by clicking the **U Loft** button, then selecting the curves in the viewport. Or, you can create a new curve based on an existing surface with the **U Iso Curve** or **V Iso Curve** buttons. There are many possibilities for projecting, trimming, and blending curves and surfaces.

The tools for building NURBS sub-objects are also found in the **NURBS Creation Toolbox**, which is accessed by clicking on a button ⊞ on the Modify panel. See the following illustration.

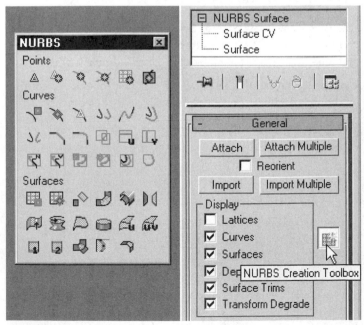

Figure 9-20: **NURBS Creation Toolbox**

EXERCISE 9.3: Flying Saucer

In this tutorial, you'll create a classic 1950s flying saucer using NURBS.

1. Reset 3ds Max.

2. In the Top viewport, create a sphere with a **Radius** of about **40** units.

3. With the sphere selected, right-click to access the Quad Menu, and choose **Convert to NURBS**. The sphere has been changed to a NURBS object. It no longer has sides and segments. View the sphere in Wireframe mode in the Perspective viewport.

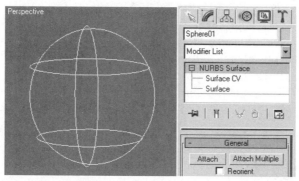

Figure 9-21: NURBS sphere in Wireframe display mode

4. Press **<F3>** to switch the Perspective view back to **Smooth+Highlights**. Open the Modify panel and choose **Surface CV** as the sub-object level. A series of control points appear around the sphere. They are joined by lines called a *control lattice*.

5. On the Modify panel, click the **Column of CVs** button. Select one of the control vertices at the center of the sphere. All CVs around the equator are selected.

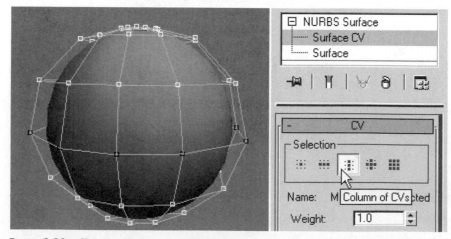

*Figure 9-22: Click the **Column of CVs** button and select one of the center CVs*

6.   With the equator CVs still selected, scale them in the world XY plane. The scaling is the same in both axes, so the saucer remains symmetrical.

NOTE: If the scaling is not uniform, go to the Main Menu and choose **Customize > Preferences**. In the **Preference Settings** dialog, select the **Gizmos** tab. In the **Scale Gizmo** section, check the box labeled **Uniform 2-Axis Scaling**.

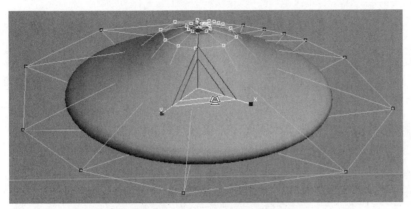

Figure 9-24: Symmetrical flying saucer

10.   In the Front viewport, select one of the control vertices in the next column up on the flying saucer. All of the vertices in that column are selected. On the Modify panel, increase the **Weight** parameter to **200**.

The surface is drawn toward the selected control vertices.

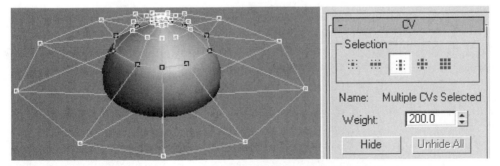

*Figure 9-25: Increase the **Weight** of the selected control vertices to **200***

11. Select the control vertices on the equator again, and increase the **Weight** parameter to **200**. The surface is drawn toward the equator.

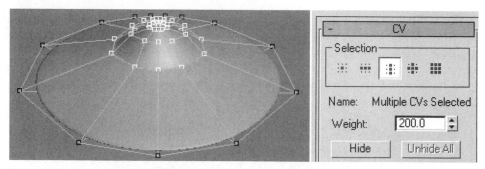

*Figure 9-26: Increase the **Weight** of the CVs at the equator*

12. Continue editing the scale, position, and Weight of the CVs to produce a finished NURBS model. Save your scene as **FlyingSaucer.max** in your folder.

Figure 9-27: Flying saucer: "Klaatu barada nikto!"

PAINT TOOLS

One of the most exciting, intuitive features in the 3ds Max modeling arsenal is the Paint toolset. Using a circular brush icon, you can perform several wonderful functions, such as selecting sub-objects, painting colors of vertices, setting weights for animated deformation with the Skin modifier, or sculpting Editable Poly objects by pushing or pulling.

All of these functions share common options that are controlled with the **Painter Options** dialog. For example, to paint soft selections within Editable Poly, enable the **Use Soft Selection** option, and click the **Brush Options** button.

In the Painter Options dialog you'll find many controls for altering the size, strength, and display settings of the brush. If you have a pressure-sensitive graphics tablet, you can use the pressure to alter the brush parameters, just as you would in an advanced 2D paint program. The Painter Options dialog also features a curve editor to determine the falloff of brush strength. If you want a hard or soft edge to the brush, edit the curve control points, much in the same way that you would edit the curve in the Loft Deformation dialogs. The left side of the graph corresponds to the center of the brush, and the right side is the edge of the brush. You may also simply choose among five preset curves by clicking the curve icons at the bottom of the editor area.

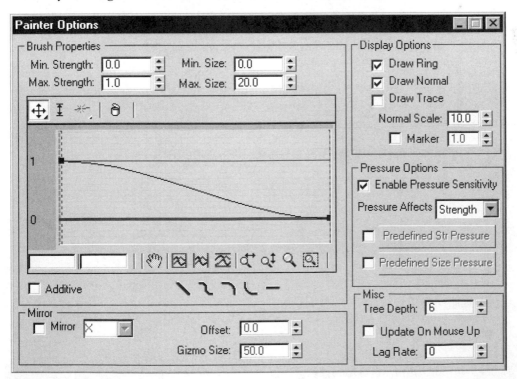

Figure 9-28: **Painter Options** *dialog*

In addition, you can define your own brush presets, so you don't have to keep changing the Painter Options manually. To use and edit brush presets, use the **Brush Presets** toolbar. Right-click an empty area of a toolbar, and choose Brush Presets from the pop-up menu. Then you can dock the toolbar if you wish, by dragging it to a toolbar area. (Remember that you can't dock toolbars if you have the **Customize > Lock UI Layout** option active.)

Figure 9-29: **Brush Presets** *toolbar*

The Brush Presets icons are grayed out unless you have one of the paint tools active. If you click on one of the circular icons while a paint tool is active, then that brush preset is loaded.

To create a customized brush preset, click the **Add New Preset** button on the Brush Presets toolbar. The current settings from the Painter Options dialog are used to create a new preset, and a new button appears on the toolbar. The appearance of the new button is determined by the current Painter Options settings. This is a great feature, because it gives you immediate visual feedback about the preset parameters. You can quickly create your own brush presets, and not have to bother with the Painter Options dialog every time you want to change brush settings.

In addition, you can edit brush presets using the **Brush Preset Manager**. Click the button on the Brush Presets toolbar to launch a dialog that gives you control over all of your brush presets. When a preset is selected in the Brush Preset Manager or on the Brush Presets toolbar, you can edit that preset using the Painter Options dialog.

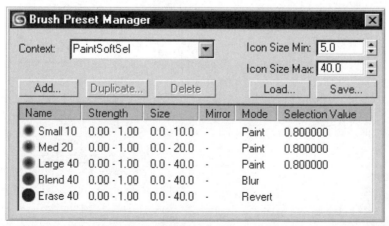

Figure 9-30: **Brush Preset Manager**

NOTE: Brush Presets are stored in the 3ds Max user interface preference files, not in your **.MAX** scene file. If you wish to save and load brush presets, use the Brush Preset Manager to store these settings on disk in a special file with the extension **.BPR**.

PAINT SOFT SELECTION

One of the most useful applications of the paint tools is **Paint Soft Selection**. This is only available from within Editable Poly, Edit Poly, or Poly Select. Enter any sub-object mode, open the **Soft Selection** rollout, enable **Use Soft Selection**, and then click the **Paint** button. Now when you drag your cursor over the model, sub-objects are soft selected according to the settings in the Paint Soft Selection section of the Soft Selection rollout, and in the Painter Options dialog.

The strength of the selection is determined by both the **Selection Value** and the **Brush Strength**. Selection Value determines the maximum amount that a sub-object can be selected. Brush Strength controls the influence of the brush cursor. If the Brush Strength is set to a low value, you'll need to drag the brush over the sub-objects many times to reach the full Selection Value.

The **Revert** button lets you use the brush to unselect sub-objects. In this mode, the Selection Value spinner has no effect, and only the Brush Strength setting is recognized. To fully unselect sub-objects, the Brush Strength must be set to **1.0**.

Click the **Blur** button to enter a mode in which you can soften the edges of existing painted soft selections.

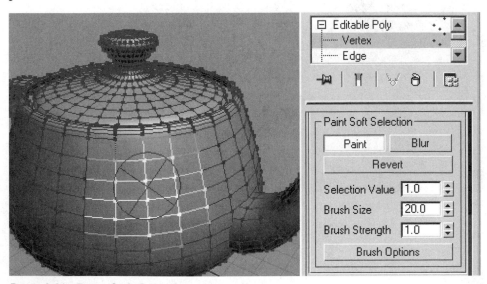

Figure 9-31: **Paint Soft Selection**

EXERCISE 9.4: Sculpted Landscape

In this exercise, you'll use Paint Soft Selection in combination with a Push modifier to create a simple landscape.

1. Reset 3ds Max.

2. Maximize the Perspective view. Create a **Plane** primitive with a **Length** and **Width** of **200** units. Give the Plane **32 Length Segments** and **32 Width Segments**. Press **<F4>** to view Edged Faces.

   This will be a rough landscape; if you were doing a final production model, you'd use more segments, or perhaps add a subdivision modifier.

3. Add a **Poly Select** modifier to the plane. Enter Vertex sub-object mode. Open the Soft Selection rollout and enable **Use Soft Selection**. In the Paint Soft Selection section of the rollout, click the **Paint** button. Drag the cursor across the plane to soft select some of the vertices.

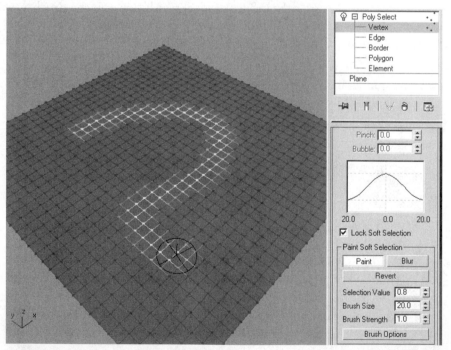

Figure 9-32: Paint a soft selection of vertices

4. Without exiting Vertex sub-object mode, add a **Push** modifier to the object. Increase the **Push Value** to approximately **30** units. The vertices you soft selected earlier are pushed up along their normals, resulting in a raised area on the plane.

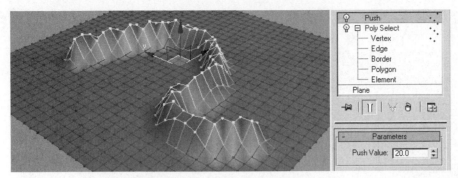

Figure 9-33: **Push** *modifier applied to painted soft selection*

5. Go back to Vertex sub-object mode within the Poly Select modifier. Enable Show End Result so you can see the effects of your painted soft selection. Change the **Selection Value** to **0.1** and paint on the plane again. Note the subtlety of the effect: the amount of displacement applied by the Push modifier is much less. If a vertex is not selected very much, then it won't be pushed very much.

6. With the **Paint** button still pressed, right-click a toolbar and choose **Brush Presets** from the pop-up menu. Click the **Add New Preset** button. The Brush Preset dialog pops up. In the **Preset Name** field, type the name **SoftSelect 0.1**, and click OK. When the dialog closes, a new button appears at the right edge of the Brush Presets toolbar. This is the preset you just created. The button is highlighted to indicate that the preset is active. Click the button to deactivate the preset. If you don't do this, then any changes you make in the following steps will affect this preset, and you don't want that.

7. In the Paint Soft Selection section of the Soft Selection rollout, change the Selection Value to **0.5**. Click the Add New Preset button on the Brush Presets toolbar. Give the new preset the name **SoftSelect 0.5**. Click the new button on the Brush Presets toolbar to deactivate the new preset you just created.

    Repeat this process once more, changing the Selection Value to **1.0** and creating a new preset with the name **SoftSelect 1.0**. You now have three new brush presets, with Selection Values of **0.1**, **0.5**, and **1.0**. Open the **Brush Preset Manager** to double-check your work. See the following illustration.

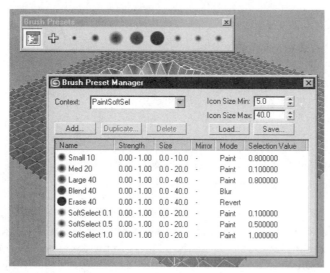

Figure 9-34: Three new brush presets shown in the Brush Preset Manager

8. Use the three new brush presets to paint different amounts of selection on the plane. This gives you three different levels of elevation for the landscape. If you wish, create additional brush presets, or edit the existing ones. For example, you may want to make the **Brush Size** larger or smaller. Simply select the preset in the Brush Preset Manager, and change the Brush Size in the Poly Select modifier. If you need finer control, try using different **Brush Strength** values.

9. Continue painting the soft selection to create your landscape. To change the overall height of the landscape, adjust the Push Value. If you wish to completely flatten an area, use the **Revert** tool with a Brush Strength of **1.0**. When you're happy with the results, save your scene as **SculptedLandscape.max** in your folder.

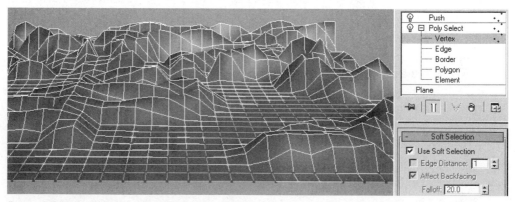

Figure 9-35: Sculpted landscape

PAINT DEFORMATION

Another terrific application of the 3ds Max paint tools is **Paint Deformation**. This is a feature of Editable Poly that lets you push or pull vertices to sculpt geometry. It's more free-form and less procedural than the technique shown in the last exercise. If you have experience sculpting with real-world materials such as clay, then you'll feel right at home with Paint Deformation.

To use this tool, simply open the Paint Deformation rollout within an Editable Poly object or an Edit Poly modifier. Click the **Push/Pull** button, and drag the brush cursor over your model to sculpt. Change the brush settings as desired by changing the **Push/Pull Value**, **Brush Size**, or **Brush Strength** spinners. A positive Push/Pull value will move vertices outward from the surface, and a negative value will move them inward. As with Paint Soft Selection, additional brush parameters can be accessed by clicking the **Brush Options** button to launch the Painter Options dialog, or by choosing presets from the Brush Presets toolbar or Brush Preset Manager.

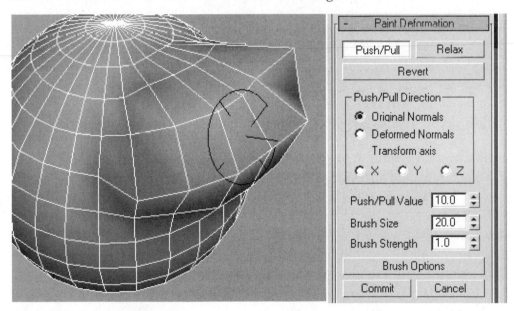

Figure 9-36: **Paint Deformation**

When you use the Push/Pull tool, you can choose what **Push/Pull Direction** you wish the vertices to move. By default, they move along the direction of the **Original Normals**, so the initial state of the surface determines the direction of deformation. A vertex normal is a line projecting from a vertex, perpendicular to the surface. If desired, you can choose to move vertices along their **Deformed Normals**, so they move in the direction of their current normals, instead of the normals of the undeformed object. This gives a different "feel" than using the default setting of Original Normals. You may also simply move vertices along the **X, Y,** or **Z Transform Axis** by choosing the axis. These transforms use the World coordinate system.

If you wish to partially or completely undo your sculpting, click the **Revert** button and drag your brush cursor over the model. As long as you haven't changed the topology of the model by adding or removing sub-objects, you can restore the shape of the model to what it was before you used the Push/Pull tool.

When used in combination with other 3ds Max tools, Paint Deformation gives you a great deal of control over the form of your model. For example, you can choose to only deform a sub-object selection. Simply select vertices, edges, or polygons, and use the brush tool to push or pull them. Unselected sub-objects are not affected.

The Painter Options dialog also features a **Mirror** option. This is very effective for sculpting bilaterally symmetrical objects such as character heads. Simply click the Brush Options button, and enable the Mirror checkbox at the bottom left of the Painter Options dialog. You can also choose which axis you which to mirror, change the **Gizmo Size** of the mirror plane displayed in the viewport, and define an **Offset** value to move the mirror plane gizmo.

The Paint Deformation tool only works on Poly objects. It doesn't work with Patches or NURBS. This means that you should have your object set to the appropriate level of detail before beginning to sculpt. It's also best to first get the overall structure of the object established, with edges running along the general contours of the surface.

As you'll see in the next section, *subdivision surface* modeling is very effective for creating highly detailed, organic polygonal objects. The Paint Deformation tool, when used in combination with subdivision surfaces, gives you the freedom to model almost anything you can imagine.

SUBDIVISION SURFACES

At last we come to the crown jewel of 3ds Max's modeling tools: *subdivision surfaces*. Many modelers view this as the ideal technique for creating curved, organic objects, because it combines the power of mesh editing with the subtlety of Surface tools or NURBS. It is widely considered to be easier to use than the Surface modifier, but it still takes considerable time to master the technique.

Like standard polygonal modeling, subdivision surface modeling starts with a box or other primitive. The difference is that in a subdivision surface model, the vertices of a mesh object become control points for an underlying, higher density, smoothed mesh. In some ways it's similar to working with the Surface modifier, because the model is edited by moving the components of a *control cage*. However, a spline cage always sits directly on the final surface, whereas a subdivision mesh cage usually lies off the surface by quite a bit.

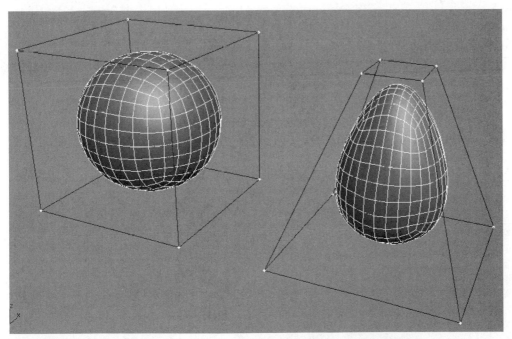

Figure 9-37: A cube becomes an egg by scaling the top polygon of the control cage

The easiest way to create a subdivision surface model is to start with a Box primitive, and convert it to an **Editable Poly**. Scroll the Modify panel and you will find a rollout labeled **Subdivision Surface**. Click the check box labeled **Use NURMS Subdivision**, and then increase the number of **Iterations** to **2** or **3**. The model is subdivided into smaller polygons, and angles among polygons are smoothed, resulting in a softly curved polygonal mesh.

WARNING: Use extreme caution when changing the number of Iterations of a subdivision surface model! Each iteration multiplies the number of quadrilateral polygons by four. So an Iterations value of 7 converts a single polygon to 4^7, which is 16,384 polygons! You should never take the Iterations value above 5, unless you really want to crash your computer. As a general rule of thumb, if you need more than 3 Iterations, there's something wrong with your model that subdivisions won't fix.

To edit the shape of a subdivision surfaces model, simply enter a sub-object mode and use the tools in Editable Poly, which you've already learned. Scale a polygon or move a vertex, and the underlying subdivision surface changes shape. Add more vertices to the control cage with **Slice** or **Cut**, and the underlying surface becomes more detailed and also changes shape.

In addition, you can increase the **Weight** of vertices or edges. Enter Vertex or Edge sub-object mode, select a vertex or edge, and look in the **Edit Vertices** or **Edit Edges** rollout. Increasing the Weight of a vertex draws the surface toward the vertex; increasing the Weight of an edge pushes the surface away.

The Edit Edges rollout also features a **Crease** parameter. Select an edge and increase the Crease value, and the underlying surface will be drawn toward towards the edge. A sharp crease in the surface is formed.

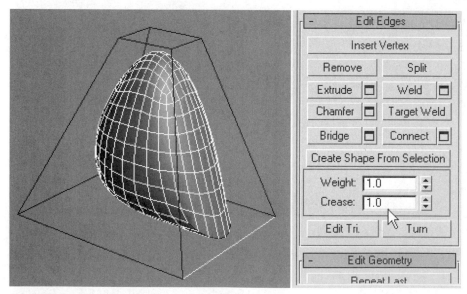

Figure 9-38: A creased edge

You will find that once you get the hang of subdivision surface modeling, you can make almost anything with this technique. It's powerful, easy, and *FUN!*

EXERCISE 9.5: Subdivision Surfaces Goldfish

In this tutorial, you'll create a simple goldfish cracker from a Box primitive. We will explore subdivision surfaces more thoroughly in the End-of-Chapter Exercise.

1. Reset 3ds Max.

2. In the Perspective viewport, create a Box with the following parameters:

Length	5
Width	25
Height	16
Length Segs	1
Width Segs	4
Height Segs	3

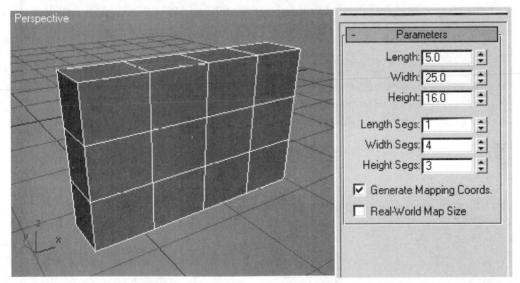

Figure 9-39: A box that will soon be a fish

3. Select the Box. Right-click and choose **Convert To > Convert to Editable Poly** from the Quad Menu.

4. Open the **Subdivision Surface** rollout on the Modify panel. Check the option labeled **Use NURMS Subdivision**. Increase the number of **Iterations** to **2**. The box is smoothed.

5.   Access the **Vertex** sub-object level. The control cage of the box appears.

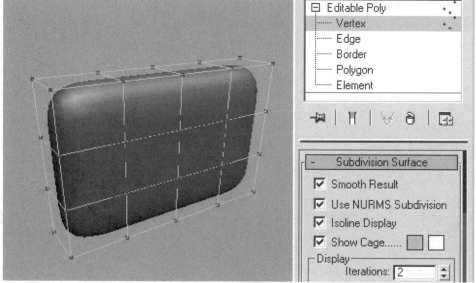

Figure 9-40: Subdivision control cage

6.   In the Front viewport, click and drag to select all of the vertices at the upper right corner of the box. Move the vertices toward the center of the box to start the rounded shape of the fish head.

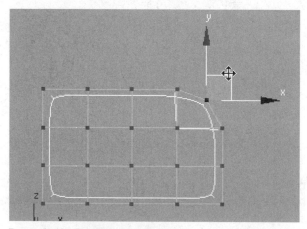

Figure 9-41: Shaping the fish head

7.   It's easier to see what you're doing if you view the model in shaded mode. Press **<F3>** to change the Front viewport to **Smooth+Highlights** rendering mode. Move the vertices at the lower-right corner of the box in the same way that you did in step 6.

8. Next, you'll shape the fish's tail. In the Front viewport, select the second column of vertices. Select the Scale tool from the Main Toolbar and scale the vertices in the Y axis.

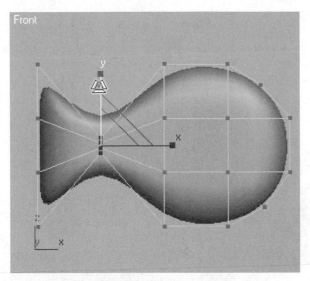

Figure 9-42: Pinching the tail

9. Exit sub-object mode and view the goldfish cracker in the Perspective viewport. Press the **<F4>** key to view Edged Faces. Now you can see the isolines, which are the edges on the subdivided mesh that correspond to the original edges of the control cage. To see all of the edges in the subdivided mesh, disable the **Isoline Display** option in the Subdivision Surfaces rollout.

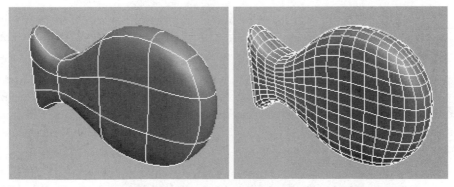

*Figure 9-43: Finished goldfish cracker before and after disabling **Isoline Display***

10. This is a very simple model, but much more is possible. Experiment with the tools in Editable Poly, such as **Cut**, **QuickSlice**, and **Extrude**, to add detail to the model and make it look more like a real fish. Save the scene as **Goldfish01.max** in your folder.

SUBDIVISION SURFACE TECHNIQUES

Editable Poly Tools

All of the Editable Poly tools you learned about in Chapter 3, *Modeling*, can be brought to bear on the creation of a subdivision surfaces model. For example, you can use **Extrude** or **Bevel** to quickly and easily create branches in your model. In this way, it is very easy to build the arms of a character or the limbs of a tree. The same soft, organically curved result can be achieved with the Surface modifier, but it is much more difficult to create branching architecture with spline tools.

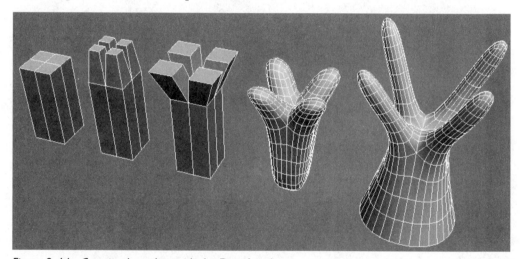

*Figure 9-44: Creating branches with the **Bevel** tool*

Use other Editable Poly tools, such as **Cut, Slice,** and **Tessellate**, to create detail within your subdivision models. More polygons and vertices in the control mesh will result in more subdivisions, which in turn results in a higher-density model.

In general, subdivision surfaces produce the best results when the control cage has only four-sided polygons, known as *quads*. Often, this is not possible, and you'll be forced to use triangular polygons. That's OK, but be certain not to use polygons with five or more sides, or your subdivided surface will have problems such as bizarre nodules in the mesh.

Be on the lookout for extra vertices that aren't needed. These extra vertices are often left over after removing edges. The subdivision algorithm is affected by all vertices, even those which are not necessary to the shape of the control mesh. The result of extra, "orphan" vertices is a mesh that looks strange and will deform poorly when animated. To solve this problem, use the **Remove** or **Target Weld** tools to get rid of the unnecessary vertices.

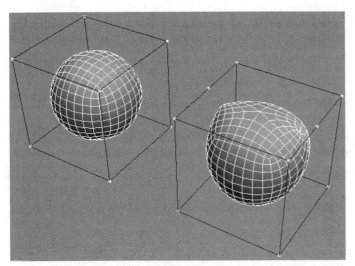

Figure 9-45: The mesh on the right has extra vertices that must be removed

MeshSmooth Modifier

The **MeshSmooth** modifier lets you apply subdivision smoothing to any mesh object, including primitives. MeshSmooth gives the artist more control than is possible with the Subdivision Surface rollout in Editable Poly.

Three types of smoothing arc available within the MeshSmooth modifier: **Classic**, **Quad Output,** and **NURMS**. Try each one to see which gives the best results for the particular model you're creating.

NURMS is the most sophisticated smoothing type, and it's the same one found in the Subdivision Surface rollout of Editable Poly. It stands for *Non-Uniform Rational MeshSmooth.* This name was invented in an attempt to associate the subdivision surface algorithm with NURBS modeling, when in fact the two have very little in common.

The **Smoothness** parameter in MeshSmooth collapses polygons that meet at shallow angles, in order to reduce the overall polygon count. This is very similar to the **Optimize** modifier. Both the Smoothness parameter and the Optimize modifier should be used with caution; sometimes they can do more harm than good, by radically changing the structure of the model.

With the **Classic** and **Quad Output** types, the **Strength** parameter increases or decreases the smoothing effect. The default Strength is **0.5**; greater or smaller values can change the original object's shape a great deal. The **Relax** parameter can be increased to soften any hard angles on the object. A negative Relax value will enhance the angles of the object, causing its edges to stick out.

If the control mesh vertices have been moved in such a way that the object is pinched or uneven, MeshSmooth will smooth these areas as much as possible, but it can only do so much. MeshSmooth is designed to smooth out a low-density model, not to correct poor modeling.

WARNING: The **Local Control** rollout within MeshSmooth lets you edit vertex and edge positions and weights *after* the subdivision has been applied. This is a powerful feature, but it is not necessary to achieve professional results. The Local Control tools are not recommended for beginners, because it's easy to get confused about whether one is editing the control mesh or the subdivided mesh. Instead of editing sub-objects within the Local Control rollout of MeshSmooth, you should edit the sub-objects of the Editable Poly object at the bottom of the Modifier Stack.

Sometimes you need finer control, but don't wish to weigh down the model with excessive detail in the control mesh. For example, you may need to edit small features such as eyelids, but if you refined the control mesh, the subdivision algorithm would create too much polygonal detail. In this case, work on your model until it's finished except for those fine features that require editing of individual polygons. Then convert the model to Editable Poly, collapsing the Stack and making the MeshSmooth subdivision permanent. You can then edit sub-objects of the collapsed mesh without adding too many polygons or risking the confusion that the Local Control tools can cause.

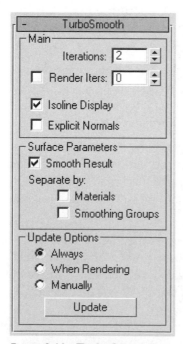

Figure 9.46: *TurboSmooth*

TurboSmooth Modifier

As the name implies, the **TurboSmooth** modifier is a faster, more streamlined version of MeshSmooth. It performs the basic functions of NURMS subdivision more quickly and simply than MeshSmooth. It's also more efficient than the NURMS subdivision option built into the Editable Poly object.

The performance improvement of TurboSmooth over MeshSmooth is very significant, so in most cases you should use TurboSmooth instead. One of the only times you might use MeshSmooth is if you wish to subdivide only part of a model. TurboSmooth doesn't have the **Apply to Whole Mesh** option, so you can't disable it like you can in MeshSmooth.

Symmetry Modifier

In the Mask tutorial, you used the **Mirror** modifier to save time and effort. You only needed to model half of the mask, and the Mirror modifier did the other half. When working with subdivision surfaces, the **Symmetry** modifier is superior to Mirror. Symmetry automatically welds the vertices along the seam, so you don't get any creases in the model. This is perfect for characters and other objects that are symmetrical.

Generally speaking, it's best to place the MeshSmooth or TurboSmooth modifier above the Symmetry modifier in the Stack. That way, the welding across the seam is done before the subdivision, which yields a result that is both cleaner and has fewer polygons.

Turn **Show End Result** on and off as needed to view the control cage and the finished, smoothed, and reflected model.

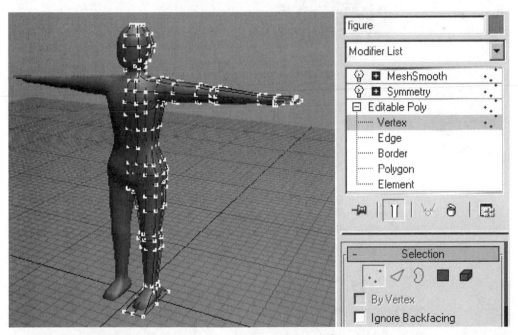

*Figure 9-47: A simple figure modeled with **MeshSmooth** and **Symmetry***

DeleteMesh Modifier

The **DeleteMesh** modifier deletes a selected portion of a mesh object. It is extremely useful for subdivision surface modeling, because you will often experience system slow-downs with a complicated model. DeleteMesh lets you "hide" part of the control cage from the MeshSmooth modifier, so you can focus in on a certain area of the model. You'll experience much better viewport performance this way.

DeleteMesh works in conjunction with a selection modifier. **Volume Select** works best because it is not affected when you add or delete sub-objects. (Remember, **Mesh Select** relies on vertex numbers, and when the number of vertices changes, the selection also changes.) Add a Volume Select modifier below the DeleteMesh in the Stack. Choose **Face** in the **Stack Selection Level** section. Move and scale the Volume Select **Gizmo** to select what part of the model you wish to delete. When your model is finished you can remove the Volume Select and DeleteMesh modifiers.

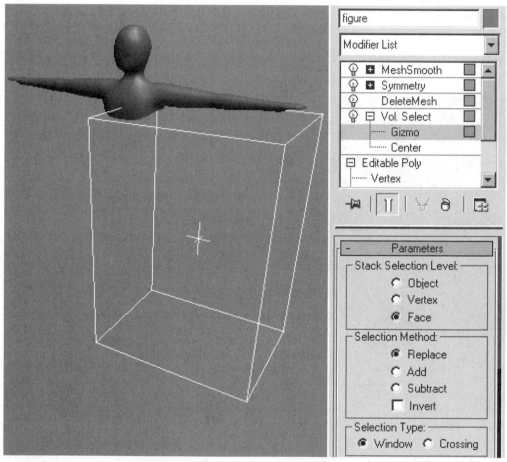

*Figure 9-48: Using **Volume Select** and **DeleteMesh** to speed up viewport performance*

END-OF-CHAPTER EXERCISES

EXERCISE 9.6: NURBS Tablecloth

In this exercise, you'll use NURBS to create a tablecloth.

1. Load the file **TableScene10.max** from the disc that comes with this book.

2. In the Top viewport, create an **NGon** shape to fit the top of the table. On the Command Panel, check the **Circular** check box, and set **Sides** to **12**. Position the circle just over the table.

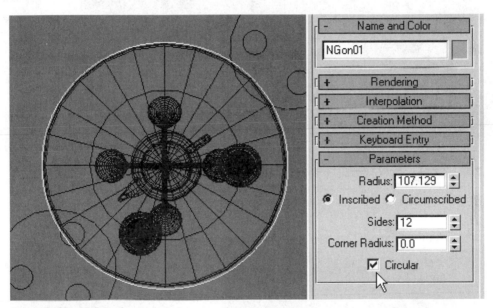

Figure 9-49: **NGon** *for tablecloth*

3. In the Front viewport, hold down the **<SHIFT>** key and move the NGon down to create a clone that almost touches the floor. In the **Clone Option**s dialog, choose **Copy**.

4. Activate the Perspective viewport. Go to the Modify panel. Increase the **Radius** of the second NGon slightly.

5. With either NGon selected, right-click and choose **Convert To > Convert to Editable Spline** from the Quad Menu. In the **Geometry** rollout of the Modify panel, click the **Attach** button. Then click the other NGon to attach it. Now both circles are part of the same Editable Spline object.

6. In the Top viewport, move the vertices of the outer circle to create folds in the table-cloth. In the Left viewport, move some vertices slightly up or down.

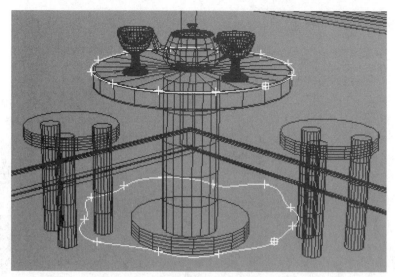

Figure 9-50: Perspective view of Editable Spline object

7. Exit sub-object mode. With the Editable Spline selected, right-click and choose **Convert To > Convert to NURBS** from the Quad Menu.

The object is converted, and the **NURBS Creation Toolbox** appears.

8. On the NURBS Creation Toolbox, click the **Create U Loft Surface** button.

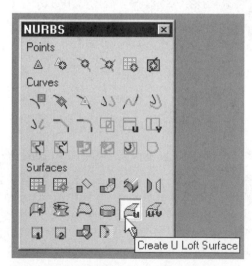

*Figure 9-51: **Create U Loft Surface** button on the **NURBS Creation Toolbox***

Move the cursor over one of the NGons until it turns blue, then click. Move the cursor over the other NGon until it turns blue, and click again. Right-click to end the U Loft creation process. Click the Create U Loft Surface button again to turn it off.

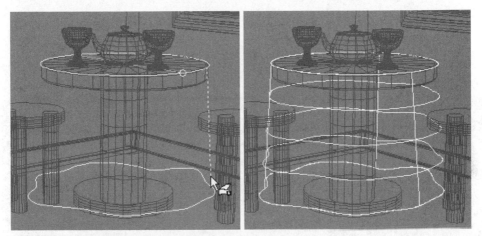

*Figure 9-52: Click and drag to create the **U Loft** surface*

9. A surface has been created, but it might be inside out. If you need to flip the normals of the surface, enter **Surface** sub-object mode in the Modifier Stack. Click the surface in the viewport to select it, and scroll to the bottom of the **Surface Common** rollout. Click the check box labeled **Flip Normals**.

    The tablecloth now needs a top surface.

10. On the NURBS Creation Toolbox, click the **Create Cap Surface** button.

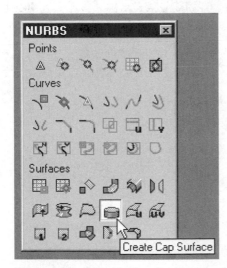

*Figure 9-53: **Create Cap Surface** button on the NURBS Creation Toolbox*

Move the cursor over the topmost circle until it turns blue, then click. If the hole is not capped, scroll to the bottom of the command panel and check the **Flip Normals** check box.

You now have a tablecloth for your table, as shown in the following illustration.

Figure 9-54: Tablecloth

11. Save the scene with the filename **TableScene11.max** in your folder.

Hints

- You might find it helpful to hide all objects except the two NGons to see how they look before performing the U Loft.

- Once the tablecloth has been created, you can delete or hide the original table.

Advanced Exercises

- Create a tablecloth with a filleted edge by using more circles to define the NURBS object. Three circles define the filleted edge, and one defines the lower edge of the tablecloth. The topmost circle will still have to be capped with the **Create Cap Surface** tool.

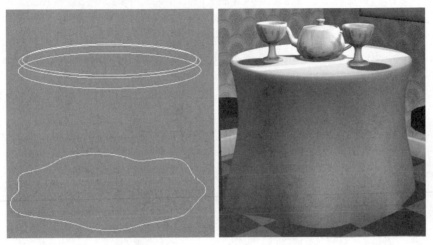

Figure 9-55: Filleted tablecloth

- Edit the tablecloth after it has been created. The U Loft and Cap surfaces are dependent on the shape of the curves. You can edit the surface by editing the curves.

  To do this, enter **Curve** sub-object mode. Select curves and move or scale them. Choose the **Curve CV** sub-object level. A series of control points appear around each curve. Move control points to alter the shape of the curve.

- Create curtains over the windows in the room using the same method you used to create the tablecloth.

EXERCISE 9.7: Character Head

In this exercise, you'll use Paint Deformation and TurboSmooth to create a simple character head.

1. Reset 3ds max.

2. Maximize the Perspective view. Create a Box primitive with a **Length**, **Width**, and **Height** of approximately **100** units. Give the box **12 Length**, **Width**, and **Height Segments**. Press the **<F4>** key to view Edged Faces. Rename the object **Head**.

3. Add a **Spherify** modifer to the object. With the default Percent value of 100, the object is completely spherical.

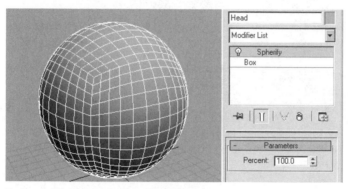

*Figure 9-56: Add a **Spherify** modifier*

4. With the **Head** selected, right-click in the viewport and choose **Convert To > Convert to Editable Poly** from the Quad Menu.

5. Enter **Element** sub-object mode. Click the object in the viewport. All of the polygons are highlighted. Use the **Scale** tool to stretch the polygons vertically in the world Z axis. Now the **Head** is more oval in shape. Exit sub-object mode.

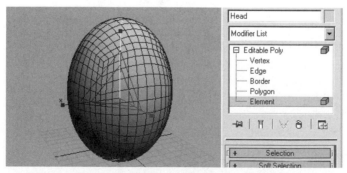

*Figure 9-57: Scale the polygons in **Element** sub-object mode*

6. Scroll down to the **Subdivision Surface** rollout. Enable the **Use NURMS Subdivision** check box. Leave the number of **Iterations** at the default value of **1**. Disable the **Isoline Display** option, so you can see that the object actually has a much higher level of detail. Press the **<F4>** key again to disable Edged Faces. Now your display will not be as cluttered, and it will be easier to see what you're doing during the sculpture process.

7. Open the **Paint Deformation** rollout. Click the **Brush Options** button. In the Painter Options dialog, enable the **Mirror** option at the bottom of the dialog. Leave the axis selection at the default of **X**. Close the dialog.

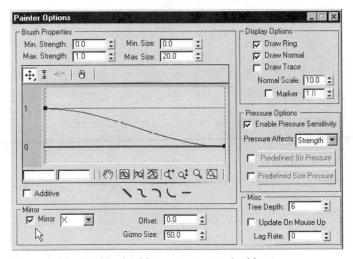

*Figure 9-58: Enable the **Mirror** option in the X axis*

8. Click the **Push/Pull** button in the Paint Deformation rollout. As you drag the brush cursor over the **Head**, you see two brushes. Move the cursor to the center of the object. Click and drag to create a nose for the character.

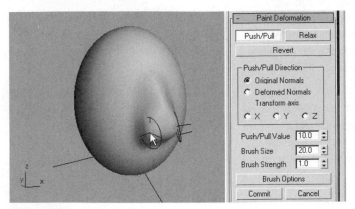

*Figure 9-59: Use the **Push/Pull** tool with the Mirror option to create a nose*

9. Change the **Push/Pull Value** to **-10**. Click and drag on the Head to create eye sockets and a mouth.

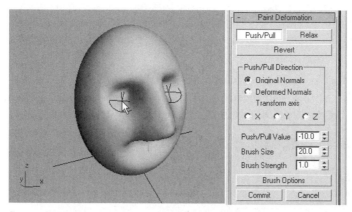

*Figure 9-60: Use a negative **Push/Pull Value** to create a mouth and eye sockets*

10. Continue using the Push/Pull tool to sculpt your character. Try different brush settings. If you find you can't push or pull vertices any further, click the **Commit** button to reset the brush.

If you wish to add features that change the mesh topology, such as ears, you may use the other tools within Editable Poly, such as **Extrude**. Just be aware that if you do so, you won't be able to use the **Revert** tool to go back to the state of your model before you started using Push/Pull.

11. If you wish to achieve finer detail, you can convert the object to Editable Poly again. This makes the subdivision surface **Iterations** permanent, and you can edit the subdivided mesh using any of the Editable Poly tools, including Paint Deformation.

12. When you're close to being finished, add two spheres for eyes, and edit the model to fit around the eyeballs. Save your scene as **SculptedHead.max** in your folder.

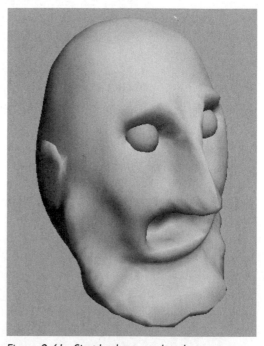

Figure 9-61: Simple character head

EXERCISE 9.8: Subdivision Surfaces Character

About This Tutorial

In this tutorial, you'll create a complete character using box modeling techniques. In the end you'll have a single object that constitutes the character's body and his clothes. This will make it a little easier to animate this character with control objects such as **Bones**. You'll learn about Bones in Chapter 10, *Advanced Animation*.

There are many ways to model, and this is just one of the possible approaches. You could also build each component of the character as a separate object and still control them with Bones. The approach we take in this tutorial is to create a single mesh object, which is a little more challenging to model, but a little easier to animate with Bones. Also, a single mesh object such as this is required by most 3D game engines.

Reference Pictures on Image Planes

Before starting to model, you should have reference drawings or photos of the character you want to make. Taking a little time to sketch out the character, even roughly, will save you hours of time in modeling. The ideal pose for your character is the classic "Da Vinci" pose: legs slightly apart and arms held out to full length. This will make it a lot easier to model the figure.

Figure 9-62: Reference drawing

You can scan in reference pictures and display them in the viewports when modeling. The best way to do this is to create *image planes*. Image planes are objects that give you a point of reference when building the model. You'll need at least one reference image. Many modelers use front and side reference images, and sometimes a top image is useful as well.

Tips

- Scan your images at fairly low resolution: 72 ppi is fine for a letter size drawing. Make sure you use the same settings for all scans.

- Using a bitmap editor such as Photoshop, crop the front and side images so they line up vertically. The tops of the heads, the eyes, and the feet must line up on both images. Write down the resolutions of these images; you'll need that information later. Don't create files with a resolution higher than 512 x 512, or you will end up slowing down viewport performance. Reduce the contrast of the image so you'll be able to see 3ds Max interface elements better when they are superimposed. Save as `.TGA` or `.TIF`.

- Use Box primitives for your image planes. Plane primitives are not good for this usage, because you can only select one side of a plane. Make the Box dimensions the same as your image resolution. For example, if the image is 350 x 500 pixels, make the Box **350** units wide and **500** units tall. This way, the reference bitmap will not be distorted when you apply it to the Box surface. You can give the box a thickness of 0.

  When you are finished with your model, you can scale it to the appropriate real-world size, then use the **Reset Scale** command on the Hierarchy panel.

- If you are using both a side and front view for reference, create a second Box object perpendicular to the first one. Adjust the Length, Width, or Height if necessary to match the resolution of the side view image file.

  Position the boxes as two intersecting planes precisely at the world origin. Some modelers prefer to position the image planes to form two sides of a large box. Either way, set up your reference image planes so you can build the model at the world origin.

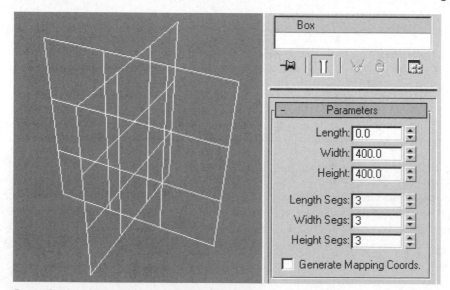

Figure 9-63: Intersecting image planes

Create an Image Plane

1. Create a folder on your hard drive for this project. Locate the file **SkaterFront.tif** on the disc that comes with this book, and copy it to your project folder.

2. Reset 3ds Max. Create a **Box** object in the Top viewport. Give the Box the following dimensions:

Length	0
Width	400
Height	400

3. Select the Move tool. Use the **Transform Type-In** in the Status Bar area to move the Box object to the origin, at coordinates **(0,0,0)**.

4. In the Material Editor, create a material with an Opacity of **50%**. Assign **SkaterFront. tif** as the Diffuse Color map. Apply this material to the image plane. Turn on the **Show Map in Viewport** button in the Material Editor.

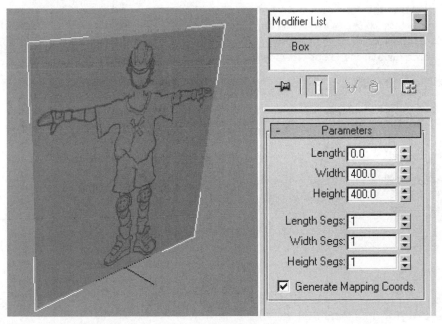

Figure 9-64: **SkaterFront.tif** *applied to image plane*

5. It's very helpful to have semitransparent image planes in the Perspective view, but not in the orthographic views. Right-click the Perspective viewport label and select **Transparency > Best**. Right-click the Front viewport label and choose **Smooth+Highlights**, and then select **Transparency > None**.

Start with a Box

Most subdivision surface models start with a Box primitive. The Box will be smoothed later, so very few polygons are needed at first.

1. In the Top viewport, create a **Box** primitive. Set the parameters as follows:

Length	70
Width	90
Height	90
Length Segs	3
Width Segs	4
Height Segs	2

2. Use Transform Type-In to move the Box to the origin.

3. In the Front viewport, move the Box so the top of the Box aligns with the skater's shoulders in the reference picture.

4. Notice that the drawing is not perfectly aligned with the world axes. Move the image plane slightly to the left in the Front viewport, so that the world axis, the image plane, and the new Box are all precisely aligned. When you are finished, press the **<G>** key to hide the Grid in the Front viewport.

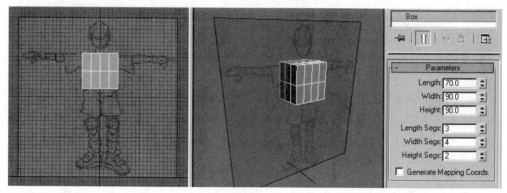

Figure 9-65: Box created and image plane aligned

5. Convert the new Box to an Editable Poly. Right-click and choose **Convert To > Convert to Editable Poly** from the Quad Menu.

    To save time, we will only model half of the figure, and let 3ds Max mirror the other half for us.

6. In the Front viewport, select all polygons on the side of the box that is on the left side of the screen. Press the **<DELETE>** key.

7. Exit sub-object mode. Apply a **Symmetry** modifier to the box.

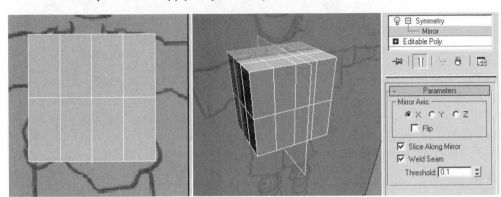

*Figure 9-66: Left side of box deleted; **Symmetry** modifier added*

Shaping Polygons

Once the primitive is converted to an Editable Poly, the first step is to form limbs off the main box. This is accomplished by extruding and shaping polygons.

1. Enter **Polygon** sub-object mode. Turn on **Show End Result**. In the Perspective view, select the polygon located at the shoulder of the figure. Extrude the polygon about **40** units.

Because the Symmetry modifier is applied, the other side of the box is also extruded.

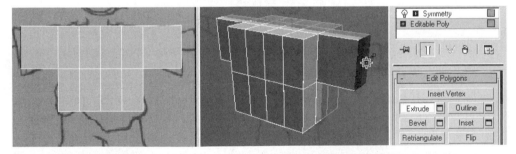

Figure 9-67: Extrude from the shoulder

You can change the sizes of extruded polygons by using the **Bevel** command instead, or by using the **Outline** command after the extrusion is done. You can also simply scale the polygons.

2. With the extruded polygon selected, scale it up a little. Then move it down so it approximates the sleeve.

Don't be concerned if the extruded sleeves don't match exactly with the reference picture. We're using the picture as a guide for size, not necessarily for an exact match.

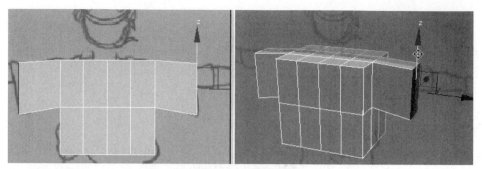

Figure 9-68: Scale and move polygon at the end of the sleeve

3. To create the fold in the sleeve, we'll use the **Inset** command. Click the Inset button on the Modify panel, then click and drag on the polygon at the end of the sleeve.

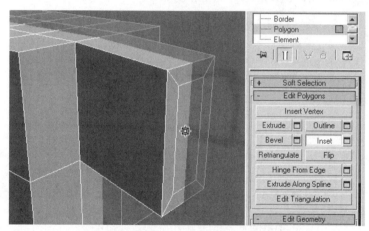

*Figure 9-69: **Inset** the polygon*

4. Move the polygon back into the sleeve. This will help create a fold in this area. Move and scale the polygon non-uniformly to match the proportions of the arm seen in the image plane.

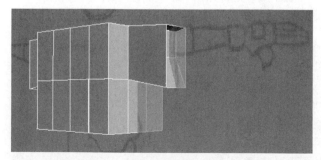

Figure 9-70: Move and scale the inset polygon to match the image plane

5. With the inset polygon still selected, extrude it slightly. You now have a fold in the sleeve, and the beginnings of an arm.

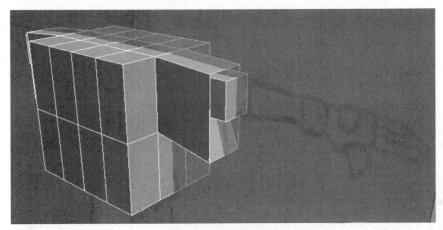

Figure 9-71: Extrude the inset polygon to create an elbow joint

6. Continue to scale, extrude, and move polygons until you have created a forearm and palm that looks similar to the following illustration.

   You can also select multiple edges or vertices in the Front viewport and scale them to change the basic form of the arm after you have made the extrusions.

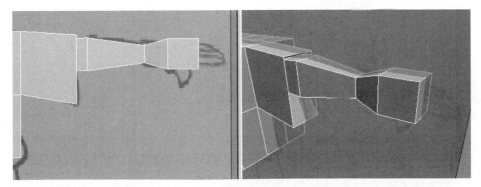

Figure 9-72: A rough approximation of an arm

7. Save your model as **Skater01.max**.

TIP: Always save successive versions of your work. That way, you can always return to an earlier version if something goes wrong or if you simply change your mind about a modeling decision. The amount of time you spend redoing a project is determined by how long ago you last saved a new scene. So, if you create a new scene file every 30 minutes, you'll never lose more than 30 minutes of work. Never save over a scene file if you've made changes to it!

Use what you've learned to extrude and shape the legs and feet.

8. Select the polygons at the bottom of the figure and extrude them down. Don't move or scale them. Bring them down to about the level of the hips.

9. Click the light bulb icon next to the Symmetry modifier to turn it off for a moment. Arc Rotate so that you can see the area at the center of the model. Notice that when you extruded polygons along the seam, unwanted polygons were created. Select these polygons and delete them.

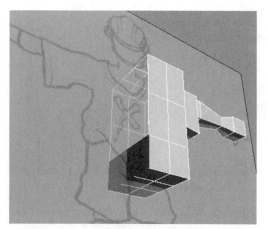

Figure 9-73: Unwanted polygons created by extrusion

10. Select the polygon shown in the following illustration. Move and scale the polygon to create a larger surface for extrusion.

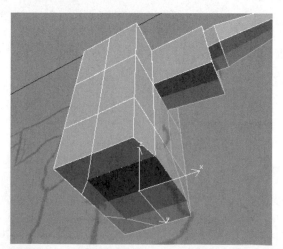

Figure 9-74: Scale and move the polygon at the bottom of the figure

11. Extrude the legs. Use the same techniques you did for the arms, creating a fold for the trousers and following the general shape of the reference image.

Your model should look similar to the one shown in the following image. Don't worry if your model doesn't look exactly like the one in the picture. Box modeling takes practice! If your model becomes hopelessly mangled, or if you simply don't like the result, reload the file **Skater01.max** and try again. When you're satisfied with the model, save your work as **Skater02.max**.

TIP: To do this easily, choose **File> Save As** from the Main Menu, then click the plus sign **[+]** on the **Save File As** dialog.

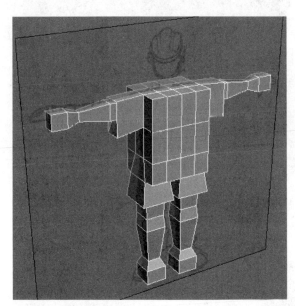

Figure 9-75: **Skater02.max**

Shape the Model with Vertices

Extruding and outlining polygons will get you only so far in box modeling. At some point, you will have to start moving vertices around to get the model into shape.

1. Extrude a polygon for the neck, and delete the unwanted polygon along the seam.

2. Work with the vertices on the model until the model looks similar to the following illustration.

3. Save your work as **Skater03.max**.

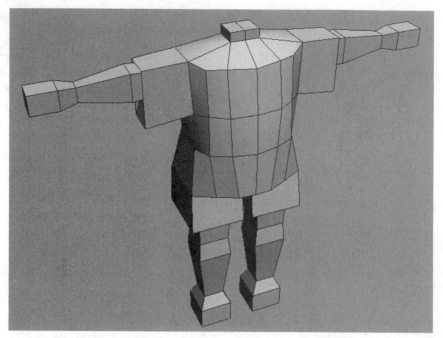

Figure 9-76: **Skater03.max**

Tips

- To shape your model, work on one part at a time, such as the shoulders, chest, hips, or arms.

- Select vertices as best you can in one viewport, then use other viewports to add or subtract from the selection. Move and scale vertices as necessary.

- Always check more than one viewport to ensure you have the correct vertices selected before moving or scaling.

- Use the **Ignore Backfacing** feature if you need to avoid selecting vertices on the far side of the model, facing away from you.

- Never move the vertices at the seam of the model in the world X axis! If you do, you'll create a gap or an overlap in the model where it is mirrored by the Symmetry modifier. Only move vertices along the seam in the world Y or Z axes, so the seam will remain straight.

- Characters are usually more round than boxy, so the first thing you'll want to do is smooth out the sharp corners. Then you can start shaping the body, arms, and legs to more closely match the reference picture.

- When scaling vertices, it is often useful to change the transform center to **Use Transform Coordinate Center**, and choose the **World Coordinate System**. This way you can scale vertices toward or away from the center of the model.

Subdivision Surface

Now you're ready to see what the model looks like as a subdivision surface.

1. Apply a **MeshSmooth** modifier to the top of the stack, and increase the **Iterations** parameter to **1**.

2. Turn off the **Edged Faces** display option if it is on. Press **<F4>** to show and hide polygonal edges when in a shaded view.

3. Select the **Vertex** sub-object level of the Editable Poly.

    Turn on **Show End Result** . Now you can see the control mesh (the original polygons) and the subdivided surface at the same time. The edges of the control mesh display in orange, and vertices display in blue.

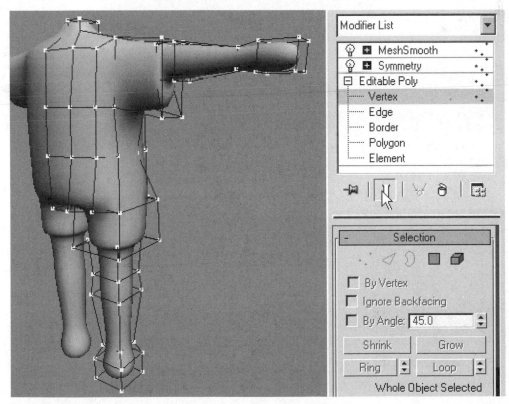

Figure 9-77: Control mesh superimposed over subdivision surface

WARNING: Remember to take care when increasing the value of the **Iterations** parameter. Increasing it beyond **3** may slow down your system, or cause 3ds Max to freeze. If you need more than **3** iterations to smooth your model, the model probably needs more detail. This is best fixed by adding polygons to the control cage.

Currently, some areas are not well defined. This can't be solved by adjusting vertices, as there aren't enough edges and vertices to make the necessary detail. This problem can be solved with the tools found in Editable Poly.

Slice Polygons

Now we will add detail to the entire model with the **Slice** tool.

1. Turn off the MeshSmooth modifier by clicking the light bulb icon in the Modifier Stack.

2. Access the Editable Poly level of the stack, then access the **Polygon** sub-object level. Select all of the polygons in the model.

3. Turn on **Angle Snap** if it is not already active.

4. Click the **Slice Plane** button in the **Edit Geometry** rollout.

5. Rotate the Slice Plane exactly **90** degrees around the world X axis. Then move it slightly toward the front of the model. You should see a new series of edges appear along the sides, top, and bottom of the model. This is called an *edge loop*.

When your screen looks similar to the illustration below, click the **Slice** button to complete the operation. Then click the Slice Plane button again to turn it off.

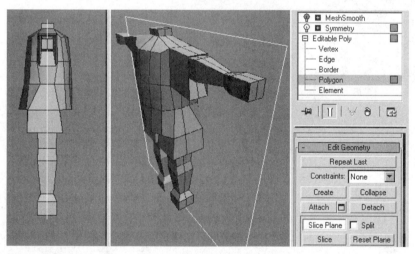

Figure 9-78: Creating an edge loop with the Slice tool

Chamfer Edges

Notice that the shirt does not have a constant fold around the entire body. On the side of the model, the shirt joins directly to the trousers. We'll fix that with the **Chamfer** tool.

1. Activate the **Edge** sub-object level.

2.   Select the two edges shown in the following illustration.

3.   On the **Edit Edges** rollout, click the **Chamfer** button. Click and drag the selected edges. The edges are chamfered, converting them into polygons.

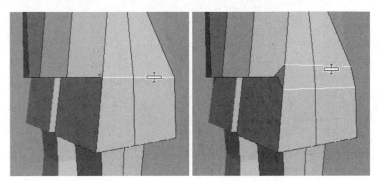

Figure 9-79: Chamfer two edges

Target Weld Vertices

A by-product of the Chamfer operation is that new polygons are created at the sides of the chamfered edges. Also, nearby polygons have five sides. The MeshSmooth algorithm works best if all polygons are quadrilateral. We'll need to clean up this area by welding vertices.

1.   Activate **Vertex** sub-object level.

2.   Zoom in on the chamfered area. Note the irregularly shaped polygons created on either side of the chamfer.

3.   Click the Target Weld button. Click the far left vertex of the irregular polygon, and drag to the vertex at the top right. Release the mouse button. The two vertices are welded.

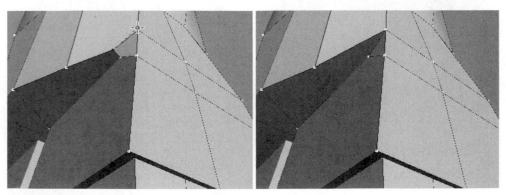

*Figure 9-80: **Target Weld** to collapse vertices*

4.   Repeat the Target Weld operation for the bottom two vertices.

5. Arc Rotate around to the back of the model and Target Weld the vertices as you did on the front.

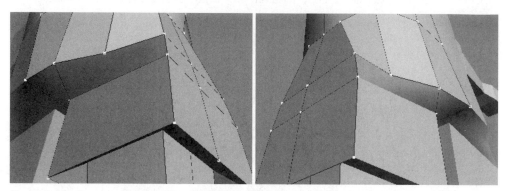

Figure 9-81: Front and back views of the model after Target Weld

Cut Tool

There is not enough detail in the front and back of the legs. Notice that there are now six-sided polygons under the shirt. This gives us an opportunity to extend the unshared edges down along the front and back of the legs.

1. Arc Rotate to the front of the model. While in Vertex sub-object mode, activate the **Cut** tool. Click the middle vertex of the six-sided polygon and move the mouse downward and slightly to the left. Click again when you reach the edge at the bottom of the six-sided polygon. Right-click to exit the Cut tool.

    The goal is to cut the six-sided polygon into two. The new vertex you create should be lined up with the approximate center of the leg.

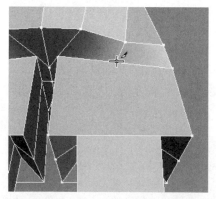

Figure 9-82: Cut the six-sided polygon in half

2. Zoom out so you can see the entire leg. Arc Rotate so you are looking up at the model from below. With the Cut tool active, click the new vertex you just created. Drag down to the bottom of the foot and click again. A series of edges and vertices is created on the front of the leg.

    Sometimes the Cut tool won't cut across a polygon, depending on your angle of view. You may need to create the new line of edges in several separate Cut operations.

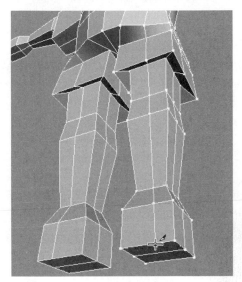

Figure 9-83: Cut a new line of edges down the front of the leg

3. Repeat the Cut commands at the back of the model, so you end up with a new line of edges down the back of the leg.

4. Use the Cut tool to connect the new line of edges on the front of the model with the new line on the back. Cut across the bottom of the foot to complete an edge loop that runs all the way through the model.

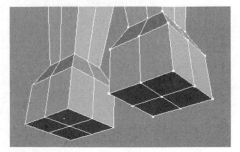

Figure 9-84: Cut the bottom of the foot

5. Save the scene as **Skater04.max**.

Shape Clothing

Now that you have more detail to work with, you can start to shape the clothing. This is accomplished by moving vertices near the clothing's openings inward past the sleeve or pant leg. You began this process by extruding polygons at the beginning of the tutorial; now you'll fine-tune your work.

Ordinarily, you do not want to move vertices past neighboring vertices. This can create polygons that intersect one another, a situation you should normally avoid. However, subdivision surfaces behave differently than standard mesh objects. Remember that the Editable Poly object is just a control mesh, and what really matters is the resulting subdivided surface. So, in this case, it's OK for you to create situations where polygons overlap or intersect, as long as the resulting subdivision surface does not intersect with itself.

When fine-tuning geometry, turn on **MeshSmooth** from time to time to make sure you're getting the shape you want.

1. Select the row of vertices that will define the bottom seam of the shirt. Move them into position so they form a rough circle around the figure. Remember, never move the vertices at the center of the model away from the line of symmetry, or you'll create a gap or an overlap in the surface.

2. Choose the World coordinate system. On the Main Toolbar, select **Use Transform Coordinate Center** . Scale the vertices in the X and Y axes to make them stick out beyond the trousers.

3. Move the vertices down. You want to create an area of overlap so the shirt looks like it's hanging down over the trousers.

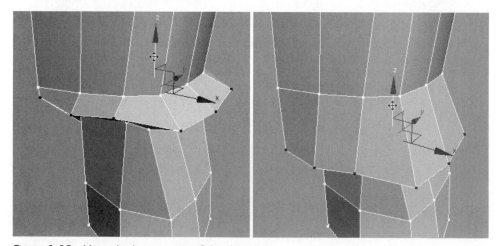

Figure 9-85: Move the lower seam of the shirt down

4. To make the outer seam of the clothing more defined, enter **Edge** sub-object mode, select the seam, and increase the **Crease** parameter.

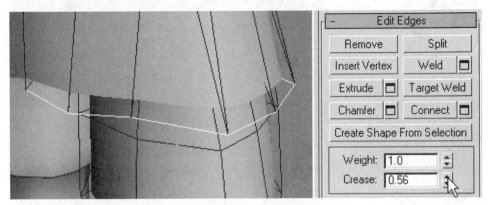

Figure 9-86: Increase the Crease parameter

5. Extrude a foot and shape it by moving and scaling sub-objects.

6. Use Slice to create new edge loops for shoes and socks.

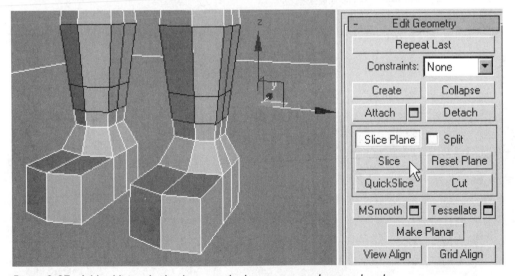

Figure 9-87: Add additional edge loops to the legs to create shoes and socks

7. Continue using these same techniques to fine-tune the clothing and the figure. Use Slice and Cut to create new edge loops if necessary. Use Crease at the outer seams of all clothing.

8. Save the scene as **Skater05.max**.

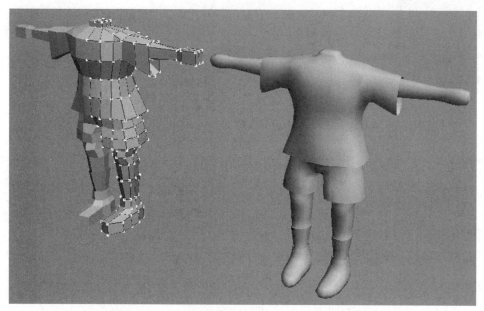

Figure 9-88: Details added to clothing and figure

Give Him a Hand

Making the thumb is easy; just Bevel a polygon and shape it by moving and rotating.

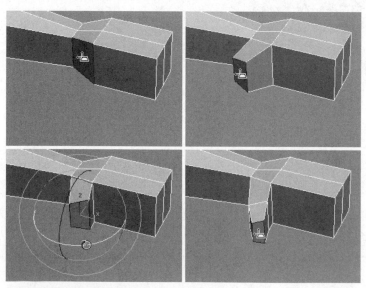

Figure 9-89: Creating a thumb

I. Access the **Polygon** sub-object level.

2. Select the thumb polygon on one hand, as shown in the previous illustration. Use the **Bevel** tool to extrude and outline the polygon in one step.

3. Move and rotate the polygon.

4. Create another Bevel for the second thumb joint.

   To create the fingers, we'll need to add additional detail at the end of the palm.

5. Use the Cut tool to create two additional edges at the end of the hand.

   While you cut in the Perspective view, it's helpful to also look in an orthographic view to check your work. Activate the Right viewport and zoom in on the hand. In the Perspective view, try to cut edges that are close to vertical.

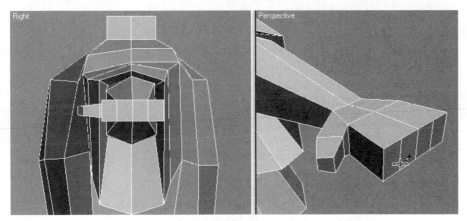

Figure 9-90: Use Cut to create new edges at the end of the palm

6. Shape the vertices at the end of the palm into a more rounded form. When we extrude the fingers, this will cause the fingers to extrude outward from one another.

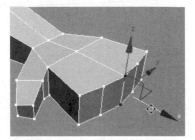

Figure 9-91: Move the vertices to round out the palm

7. To extrude the fingers separately, select the polygons at the end of the palm. Open the **Extrude Polygons** dialog by clicking on the **Settings** button next to the Extrude button. Change the Extrusion Type to **By Polygon**. Adjust the **Extrusion Height**. Don't click Apply or OK just yet.

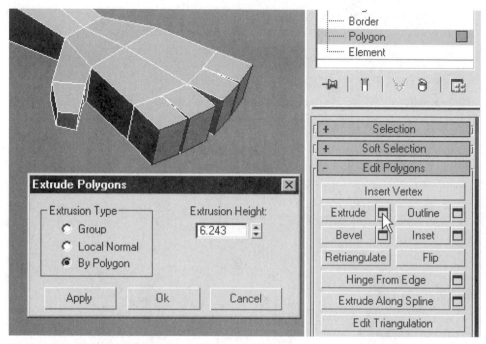

Figure 9-92: **Extrude Polygons** *dialog*

The tool dialogs within Editable Poly allow you to see a preview of the operation, such as Extrude, before committing to it. If you click OK, the operation is completed, and the dialog closes. If you click Apply, the operation is completed, and another operation is begun. You see the first operation finished, and a new operation previewed.

8. Click the **Apply** button. A second extrusion appears. This is a preview. Adjust the Extrusion Height if you like.

9. Click the Apply button a second time to finish the second extrusion. A third preview extrusion appears. Adjust Extrusion Height if needed, and click the **OK** button. The third operation is completed, and the dialog closes.

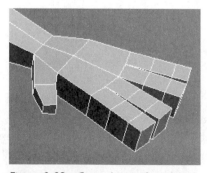

Figure 9-93: Control cage for a hand

Now we will adjust the vertices of the hand to create a better shape. Since it is difficult to select the entire finger, we will use the Grow command.

10. Select the polygon at the end of a finger. Then click the **Grow** button on the Selection rollout. The polygons that border on the current one are selected. Click the Grow button until the entire finger is selected.

11. Transform the fingers to move them into shape. Scale them down or up, and rotate them into position. When scaling polygons, it's useful to use the **Local** coordinate system, so the selection doesn't get distorted.

Don't spend too much time on the hand, because this character is destined for a wide shot, and we'll never see fine details of the fingers.

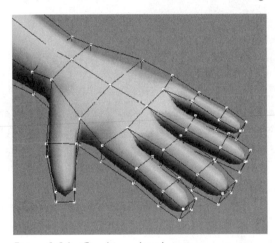

Figure 9-94: Give him a hand

12. Save the scene as **Skater06.max**.

Create the Collar

1. Use the Cut tool to create an edge loop around the upper chest and back. This will be the collar of the shirt.

Start from the center of the model and work your way to the outside. Instead of cutting across several polygons at once, click each edge as you go.

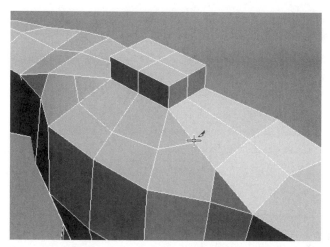

Figure 9-95: Use the Cut tool to begin making a collar

2. Arc Rotate around to the back and continue to create new edges with the Cut tool.

3. When the edge loop is complete, enter Edge sub-object mode. The edges you just created are still selected. Use the Chamfer tool to create a second edge loop based on this one. This will allow you to create a fold, just as you have done before.

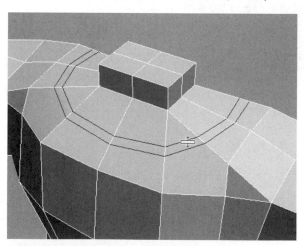

Figure 9-96: Chamfer the collar edges

4. Select just one edge of the outer edge loop of the collar. Then click the **Loop** button on the Modify panel. The entire edge loop is selected automatically. IT'S MAGIC!

5. Use the Scale and Move tools to create a fold. Remember to take care with the Scale tool. If you Scale using the **Use Pivot Point Center** transform option, the vertices will not line up in the center of the model. Instead, Scale the edge loop relative to the World coordinate system, with **Use Transform Coordinate Center**.

6. Increase the Crease parameter for the edge loops of the collar.

Figure 9-97: Creasing the collar

7. Save the scene as **Skater07.max**.

Model the Head

There are many ways to create models, and many ways to create heads and faces. As you learned earlier, you can model a face with spline curves and the Surface modifer. You could build the head separately, and then simply link it to the body, but then you'd have two separate objects, and a visible seam where they overlap.

If you had a lot of time, you could weld the head to the body. You'd need to "massage" (adjust) both models so they were compatible. That would take a while, and the results might not be ideal.

So, in this situation we will simply extrude the head out from the body and progressively refine it using tools within Editable Poly. Since our goal is to end up with a single-mesh character, this is really the best approach.

1. Turn off Show End Result. Arc Rotate so you can see inside the center of the model.

2.   Select the polygons at the top of the neck. Go into the **Extrude Polygons** dialog and reset the **Extrusion Type** to **Group**. Close the dialog. Extrude the polygons upward to the full height of the head.

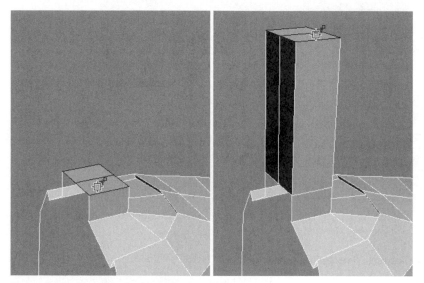

Figure 9-98: Extrude the head upward from the neck

3.   Notice that, once again, unwanted polygons are created on the inside of the model. Select the polygons and delete them.

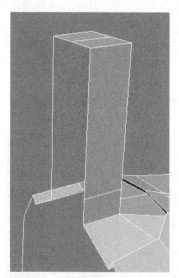

Figure 9-99: Delete the polygons on the inside of the extrusion

4.   Select the polygons of the head and use the Slice Plane to create four new edge loops running horizontally through the face.

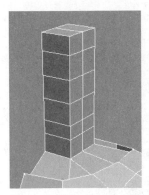

Figure 9-100: Slice four new edge loops through the face

5. Begin to pull the polygons and vertices into a rough approximation of a head shape. Turn on Show End Result, and turn MeshSmooth on and off as needed to see the results of your edits.

6. Using the Cut tool, create new edges running vertically on the front and back of the head.

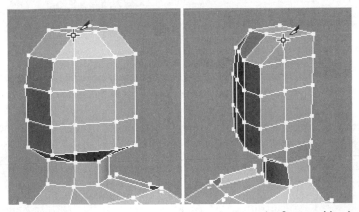

Figure 9-101: Cut new edges running vertically on the front and back of the head

TIP: Remember that you can remove edges with the **<BACKSPACE>** key and create new ones with the **Create** command. This way you can restructure the mesh to more closely follow the contours of the head.

7. Now that you have more detail to work with, continue to pull vertices into a better head shape.

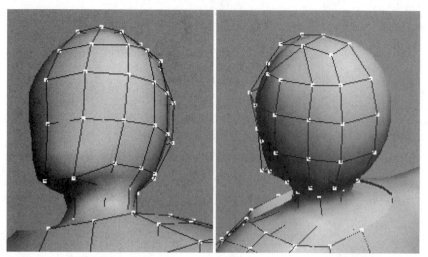

Figure 9-102: A rough head

8. Save the scene as **Skater08.max**.

Hide the Body

Focusing in on the facial features requires close work, and you'll need to see both the control mesh and the resulting surface at the same time. This will probably slow down your computer as 3ds Max tries to calculate the MeshSmooth for the entire model. Use Volume Select and Delete Mesh to reduce the amount of subdividing, and speed up viewport performance.

1. Add a **Volume Select** modifier above the Editable Poly in the stack, but below the Symmetry modifier.

2. In the **Stack Selection Level** section of the Volume Select modifier, choose **Face**.

3. Enter **Gizmo** sub-object mode, and move the Volume Select Gizmo down in the world Z axis, and scale it until the box encloses the body up to shoulder level. Turn off the **Auto Fit** option in the Alignment section. See the following illustration.

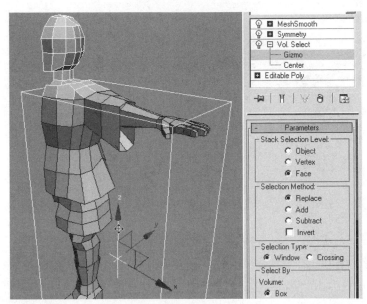

Figure 9-103: Volume Select Gizmo encloses the body, but not the head or neck

4.   Add a **DeleteMesh** modifier above the Volume Select modifier.

5.   Go back to Vertex sub-object level within the Editable Poly control mesh. Only the head is reflected and subdivided, making it much easier to focus on modeling the head.

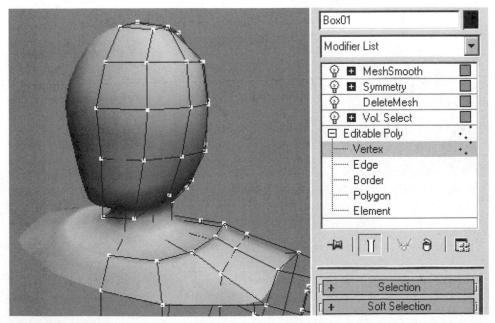

Figure 9-104: Only the head is reflected and subdivided

Refine the Features

1. Use the Cut tool to create a new edge for the side of the nose.

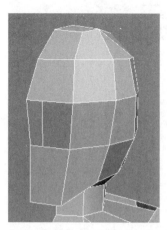

Figure 9-105: Cut an edge for the nose

2. Select the polygon that will be the nose. Activate the **Hinge From Edge** tool on the Edit Polygons rollout. Click the top edge of the polygon and drag downward.

    Hinge From Edge is similar to Extrude, except that the selected edge is not extruded.

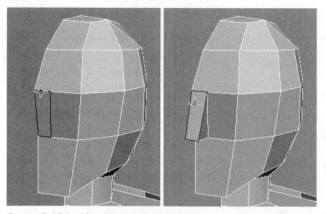

*Figure 9-106: Use **Hinge From Edge** to create a nose*

3. With Show End Result turned off, Arc Rotate to view the center of the model. Delete the extra polygon in the middle of the nose.

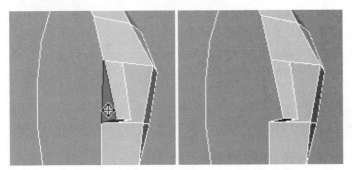

Figure 9-107: Delete the extra polygon in the center of the nose

4. The Hinge From Edge operation created a gap in the model, which you can see if you turn on Show End Result and MeshSmooth. To close the gap, select the vertex at the center tip of the nose, and move it to the center of the model in the Front viewport.

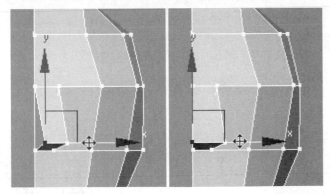

Figure 9-108: Close the gap in the nose

5. Use Target Weld to collapse the two vertices at the top of the nose, giving it a pyramid shape. Weld the outer vertex to the inner vertex.

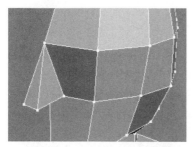

Figure 9-109: Use Target Weld to create a pyramidal nose

6. Use the Cut tool again to create two edges horizontally across the mouth area.

7. Select the three edges that border on the mouth. Don't select the edge at the center of the model.

8. Click the **Settings** button next to the **Chamfer** button. Adjust the **Chamfer Amount** in the **Chamfer Edges** dialog, and click OK.

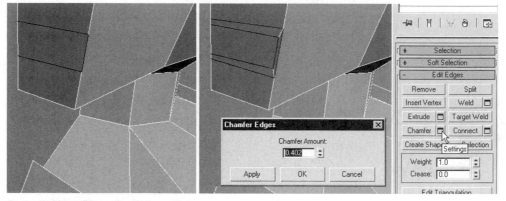

*Figure 9-110: **Chamfer Edges** dialog*

9. Select the inner edges of the mouth, and repeat the Chamfer operation.

Now you have three edge loops. The two outer ones will be the lips. The innermost edge loop will define the back of the mouth.

10. Select the polygon at the center of the mouth and move it back in the world Y axis. This creates the mouth cavity.

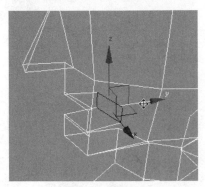

Figure 9-111: Move the center polygon back to create a mouth cavity

NOTE: It is a common practice to model a character with its mouth open, so it may be easily animated later. If the lips are closed, it will be difficult for animators to separate and animate them.

11. Continue to use the tools within Editable Poly to shape the eyes, eyebrows, and ears.

12. Save the scene as **Skater09.max**.

Hints

- Create or remove edges as you go. Your goal is to make the edges of the control mesh follow the contours of the surface. That way, the subdivided surface will also follow the contours, and the model will deform and animate well.

- Four-sided polygons work best with MeshSmooth. Three-sided polygons are also OK, but they don't work quite as well. Try to avoid polygons with more than four sides, or places where more than four polygons meet at a single point. It's nearly impossible to build a model strictly from quads, but the closer you can get, the better off you'll be.

- It might take a while to get the results you want. Don't expect instant gratification; modeling something worthwhile takes time. The head shown here took an hour to create, and it's still missing details such as eyelids. The sense of satisfaction when you're finished is very strong. Don't give up! Your persistence will be rewarded.

- Your model won't look exactly like the examples pictured here. You might create a head with more or less detail. These images are only provided to give you an idea of how to go about the process. Be creative! Make a unique character.

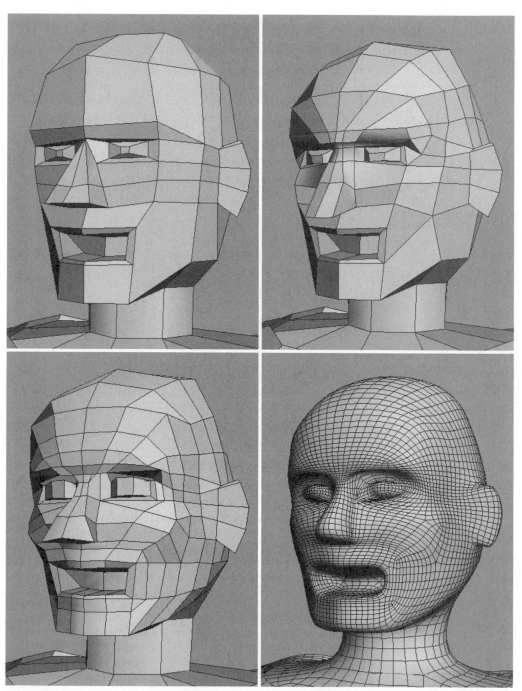

Figure 9-112: Stages in modeling a head

Collapse Symmetry and Weld the Seam

The Symmetry modifier is an excellent modeling aid, but you can't leave your model in this state. You need to create a single mesh, and you also need to be able to add asymmetrical details such as folds in clothing.

1. Delete the **Volume Select** and **Delete Mesh** modifiers. They have served their purpose and are no longer needed.

2. Disable the **MeshSmooth** modifier. You should see a complete control mesh in the viewport.

3. Right-click the **Symmetry** modifier in the Stack. Choose **Collapse To** in the pop-up menu. A warning dialog appears. Click OK.

   Collapse To allows you to collapse part of the Stack, up to and including the selected modifier. Modifiers above the current level are preserved.

   Before the Symmetry modifier was introduced, it was necessary to manually weld the seam. If you've been careful not to move the seam vertices off axis, then you won't need to weld them. To check the model for problems, use the STL Check modifier.

4. Exit sub-object mode if necessary. Add an **STL Check** modifier above the Editable Poly. Activate the **Check** option at the bottom of the Modify panel. If there are problems with the object, they will be highlighted in red.

   You'll only need to do the following steps if the STL Check modifier reported that you have problems.

5. Enter **Vertex** sub-object mode in the Editable Poly object. Make sure **Ignore Backfacing** is turned off.

6. Use the Front viewport in Wireframe mode. Select all vertices along the seam where the left half of the model joins the right half. Arc Rotate around the Perspective view in Wireframe mode to make sure you've selected all of the seam vertices, and only the seam vertices.

7. Open the **Weld Vertices** dialog by clicking the **Settings** button next to the **Weld** button. Increase the **Weld Threshold** until the problem vertices are collapsed. Click OK.

8. Delete the STL Check modifier.

9. Save the scene as **Skater10.max**.

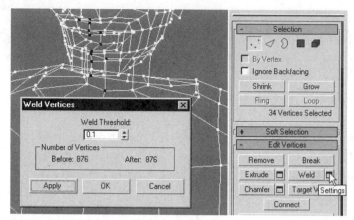

Figure 9-113: Welding the seam

Make Hair

Hair is one of the hardest things to model in 3D. If you think about it, your hair is made up of thousands of tiny objects. It's not at all practical to even consider modeling hair strand by strand. However, there are several techniques you can undertake.

- Use the **Hair and Fur** modifier. This produces the most realistic result, but is only a viable option if you are producing a rendered image. It simply won't work in real-time situations such as games. Hair and Fur is covered in Chapter 10, *Advanced Animation*.

- Create flat, 2D polygons or patches and drape them over the head. Use an opacity map to simulate many clumps or strands of hair. The advantage to this approach is that it is relatively easy to set up, and it's animatable. The disadvantage is that the end result is not a single mesh object, and this may pose problems if you're exporting to a game engine.

- Model a "hair helmet" that simply sprouts from the character's head. This is the most traditional approach, but it's also the least realistic. It works best if the hair is designed to be short. We'll use this technique for the tutorial.

1. Select all of the polygons within the character's hairline. Hopefully your control mesh roughly conforms to the places where hair should be. If you've modeled the head so that its edge loops follow natural contours, then you'll have no problem.

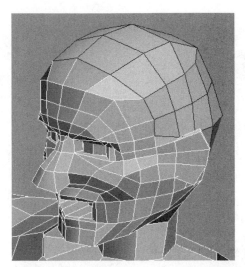

Figure 9-114: Select the polygons within the hairline

2. Use the **Bevel** tool to extrude and outline the hair.

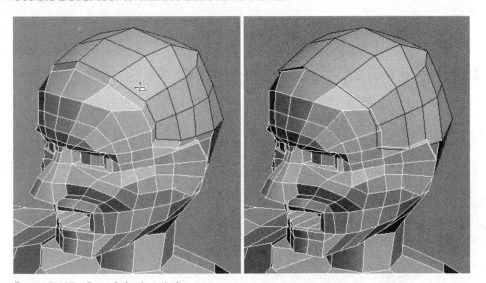

*Figure 9-115: **Bevel** the hair helmet*

3. Push and pull the vertices to make the hair more natural. Use Cut and other tools if necessary to add or remove detail.

   You'll probably need to add another **Volume Select** and **Delete Mesh** while you're working on the hair. Remember to turn off the **Auto Fit** option in Volume Select.

4. Save the scene as **Skater11.max**.

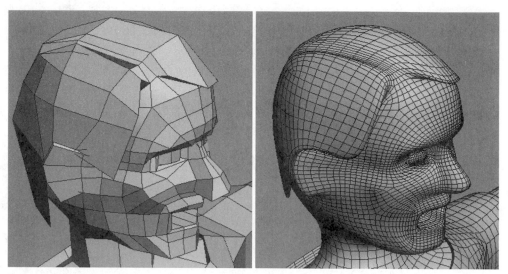

Figure 9-116: Finished hair

Conclusion

1. Add any finishing touches. Randomize the edges of the clothing a little so that the model is not perfectly symmetrical. Increase the MeshSmooth **Iterations** to **2**.

2. Congratulations! Save the model as **Skater12.max**.

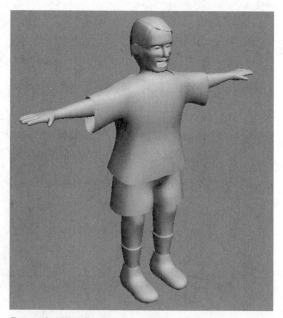

Figure 9-117: Completed model

SUMMARY

Boolean Compound Objects create new forms based on the overlapping volumes of two existing meshes. The three Boolean operations are **Union, Intersection**, and **Subtraction**.

Bezier **Patches** create soft, smooth surfaces based on Bezier curves. The **Surface** modifier automatically generates Bezier Patches from intersecting splines, called a *spline cage*. The spline cage must create three- or four-sided areas for the Surface modifier to work.

Subdivision surface modeling consists of using the tools within **Editable Poly**, often in conjunction with the **MeshSmooth** or the **TurboSmooth** modifier. Subdivision surface modeling uses the powerful toolset of Editable Poly to create smooth curves. The resulting surfaces are similar to Patches, but easier and more flexible to work with. Other modifiers, such as **Symmetry, Volume Select, Delete Mesh,** and **STL Check,** can be employed temporarily during the subdivision modeling process.

NURBS objects are also made of curves, similar to Bezier Patches. NURBS surfaces, such as **U Loft,** can be built from NURBS curves. NURBS is different from Bezier Patches in a variety of ways. For example, NURBS points can be weighted, and there is no limitation on the number of control points surrounding a surface.

Objects and scenes can be combined with grouping, merging, and other tools.

REVIEW QUESTIONS

1. What modifiers are used in *subdivision surface* modeling?

2. What is the purpose of using **Volume Select** and **Delete Mesh** when working on a subdivision surface model?

3. What is a *spline cage*?

4. What must happen for the **Surface** modifier to work correctly?

5. How are Bezier **Patches** and **NURBS** similar? How are they different?

6. Name two ways to make a NURBS object.

7. What is a **Group**? When would you use it? When would you want to avoid it?

8. What is a **Boolean**?

9. What is the advantage of the **Symmetry** modifier over the **Mirror** modifier?

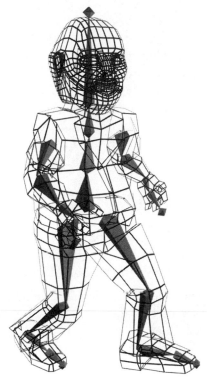

Chapter 10
Advanced Animation

OBJECTIVES

In this chapter, you will learn about:

- Advanced coordinate systems
- Controllers and Constraints
- Parameter Wiring and Manipulators
- Expressions
- Dynamic simulations such as reactor, Hair and Fur, and Cloth
- Inverse Kinematics
- Animation modifiers such as Skin and Morpher
- Character animation

ABOUT THIS CHAPTER

In Chapter 6, *Keyframe Animation*, you learned how to make objects move and parameters change over time. Now you'll delve deeper into animation, and learn some of the core concepts of how animation works in 3ds Max. You'll also work with alternate methods of creating animation, such as how to place an object on a path, or change parental linking over time. We will discuss the basics of character setup as well as simulations such as reactor dynamics, Hair and Fur, and Cloth.

COORDINATE SYSTEMS

PARENT SPACE

When you position, rotate, or scale an object, the transforms are *always* made relative to the parent of that object. Even if you choose World or Local as the Reference Coordinate System, 3ds Max always internally stores the transforms in parent space.

This can cause some confusion if you're not used to the idea. For example, you may rotate an object around its local Y axis, then go to the Curve Editor, only to find that the rotation information is stored in the Z Rotation track. This happens quite often when objects are linked. If an object isn't linked, then the World is its parent.

EXERCISE 10.1: Parent Space

1. Reset 3ds Max. Create a **Teapot** in the Top viewport. Create a second Teapot in the Front viewport.

2. Choose **Local** as the Reference Coordinate System. Maximize the Perspective viewport. Select one teapot, then select the other. Notice that the axis tripods of the objects don't match. The first teapot's Z axis is pointing up, and the second teapot's Y axis is pointing up. This is because they were created in different viewports.

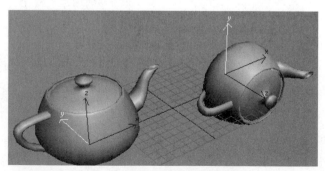

Figure 10-1: Two teapots with different orientations

3. Activate the **Rotate** tool. Notice that the Reference Coordinate System is set to View by default. Each selection or transform tool has its own setting for the Reference Coordinate system. Choose **Local** from the pulldown.

4. Parent the second teapot to the first teapot with the **Select and Link** tool. The teapot sitting on its side should be the child of the teapot sitting upright.

5. Animate the child rotating around its local Y axis.

6. Open the **Curve Editor**. The animation you created is in the Z Rotation track, because it was recorded relative to the parent object.

INHERITING TRANSFORMS

It is possible to set up a linked object so it only inherits certain transforms from its parent. For example, you can limit the link so it follows only the rotation of the parent, or just the position.

To limit the inheritance of linked transforms, select a child object. Go to the **Hierarchy** panel and click the **Link Info** button at the top of the panel. By default, all check boxes under the **Inherit** rollout are checked. Uncheck the boxes under **Move**, **Rotate**, or **Scale** to prevent the motion, rotation or scaling along that axis from being passed from the parent to the child. The World coordinate system is used.

EXERCISE 10.2: Unicycle Pedals

In this exercise, you'll animate a unicycle pedaling itself.

1. Load the file **Unicycle.max** from the disc that comes with this book.

    This is a 100-frame animation of a unicycle with a rotating wheel and pedals. The pedals are linked to the bar assembly that comes out of the center of the wheel. The pedals rotate along with the bars. We would like the pedals to follow the bars, but not rotate.

2. Select one of the pedals.

3. Go to the **Hierarchy** panel and click the **Link Info** button. Under the **Inherit** rollout, uncheck the **X** check box under **Rotate**.

    This will prevent the pedal from rotating on the X axis.

4. Pull the time slider. The pedal still moves with the assembly, but does not rotate as it moves.

5. Select the other pedal, and do the same to keep the pedal from rotating.

Figure 10-2: Pedals remain parallel to the ground

6. Save the scene with the filename **UnicyclePedaling.max** in your folder.

WARNING: Because the controls in the **Inherit** rollout work in world space, not parent space, disabling transform inheritance won't work in all situations. For example, turning off just one or two rotation axes may lead to unpredictable behavior.

You probably noticed that the Hierarchy panel also has a **Locks** rollout. Use these check boxes to prevent an object from being directly transformed with the Move, Rotate, or Scale tools. The Lock controls do not affect how transforms are inherited from parent objects. Locks are all relative to the object's **Local** coordinate system.

CONTROLLERS AND CONSTRAINTS

All animation in 3ds Max is managed by program modules called **Controllers**. Some Controllers are called **Constraints**, because they constrain an object to some other object. This terminology was adopted to make 3ds Max more in line with other 3D packages, but at their heart there is no difference between Controllers and Constraints.

Controllers and Constraints determine how an object is animated. So far, you have only used the default Controllers, such as **Position XYZ**. This Controller lets you set the location of an object with the **Select and Move** tool. However, there are many other ways to set the values of an object's animation track, such as Position.

For example, the Position track could be determined by a spline path, or by random noise, or by a combination of these effects. Or, Rotation could be set by making the object always point at another object.

Different Controllers are available for different animation tracks. Some Controllers are designed to control multiple parameters at once. So, Controllers can be assigned to all three transforms as a unit, or to each Position, Rotation or Scale track separately. You may also assign Controllers to parameter values such as the **Angle** of a Bend modifier.

ASSIGNING A CONTROLLER

Controllers are assigned in numerous ways in 3ds Max. You can use the **Animation** menu, the **Motion** panel, or a **Track View**.

The Animation menu is the quickest and easiest way to assign a transform Controller or Constraint. Simply select the object, and choose an item from the Animation menu. For example, to assign a **Path** Constraint, choose **Animation > Constraints > Path Constraint**.

To assign a transform Controller on the **Motion** panel, select the object and expand the **Assign Controller** rollout. You'll see a hierarchy of the various transform tracks. Each track has its current Controller type listed after a colon.

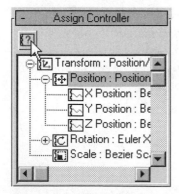

Figure 10-3: **Assign Controller** *rollout on the Motion panel*

Select one of the tracks listed on the rollout window. Click the name of the track, not its icon. You'll know that the track is selected when it is highlighted.

Click the **Assign Controller** button 🖸 at the top of the rollout. A list of Controllers available for that transform is displayed. Choose a Controller from the list. The Motion panel changes to display the available tools for the active Controller.

A Controller can also be assigned in a Track View. Here, you can assign controllers to any track, including modifier or object parameters, materials, lights, and so on.

To assign a Controller in a Track View, open a Curve Editor or Dope Sheet. Select the desired track in the **Controller Window** on the left side of the Track View. Right-click and choose **Assign Controller** from the Quad Menu, or choose **Controller > Assign** from the Track View menu.

PATH CONSTRAINT

A **Path** Constraint is used to make an object follow a spline curve. To use this Controller, create a line to use as a path, then select the object that will follow the path. Choose **Animation > Constraints > Path Constraint** from the Main Menu. Then immediately click the spline curve in the viewport. The object jumps to the start of the path.

Play the animation to see the object follow the path. By default, the object starts at one end of the path and goes to the end over the entire animation range.

Open the **Motion** panel. Check the **Follow** check box to make the object rotate as it moves along the path. You can still rotate the object normally if you wish. If the object is not oriented correctly on the path, go to frame 0 and rotate it so it's pointing in the right direction.

The **% Along Path** parameter determines how far along the path the object has traveled at the current frame. The % Along Path value is automatically animated when you pick the path, so that the object is **0%** along the path on the first frame of the animation, and **100%** along the path on the last frame of the animation.

You can change the % Along Path by turning on the **Auto Key** button, going to a certain frame in time, and changing the spinner value on the Motion panel. Of course, you can also control the keys and function curve for the **Percent** parameter in a Track View. A Percent over **100** or less than **0** causes the object to loop around to the beginning or end of the path. A value of **101%** is considered to be the same as a value of **1%**.

The object following the path is oriented to the path by its Pivot Point. When the object moves along a path, its Pivot Point is always on the path. Before setting up a Path Constraint, change the object's Pivot Point if necessary so it's at the correct spot.

EXERCISE 10.3: Flying Along a Path

1. Load the scene **SpaceScene05Blur.max** from the disc.

2. Delete the Spaceship object from the scene.

3. Choose **File > Merge** from the Main Menu. Merge the NURBS spaceship from the file **FlyingSaucer.max** on the disc. Don't merge the lights or camera, just the NURBS ship, which is probably named Sphere01.

    It would have been easier to know what object to merge if we had named the NURBS ship intelligently!

4. Rename the NURBS ship to **Spaceship**.

5. Scale the NURBS ship to the scene. Then go to the **Hierarchy** panel and look for the **Adjust Transform** rollout. Click the **Scale** button to reset the scale of the object to 100% without changing its size in the scene.

6. Click **Time Configuration** 🗔 and set the animation **Length** to 250.

7. Select the camera in the **Top** viewport. On the **Modify** panel, check the **Show Cone** check box. This causes the camera cone to be visible at all times, making it easier for you to create a path that will pass in sight of the camera.

8. If necessary, increase the **Target Distance** parameter, so the camera cone encloses all geometry in the scene.

9. In the Top viewport, create a curved line to be used as the path for the object. Draw the path so it comes into the camera's cone, flies around a little, then goes back out again near the camera.

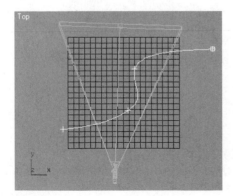

Figure 10-4: Path curving into then out of camera cone

10. In the **Front** or **Left** viewport, move the path higher so it traverses above the ground.

11. Adjust the vertices of the path so its height varies over the course of the path, while still staying in the camera view.

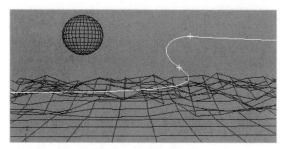

Figure 10-5: Path with varying height

If the line doesn't have enough segments to make it smooth, expand the **Interpolation** rollout and increase the **Steps** value, or use the **Adaptive** option.

12. Select the **Spaceship**, and choose **Animation > Constraints > Path Constraint** from the Main Menu. Click the spline object.

    The spaceship jumps to the start of the path.

13. Play the animation. The spaceship moves along the path over the entire **250** frames.

Figure 10-6: Spaceship following path

14. Open the **Motion** panel. Scroll down to the **Path Parameters** rollout. Check the **Follow** and **Bank** check boxes. The object turns to follow the path, and banks at the corners.

    If the spaceship banks too much, decrease the **Bank Amount** parameter. If the spaceship jitters, increase the **Smoothness** parameter.

15. Edit the vertices and Bezier handles of the path object to fine-tune the movement.

16. Select the spaceship and move the keyframe at frame **250** to frame **100**. The spaceship reaches **100%** along the path at frame **100**; therefore, it moves much more quickly.

17. With the spaceship selected, open the **Curve Editor**. The function curve of the Path Constraint's Percent parameter is visible.

To change the speed of the spaceship as it travels along the path, you need to edit the function curve. However, the default type of Controller for the Percent track is **Linear Float**. This Controller uses only linear interpolation between keyframes, and it doesn't allow you to edit Bezier handles.

18. Make sure the Percent track is selected and highlighted. In the Curve Editor menu, choose **Controller > Assign**. In the Assign Float Controller dialog, click **Bezier Float**. Click OK to close the dialog.

Now the Percent track uses Bezier interpolation, and you can change the shape of the function curve to suit your needs.

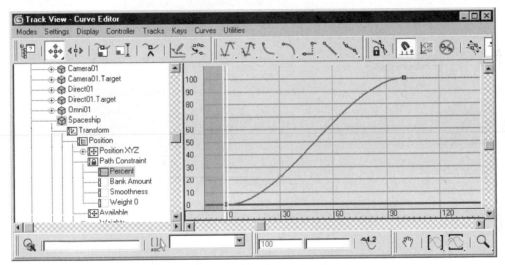

Figure 10-7: **Bezier Float** Controller assigned to Percent track

19. Fine-tune the animation. Change the length of the animation. Add more keys to the Percent track to make the spaceship speed up or slow down. You may also add rotation keys to the spaceship to make it spin or to manually animate the banking.

20. Save the scene as **SpaceScenePath.max** in your folder.

To view a finished animation, open the file **SpaceScenePath.avi** on the disc.

LOOKAT CONSTRAINT

A **LookAt** Constraint causes an object to rotate so that one of its local axes always points to another object. To set it up, create two objects, one to point at another. Select the object that will "look at" the other, and assign the LookAt Constraint from the Animation menu.

Click the object to look at. If necessary, choose the axis to point at the object from the Motion panel. Check **Flip** if necessary to flip the object the other way along the selected axis.

The LookAt Constraint is handy in many situations, such as making a character's eyes follow an object, or for making a gun follow a moving target. In these cases, the object being followed is usually a **Point** Helper.

LINK CONSTRAINT

When you use **Select and Link** to parent one object to another, the objects are linked throughout the entire animation. There will be times when you want an object to be linked first to one object, then to another. To accomplish this, you can use the **Link Constraint**.

The Link Constraint is assigned as a Transform Controller. Don't parent the object to anything. Select the object and open the Motion panel. On the **Assign Controller** rollout, highlight **Transform**, then click the **Assign Controller** button. Choose **Link Constraint** from the pop-up dialog.

The **Link Params** rollout is displayed on the Motion panel. Click **Add Link**, then click the object that you want to be the initial parent. The parent's object name appears in the Link Params rollout.

Links assigned with Link Constraint do not shift gradually, but change abruptly at a particular frame. Note that the **Start Time** parameter in the Link Params rollout is set to the current frame number. This is the frame at which the link will take effect.

To animate the transfer of a link to a new parent object, position the Time Slider at the frame you wish the transfer to take place. Then click the **Add Link** button, and click the new parent object in the viewport. The Start Time for the new parent is automatically assigned to the current frame.

At any time, you can highlight an object in the Link Params list and change its Start Time. You may also use the **Delete Link** button to remove parent objects from the list.

EXERCISE 10.4: Passing a Ball with Link Constraint

In this tutorial, you'll use the **Link Constraint** to cause an object to be linked to several different objects over the course of an animation.

1. Load the file **Sticks.max** from the disc. Play the animation.

    This animation consists of six sticks rotating in unison over **150** frames. There is a ball out of view of the camera sitting at the top of the rightmost stick in its starting position at frame **0**. You will use the **Link Constraint** to make the sticks pass the ball along.

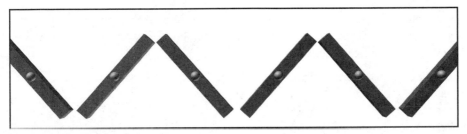

Figure 10-8: Sticks

2. Select the ball in the Front viewport.

3. Go to the **Motion** panel. Expand the **Assign Controller** rollout.

4. Highlight the **Transform** track and click **Assign Controller** [?]. On the **Assign Transform Controller** dialog, choose **Link Constraint**.

    The Motion panel changes to display the **Link Params** rollout.

5. Go to frame **0**. Click **Add Link**, and click the stick upon which the ball is currently sitting.

    The picked object name, **Pass01**, is placed on the list on the **Link Params** rollout with a **Start Time** of **0**.

6. Pull the time slider. The ball is now linked to the stick you picked, so it moves along with the stick.

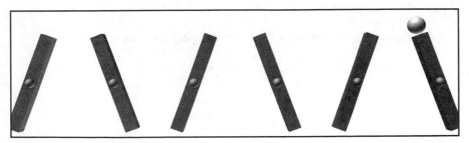

Figure 10-9: Linked ball moves with stick

7. Go to frame **25**. This is the frame at which you want the ball to be linked to the second stick.

8. The **Add Link** button should still be pressed. Click the second to last stick, named **Pass02**. The object appears on the list with a **Start Time** of **25**.

   Instead of picking the object from the scene, you can press the **<H>** key on the keyboard to select the object from a list.

9. Go to frame **50**, and click the next stick, **Pass03**. This object appears on the list with a **Start Time** of 50.

10. Continue to move another **25** frames ahead, and choose the appropriate stick. When you have finished, the **Link Params** rollout should show the following objects, each with a different **Start Time**:

Object	Start Time
Pass01	0
Pass02	25
Pass03	50
Pass04	75
Pass05	100
Pass06	125

11. Click **Add Link** to turn it off.

12. Activate the **Camera01** view and play the animation. The sticks appear to pass the ball to the left.

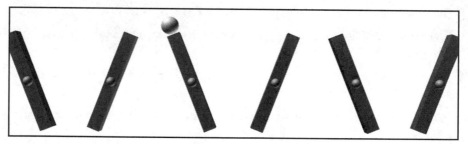

Figure 10-10: Sticks passing ball

13. Save the scene with the filename **SticksPassingBall.max** in your folder.

PROCEDURAL ANIMATION

Just as a map can be generated by a procedure or algorithm, animation can be generated with a procedural Controller. For example, a **Noise Position** Controller can be applied to a Position track to randomize the position of an object. Or, a **Waveform Float** Controller can animate a repeating pattern such as a sine wave.

All procedural Controllers have their own controls and settings. There are many possibilities for natural, fantastic, or abstract animation effects. The best way to learn about these Controllers is to simply experiment with them. Select a track name in the Dope Sheet or Curve Editor, right-click and choose Properties from the Quad Menu. A dialog that has settings specific to that Controller appears. Some of these parameters are animatable, and some are not.

NOTE: Not all Controllers work with all animation tracks. Some Controllers are designed to output three parameter values, such as **XYZ Position**. Others only output one parameter value, such as **Waveform Float**.

LIST CONTROLLER

Often, it is useful to stack Controllers on top of one another. For example, you may wish to send an object down a Path, and also add a little bit of Noise to the motion to randomize it. In this case, use a **List** Controller. A List Controller lets you combine the effects of other Controllers, and balance their relative influence with an animatable **Weight** parameter.

To use a List Controller, select a transform from the Assign Controller rollout in the Motion panel, or select any track in a Track View. Assign a List Controller, such as **Position List**. The old Controller now becomes the first in a new list of Controllers.

To add additional Controllers, look for the word **Available** in the Track View or Motion panel. Assign a new Controller to the Available track. If desired, use the Weight parameters to control the influence of the various controllers.

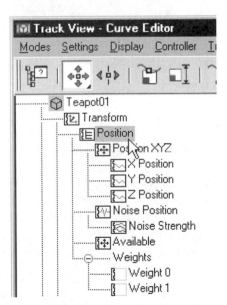

Figure 10-11: **Position List** *Controller in the Curve Editor*

You can also remove controllers from the List, or change their order within the list. Open the List rollout in the Motion Panel, or right-click the List Controller in a Track View and choose **Properties** from the Quad Menu.

PARAMETER WIRING AND MANIPULATORS

Another way you can automate the animation process is through **Parameter Wiring**. This allows you to send animation data from one track to another over a virtual "wire." You can make a wheel turn automatically as a car moves, or control multiple objects from a virtual slider, called a **Manipulator**.

PARAMETER WIRING DIALOG

Animation data can be transferred from one track to another if the two tracks have been connected in the **Parameter Wiring** dialog. To open it, choose **Animation > Wire Parameters > Parameter Wire Dialog** from the Main Menu.

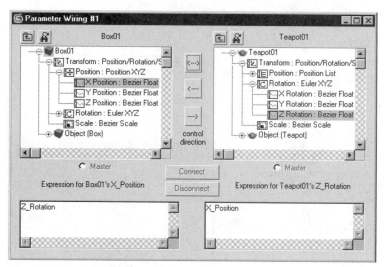

*Figure 10-12: **Parameter Wiring** dialog*

The Parameter Wiring dialog has two sides. You can choose an animation track from one side and send its data to another track on the other side. You may also create a two-way wire, so that each track will influence the other. Highlight one track on each side, and then press one of the buttons labeled **control direction**. If you click a one-way arrow, then one of the tracks becomes the **Master** for the other.

Below each track listing is a window that lists an **Expression**. Expressions are like mathematical equations. When you first make a connection by clicking the **Connect** button, the name of the track on the opposite side of the dialog is listed in each Expression window. In the previous illustration, there is a two-way connection between the X Position of one object and the Z Rotation of another object.

You can type in mathematical expressions in the Expression windows, to alter the data coming in over the wire. An example is shown in the following exercise.

EXERCISE 10.5: Turning a Wheel

In this exercise, you will use Parameter Wiring to make a wheel spin automatically as a car moves.

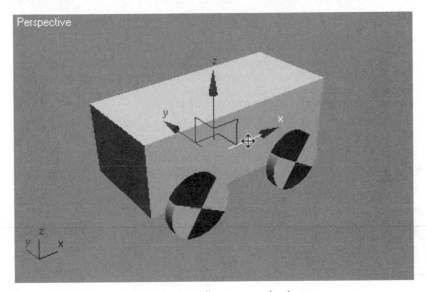

Figure 10-13: Boxcar with automatically rotating wheels

1. Open the file **Boxcar.max** from the disc.

2. You should see a checker pattern on the wheels. If not, go into the Material Editor and click the **Show Map in Viewport** ![icon] button.

3. Select the Box and move it in the global X axis. The wheels are parented to the box, but they don't rotate.

4. From the Main Menu, choose **Animation > Wire Parameters > Parameter Wire Dialog.**

5. In the Parameter Wiring dialog, scroll the left window and highlight the track **Box01 > Transform > Position > X Position**.

6. Scroll the right window and expand the **Box01** hierarchy to see its children, the cylinders. Highlight the track **Cylinder01 > Transform > Rotation > Y Rotation**.

7. Click the right-facing arrow to make the **Box01** X Position the Master track.

8. Click the **Connect** button. Minimize the dialog.

9. Select the Box in the Perspective view and move it in the global X axis. The **Cylinder01** object spins wildly.

For each unit the Box moves, the Cylinder spins **30** times. This is because the Cylinder is **30** units in Radius.

10.   Restore the Parameter Wiring dialog. In the right-hand Expression window, change the expression to read **X_Position/30**. This divides the data coming from the X Position track by **30**, which is the Radius of the Cylinder. Click the **Update** button.

11.   Move the Box in the X axis. The wheel now spins correctly.

12.   Use the Parameter Wiring dialog to wire the Y Rotation of the other wheels to the X Position of the Box.

13.   Save the scene as **BoxcarWireParameters.max** in your folder.

NOTE: This is a simple setup that will not work if you move the Box in any direction other than the global X axis. You could make the wheels turn wherever you drove the car by using Constraints and MAXScript, but that is beyond the scope of this book.

Another way to solve the problem would be use a series of Dummy or Point Helpers to control the transforms. To see how this can be done, load the file **WheelRoll-Dummies.max** from the disc. This method does not require scripting, but it's not as flexible or elegant as the scripting method.

MANIPULATORS

One button on the Main Toolbar we haven't discussed yet is **Select and Manipulate**. This enables display and control of **Manipulators**, which are Helper objects. They're designed to make changing parameters easier with controls within the viewports.

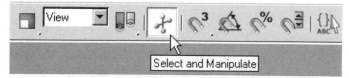

Figure 10-14: ***Select and Manipulate*** *button on the Main Toolbar*

Some primitive objects have parameters, such as Radius, which can be controlled with built-in Manipulators. You can also change the Hotspot and Falloff of a Spot light with its built-in Manipulators. But Manipulators really show their usefulness when used in conjunction with Parameter Wiring.

Manipulators are very handy when you wish to change more than one parameter at a time, or if the parameters are commonly used, or hard to get to. You can create a Manipulator Helper object, such as a **Slider**, and use it to control one or more animatable parameter in 3ds Max.

To create a Slider, open the Create panel and select Helpers. Choose Manipulators from the pulldown list, and click the Slider button. Then click in the viewport to create a Slider. Right-click to end creation mode.

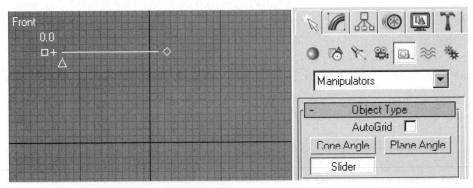

*Figure 10-15: Creating a **Slider** Manipulator*

Sliders will appear in any active viewport. To move the Slider around on the viewport, activate the Select and Manipulate button on the Main Toolbar. Then click the square at the left side of the Slider, and drag to place the Slider on the viewport. To minimize a Slider, click the small cross at the left side of the Slider.

Sliders have parameters just like any object, including the Slider Values, Label, placement, and length. To access these parameters, you must turn *off* the Select and Manipulate button, click the Slider, and open the Modify panel.

Once you have created a Slider, you can use Parameter Wiring to send its Value to any other parameter in 3ds Max.

EXERCISE 10.6: Butterfly Wings

In this exercise, you'll combine Manipulators and Parameter Wiring to make a butterfly's wings flap together.

1. Open the file **Butterfly.max** from the disc. This scene consists of a simple butterfly body and two wings. The wings are parented to the body, and the body is parented to a **Point** Helper, simply to make it easier to select and transform.

2. On the Create panel, choose the **Helpers** panel . Select **Manipulators** from the pulldown list. Click the **Slider** button, then click in any viewport to create a Slider.

3. On the Create panel, change the Slider's parameters to the following:

Label	wing angle
Minimum	-89
Maximum	10

    Right-click in the viewport to exit Slider creation.

    Now we'll wire the Slider Value to the Y Rotation angle of the wings.

4. From the Main Menu, choose **Animation > Wire Parameters > Parameter Wire Dialog**.

5. On the left side of the dialog, expand the hierarchy and highlight the track **Slider01 > Object (Slider) > Value**. On the right side, expand the hierarchy and highlight the track **wing_right > Transform > Rotation > Y Rotation**.

6. Click the arrow facing right. Then click the Connect button to wire the output of the Slider value to the input of the Y Rotation track.

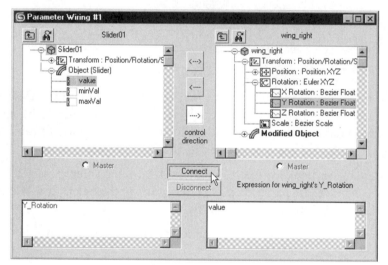

*Figure 10-16: Connect the **Slider Value** to the **Y Rotation** angle*

7. Minimize the Parameter Wiring dialog. Click **Select and Manipulate**  on the Main Toolbar. Click and drag the triangle at the bottom of the Slider to move it left and right.

Figure 10-17: Click and drag the Slider triangle to change the value

The wing spins wildly out of control. This is because the Rotation values are calculated in radians, not in degrees. There are 2π radians in one circle.

To correct this situation, we must use a simple expression to convert degrees into radians.

8. Restore the Paramater Wire dialog. In the expression window at the bottom right of the dialog, enter the following expression:

    degToRad(value)

    Click the **Update** button.

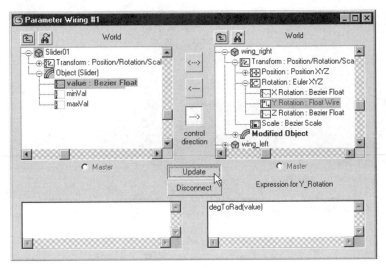

Figure 10-18: **Update** *the expression*

9. Minimize the Parameter Wiring dialog. Try the Slider again. Now the wing rotation is the same as the Slider value, measured in degrees.

10. Restore the Parameter Wiring dialog. On the right side of the dialog, highlight the track **wing_left > Transform > Rotation > Y Rotation**. Enter the following expression into the expression window:

    degToRad(-value)

    We are using a negative value here so that the left wing will rotate in the opposite direction of the right wing.

    Click the right-facing arrow, and then click the **Connect** button. Close the Parameter Wiring dialog.

11. Move the Slider triangle. The wings move together, in opposite directions. Animate the Point Helper flying around, and animate the Slider value with the Auto Key button.

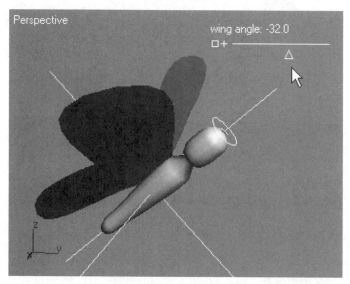

Figure 10-19: Slider value controls wing rotation in degrees

12. Save the scene as **ButterflySlider.max** in your folder.

DYNAMICS

Some types of animation are simply too complex or difficult to be feasibly keyframed by hand, or even through particles and Atmospheres, procedural Controllers or Wire Parameters. When a project calls for realistic simulation of moving objects, sometimes the best solution is to use a form of *dynamics*.

Dynamics are physically accurate simulations of objects interacting in some way with one another, or with Forces such as gravity, friction, and inertia. They work by calculating the movements of objects as if they had real mass, volume, elasticity, and other properties. The laws of Newtonian mechanics are used to create representations of how objects would move and deform based upon their physical characteristics.

REACTOR

There are many different types of dynamic simulations. 3ds Max's core dynamics engine is called **reactor**. This is a plugin licensed from another company, called Havok. This same simulation technology is widely used in real-time action games.

NOTE: 3ds Max also has a feature called simply **Dynamics**, which can be accessed via the Utilities panel by clicking the More button. You'll also find it in the Material Editor, in the Dynamics Properties rollout. This is a "legacy" feature, which is left over from previous versions of 3ds Max. It never worked very well, and it has been effectively replaced by reactor. The only reason it's still in 3ds Max is for compatibility with old scene files.

Figure 10-20: ***reactor*** *toolbar*

There are many tools in the reactor plugin. They are accessed from the reactor toolbar, which is located on the left side of the 3ds Max interface by default. In this book, we only have space to cover the most basic dynamic simulation, the **Rigid Body Collection**. If you wish to learn about simulations of other phenomena using features such as **Soft Body Collection, Water,** and **Fracture**, consult the 3ds Max User Reference.

In addition to the specific tools found in the reactor toolbar, the

Utilities panel ![icon] provides access to global settings for the reactor dyamics engine. Here you can define parameters for the dynamics in the scene as a whole, such as the **Start Time** and **End Time** of the simulation, accuracy of calculations, strength of **Gravity**, amount of **Drag Action**, and so on.

If you get deep into dynamic simulations, you'll find that the reactor toolset is quite comprehensive. Even so, it's important to realize that reactor is a plugin added to 3ds Max. It's not part of the core application programming architecture, and is really a separate program within 3ds Max. So the reactor plugin does work with object transforms that are animated by hand using traditional keyframing, but in general it doesn't work with other animation techniques, such as procedural Controllers, Constraints, particle systems, and Forces. However, in most cases reactor has built-in features that perform the same functions. For example, reactor has its own **Wind, Gravity,** and **Constraints,** such as **Point-to-Point, Hinge,** and **Rag Doll.**

Figure 10-21: ***reactor*** *panel*

Reactor is entirely physically based. In order to get anything like accurate results, you must model your scene to scale. It's absolutely critical that you use real-world units of measurement when building your scene. Before you begin modeling or animating, choose a system of measurement from the **Units Setup** dialog, accessed through the **Customize** menu. It doesn't matter whether you choose **Metric** or **US Standard** as long as your objects are the right size in your scene. Also remember that if you must scale your objects, use the **Reset: Scale** tool in the **Hierarchy** panel to restore your objects to scale values of 100% in all three dimensions.

In addition to size, certain physical properties of objects must be defined in order to simulate their movement. Most important of these properties is mass, which is proportional to weight. Under the same gravitational conditions, an object with higher mass will weigh more than an object with lower mass. This determines how the objects will move and collide. If you roll a balloon to collide with a bowling ball, the bowling ball won't move much, if at all. But if you roll a bowling ball to a balloon, the balloon will bounce away. Because of the difference in mass, the objects behave very differently, even though they are about the same size. An object with a great deal of mass has more *inertia* than an object with very little mass. Inertia is the tendency of a moving object to remain moving, and a static object to remain stationary, unless acted upon by a force such as gravity or collision.

Other important physical properties to consider when designing a dynamic simulation are *friction* and *elasticity*. Friction is the effect that the roughness of a surface has on inertia. Ice skates work well on ice, but not so well on gravel, because of the low friction of slippery ice. Elasticity is the "bounciness" of objects. A rubber ball bounces well because its shape is highly elastic.

Each of these three properties must be defined for each object in a simulation. They are accessed through the **Properties** rollout at the bottom of the reactor panel, or through a modeless dialog called

the **Property Editor**, launched from a button on the reactor toolbar. **Mass** is measured in kilograms. An object with **0** Mass (the default) will not move at all during the simulation. Normal values for **Friction** and **Elasticity** are between **0** and **1**.

Figure 10-22: Reactor Property Editor

Creating a Reactor Simulation

To create a reactor simulation, set up the initial conditions, which includes modeling your scene and keyframing any objects that will influence the simulation. Add reactor scene elements, such as **Rigid Body Collection**, **Motor**, or **Spring**. For Collections of objects such Rigid Body Collection, add the necessary objects to the Collection. Define the reactor Properties of all necessary objects, such as their Mass, Friction, and Elasticity. If desired, adjust the simulation's global parameters in the reactor panel, found in the Utilities panel.

Once you've got the scene ready, it's time to preview the simulation. Activate a Perspective or Camera view, and adjust your point of view as desired to test the simulation. Click the **Preview Animation** button ▣ on the reactor toolbar, or click the **Preview in Window** button on the reactor Utilities panel. The **reactor Real-Time Preview** window opens. Press the **<P>** key to see the simulation run. To see it run again, choose **Simulation > Reset** from the menu at the top of the preview window, and press **<P>** again.

WARNING: While the reactor Real-Time Preview window is open, reactor uses all of your computer's available processing power to calculate the simulation, *even if the simulation is not currently playing*. All other functions of 3ds Max are disabled while the window is open. Performance in any other open application is severely degraded.

If the preview looks good, you can close the window and complete the process by clicking the **Create Animation** icon button ▣ on the reactor toolbar, or the **Create Animation** text button on the reactor panel. A dialog appears, warning you that this action cannot be undone. If you click OK to proceed, reactor generates keyframes for all of the objects in the simulation. Once this is accomplished, you can edit the keyframes and function curves just as you would any other animation.

Generally speaking, it's unlikely that the first preview you perform will be satisfactory. Then you need to adjust object properties and global parameters in an attempt to get better results. Working with dynamics often involves changing parameter values, previewing the simulation, changing the same or different parameters, previewing again, and so on until you get more or less what you want. It's never perfect, but in some situations it can be faster and more effective to create a convincing illusion using dynamics than through manual keyframing.

There are many idiosyncrasies to working with dynamics. Working around these problems takes an awareness of how things work in the real world, careful study of the product documentation, plus a lot of trial and error. For example, reactor is designed to work with objects near the scale of a human body. It doesn't work well with small objects unless certain parameters are adjusted, as we will see in the following exercise.

EXERCISE 10.7: Reactor Billiards

In this exercise, you'll use reactor to animate a simple scene of billiard balls bouncing

1. Open the scene **PoolTable.max** from the disc that comes with this book. Note how the scene is constructed. The sides of the billiard table are built from simple boxes. This simplifies the calculations of collision detection that reactor must make.

2. Enable **Auto Key**. Move the Time Slider to frame **5**. Select the **cue ball** object and move it forward in the world Y axis until it is almost touching the other three balls. To put some spin on the **cue ball**, rotate it approximately **-2000** degrees around the world X axis. Disable Auto Key.

NOTE: If you view the rotation transform values using World as your Reference Coordinate System, what you'll see is the remainder after the actual rotation value is divided by 360 degrees. This is one of the weird quirks in 3ds Max. If you view transform values using the Local coordinate system, everything will read **0** unless you're currently adjusting the transform. To see the true rotation values of an object, choose **Parent** from the Reference Coordinate System pulldown on the Main Toolbar, or look in the Curve Editor.

3. With the **cue ball** selected, open the Curve Editor. Select all of the keyframes and click **Set Tangents to Linear**. Now the ball's position and rotation do not ease out of or into their keyed values. Click in the viewport to deselect the **cue ball**.

4. If the **reactor** toolbar is not visible, right-click an empty area of a toolbar and choose reactor from the pop-up menu. Click **Create Rigid Body Collection** on the reactor toolbar, then click anywhere in the Perspective view to create the reactor Collection. The location of the Collection object does not matter.

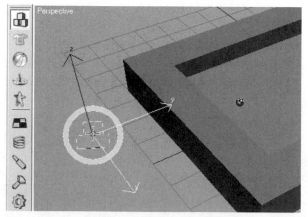

Figure 10-23: **Create Rigid Body Collection**

5. With the Rigid Body Collection object selected, open the Modify panel. Click the **Add** button. The **Select Rigid Bodies** dialog appears. Choose all of the objects in the list and click **Select**.

6. Now we will set the physical properties of the ball objects. Select the **cue ball** and

    click the **Open Property Editor** button  on the reactor toolbar. In the Rigid Body Properties dialog that appears, set the **Mass** of the ball to **0.17** kilograms and the **Elasticity** to **1.0**. Leave the **Friction** at its default value of **0.3**.

NOTE: To help ensure the accuracy of the simulation, this scene has been built to scale, using regulation sizes and weights for the balls.

7. Select the other three balls and set their **Mass** to **0.167** and their **Elasticity** to **1.0**.

    It's not necessary to change the properties of the objects that make up the pool table. The default values will work just fine. Specifically, the default Mass of **0** prevents an object from moving, but allows other objects to react to it. That's just what we want for the pool table.

8. Click the **Preview Animation** button 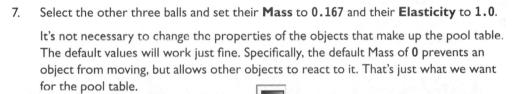 on the reactor toolbar. The **reactor Real-Time Preview** window opens. Press the **<P>** key to play the simulation. Unfortunately, the initial results of the simulation are completely useless. The balls instantly fly off into infinity.

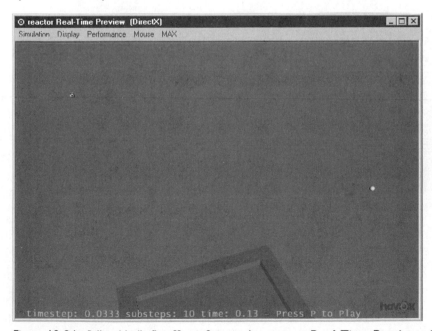

*Figure 10-24: Billiard balls fly off to infinity in the **reactor Real-Time Preview** window*

The reason there's a problem here is that the global parameters of the simulation are not optimized for this scene. Reactor's default settings don't work well for objects of such small size and low mass.

9.   Close the reactor Real-Time Preview window. Open the **Utilities** panel and click the **reactor** button to open the reactor panel.

To allow reactor to take the animation of the cue ball into account, the start time of the animation must be changed to a frame in which the ball is in motion. This means that the start time should be somewhere between frame **0** and frame **5**.

Open the **Preview & Animation** rollout. Change the **Start Frame** to 4. Click the **Preview in Window** button. Press **<P>** to preview the animation. Now the balls are colliding, but they are still flying off to infinity. Close the preview window.

10.   In the Preview & Animation rollout, set the number of **Substeps/Key** to 60. This increases the accuracy of the result by making more calculations for each keyframe of the simulation. Preview the animation again. This time, the balls collide but don't fly off to infinity. However, if you look closely, you'll see that the balls aren't touching the table at any time. Move your viewpoint around in the preview window using **<ALT> + left mouse** to orbit and **<ALT> + middle mouse** to pan.

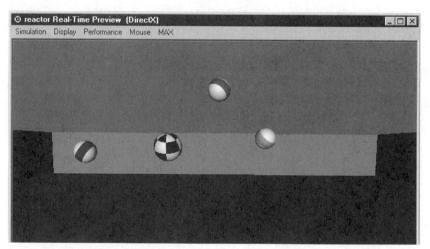

Figure 10-25: Balls hover above the table

11.   This happens when the **Collision Tolerance** of the simulation is set too high for the size of the objects. Objects that react to the simulation can never be closer together than the Collision Tolerance value. In the **World** rollout of the reactor panel, set the **Col. Tolerance** to 0'0 2/8". Now when you preview the animation, the balls still float above the table, but only by a quarter of an inch. If this becomes a problem for the final animation, we can generate the keyframes using a Col. Tolerance of **0'0 2/8"**, and then move the table up by a quarter inch so that it touches the billiard balls.

12. Look very closely at the balls' movement in the preview window. They are colliding, but not rolling. This is because there is too much drag on them. Reactor has a default function that applies a drag factor to the entire simulation, to prevent objects from moving forever. This is in addition to the Friction parameters for objects. The default amount of drag is too much for such small objects, and it completely prevents them from rolling.

    In the World rollout, look for the section labeled **Add Drag Action**. Set the **Lin** and **Ang** values to **0.001** each, and preview the simulation again. Now the balls are rolling properly.

13. If you wish, make any additional adjustments to the properties and parameters discussed in this exercise. You can also change the speed of the **cue ball** by moving its second keyframe to an earlier or later frame in time. If you do so, just make certain that the simulation **Start Frame** is set to a frame in which the cue ball is in motion.

14. When you're happy with the simulation, save the scene as **PoolSimulation.max** in your folder. Click the **Create Animation** button on the reactor toolbar, or click the **Create Animation** text button on the reactor panel. A dialog appears, warning you that the action cannot be undone. Click **OK** to proceed, and position and rotation keyframes are created for all four balls.

    Play the animation in the 3ds Max Perspective viewport. If there is any problem with the animation, you can reload the **PoolSimulation.max** file, make adjustments, and create the animation again.

    Select one of the balls and look at the Track Bar. Notice that there are keyframes for every frame of the simulation. This is because the default number of **Frames/Key** is **1**. This can result in an unwieldy amount of animation data, but if you increase the Frames/Key value, the simulation will be much less accurate. If you need to edit the keyframe data, you can use the **Keys > Reduce Keys** menu item in the Curve Editor, which will streamline the key data while maintaining the overall shape of the function curves.

15. Save the scene as **PoolAnimation.max** in your folder.

CLOTH

One of the most difficult things to animate by hand is cloth. Since realistic cloth is such a common requirement of 3D computer animation, 3ds Max offers a way to simulate cloth interacting with character bodies and other objects.

In fact, there are two completely different ways of simulating cloth in 3ds Max. The older method uses Havok's reactor engine. However, with 3ds Max 8, a new and better cloth system is available.

This new cloth feature set gives the artist much more freedom than reactor. It works with standard 3ds Max Forces such as Wind and Gravity. You can even use 3ds Max to simulate clothing made from 2D garment patterns, just as fashion designers and tailors do in the real world.

The new cloth system consists of two modifiers: **Garment Maker** and **Cloth**. Garment Maker is applied to Editable Spline objects. It converts the 2D spline to a special type of mesh object that deforms naturally during the cloth simulation. Using Garment Maker to build clothing such as dresses and shirts is beyond the scope of this book, but it's relatively straightforward to use. You can even import data from pattern-making software used in the fashion industry.

The Cloth modifier is applied to all objects that take part in the simulation, whether they are cloth objects or objects that will collide with the cloth. Each Cloth modifier is its own simulation. This means that any objects that will potentially touch must have the same instanced Cloth modifier. For simple scenes, this means you only have one Cloth modifier applied to many objects. You can select multiple objects and apply the Cloth modifier to all of them, or you can add and remove objects from the simulation later, using the **Object Properties** dialog found on the Modify panel.

Figure 10-26: Cloth modifier

The general workflow of a cloth simulation is as follows. Model the scene, including characters and props. Use Garment Maker to create clothing or other fabric. Use the Cloth modifier to define properties and parameters for the simulation, and test the simulation many times along the way, just as you do with reactor. However, unlike reactor, the processes for previewing and creating animation are not separate. In the Cloth modifier, click **Simulate**, and keyframes are created for the fabric objects' mesh vertices. To alter the simulation, simply press the **Erase Simulation** button, make your changes, and click Simulate again.

The 3ds Max cloth system relies on real-world measurements, just as reactor does. So, once again, it's very important that you build your scene to scale. In addition, don't change the **System Unit Setup** in the **Customize > Units Setup** dialog. If you do, the simulation will not be accurate, and then you must perform manual calculations and enter the results in **cm/unit** field in the Cloth modifier. Generally speaking, it's a bad idea to change the System Unit Setup anyway.

EXERCISE 10.8: Cloth

In this exercise, you'll learn the basics of working with the 3ds Max cloth system, including collision with objects and the effects of Forces on cloth.

1. Open the file **ClothRobot.max** from the disc that comes with this book. Play the animation. At frame **40**, the robot arm suddenly moves.

2. Select the Editable Spline object called **Fabric**, which is hovering over the robot arm. Enter Vertex sub-object mode. Select all four vertices of the object, and click the **Break** button on the Geometry rollout. This step splits the spline into four separate Spline sub-objects, which helps the Garment Maker modifier work more accurately.

3. Exit sub-object mode and add the **Garment Maker** modifier to the fabric object. In the Perspective view, press **<F3>** to see the **Fabric** in Wireframe mode. Note the irregular triangulation. This is called a Delaunay mesh, and it produces a much more natural result than a regular tessellation. Press **<F3>** to return to Smooth+Highlights mode.

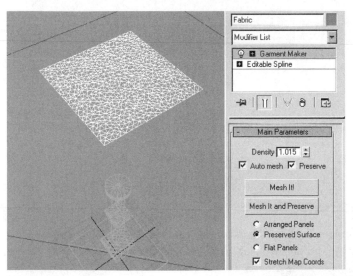

Figure 10-27: **Garment Maker** *creates a Delaunay mesh from an Editable Spline*

NOTE: Increasing the **Density** parameter in the Garment Maker modifier would result in a more finely triangulated mesh, but it would increase calculation time for the simulation. If you needed more detail, you could add a **HSDS** modifier, which is a type of subdivision surface algorithm that works better than MeshSmooth in this case.

4. Add a **Cloth** modifier to the **Fabric**. In the **Object** rollout of the Modify panel, click the **Object Properties** button. In the Object Properties dialog, click the **Add Objects** button. The **Add Objects to Cloth Simulation** dialog appears click the **All** button, then click **Add**. All of the objects in the scene are added to the current simulation, which means they all share the same instanced Cloth modifier.

5. In the Object Properties dialog, all of the objects in the list are highlighted except the **Fabric** object. Choose the **Collision Object** option near the bottom of the dialog. All of the objects except for **Fabric** are now part of the simulation, and they are designated as objects that will collide with the cloth object.

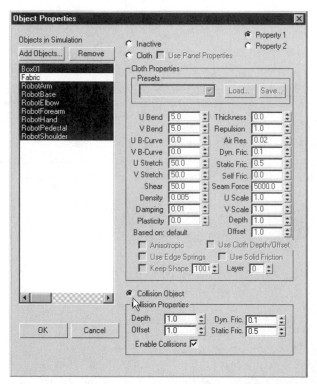

Figure 10-28: Define all objects except **Fabric** *as* **Collision Object**

6. Select the **Fabric** object from the list in the Object Properties dialog. Choose the **Cloth** option near the top of the dialog. In the **Cloth Properties** section, click the **Presets** pulldown, and choose **Cotton**. Many of the fields in the Cloth Properties section update to reflect the Cotton preset. Click OK to accept the changes and close the Object Properties dialog.

7. Click the **Simulate** button in the Cloth modifier's **Object** rollout. 3ds Max calculates the simulation, and updates the viewport frame by frame. Depending on the speed of your computer, it can take several minutes to calculate the 100 frame animation.

    When the simulation is completed, play the animation in real time using the Play Animation button in the 3ds Max animation controls, or by using the hotkey, which is **</>** (forward slash). Notice how the **Fabric** slides across the floor in an unrealistic manner.

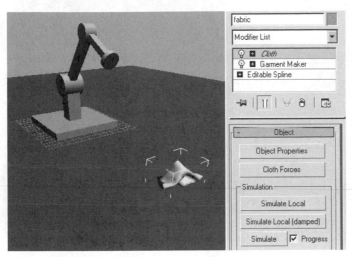

Figure 10-29: **Fabric** *slides across the floor*

8. You might think that increasing the Friction properties of the Fabric or the plane object would help the sliding problem, but in fact the problem is one of accuracy. In the Simulation Parameters rollout of the Cloth modifier, increase the **Subsample** parameter to **5**. This causes 3ds Max to calculate the solution for the cloth deformation five times per frame, resulting in a much more accurate result.

9. Click the **Erase Simulation** button in the Object rollout, and then click **Simulate** again. The new simulation takes five times as long to calculate. This time, the **Fabric** does not slide across the floor, but stays put at the base of the robot arm.

10. Now we'll add a gust of wind to the simulation. Click the Erase Simulation button again. Right-click in the viewport and choose Unhide All from the Quad Menu. Two objects appear in the scene: a Wind Space Warp and a Spray particle system. The Spray is only in the scene so that we can more easily see the effects of the Wind.

    Play the animation. The **Spray01** particles are blown by the Wind.

11. Select the **Wind01** object. Change its **Turbulence** parameter to **10** and play the animation. The particles are blown around more chaotically.

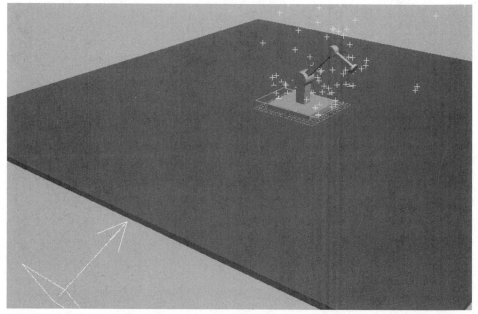

*Figure 10-30: Wind **Turbulence***

12. Now we'll animate the wind to provide a gust of wind at a specific time. Activate Auto Key. From frame **0** to **60**, set the **Wind01** object's **Strength** parameter to **0**. To do this, go to frame **0** and enter a **0** in the Strength field. Then go to frame 60 and click the Strength spinner up once, then down once, to set a keyframe for that value. This is a little tricky. Play back the animation to make certain that the Wind Strength remains at a constant value of **0** from frame **0** to **60**.

    Now go to frame **65** and set the Strength parameter to **10**. Disable Auto Key and play back the animation. The particles drift randomly until frame 60, and then they are blown away by a sudden gust of wind.

13. Select the **Fabric** object. In the Object rollout, click the **Cloth Forces** button. The Forces dialog opens. On the left side is the Forces in Scene column. Select **Wind01** and click the right-facing arrow to send **Wind01** into the Forces in Simulation column. Click OK to accept the changes and close the dialog.

14. Click the **Simulate** button again. This time the **Fabric** falls to the floor, and is then blown away by the wind.

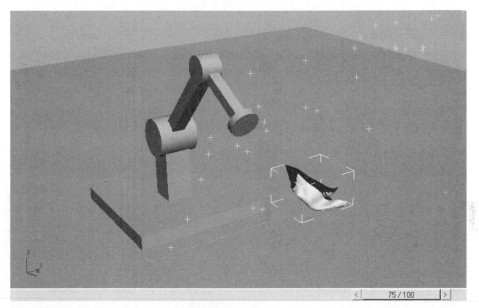

Figure 10-31: **Fabric** *is blown away by the wind*

15. Save the scene as **ClothWind.max** in your folder.

HAIR AND FUR

The last form of dynamic simulation we'll look at is **Hair and Fur**. As mentioned in Chapter 9, *Advanced Modeling*, it's not practical to manually model thousands of strands of hair. 3ds Max provides a relatively simple and intuitive way to add hair and fur to your characters. This system can also be used to create grass. With the **Instance Node** feature, you can even use a source object to generate complex geometries. For example, by assigning a flower model as an Instance Node, you can grow flowers instead of simple strands of hair.

The 3ds Max Hair and Fur interface consists of a modifier and a Render Effect. Hair properties such as color and length are defined with the **Hair and Fur** modifier. The modifier also includes a sophisticated yet simple dialog called **Style Hair**, in which you can comb, brush, and cut the hair. The thousands of individual strands are drawn at render time and composited with the standard scene geometry. Because of this, you can't see the hair rendered in the viewports; you can only see a preview of a few hairs displayed as wires.

The Hair and Fur modifier can be applied to a mesh object, or to an Editable Spline with multiple Spline sub-objects. If applied to an Editable Spline, the Hair and Fur modifier generates strands of hair between the Spline sub-objects. If applied to a mesh object, strands are created over the entire object, or over desired Faces or Polygons selected in the Hair and Fur modifier.

Many of the parameters in the Hair and Fur modifier can be mapped. For example, you can use a grayscale map to define the length of hair. Darker values in the map image result in shorter hair on the model. If you use the **VertexPaint** modifier, you can paint directly on the model to define areas of differing color, length, density, and so on.

EXERCISE 10.9: Hair

In this exercise, we will explore the basic features of the 3ds Max Hair and Fur system.

1. Reset 3ds Max. Create a Sphere with a Radius of about **30** units.

2. Assign the **Hair and Fur (WSM)** modifier to the sphere. Red wires appear growing out of the sphere.

3. Render the scene from the Perspective viewport. The ball appears covered in brown fur.

4. Open the **Dynamics** rollout of the Hair and Fur modifier. In the Mode section, activate the **Live** option. Use the Select and Move tool to move the sphere around in the viewport. The hair wires flow in response to the movement.

5. Press the Play Animation button and let the animation play through for a few seconds. The hair wires droop under the inflence of gravity.

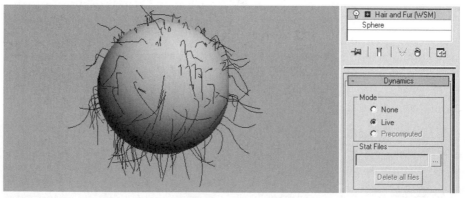

*Figure 10-32: **Live** Dynamics in the Hair and Fur modifier*

6. Note how the hair drifts and moves too much. Increase the **Dampen** parameter to **0.1**. Play the animation again. The hair settles down and doesn't drift.

7. Enter **Polygon** sub-object mode within the Hair and Fur modifier. Make a selection of polygons that roughly corresponds to where hair would grow on a human head. Click **Update Selection**. Now the hair only grows from the selected polygons.

Figure 10-33: Hair grows from selected polygons only

8. Exit sub-object mode. In the **Tools** rollout, click **Regrow Hair**, and then play the animation. The hair settles down in its new configuration.

9. In the Tools rollout, click **Load** in the **Presets** section. In the **Hair and Fur Presets** dialog, double-click **fuzzybrown.shp**. Many parameters in the modifier are now changed. Render the Perspective view.

10. Create a Target Spot light and shine it at the sphere. Enable shadows, using the default Shadow Map parameters. Render the Perspective view again.

Figure 10-34: Hair rendered with a Spot light

11. Save the scene as **Hair.max** in your folder.

ANIMATION MODIFIERS

Several modifiers are designed for animating soft, organic objects such as characters.

SKIN

The **Skin** modifier associates a mesh with a skeletal structure, so the skeleton can control the organic deformation of the mesh. The skeleton can be built from simple objects such as boxes and splines, or it can be built with specialized **Bones** objects. Bones objects have numerous advantages over boxes. Either way, the control objects that make up the skeleton are categorically referred to as *bones*.

The skeleton can be arranged in any configuration that suits the mesh. Bones are usually linked together so they can be easily animated. Since the mesh is controlled by the transforms of the bones, you don't need to parent the mesh to the bones.

Apply the Skin modifier to the mesh, not the bones. On the Modify panel, click the **Add** button to choose the bone objects from a list. This associates the selected bones with the mesh.

When a mesh is animated with Skin and bones, you animate the bones, not the mesh. In a Track View, you should see keyframes for one or more bones, but no keyframes for the mesh. To prevent accidentally animating the mesh, it's a good idea to freeze it.

In 3ds Max, there are several ways ways to control how a particular bone will influence the mesh. This is called *vertex assignment*, because each vertex in the mesh is assigned to be influenced by one or more bones. The easiest way to assign vertices to bones is with bone **Envelopes**, which are volumes of influence around the bones. If a mesh vertex falls within a particular bone's envelope, then that vertex is influenced by that bone. Another way to assign vertices is to use the **Paint Weights** tool. This is done with the same paint interface that is used for Paint Deformation and Paint Soft Selection. Finally, you can adjust vertex weights directly by selecting them and entering a value from **0** to **1** in the **Abs. Effect** spinner field.

EXERCISE 10.10: Mr. Puppet Comes Alive

In this exercise, you'll animate a puppet with a simple Skin and Bones setup.

Create Bones

1.   Load the file **Puppet.max** from the disc.

2.   Press the **<S>** key to activate **3D Snap**. Maximize the Front viewport.

3.   Go to the **Create** panel and open the **Systems** panel. Click the button labeled Bones.

4. Now you are going to create a series of Bones objects for the skeleton of Mr. Puppet. Click at the base of Mr. Puppet's body, making sure that the cursor snaps to the origin. Click again at the midpoint of Mr. Puppet's body, then at his neck, and finally at the tip of his hat. Right-click to finish creating this particular system of Bones. An extra Bone is created above his head; this is normal. Since Bones don't normally render, it's OK to leave this one where it is.

Right-click again to exit Bone creation mode.

You should now have a series of three bones inside Mr. Puppet, plus the extra Bone created over the top of his head. All of the bones should be precisely aligned with the origin, and in a perfectly straight line. See the following illustration.

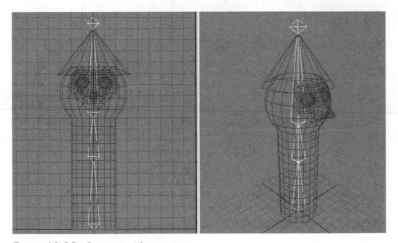

Figure 10-35: Bones inside puppet

The Bones have been parented to one another automatically. The first Bone you create is the parent of the second Bone, and so on. There is no need to use the Select and Link tool.

After creating Bones, it's usually necessary to adjust their positions. This is so that the Pivot Points of the Bones will be placed correctly.

Do not move, rotate, or scale the Bones. Don't use the tools in the Hierarchy panel to adjust the Pivot Points either. Instead, use Bone Edit Mode.

5. Disable 3D Snap. Go to the Main Menu and choose **Character > Bone Tools**. The Bone Tools dialog appears. Click the **Bone Edit Mode** button.

Now you may edit the position of the Bone objects. Move the Bones up or down in the world Z axis to adjust them. In Bone Edit Mode, moving a Bone causes the parent Bone to change length.

Move the second Bone upward in the Z axis to make Mr. Puppet's neck Bone shorter. When you are finished, turn off Bone Edit Mode and close the Bone Tools panel.

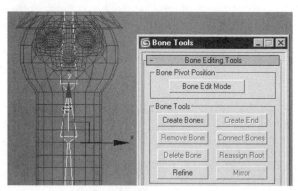

*Figure 10-36: Adjusting the initial position of Bone objects in **Bone Edit Mode***

Add Bones to the Skin

1. Select the puppet mesh. Apply the **Skin** modifier to it.

2. On the **Parameters** rollout, click the **Add** button. The **Select Bones** dialog appears. Select the first three Bone objects and click the Select button. The selected Bones appear inside the list on the Parameters rollout.

   You don't need to select the last Bone, which is Bone04, at the top of the hat. This Bone may be needed for a type of animation called **Inverse Kinematics**, but it is not needed to deform the mesh. Therefore, don't add it to the Skin modifier list.

3. Select the mesh object. In the Modify panel, select **Bone01** from the list and click the **Edit Envelopes** button.

   Two capsule-shaped envelopes appear in the viewport, one inside the other. Each capsule has two **Cross Sections**, with four control points on each Cross Section. The control points are indicated by square boxes.

4. Select one of the control points on the outer envelope, and use the Move tool to drag it until the envelope encloses the mesh.

   Select a control point on the other Cross Section and move it outward in the same way. All of the vertices in the neighbohood of **Bone01** should be highlighted in red.

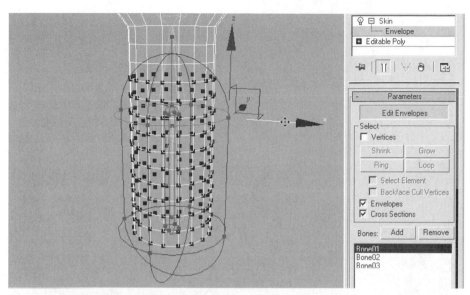

Figure 10-37: Expand the Envelope for Bone01 until it entirely encloses the mesh

5. Expand the envelopes for **Bone02** and **Bone03** in the same way. Select the Bone from the list on the Modify panel, then click one of the control points on a Cross Section, and drag it outward. Then do the same for the second Cross Section of that Bone.

    Notice that where envelopes overlap, the vertex colors change. This is simlar to Soft Selection: red vertices are most influenced by the selected Bone. Orange vertices are influenced less, and blue vertices are influenced very little.

    You can also edit the position of the entire envelope, to change the volume of influence. By default, the envelope is centered on the Bone, but that might not give the effect you need. If you zoom in very closely to the center of an envelope, you'll see a yellow line. That is the center of the envelope. At either end of the yellow line are two control points. You can move these to place the envelope wherever you need it, no matter where the Bone is.

6. Select **Bone03** from the list. In the Left viewport, zoom in very close to the Bone. Look for the yellow line. If you can't see it, try looking for it in the Perspective view. Select the control point at the bottom of the yellow line, and move it upward. This reduces the influence **Bone03** has in the neck area, allowing the neck bone more control and giving a sharper transition in this area when Mr. Puppet is animated.

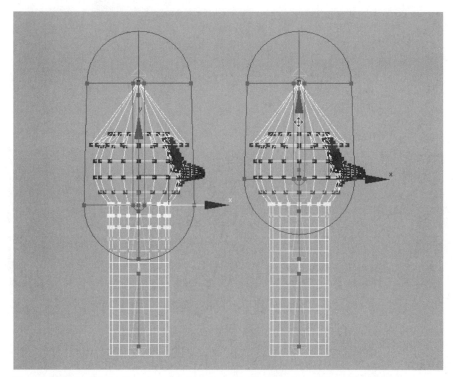

Figure 10-38: Move the control point at the bottom of the yellow centerline

11. Exit Edit Envelopes mode. Select all of the objects in the scene, and choose **Character > Set Skin Pose** from the Main Menu. Click OK in the subsequent pop-up dialog to confirm the Skin Pose. Now you can always return Mr. Puppet's bones to their original setup.

12. Test your setup by rotating the Bones in the viewport. If any vertices are left behind and the mesh stretches, or you don't like the way Mr. Puppet bends, undo the rotations and go back into Edit Envelopes mode. You can also choose **Character > Assume Skin Pose** to undo all rotations. Adjust the Cross Section radii or envelope positions as needed, and test the results.

12. When you are satisfied that Mr. Puppet's skin is working correctly, freeze the mesh. Select the mesh, and go to the **Display** panel. Expand the **Freeze** rollout, and click **Freeze Selected**. Now you can't select the mesh at all, and can work with the Bones more easily.

13. Create a simple animation by rotating the Bones in the Local coordinate system.

14. Save the scene with the filename **PuppetSkin.max**.

Tips for Working with Skin

If you find that a character mesh doesn't have enough segments to deform smoothly, you can add a TurboSmooth modifier to all or part of the mesh above the Skin modifier. It's OK to add a subdivision surface modifier, such as TurboSmooth, above the Skin, but don't add or remove vertices below the Skin modifier in the stack! If you do, any direct or painted vertex weighting will be scrambled because the vertex ID numbers have changed.

A single bone object can control more than one mesh. In this way, you can animate segmented characters with Skin and bones. Model your character in several pieces, and apply the Skin modifier to each piece. Add the relevant bones to each Skin modifier, and animate the bones normally. Knowing that you can animate in this way may make modeling easier if you aren't exporting to a game engine that requires single-mesh characters.

Bilaterally symmetrical characters, such as human figures, have identical skeletons on either side of their bodies. During character setup, a great deal of time can be saved by using mirroring. Create Bones and other control objects for the character's spine and just one arm and one leg. Use the **Mirror** button on the Bone Tools dialog to create the Bones for the other arm and leg. Of course, this only works for Bones System objects, and not for non-Bones objects. For Helpers or other objects, use the Mirror tool on the Main toolbar. Since the Mirror tool on the Main Toolbar tool works by inverting scale values along one axis, you must also use the **Reset: Scale** command on the Hierarchy panel.

After you've set the Skin vertex weights for one side of the figure, use the tools in the **Mirror Parameters** rollout of the Skin modifier to copy the vertex weighting to the other side of the body. Vertex weighting is a time-consuming process, and this procedure is very valuable.

MORPHER

Morphing is a term used to describe the seamless transition of one image or shape into another. The **Morpher** modifier allows you to morph all or part of one object to another. The objects to which an object is morphed are called *morph targets*. By animating the strengths of various morph targets, you can blend many different objects together. This gives the illusion that there is one object that is changing shape in complex ways.

Morphing is perfect for facial animation. Create several variations of a face or head model, with different expressions, and use the Morpher modifier to animate the face changing expression.

All of the morph targets must have the same number of vertices. So, you should model one object, such as a face, and then make copies of it. Then alter the vertex positions of the copies to create the various expressions. Do not add or remove vertices!

Apply the **Morpher** modifier to the face with the most neutral expression. On the Modify panel, click **Load Multiple Targets** and choose the morph targets from the list. Once you've added all of the targets, you can hide them so that only the morphing face is visible in the scene.

Each morph target appears on the **Channel List** rollout. Move the time slider to a frame other than **0** and turn on the **Auto Key** button. Change the percentages to the right of one or more morph targets to morph the object.

EXERCISE 10.11: Changing a Mood with Morpher

In this exercise, you'll use the **Morpher** modifier to animate facial expressions.

1. Load the file **FacialExpressions.max** from the disc.

    This file contains a girl's head, plus four versions of her face with different expressions. Each head was created as a variation on the first head, by moving vertices. All the heads have the same number of vertices.

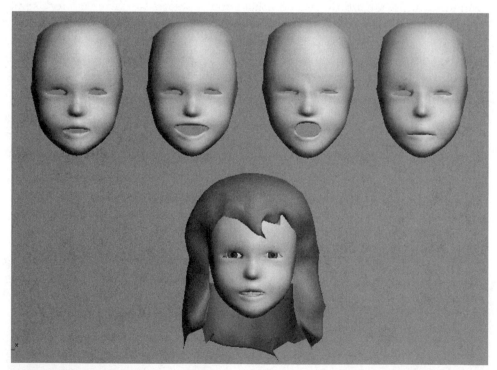

Figure 10:39: Morph targets

2. Select each of the face objects and note its name. Each one is named for the facial expression it represents.

3.   Select the object **Face – Base**, which is the face that has eyes, teeth, and hair.

4.   Go to the **Modify** panel. Apply the **Morpher** modifier to **Face – Base**.

5.   At the bottom of the **Channel List** rollout, click the **Load Multiple Targets** button. All objects in the scene with the same number of vertices as **Face – Base** appear on the list. Choose all the objects. The morph targets appear on the **Channel List** rollout.

Figure 10-40: Channel List rollout with morph target names

The number to the right of each morph target name indicates the percentage of the object that is used for the current facial expression. Right now they are all set to **0**, so there is no change to the **Face – Base** object.

6.   Move the time slider to frame **20**, and turn on the **Auto Key** button.

7.   Next to the object name **Face – Happy** on the **Channel List** rollout, change the percentage number to **100**. The face changes to a look like the **Face – Happy** morph target.

8.   Move the time slider to frame **40**. Change the percentage next to **Face – Happy** to **0**, and change the percentage next to **Face – Angry** to **100**. The face changes to look like the **Face – Angry** morph target.

You can quickly change a parameter to its lowest possible value by right-clicking on either one of the spinner arrows. You can use this method to change a percentage back to **0** after it has been increased to a higher number.

9. Move the time slider back to frame **20**.

   Note that the percentage for **Face – Angry** is **50**. This is because the last key for this morph target is at frame **0**, so the percentage goes gradually from **0** to **100** from frames **0** to **40**.

10. Change the percentage for **Face – Angry** to **0**.

    Now you're ready to play the animation and see what the face is doing so far.

11. Play the animation. The face smiles, then appears angry.

12. Continue to animate the facial expression on various frames by changing the percentages on the Channel List rollout.

13. Save the scene with the filename **FacialMorphing.max** in your folder.

USING MORPHER AND SKIN TOGETHER

When animating a character, you'll probably want to use morphing for the facial expressions, and skeletal animation for the rest of the body. This is easy to set up if the head and body are different objects. Just model the head and neck as one object, and model the shirt or jacket as a separate object. Apply the Morpher modifier to the head, and apply Skin to the rest of the body.

However, if you're working with single-mesh characters (such as creatures who don't wear clothes), the problem becomes more challenging. If you need to combine Morpher and Skin on a single-mesh character, follow this procedure.

1. Model the character as a single polygonal mesh object. Then cut off the head. Use the **Detach** command to separate the head from its body. Make sure it's a clean break.

2. Make copies of the head and create morph targets for the facial expressions. Leave the original head where it is, placed as a separate object on top of the body.

3. Apply the **Morpher** modifier to the head, and assign the copies as morph targets.

4. Apply the **Edit Mesh** modifier to the head, above the Morpher in the stack. Use the **Attach** command to re-attach the body. **Weld** the seam between the head and body.

5. Add the **Skin** modifier above the Edit Mesh. Add bones and adjust envelopes.

6. If desired, place a **TurboSmooth** modifier at the top of the stack.

INVERSE KINEMATICS

The usual way that linked objects work together is called *forward kinematics* (FK). In forward kinematics, a child object always inherits the transforms of the parent. *Inverse kinematics* (IK) is an alternate method of interpreting the parent-child relationship between linked objects. With IK, when you move a child, the parent objects rotate accordingly. It's the opposite of FK: the parent inherits the transforms of the child.

IK makes character animation much easier, because you only have to move a hand or foot to its destination, and the arm or leg joints will follow automatically. Animating a simple motion such as picking up a drinking glass can be quite difficult with FK, but with IK, it's very easy.

You can use IK and Bones together to animate organic characters. However, IK will work with any linked objects, and you don't need to use Bones to work with IK.

IK SOLVERS

IK differs from FK, because IK has multiple possible solutions for any movement or rotation of the child object, while FK only has one. The determination of how the parents should be transformed is a job done by a program module called an **IK Solver**.

3ds Max has several IK solvers, found in the Animation menu. Each solver has different characteristics. The most versatile one is called the **History-Independent Solver**, or **HI Solver** for short.

The **IK Limb Solver** is a variation on the HI Solver. As the name indicates, the IK Limb Solver is designed for animating character limbs such as arms and legs. The difference between it and the HI Solver is that the IK Limb Solver that can only have three objects in the chain, whereas the HI Solver can have any number of joints. The IK Limb Solver was created so that it may be customized by game programmers. You should use the HI Solver instead.

The **SplineIK Solver** is primarily used for animating objects which are long and flexible, such as snake bodies and animal tails. It uses a spline object and several control objects to influence the rotations of a series of bones.

The **History-Dependent** or **HD Solver** is the form of IK used in earlier versions of 3ds Max. It is primarily included in the current version for compatibility with old scene files. In most cases, you should not use the HD Solver, and this book does not discuss it.

History-Independent IK

To use the HI Solver, you must have at least three objects linked to one another in a hierarchy. When you assign an HI Solver to a series of linked objects, an **IK Chain** is created. An IK Chain is indicated by a line that connects the first parent in the chain to the last descendent in the chain.

To assign an HI Solver, select a parent object. Then choose **Animation > IK Solvers > HI Solver** from the Main Menu. Finally, click the descendent object to create the IK Chain. The descendent can be any number of steps away from the parent in the hierarchy.

The HI Solver is designed so that objects in the chain can only rotate in one direction. This means that before you assign an IK solver, you need to rotate the objects a little bit, to give the IK solver a "hint." That way, it will "know" which direction you want a knee to bend, for example.

When using the HI Solver, you will also see a cross-shaped point helper at the end of the chain. This is called the **Goal**. To animate the entire chain of linked objects, simply animate the Goal.

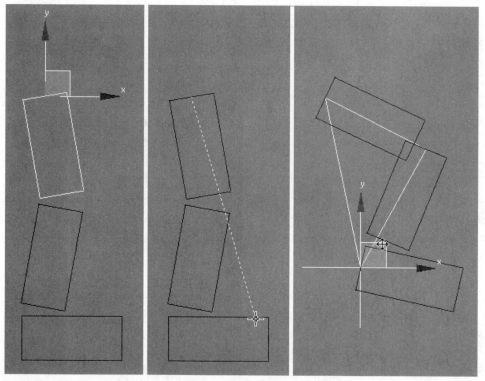

Figure 10-41: Rotate objects, assign the IK solver, and move the Goal

Just as with forward kinematics, you must take care to set up your objects' Pivot Points very carefully before attempting to animate them. When you create an IK Chain, make sure to test it thoroughly before trying to use it for animation. Move the Goal all around and observe how the IK Chain reacts. If there are problems, it's best to simply delete the IK Chain, fix the problems, and then create a new IK Chain.

When you move the Goal, all of the objects included in the IK Chain will rotate to follow the Goal. However, you can't make a joint bend backward, away from its initial rotation. This is perfect for animal limb joints, which can only bend in one direction.

If you need to rotate an entire HI IK Chain, you can use the **Swivel Angle**. This will simulate motions such as that of a hip joint. To use the Swivel Angle, turn on the **Select and Manipulate** button . A Swivel Angle Manipulator appears at the start of the IK Chain. Move the Manipulator to adjust the rotation of the entire chain.

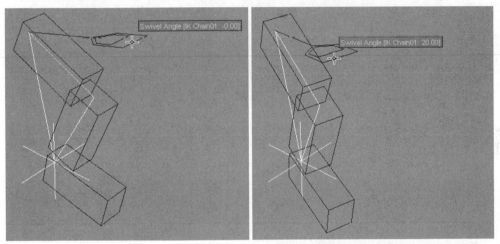

*Figure 10-42: Adjust the **Swivel Angle Manipulator** to rotate the IK chain*

You can also adjust the Swivel Angle, and other parameters, by selecting the IK Chain and opening the **Motion** panel. One useful technique is to assign a **Target** for the Swivel Angle. If the Swivel Angle has a Target, the chain will always swivel to follow the Target. It is often easier to move a Target, such as a **Point** Helper, than it is to move the Swivel Angle Manipulator.

NOTE: A **Point** is a general-purpose Helper object that does not render, and it is very useful in many situations. In some programs, it is called a *locator*, a *null*, or a *dummy* object. 3ds Max also has a Helper called **Dummy**. The Point Helper is an improved version of the Dummy that gives you better control over the display of the object.

EXERCISE 10.12: History-Independent IK

In this exercise, you'll set up a simple IK system to animate the legs of a character built of boxes.

1. Open the file **BoxLegs.max** from the disc that comes with this book.

2. Begin by parenting the boxes to one another with the **Select and Link** tool. Parent **FootLeft** to **CalfLeft**, parent **CalfLeft** to **ThighLeft**, and parent **ThighLeft** to **Torso**. Repeat this process for the right leg.

3. To assign the first IK chain, select the **ThighRight** object. Then choose **Animation > IK Solvers > HI Solver** from the Main Menu. Click the **FootRight** object to create the chain. Test the IK by moving the **Goal** of the newly created chain.

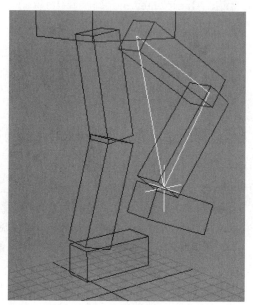

*Figure 10-43: Move the **Goal** of the new IK Chain*

4. Undo the movement of the Goal. Create a second chain between the **ThighLeft** object and the **FootLeft** object. Test it by moving the Goal.

5. Move the **Torso** object. Notice that the legs stretch out, because the Goals have been left behind. The IK chains themselves are not parented to anything in the scene.

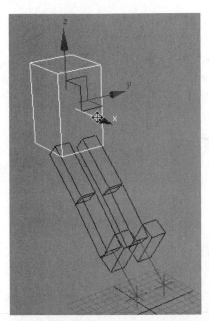

Figure 10-44: Moving the **Torso** *object*

6.   Undo the movement of the **Torso** object. Turn on the **Select and Manipulate** tool. The **Swivel Angle Manipulators** are now visible. Adjust one of them to pose the figure in a more natural position.

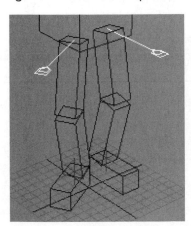

Figure 10-45: Adjusting the ***Swivel Angle Manipulator***

7.   Undo the Swivel Angle adjustment. Go to the **Display** panel and click **Unhide All**. Two Point Helper objects appear in front of the figure. These will be Targets for the Swivel Angles.

8.   Select **IK Chain01**, which is on the right side of the figure. Rename this object to **IK Chain LegRight**. Rename **IK Chain02** to **IK Chain LegLeft**.

9.   With **IK Chain LegRight** selected, open the **Motion** panel. Scroll down to the **IK Solver Properties** rollout. In the **IK Solver Plane** section, click the **Pick Target** button labeled **None**. In the viewport, click the Point Helper labeled **KneeSwivelRight**.

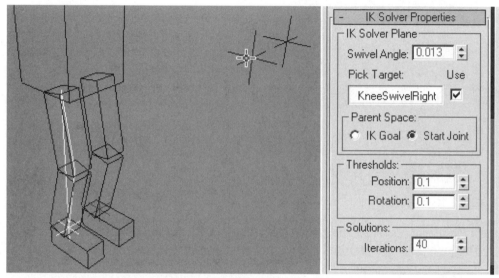

*Figure 10-46: Pick the KneeSwivelRight Point Helper as the **Target** of the Limb Solver*

10. Test the swivel target by moving the Point Helper.

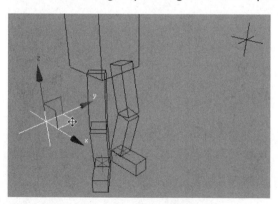

Figure 10-47: Move the swivel target

11. Undo the movement of the swivel target. Select the other IK chain, **IK Chain LegLeft**, and assign the **KneeSwivelLeft** Point Helper as its swivel target.

12. Sometimes it is difficult to select the IK Goals. Select each Goal in turn and increase its **Size** on the **IK Display Options** rollout.

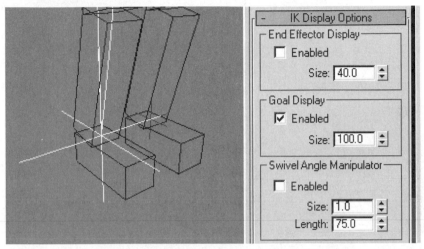

Figure 10-48: Increase the **Size** of the **Goal Display**

13. Finally, parent the swivel targets to the **Torso**. That way, when you move the **Torso**, the Swivel Angle won't change.

14. Create a simple animation of the torso and legs jumping, walking, or shifting position. Save the scene as **BoxLegs_IK.max** in your folder.

Hints

- Animate the torso, the Goals, and perhaps the swivel targets. If you're animating a walk, it's easier to use the Front viewport in this case.

- If you're making a repetitive motion, such as a walk cycle, remember that you only need to animate one cycle. You can repeat the cycle with the **Parameter Curve Out-of-Range Types** options in Track View. **Relative Repeat** works well for IK walk cycles.

- You don't need to animate the legs directly. They will probably only get in the way, so it's a good idea to **Freeze** the legs when you're trying to select the IK Goals.

- It helps to turn on **Trajectories** and select various objects, such as the Torso. Even if an object is not directly animated, its Trajectory will be displayed in the viewport. This will greatly help in visualizing smooth motion.

CHARACTER ANIMATION

When you plan to work a great deal with a character, you can make animation easier by setting up a character rig. A *character rig* is a collection of Bones, IK Chains, Helper objects, and other tools to help automate the animation process. A character rig helps minimize the time an artist must spend on technical considerations, allowing him or her to focus on creativity.

A character rig can be quite simple, consisting of a just a few controls. When a character will be animated a great deal, as with a weekly animated series or a feature film, a more complex rig is used.

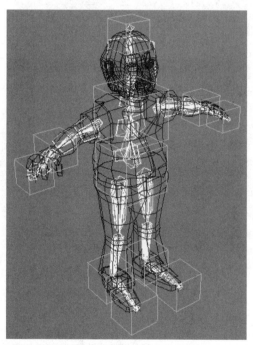

Figure 10-49: A simple character rig

If you are interested in learning more about character rigging, you will find a supplemental exercise on the disc. Look in the folder for Chapter 10, and you will find a filenamed **IK-Rig.PDF.** This is an Adobe Acrobat file. To open it, you need the Acrobat Reader software, which is available from the Adobe website: **www.adobe.com**.

This exercise takes you through the process of building a simple character rig with History-Independent IK and Helper objects. This is essentially the same rig shown in the accompanying illustration. You will use this character setup in Exercise 10.13: *Walk Cycle with a Character Rig.*

POSE-TO-POSE ANIMATION

Many character animators use a technique called *pose-to-pose animation*. With this method, you create poses of the entire character body. For each pose, keyframes are created for every single animatable object in the rig. This technique prevents unexpected keyframe interpolation, and therefore keeps body parts from moving out of control. It also lets animators try out various character poses before committing to them.

SET KEY MODE

Set Key mode is necessary because it is very difficult to use Auto Key mode for pose-to-pose animation. Set Key mode provides a quick way to create keys for a specific set of objects and/or animation tracks on the current frame. With one mouse click, you can create keyframes for multiple objects, such as Bones and Helpers.

To use Set Key mode in its simplest form, activate it by pressing the **Toggle Set Key Mode** button on the Status Bar. Go to a frame in time and transform objects in the viewport. When the pose is correct, click the **Set Keys** button, which is the big button that has a picture of a door key on it. By default, keyframes are created for all objects that are currently selected in the viewport.

Figure 10-50: **Toggle Set Key Mode**

If you don't press the big Set Keys button, no keyframes are created. This lets you experiment with different poses without disturbing poses at other points in time.

Set Key Selections

One great advantage of Set Key mode lies in the ability to create keyframes for Named Selection Sets of objects, even if they aren't selected in the scene. That way, you can set a keyframe for every Bone or Helper object of a character rig with a single mouse click. This is critical to effective pose-to-pose animation.

To do this, create a Named Selection Set containing the objects you wish to keyframe. Then select the Selection Set from the **Set Key Selection List** pulldown. When you click the big Set Keys button, 3ds Max creates keyframes on all objects in the currently active Set Key Selection List, even if the objects aren't selected in the viewport.

Figure 10-51: **Set Key Selection List**

Set Key Filters

By default, 3ds Max creates keyframes on all transform tracks for objects in the current Set Key Selection. Often, this is not necessary. For example, it is common to create Position keyframes for Helper objects in a character rig, but usually you don't need to create Scale keys.

To designate which tracks you wish to animate with Set Key mode, click the **Key Filters** button. The **Set Key Filters** pop-up dialog appears. Check the boxes for all categories of animation tracks that you wish to keyframe with Set Key mode. If a check box is off, no keys will be set.

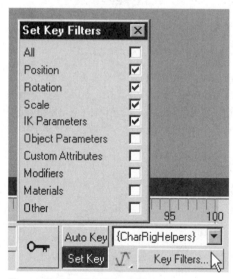

Figure 10-52: **Set Key Filters**

Keyable Icons in Track View

In addition to controlling keyframe creation by animation category with Set Key Filters, you can also designate individual animation tracks as keyable or not keyable. For example, you may wish to set keys for the **X Position** track, but not the **Y Position** or **Z Position** tracks. This is accomplished in the Track View.

On the Track View toolbar in Curve Editor or Dope Sheet, there is a button labeled

Show Keyable Icons [icon]. When this button is active, you will see door key icons next to each track name in the Controller Pane of Track View. If a Keyable icon is red, then that track is keyable. If a Keyable icon is gray with a slash through it, then that track is not keyable. Click a Keyable icon to toggle it between keyable or not keyable. By default, all tracks are keyable.

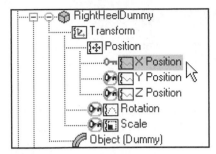

Figure 10-53: **X Position** *is keyable, other tracks are not keyable*

If a track is not keyable in Track View, it can't be altered in the scene. For example, if X Position is the only track made keyable, you can only move the object in the X axis, regardless of whether you are in Set Key mode, Auto Key mode, or no key mode at all. In this way, Keyable icons are similar to Transform Locks on the Hierarchy panel, except that Keyable icons exist for all scene parameters, including modifiers and materials.

The Show Keyable Icons button is merely a display option. If you turn it off, the icons are hidden from view, but the choices you made are still active.

The effects of Keyable icons in Track View and the Set Key Filters are cumulative. This means that in order for a track to be affected by the Set Key button, the animation category must be enabled in the Set Key Filters dialog, and the track must also be designated Keyable in Track View.

Auto Key Mode

In general, it is not a good idea to switch between Set Key and Auto Key modes. Choose one mode and stick with it. Set Key mode was specifically designed for character rigs, and it gives you far more control, and is easier to use, than Auto Key mode.

Although Set Key Filters don't work in Auto Key mode, Keyable icons in Track View do work in Auto Key mode.

EXERCISE 10.13: Walk Cycle with a Character Rig

In character animation, one of the most common needs is a walk cycle. A walk cycle consists of a step for each foot. One foot comes up, then steps in front of the body. Then the other foot comes up and steps in front.

TIP: If you've never animated a walk cycle before, it can be helpful to spend a few moments walking around and noting how your legs and feet move. A full-length mirror is very helpful. You can also have a friend walk slowly around the room while you study his motions, or study video of someone walking.

In this exercise, you'll create a very simple walk cycle. Although you could create a walk cycle by animating a character's bones directly, the task is much easier if you set up a character rig.

You will load a file that has already been set up with a simple character rig and animate it with the pose-to-pose technique. If you want to know how this rig was created, see the file **IK-Rig.PDF** on the disc.

At the end of each section of the exercise, you are instructed to save a scene file. These scene files are also found on the disc.

Test the Character Rig

1. Load the file **BoboWalkStart.max** from the disc.

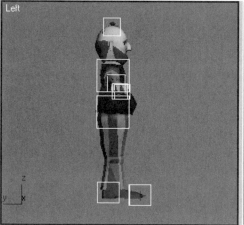
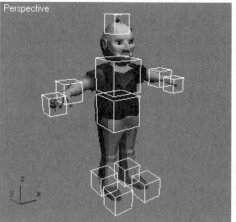

Figure 10-54: Bobo character model and rig

This file contains a character named Bobo. Bobo has been set up with Bones, IK Chains, and Dummy objects for controlling the character.

To animate this character, you need only animate the Dummy objects. For this reason, all objects except the Dummy objects have been frozen. You can't select frozen objects.

2. To test the setup, move any Dummy object. Parts of the body will move when you move the Dummy. Be sure to Undo any movements you make.

Prepare to Use Set Key Mode

With this particular character rig, you'll only animate the positions of the Dummy objects with Set Key mode. Using Key Filters, you can limit the creation of keyframes, so that only Position transform keys will be set.

1. Click **Key Filters**. Turn off all options except **Position**. Close the Set Key Filters dialog.

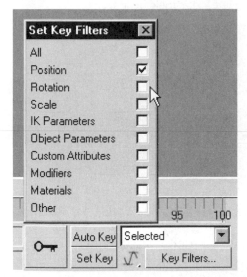

Figure 10-55: **Set Key Filters** *with all categories except Position disabled*

Next, you'll need two Named Selection Sets to help you set keys.

2. Select all of the leg and foot Dummy objects, and select the **HipDummy** object.

3. Create a **Named Selection Set** called **Leg Dummies**.

4. Select the shoulder, arm, and hand Dummy objects, and create a Named Selection Set called **Arm Dummies**.

NOTE: Some animators prefer to create a complete pose for the entire character at a specific keyframe before proceeding to another keyframe. In such a case, it would be appropriate to create a single Named Selection Set for all of the character's control objects. However, in the present exercise, you create two separate Named Selection Sets, because you are animating the arms and legs in separate stages of the exercise. This makes the animation process a little easier for artists who have never created a walk cycle before.

5. In the **Set Key Selection List** pulldown above the Key Filters button, choose the **Leg Dummies** Selection Set.

6. Turn on the **Toggle Set Key Mode** button.

    Now you're ready to animate the model.

Animate the Legs

You'll start the animation process by posing the legs.

1. In the Left viewport, on frame **0**, move the heel Dummy objects to pose the character as shown in the following illustration. The left foot should be in front.

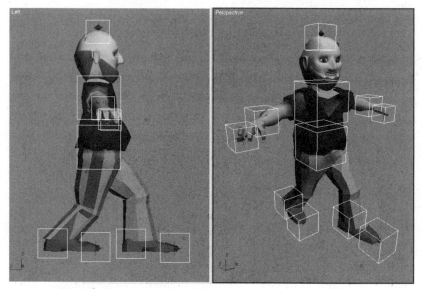

*Figure 10-56: Legs positioned at frame **0***

2. Click **Set Keys**, the big button that looks like a key. Set Key Mode must be active, as shown in the following illustration.

*Figure 10-57: Click **Set Keys** with **Leg Dummies** Selection Set chosen*

Regardless of which objects are currently selected in the scene, this sets a key for all objects in the **Leg Dummies** Selection Set.

3. Go to frame **10**. Move the right foot so it sits near the left knee, and move the right toe down slightly. Move the **HipDummy** upward and forward so the left leg is nearly straight. Don't move the left foot. Click Set Keys.

Figure 10-58: Legs and hips positioned at frame **10**

4. Go to frame **20**. Move the **HipDummy** forward and down, and move the right foot so it steps in front of the left foot. Don't move the left foot. Click Set Keys.

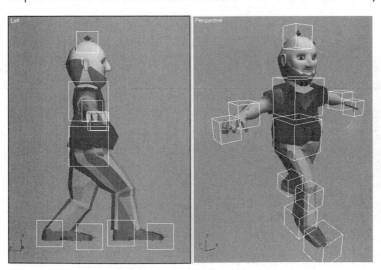

Figure 10-59: Legs and hips at frame **20**

5.   Go to frame **30**. Move the left foot near the right knee, and move the left toe down slightly. Move the **HipDummy** upward so the right leg is nearly straight. Don't move the right foot. Click Set Keys.

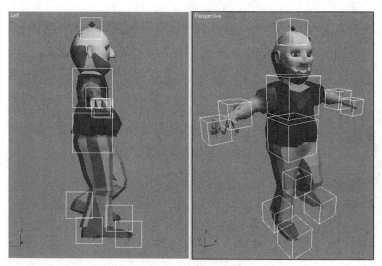

*Figure 10-60: Legs and pelvis at frame **30***

6.   Go to frame **40**. Move the **HipDummy** forward and down, and move the left foot so it steps in front of the right foot. Don't move the right foot. Click Set Keys.

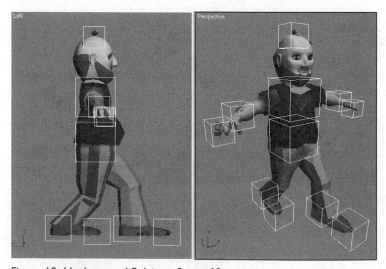

*Figure 10-61: Legs and Pelvis at frame **40***

7.   Scrub the time slider. Bobo should take two steps forward. The pose on frame **40** should look very similar to the pose on frame **0**.

8.   Save your work in the file **BoboWalkLegs.max**.

Figure 10-62: Legs and pelvis animated at frames **0, 10, 20, 30,** *and* **40**

Shift Bobo's Weight

When you practiced walking around, you might have noticed that your hips shift from side to side as you raise each leg. Each time you lift a leg, your weight shifts so it is centered over the standing leg. As you step, your hips shift again so your weight is centered between your legs. You can simulate this effect in a walk cycle by shifting the character's pelvis as he walks.

1. Activate the Perspective viewport. Press the **<F>** key to change the viewport to the Front view. Make sure nothing is selected by clicking in the viewport. Press the **<Z>** key. This **Zoom Extents All** command centers all geometry in all views.

2. At frame **10**, move the **HipDummy** to the right so the character's weight is centered over his left leg. Press the Set Keys button.

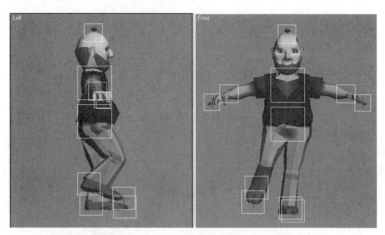

Figure 10-63: **HipDummy** *moved to the right at frame* **10**

3. At frame **30**, move the **HipDummy** to the left to center the character's weight over the standing leg. Press the Set Keys button.

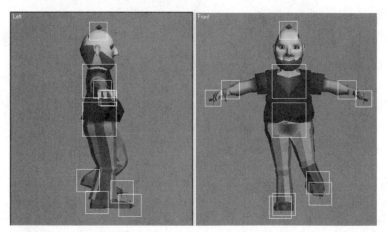

Figure 10-64: **HipDummy** *moved to the left at frame* **30**

4. Save your work as **BoboWalkHips.max**.

Animate the Arms

When you walk, your arms swing opposite your legs. When your left leg moves forward, your right arm swings forward. Then your left arm comes forward as your right leg moves to the front. When you pick up one of your feet, your arms pass by your sides. You'll animate Bobo to move his arms like this.

1. Above the Key Filters button, choose the **Arm Dummies** Named Selection Set.

2. Go to frame **10**. On this frame, Bobo's right foot is passing by his left leg. Here, the arms should pass by his sides.

3. In the Front viewport, move the hand and wrist Dummy objects to pose the hands by Bobo's sides, as shown in the following picture. Click Set Keys.

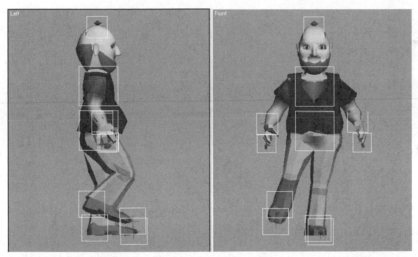

*Figure 10-65: Arms moved down at frame **10***

4. Go to frame **30**, and click Set Keys again.

5. Go to frame **0** and select the right wrist Dummy object. In the Left viewport, move the right wrist Dummy forward, and adjust the finger Dummy so the hand faces forward. Click Set Keys.

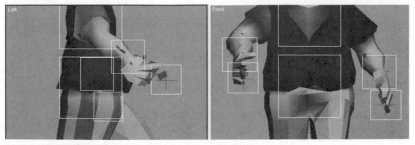

*Figure 10-66: Right wrist and finger Dummies moved forward at frame **0***

WARNING: When you move the wrist so the elbow bends, it can sometimes be tricky to move the finger Dummy so the hand faces the right way. This is one of the limitations of this simple character rig. Just adjust the hand and fingers as best you can for this exercise.

6. Right-click the Left viewport label and select **Views > Right** from the pop-up menu. The viewport changes to the Right view. You are now looking at Bobo from the reverse angle.

    This makes it easier to manipulate Bobo's left arm.

7. In the Right viewport, move the left wrist Dummy toward Bobo's back, as shown in the following illustration. Click Set Keys.

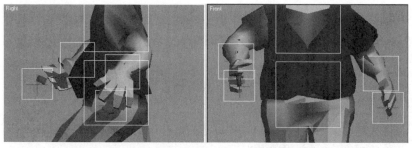

Figure 10-67: Move the left wrist Dummy in the Right viewport

8. Go to frame **20**, and move the left wrist forward and the right wrist back. Use the Right or Left viewports as needed. When you have completed the pose, click Set Keys.

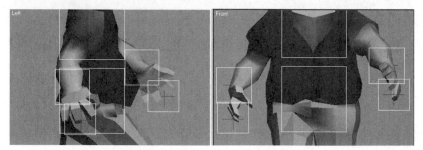

*Figure 10-68: Wrists moved at frame **20***

9. Select all the arm Dummy objects, and copy all keys on frame **0** to frame **40**.

10. Play or scrub the animation, and make sure Bobo's arms swing as he walks.

11. Save your work as **BoboWalkArms.max.**

Loop the Walk Cycle

Now you have one complete walk cycle. To keep the character moving, you'll need to loop the motion of the Dummy objects.

1. Make sure that all of the character's Dummy objects are selected. Open the Curve Editor by choosing **Graph Editors > Track View – Curve Editor** from the Main Menu.

2. On the Curve Editor toolbar, click the **Filters** button [icon]. In the **Show Only** section of the Filters dialog, turn on **Animated Tracks**. All the Position tracks for Dummy objects are now listed on the hierarchy, and the Position tracks are selected.

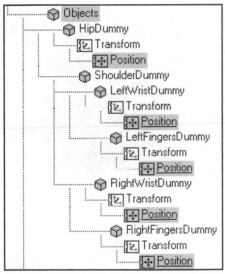

Figure 10-69: Part of the expanded hierarchy in the Controller Pane of the Curve Editor

3. On the Track View toolbar, click the **Parameter Curve Out-of-Range Types** button [icon]. In the **Param Curves Out-of-Range Types** dialog, click the large picture under **Loop**. Click OK to close the dialog.

4. Play the animation.

    The animation loops, but the entire character jumps back to its starting point when it reaches the end of each loop. You can make Bobo continue to walk forward by using Relative Repeat instead of Loop on the heels and pelvis.

5. In the Controller Window , select the Position tracks for the `HipDummy`, `RightHeel Dummy` and `LeftHeel Dummy` objects. Make sure only these three tracks are selected.

6. Click Parameter Curve Out-of-Range Types, and click the large picture under **Relative Repeat**. Click OK to close the dialog.

7. Play the animation. Bobo now walks forward.

8. Save your work as **BoboWalkLoop.max.**

There is much more you can do with a walk cycle. You could make the character's head bob, make him lean forward and back, and make his hips and shoulders turn as he walks. You can learn much more about walk cycles and other standard character motions from a book on traditional or digital character animation.

For a walk cycle with some added body motion, see the file **BoboWalkFinal.max** on the disc.

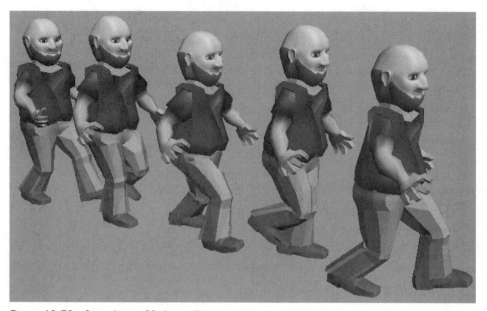

Figure 10-70: Snapshots of Bobo walking

MORE ANIMATION TOOLS

EXPRESSIONS

As we've seen, the Parameter Wiring dialog allows you to use a mathematical formula or *expression* to process the data going from one parameter to another. We can even wire an output parameter to more than one input. With different expressions on the inputs, you can achieve different results. For example, in a previous exercise, the output of the slider was converted to positive and negative values and wired to the two butterfly wings, so they rotate in opposite directions.

However, one limitation of Parameter Wiring is that you can't combine two or more parameter values and wire the result to another parameter. When such control is needed, the Expression Controller comes in quite handy. For example, you could create an expression to find the average position of two objects. This type of expression works well with Inverse Kinematics. You can find the average X and Y position of a character's feet, and apply those averages to the character's hips. Then as you move the feet, the hips automatically follow.

To use an expression, assign an Expression Controller to a track, just as you would any other Controller. Access the track in the Motion panel or Track View, and use the Assign Controller command. For example, to create an expression for the X Position track, choose a **Float Expression** Controller.

Immediately after you make the assignment, the Expression Controller dialog appears. To process existing scene animation data, create variables and assign them to other tracks in the scene. Then write an expression using those variables. The result of the expression is applied to the current track. If you need to access this dialog after closing it, right-click the track and choose **Properties** from the pop-up menu.

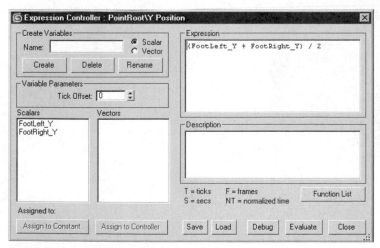

Figure 10-71: **Expression Controller** *dialog*

To begin creating an expression variable, enter a new **Name** in the dialog. There are two kinds of variables: **Scalar** and **Vector**. Scalar variables are used for tracks that have a single value, and Vector variables are needed when a track has more than one value. For example, **X Position** would use a Scalar variable, whereas **Position XYZ** would require a Vector variable. Choose which type of variable you wish to use, and click the **Create** button.

Once the variable has been created, click the **Assign to Controller** button. Browse in the **Track View Pick** window to find the scene track you wish to use. Tracks that don't correspond to the current variable type are grayed out. Repeat this process for all variables you create.

Finally, enter the expression using your new variables into the Expression window on the upper right of the dialog. In the example of finding an average between two variables, we would type **(variable_A + variable_B) / 2**. Close the dialog, or click **Evaluate**, and test the expression in the viewports.

To study an example of this type of expression, open the file **ExpressionWalk.max** on the CD that comes with this book. Notice that the expression is applied to a Point Helper that is the parent of the **Torso**. This makes it easy to move the **Torso** independently, so it's not always exactly at the precise average of the feet's X and Y positions, to animate the character's weight shifting from one leg to the other.

Figure 10-72: Expressions applied to Point Helper

LIMIT CONTROLLER

Sometimes it's useful to limit a transform or other parameter to prevent it from exceeding certain values. The **Limit Controller** is a very easy way to do this. Simply assign the Controller in the usual way, and enter upper and lower limits in the **Float Limit Controller** dialog. You may also specify a **Smoothing Buffer** amount for each limit, which causes the values to ease in and out as they approach the limit, to prevent sudden stops when the limit is reached.

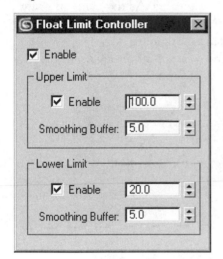

Figure 10-73: **Float Limit Controller** *dialog*

A classic example of a case in which limits are helpful is in the setup of an IK rig for a character's legs. One of the most common problems with such setups is that the IK Goals can go through the floor, causing the knees to snap in an unrealistic manner. Simply assign a Limit Controller to the Z Position track of each IK Goal, or to the Helper object that is the parent of the Goal, and set the limit values. Now the feet can't go through the floor, and the knees don't snap.

You can't assign a Limit Controller to an object that is controlled by an IK Solver. Objects in IK Chains have their own limits built in. These are set in the **IK** subpanel in the Hierarchy panel.

NOTE: Like all transform Controllers, the Limit Controller works in *parent space*. In other words, the limits you define are calculated in relation to the object's parent. If there is no parent, then world coordinates are used.

CHARACTER STUDIO

This book covers the general animation features of 3ds Max, including a little about character setup and animation. Just about any character rig can be built using the tools we've discussed so far, but you should know that there's an entirely different way of animating characters in 3ds Max. This is a set of plugins called **Character Studio**.

Character Studio consists of four components:

Biped is a System for setting up and animating character rigs. As the name indicates, it's best used with humanoid characters, although it can be adapted for four-legged creatures. Biped is a more "off the shelf" solution than the standard 3ds Max IK system. Its primary advantages are quick setup and the ability to create simple locomotion using a footstep-driven interface. The controls for Biped are found in the Motion panel.

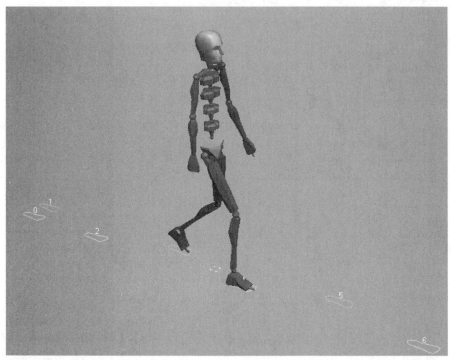

Figure 10-74: **Biped** *animated using automatic footsteps*

Physique is a modifier that serves the same essential function as Skin: it binds the vertices of a mesh object to control objects. When the control objects are transformed, the vertices move to deform the mesh. Physique can be used in conjunction with Biped, or with any other 3ds Max objects such as Bones.

Crowd is a sophisticated set of tools designed for animating large numbers of objects or characters, including Bipeds.

Motion Mixer was originally part of the Character Studio exclusive toolset, and it only worked with Biped. However, it can now be used with any 3ds Max objects. Motion Mixer works on the same principle as a nonlinear audio or video editor. You can load, save, edit, and combine animation clips in various ways. For example, you can easily speed up or slow down motion for an entire hierarchy, or make seamless transitions between different motions, such as a walk cycle and a free-form dance move.

For many years, Character Studio was a distinct software suite that required separate purchase. That's why there is some similarity in the functionality of certain features, such as Physique and Skin. Character Studio is now included with the base 3ds Max package, animators and technical directors can pick and choose which tools they prefer. In general, Character Studio is quicker, but slightly less versatile than the standard 3ds Max character tools.

END-OF-CHAPTER EXERCISE

EXERCISE 10.14: Circus Scene

In this exercise, you'll help a circus acrobat jump from a platform to a moving unicycle.

Test the Unicycle

1. Load the file `Circus_start.max` from the disc that comes with this book.

   This file contains a small circus scene, complete with a unicycle and an "acrobat"— a bouncing ball with attached "ears."

2. Activate the Left viewport. Play the animation.

   The unicycle cycles in place while the impatient acrobat bounces on the platform, waiting to jump on the unicycle.

   In this scene, you'll make the unicycle follow a circular path. The acrobat will jump on the unicycle as it passes.

3. Click **Select by Name** and look at the object names. Check the **Display Subtree** box to see the linked objects indented.

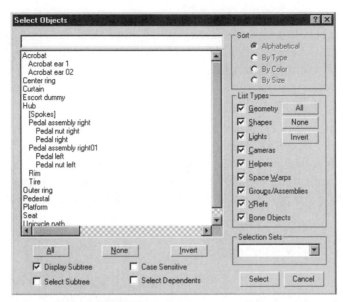

Figure 10-75: **Select Objects** *dialog*

The root parent for the unicycle is an object called **Hub**. This object is located at the center of the wheel. This is the object that rotates, spinning the wheels and pedal assembly with it.

4. Select the object **Hub**.

5. Click **Selection Lock** to lock the selection.

6. In any viewport, move the **Hub** object.

   The seat assembly doesn't go with the **Hub** when you move it. This is because the seat is not linked to the **Hub**.

7. Undo the movement to move the **Hub** back to where it began.

8. Click **Selection Lock** again to unlock the selection.

9. Select the object **Seat**, and use **Select and Link** to link it to **Hub**.

   A quick and reliable way to link the **Seat** to the **Hub** is to use the **<H>** key on the keyboard. With a selection tool active, press **<H>** to display the **Select Objects** dialog, and choose the object **Seat**. Click **Select and Link**, then press the **<H>** key to launch the **Select Parent** dialog. Select the parent object **Hub**, and click the Link button at the bottom of the Select Parent dialog.

10. Play the animation in the **Left** viewport. The seat goes round and round! This isn't exactly what we had in mind.

11. Reload the file **Circus.max** without saving the current file.

Link Unicycle Parts to Dummy Object

This time, you'll create a Dummy object at the center of the wheel that will not rotate. The **Hub** will be linked to the Dummy, and so will all objects that aren't supposed to rotate. The Dummy object will then be placed on the path so the entire unicycle moves along the path, but only the **Hub** and its children will rotate.

1. On the Create panel, click **Helpers** 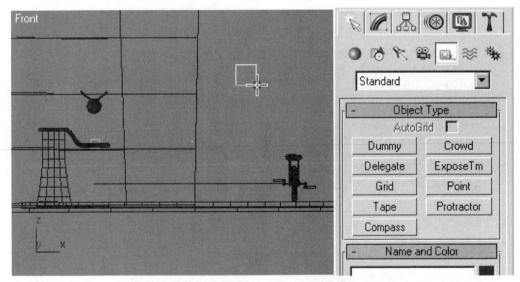. Click the button labeled **Dummy**. In the Front viewport, create a Dummy object about the same size as the bouncing ball. See the following illustration.

Figure 10-76: Dummy object in Front viewport

2. Use the **Align** tool to align the Dummy object's Pivot Point position with the **Hub** on all axes.

    After clicking **Align**, press the **<H>** key to display the **Pick Object** dialog, and choose **Hub** from the list.

3. Link **Hub** to the Dummy object.

4. Link the **Seat** to the Dummy object.

5. Move the Dummy object. The entire unicycle moves with it.

6. Play the animation. The seat stays put while the wheels turn.

Animate the Unicycle Along a Path

Next, you'll make the unicycle follow a circular path. The object `Unicycle Path` is a circle for the unicycle to follow.

1. Select the Dummy object. Go to the **Motion** panel. Expand the **Assign Controller** rollout. Highlight the **Position** track, and click **Assign Controller** [?]. Choose **Path Constraint.**

2. On the **Path Parameters** rollout, click **Add Path**, and click the `Unicycle path` in the viewport. In the **Path Options** section of the rollout, check the **Follow** check box.

   Play the animation. The unicycle follows the path, but is cycling sideways.

Figure 10-77: Unicycle pedaling sideways

3. On the Motion panel, under the **Axis** section, choose the **Y** axis.

   The unicycle is now oriented in the proper direction.

Figure 10-78: Unicycle turned in the proper direction

Acrobat Animation with Link Constraint

Next, you'll make the acrobat jump onto the unicycle. To do this, you'll use a **Link Constraint** to link the **Acrobat** first to the platform, then to the unicycle **Seat**.

1. Select the object called **Acrobat**, which is the ball bouncing up and down on the platform.

2. On the Motion panel, assign a **Link Constraint** to the object's **Transform** track.

3. Go to frame **0**. On the **Link Params** rollout, click **Add Link** and click the **Platform** object in the viewport.

   You could actually link the acrobat to any nonmoving object during the first part of the animation. However, it makes sense to link it to the **Platform**, since that is what the **Acrobat** is bouncing on.

4. Go to frame **151**.

    In the Left viewport, you can see that this is the point when the unicycle passes right underneath the platform.

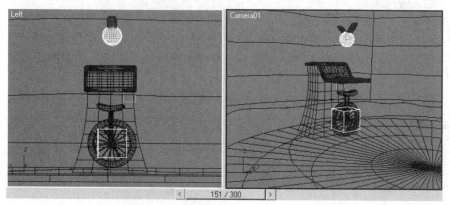

Figure 10-79: Unicycle passing under platform

5. On the Motion panel, the **Add Link** button should still be pressed. Select the **Seat** as the parent of the **Acrobat** at frame **151**.

6. Play the animation.

    The **Acrobat** jumps off the platform, but as the unicycle goes around the path, the ball is jumping up and down in the air above the seat. To solve this problem, we'll use another Dummy object, which has already been set up in the scene.

    Look closely at the platform in the Left viewport. As the unicycle passes, a Dummy object passes from the platform to the seat. It is called **Escort Dummy** because it escorts the acrobat to the unicycle.

    Up until frame **138**, the **Escort Dummy** object's Pivot Point is right at the spot where the ball hits the **Platform**. This is also the last frame at which the ball is in contact with the **Platform** before going back up the air. This is a logical time for the ball to begin bouncing forward. At frame **138**, the **Escort Dummy** begins to move to the **Seat** of the unicycle, and it arrives at frame **166**.

7. Select the **Acrobat**. Go to frame **138**. On the Motion panel, click **Add Link** if it is not already pressed, and select the **Escort Dummy** in the viewport.

    The ball is now linked to the **Escort Dummy** at frame **138**. Right-click in the viewport to exit Add Link mode.

8. On the **Link Params** rollout, select the Target object **Seat**. Change the **Start Time** for the **Seat** to frame **161**.

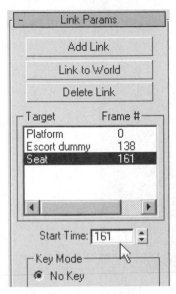

*Figure 10-80: Change the **Start Time** for the **Seat** to frame 161*

9. Play the animation.

The **Acrobat** jumps onto the unicycle as it passes by, and continues to bounce on its **Seat** as the unicycle moves along the path.

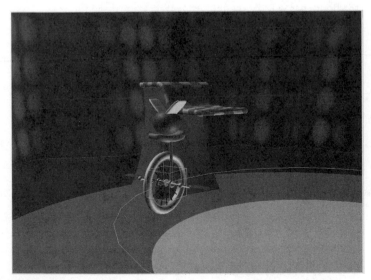

*Figure 10-81: **Acrobat** jumping onto unicycle **Seat***

Finishing Touches

Next, you'll make the Acrobat's ears flop around by applying the Flex modifier.

1. Apply the **Flex** modifier to each of the **Acrobat Ears**. Set the **Flex** parameter to **2.0** and **Strength** to **2.5**.

    The Acrobat's ears now flop as it bounces.

Figure 10-82: **Acrobat** *with floppy ears*

2. Save the scene with the filename `CircusAcrobat.max` in your folder.

    Now for some circus music.

3. Open the **Dope Sheet**. Scroll to the top of the Controller Pane. Click the track labeled **Sound** to highlight it. Then right-click the Sound track and choose **Properties** from the Quad Menu. The **Sound Options** dialog appears.

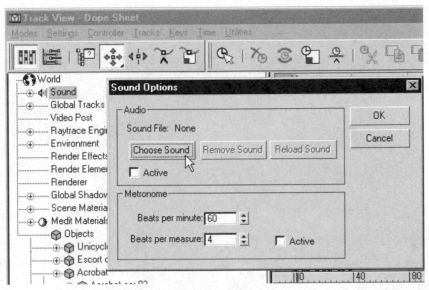

Figure 10-83: **Sound Options** *dialog*

4.   Click the button labeled **Choose Sound**, and browse to the file `Circus_loop.wav` on the disc that comes with this book. Click OK to exit the Sound Options dialog.

A sound file has now been added to the animation.

5.   Expand the **Sound** track to see the waveform of the circus music.

6.   Play the animation to hear the circus tune.

A jaunty little circus tune plays with the animation.

7.   Save the scene again.

8.   Render the animation to a 640x480 QuickTime or `.AVI` movie.

The movie renders with both animation and sound. In addition, the circus floor reflects the scene.

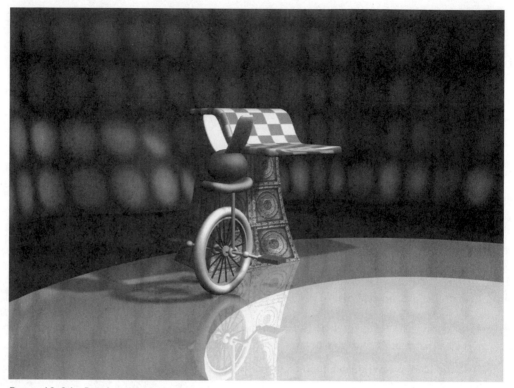

Figure 10-84: Rendered circus scene

To see the rendered animation, view the file **ACROBAT.AVI** on the disc.

SUMMARY

The transforms of an animated object are recorded relative to the local coordinate space of the object's parent. Object transforms can be disabled in the Hierarchy panel.

Animation **Controllers** and **Constraints** can achieve many effects. Controllers determine how transforms and parameters are animated. Constraints can make an object follow a path, always "look at" another object, change parental links over the course of an animation, and so on.

Parameter Wiring lets you easily control one object transform or parameter with another. You can enter *expressions* into the Parameter Wiring dialog to change the result. **Manipulators**, such as **Sliders**, are viewport controls that can be wired to object transforms and parameters.

Some types of animation are too difficult or time-consuming to keyframe by hand. Physically based *dynamic simulations* can help in these cases. The main dynamics engine in 3ds Max is called **reactor**. In addition, 3ds Max includes modifiers for simulating **Cloth** and **Hair and Fur**.

3ds Max offers many advanced modifiers for animation. The **Skin** modifier deforms a model with a series of control objects, called **Bones**. These can be specialized Bone objects, or any other objects such as Boxes or Splines. The **Morpher** modifier allows you to blend the form of one object into another. Objects that morph into one another are called *morph targets*. Morph targets must all have the same number of vertices. Morphing is very useful for facial animation.

Inverse kinematics makes animation easier, because you can animate the **Goal** of a series of linked objects, instead of animating each object individually. In IK, parent objects inherit transform information from their children. It is common to use Skin and Bones in conjunction with IK, but IK can also be used on any objects, not only Bones.

Set Key mode makes character animation more powerful by enabling *pose-to-pose animation*. You can try out different character poses before committing to them, and create keyframes for all desired tracks. This prevents poses at other points in time from being accidentally altered.

Beyond the basic animation techniques, you'll find sophisticated tools for setting up and animating characters and other scenes. The **Expression Controller** provides mathematical functionality that can't be achieved with Parameter Wiring, such as combining the values of multiple animation tracks. The **Limit Controller** is a simple method for preventing a transform or parameter from exceeding certain values.

Character Studio is an advanced suite of tools that streamline the character setup and animation process. It includes **Biped**, **Physique**, and **Crowd**. The **Motion Mixer** is a nonlinear animation interface, and it can be used with Biped or with standard 3ds Max objects.

REVIEW QUESTIONS

1. What is a **Controller**? Is it different from a **Constraint**?

2. How do you make an object travel along a path?

3. What does the **LookAt** Constraint do?

4. What does the **Link** Constraint do?

5. What is **Parameter Wiring**?

6. How can **Manipulators** make animation easier?

7. What are *dynamics*?

8. What are **Bones** objects?

9. What does the **Skin** modifier do?

10. What is a Bone **Envelope**?

11. To morph one object to another, what must the two objects have in common?

12. How is *inverse kinematics* different from forward kinematics?

13. What is *pose-to-pose* animation?

14. What is **Set Key** mode?

15. Name three ways of designating which objects and tracks are animated in Set Key mode.

16. What can you do with an **Expression Controller** that you can't do with Parameter Wiring?

Chapter 11
Advanced Materials

OBJECTIVES

In this chapter, you will learn about:

- Creating custom bitmaps as decals
- Composite maps and materials
- Transparency
- Automatic reflections, refractions, and ray tracing
- Special materials such as Ink 'n Paint
- Animated materials and maps
- Advanced mapping
- Adjusting mapping coordinates with Unwrap UVW

CREATING BITMAP LABELS

In your work with 3ds Max, you will often find that you need a custom bitmap for use as a label or logo on a 3D object. When this happens, you can use 3ds Max to create custom bitmaps. Model the label, using renderable 2D shapes, then render a bitmap image. Then you can apply the rendered bitmap as a map within a material.

Figure 11-1: 2D label applied to 3D object

To create a bitmap label, model 2D shapes of the label's text and pictures in the Front viewport. To create renderable outlines, adjust the options in the shape's **Rendering** rollout. To create solid shapes, apply the **Extrude** modifier. You won't be viewing the label from an angle, so it doesn't matter if you leave the Extrude **Amount** value at **0**.

Figure 11-2: Shapes set up for 2D label

Materials for creating bitmap textures are different from materials designed to be applied to 3D objects. Although you may apply a Diffuse texture to the label, you certainly don't want any shading or specular highlights on the label. These will be added in the 3D scene later on.

To avoid undesired shading and highlights, don't add any lights to the scene. In the Material Editor, increase the **Self-Illumination** parameter to **100**, and reduce the **Specular Level** and **Glossiness** parameters to **0**. This will produce perfectly flat shapes, with no shading or highlights.

SAFE FRAMES

Often, the logo or label you're creating won't be exactly in the proportions of the viewport. When this happens, use *safe frames*. Right-click the viewport label and choose **Show Safe Frame** from the pop-up menu.

The only safe frame you are concerned with is the yellow border around the outside edge of the viewport, so you can turn off the other ones. Right-click the viewport label and choose **Configure** from the pop-up menu. The **Viewport Configuration** dialog appears. Click the **Safe Frames** tab. Uncheck the **Action Safe** and **Title Safe** check boxes, and click **OK** to exit the dialog. Now only one line appears around the border of the viewport. This is the **Live Area**. Anything outside of this border will not appear in the rendered bitmap.

In the **Render Scene** dialog, you can set up the resolution and aspect ratio to match the proportions of your logo. The safe frame will update interactively in the viewport.

ORTHOGRAPHIC RENDERING

While it's easy to simply render the Front viewport, sometimes the orthographic view will change by itself, particularly if you change viewport configuration options. This disturbs the composition you have carefully set up. A Camera view will give you more control, and won't change unless you alter it intentionally. You can only use a camera view if you enable the **Orthographic Projection** option on the camera's Modify panel. Be careful to create the camera in the Front view, and never rotate or orbit the camera. The line of sight for the camera must be precisely perpendicular to the geometry. Adjust the camera's **FOV** parameter to zoom in and out.

*Figure 11-3: Using a camera with the **Orthographic Projection** option*

ALPHA CHANNELS

When you render a scene solely for the purpose of creating a bitmap label, it's best to save an alpha channel. This makes it easy to place the label as a decal on an object. As you remember from chapter 1, an alpha channel holds transparency in a bitmap.

Figure 11-4: Rendered label and its alpha channel

If you render your scene against an empty, black background in 3ds Max, the alpha channel will indicate what areas of the rendered image should be opaque, and which areas should be transparent. When you apply this rendered bitmap to a 3D object, the underlying color or texture will show through the transparent areas of the bitmap. This results in a very satisfying illusion of paint or ink applied to certain areas of the 3D object.

3ds Max always creates an alpha channel when rendering, but you must enable the alpha channel when you save the file. While many bitmap file formats are able to contain alpha channels, it's always safest to use the Targa (.TGA) format. The Targa format has been around for a long time, and all computer animation and bitmap editing software supports this format.

In all cases in 3ds Max, if a bitmap has an alpha channel, the alpha channel will be used to mask the bitmap. If the bitmap doesn't have an alpha channel, or if a procedural map is used, the **Alpha from RGB Intensity** option under **Output** rollout can be selected to create a fake alpha channel from the grayscale values of the map.

If you are in a tight spot, and you need an alpha channel but don't have one, you can create a grayscale image using the tools in your bitmap editor. Then you can use it to mask off another image, by using it within a **Mask** map, as described later in this chapter.

EXERCISE 11.1: Bitmap Label

In this tutorial, you'll create a bitmap label containing your name. The bitmap will be used later on in a 3D scene.

1. Reset 3ds Max.

2. In the Front viewport, create a text object of your name. Use the font **Arial Black** or a similar simple font.

3. Apply the **Extrude** modifier to the text. Leave the **Amount** at 0.

4. Press the **<F3>** key to change the Front viewport to the **Smooth+Highlights** mode. Press the **<G>** key to hide the Home Grid.

5. With the Front viewport still active, click **Zoom Extents** to place the text in the center of the Front viewport.

    Next we want to set up the safe frame to encompass just the name.

6. Right-click the Front viewport label and choose **Show Safe Frame** from the menu to turn on safe frames.

7. Right-click the viewport label again, and choose **Configure**. The **Viewport Configuration** dialog appears. Click the **Safe Frames** tab. Uncheck the **Action Safe** and **Title Safe** check boxes, and click OK to exit the dialog.

8. On the Main Toolbar, click Render Scene. In the **Output Size** section of the **Render Scene** dialog, change the **Width** parameter to 1280 and **Height** to 160. The safe frame becomes tighter around the name.

9. Zoom the **Front** viewport in or out until your name appears as large as you can make it without cutting off the edges.

Figure 11-5: Name in safe frame

10. If necessary, adjust the Height parameter on the Render Scene dialog and zoom the Front viewport until your name fits snugly into the safe frame.

11. Create a material for the name label. Choose gold as the Diffuse color. Set the **Self-Illumination** value to **100**. Change the **Specular Level** and **Glossiness** parameters to **0**. Apply the material to the name label.

12. Render the image. After it has rendered, click the **Save Bitmap** button 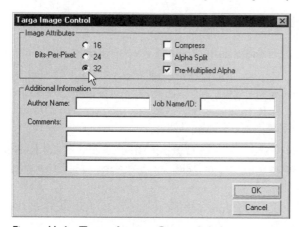 at the top left of the Rendered Frame Window. The **Browse Images for Output** dialog appears.

    Browse to your project folder on your hard drive. In the **Save As Type** pulldown, select **Targa Image File**. Enter `MyName.tga` as the filename, and click **Save**.

13. The **Targa Image Control** dialog pops up. Select **32 Bits-Per-Pixel** to save an alpha channel. You can leave the **Compress** check box on, or turn it off; it doesn't matter in this case because Targa files use lossless image compression, which does not degrade image quality. Leave the **Pre-Multiplied Alpha** option turned on.

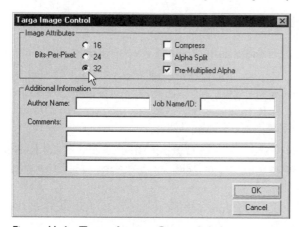

Figure 11-6: **Targa Image Control** *dialog*

If you need to access the Targa Image Control dialog in the future, or change the output settings for any rendered image file format, click the **Setup** button in the Browse Images for Output dialog.

14. Save the scene as the file `MyName01.max` in your folder.

COMPOSITOR MAPS

If you apply a bitmap that contains an alpha channel to a material attribute, such as the Diffuse color, then the alpha channel will mask off parts of the bitmap. Any areas on the alpha channel that are transparent will not affect the material attribute.

In the example of the coffee cup at the beginning of this chapter, the bitmap is saved as a 32-bit Targa file, which contains an alpha channel. When that bitmap is applied to the Diffuse color of the material, the result is a floating decal with no edges. The original Diffuse color of the material comes through in any areas where the alpha channel is transparent.

While this simple technique is easy and very useful, you will find that you need more sophisticated layering of maps. *Compositor* map types allow you to layer two or more maps, using transparency information to show one through the other. Compositors can be used to make many kinds of complex materials. If you learn these map types well, you will be able to simulate just about any real-life material.

MASK MAP

The **Mask** map uses one map and a mask. The mask determines how much of the map will be seen. The part of the map blocked out by the mask is replaced by the color or pattern underneath. For example, if a Mask map is used as a Diffuse Color map, the part of the map that is blocked out will render with the material's original Diffuse color.

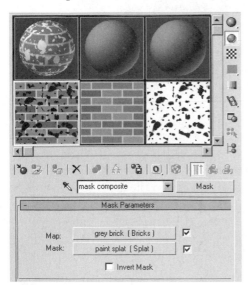 

Figure 11-7: **Mask** map applied to Diffuse color. The **Splat** map masks off the **Tiles** pattern, allowing the original **Diffuse** color to show through.

MIX MAP

The **Mix** map uses two maps and a mask. The mask determines how much of one map will be seen. The part of the first map blocked out by the mask is replaced by the colors of the second map.

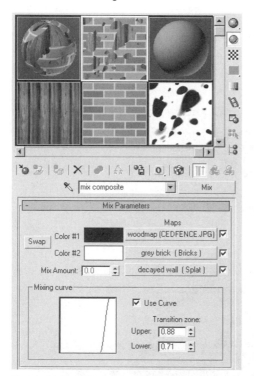

Figure 11-8: ***Mix*** *map applied to Diffuse color. The* ***Splat*** *map determines what parts of the object will render with the* **CEDFENCE.JPG** *map and what parts will render with the* ***Tiles*** *map.*

With the Mask and Mix map types, one map is specified as the mask. If the mask is a bitmap with an alpha channel, the alpha channel is used. Otherwise, the map's grayscale values are used to generate the mask. If the map being used as a mask does not have an alpha channel, it should be a grayscale image with white in the areas you wish to be opaque, and black where you wish the map to be transparent. Likewise, if you are using a procedural map as the mask, you should choose white, black, or gray colors.

COMPOSITE MAP

The **Composite** map layers two or more maps on top of each other. With Composite, each map's alpha channel or grayscale values are used to generate a mask for itself. By default, you cannot specify a separate image to be used as a mask. The alpha channel or black area of each map becomes transparent to show the map or maps underneath it.

You can set the number of maps you wish to include in a Composite map by clicking on the **Set Number** button.

Unlike other places in 3ds Max, such as the Modifier Stack, the data pipeline within the Material Editor is calculated from the top down. This means that a map that appears *lower* on the Composite map panel will actually render *in front of* the map listed above it. In the following illustration, the Tiles map is highest on the panel, but it renders *behind* the other maps.

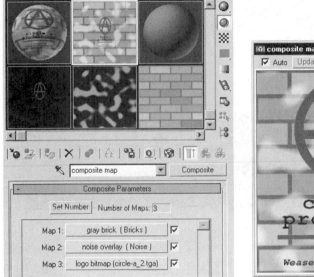

*Figure 11-9: **Composite** map applied to Diffuse color. The **Tiles** map is the first layer. The **Noise** layer is applied over the Tiles map, and finally the logo bitmap is applied over the Noise.*

Nesting Compositors

As mentioned previously, there's no way to apply a separate mask to a map within a Composite map. This would appear to be a limitation that requires alpha channels to be saved within all of the bitmaps used within a Composite map. The option to create **Alpha from RGB Intensity** on the **Output** rollout of a bitmap can help, but it can't always get you the result you want.

The answer to this problem is quite simple. You can use a **Mix** map as one of the maps within a Composite map. That way, you can apply an external grayscale bitmap, or a procedural map, to control the opacity of a map which doesn't have its own embedded alpha channel.

EXERCISE 11.2: Compositor Maps

In this exercise, you'll use a Mix map to create a nameplate for a desk. The nameplate will be created with a bitmap and its alpha channel. In this way, you'll be able to change the foreground and background colors of the nameplate quickly and easily.

Figure 11-10: Desk scene with nameplate

Diffuse Map with an Alpha Channel

1. Load the file **Desk.max** from the disc that comes with this book.

   This file contains a scene with a desk and other objects to simulate an office environment. The nameplate currently has no name on it, but mapping coordinates have already been applied so the name will lie on the nameplate correctly.

2. Open the Material Editor.

   The first material is the **Name Plate** material. This material has been applied to the nameplate in the scene. It is currently a dark gray, and has no maps.

3. Activate the first sample slot with the **Name Plate** material. Assign the file **MyName.tga**, which you created in the last exercise, as the **Diffuse Color** map.

   Turn on **Show Map in Viewport** . Your name appears on the nameplate object.

4. Render the Camera view. Notice that your name appears in gold, and the background of the nameplate object is still a dark gray. That's because you saved a 32-bit Targa file, and the effect of the Diffuse color map is masked off by the alpha channel.

Figure 11-11: Gold letters on a neutral gray background

If you wish, you can change the Diffuse color swatch at this point. However, you can't add a texture to the background of the nameplate. To do that, you must use a compositor map type, such as a **Composite** map.

Composite Map

1. In the Material Editor, you're viewing the parameter rollouts for the MyName.tga texture. Click the **Map Type** button at the lower right of the Material Editor toolbar; it currently reads Bitmap.

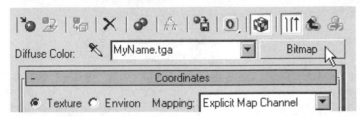

*Figure 11-12: Click the **Map Type** button to assign a different type of map*

2. The **Material/Map Browser** appears. Select **Composite** to assign a Composite map to the Diffuse color of the material.

   The **Replace Map** dialog appears. Choose **Keep Old Map As Sub-Map**, and click **OK**. This places the MyName.tga file into one of the slots within the Composite map.

3. Turn on **Show Map in Viewport** once again.

   In the Composite Parameters rollout, you now have two slots for maps. MyName.tga is currently assigned to **Map 1**, and no map is assigned to **Map 2**. Since we want the MyName.tga image to render over the top of a background map, we must place MyName.tga into Map 2.

4. Click and drag the button labeled **MyName.tga**, and drop it onto the button labeled **None**. The **Copy (Instance) Map** dialog appears. Choose **Swap**, and click OK.

5. Click the **Map 1** button, now labeled None, and choose a **Marble** procedural map from the Material/Map Browser.

6. In the **Marble Parameters** rollout, change the **Size** value to **1.0**. Change the colors if you wish. Render the camera view.

Figure 11-13: **Composite** *Map with* **Marble** *and* **Bitmap** *textures*

This looks fairly good, but we can't place a texture inside the letters. To do that, we can use a **Mix** Map.

Mix Map

1. Click the **Go To Parent** button  to move up one level in the Material Editor data flow. You are now at the level of the **Composite** map.

   If you clicked the Map Type button now to change the Composite map to a Mix map, you'd once again have the option of discarding the old map, or using it as a submap. In this case, neither option will work for us. If we keep the old map, we'll be keeping the Composite map and its submaps, resulting in a Composite map nested inside the new Mix map. That's not what we want. Instead, we need to keep the Bitmap and Marble textures, and discard the Composite map entirely.

2. In the **Composite Parameters** rollout, click and drag the **Map 1** button to one of the other sample slots in the Material Editor. The **Instance (Copy) Map** dialog appears. Choose **Copy**, and click OK. The **Marble** texture is copied to the other sample slot.

   Remember, it's OK to reuse sample slots. You are not altering the scene in any way. The material that was once in the sample slot is not lost, as long as it is still assigned to an object in the scene, or stored in a scene or a Material Library on your hard disk.

3. Click and drag the **Map 2** button to another sample slot. Now you have copied the two textures you need to create the Mix map.

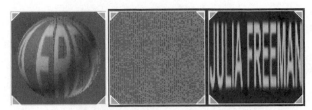

Figure 11-14: **Composite** *map in one slot; copies of* **Marble** *and* **Bitmap** *in other slots*

4. Select the sample slot with the Composite map material. Click the **Map Type** button, and choose a **Mix** map from the Material/Map Browser. This time, when the **Replace Map** dialog appears, choose **Discard Old Map** and click OK.

5. Click and drag the Marble map from the other sample slot onto the **Color #1** map button on the **Mix Parameters** rollout. The **Instance (Copy) Map** dialog appears. Choose **Instance**, and click OK.

6. Click and drag the MyName.tga bitmap from its sample slot onto the **Mix Amount** map button on the **Mix Parameters** rollout. Again, choose **Instance** in the **Instance (Copy) Map** dialog.

The Material Editor should look something like the following illustration.

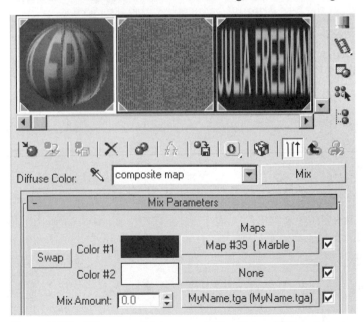

Figure 11-15: **Mix** *map with instanced* **Marble** *and* **Bitmap** *textures*

7. Render the scene. Notice how the letters are now simply a little brighter than the background, and the Marble texture is showing through.

Figure 11-16: Marble texture bleeds through

This is happening because the **Mix Amount** parameter is using the *color channels* of the MyName.tga file, and not the alpha channel of the file. Even if you used a very bright yellow when you created this image, it's still not 100% white, and therefore not 100% *opaque* (solid) when used as a mask.

8. Select the sample slot of the MyName.tga file. Since it is instanced within the Mix map, any changes made to this instance will be automatically updated in the Mix map.

   Scroll down to the **Bitmap Parameters** rollout. In the **Mono Channel Output** section, select the button labeled **Alpha**. This causes the alpha channel saved within this bitmap file to affect the **Mix Amount** parameter within the Mix map.

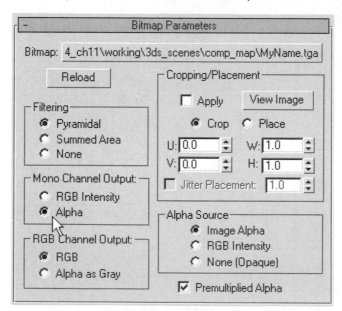

*Figure 11-17: Choose **Alpha** in the **Mono Channel Output** section*

9. Re-render the scene. The letters render in solid white, because the alpha channel within MyName.tga is now allowing **Color #2** to be completely opaque, or nontransparent.

Figure 11-18: Opaque letters

10. Select the sample slot of the Mix map. Change the color of **Color #2**, and re-render. The letters change to the color you selected.

11. Assign a procedural map, such as **Gradient** or **Smoke**, to **Color #2**. Since this is a small object, reduce the map **Size** to somewhere in the range of **1.0** to **5.0**. Re-render the scene.

12. Continue to experiment with different map types for **Color #1** and **Color #2**. Adjust the map colors and other parameters until you are happy with the result.

*Figure 11-19: **Mix** map with **Smoke** and **Gradient** submaps*

13. Save the scene with the filename **DeskWithName.max** in your folder.

OPACITY

Opacity is the opposite of transparency. This term is a variation of the word *opaque*, which means nontransparent. The term *opacity* refers to the opaque or transparent quality of the material. Opacity can be set with the **Opacity** parameter on the **Basic Parameters** rollout, or with the **Opacity** map on the **Maps** rollout.

WORKING WITH TRANSPARENCY

When first working with opacity and transparency, you will find that simply making a material transparent will not yield any usable results. For example, consider real-life glass. Everyone knows that glass is completely transparent. However, glass reflects the scene around it and usually has visible smudges and dirt.

Figure 11-20: Windows with completely transparent glass

Figure 11-20 shows a rendering of a scene with a window. A completely transparent material is applied to the glass. The resulting glass is invisible, as if it isn't even there. You could try making the glass a little less transparent, but then you end up with something that looks like tinted windows or mesh screens, as shown in Figure 11-21.

To make the glass look like ordinary windows, you'll need more than just transparency. Reflection, carefully placed dust and dirt, effective use of lights, and other elements all play a part in realistic glass. In the following image, these elements were used to create a more realistic window glass.

In this chapter, you'll learn more about how these elements come together to produce this kind of effect. In order to do this, you must first learn how opacity works.

Figure 11-21: Windows with partly transparent glass

Opacity Parameter

The **Opacity** parameter on the **Basic Parameters** rollout changes the transparency of the entire material. This parameter defaults to **100**, indicating that the material is completely opaque. When this value is low, the object appears more transparent. When it is **0**, the material is fully transparent except for specular highlights.

In practice, this parameter is rarely used to set transparency for the final material used in a rendering. Usually you will want varying degrees of transparency throughout a material, which can only be achieved by using a map to set opacity.

However, the **Opacity** parameter is useful for getting some transparency going in the material for trial renderings. Later, it can be replaced with an **Opacity** map on the **Maps** rollout.

Figure 11-22: More realistic glass with reflections, glare, and dust

Background Button

When you decrease the opacity of a material, it can be hard to see the result in the sample slot. This is because the background in each sample slot is a dark gray color. When you make a material more transparent, the material shows more of the dark background, making it appear darker rather than transparent.

To solve this problem, the dark gray background can be replaced with a background image. Click the **Background** button ▒ at the upper right of the Material Editor toolbar. This places a checkered background in the current sample slot. When the material is made to be transparent, the background image shows through the sample sphere.

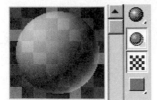

*Figure 11-23: Transparency made visible with **Background** button*

Transparency Type

In the **Advanced Transparency** section on the **Extended Parameters** rollout, you have three options for the type of transparency: **Filter**, **Subtractive**, and **Additive**.

The **Filter** color swatch alters the color of light that passes through the transparent object to reach the camera. However, the Filter color does not change the color of light that passes through the object to strike other objects in the scene.

Subtractive makes the object darker in transparent areas.

Additive makes the object brighter in transparent areas.

The type of transparency used depends largely on the type of scene you're making, the lighting around the object, and other factors. Try each one to find the best one for your material.

EXERCISE 11.3: Opacity

In this tutorial, you'll practice working with opacity to make a glass paperweight.

Figure 11-24: Paperweight on desk

1. Load the file **Paperweight.max** from the disc that comes with this book. This is a desk scene showing a paperweight on the desk. The glass has no transparency yet.

2.   Open the Material Editor. Look at the first material, which is named **Paperweight Glass**. This material has a light gray Diffuse color, but no transparency yet.

3.   On the **Blinn Basic Parameters** rollout, reduce the **Opacity** parameter to **20**.

In the Camera view, the displayed material has become partially transparent. However, the material appears dark in the sample slot.

4.   If necessary, click the **Background** button  to turn on the background in the sample slot.

5.   On the **Blinn Basic Parameters rollout**, change **Specular Level** to **80**. This will make the glass shiny, and produce specular highlights on the glass.

6.   Render the Camera view. The glass appears dingy.

Next, you'll work with the transparency type to brighten up the glass.

7.   Expand the **Extended Parameters** rollout. Note that the transparency **Type** is set to **Subtractive**. Change the transparency Type to **Additive**.

Additive transparency makes the transparent object appear brighter, as if it's catching the light.

8.   Render the Camera view. The glass appears brighter.

Figure 11-25: Brighter glass

Currently, the paperweight glass is uniformly transparent. To make it look more realistic, you'll use a reflection map.

9. On the **Maps** rollout, assign a bitmap to the **Reflection** map slot. Browse to the `\3dsMax8\maps\Reflection` folder. Choose the file `CHROMIC.JPG`.

   In the sample slot, you can see that the `CHROMIC.JPG` reflection map has pretty much taken over the material. To reduce its effect, you'll reduce the reflection Amount on the Maps rollout.

10. Click **Go to Parent**  to return to the root level of the material. Go the **Maps** rollout. Next to **Reflection**, change the **Amount** value to **10**.

11. Re-render the Camera view. Save the rendering as `PaperweightGlass.tga` in your folder.

Figure 11-26: Paperweight with reflection

12. Save the scene as the file `PaperweightGlass.max` in your folder.

TIP: You can add a TurboSmooth modifier to the `Paperweight Glass` object to soften its harsh edges.

Opacity and 2-Sided Materials

When you reduce opacity on a material, it is often desirable to check the **2-Sided** check box on the **Shader Basic Parameters** rollout. This makes the other side of a transparent object show up in a scene. If you're not sure if you want the backside of an object to show, try rendering the scene with 2-Sided on, then try it again with 2-Sided off.

OPACITY MAPPING

So far, you have learned how to change the opacity of the entire material with the Opacity parameter. To set varying degrees of transparency or opacity for different parts of the material, you can use a map to vary the opacity over the object's surface. Brighter areas of the map make the material appear more opaque, and darker areas make the material appear more transparent.

When a map is used for the Opacity map attribute, and the **Amount** is set to **100**, the transparency of the material is entirely determined by the map. When the Amount for the Opacity map is less than **100**, the transparency is determined by mixing the map's brightness levels with the Opacity value on the Basic Parameters rollout.

TIP: Color images are converted to grayscale before they are applied to the **Opacity** map. For this reason, you will get the most predictable results if you use a grayscale image to begin with. Save your opacity bitmap as a grayscale image, or use procedural maps with zero saturation. In fact, this rule applies to all mapped parameters that do not have their own color swatch, such as **Specular Level** and **Bump**. Only use color maps when the map channel has a color swatch, and use grayscale maps for all others.

Opacity Mapping and Shadows

In life, when an object is transparent, the object's shadow has varying dark and light areas corresponding to the transparency of the object.

In 3ds Max, **Shadow Map** shadows do not work with opacity mapping. If Shadow Maps are used, shadows from transparent objects appear solid. If **Ray Traced Shadows** are used, the resulting shadow *does* respect opacity mapping, and more transparent areas of the object will cast lighter colored shadows.

Unfortunately, ray traced shadows take longer to render than shadow maps. In addition, standard ray traced shadows always produce sharp edges to the shadow. This might not be the desired effect. However, 3ds Max has advanced lighting features that will allow you to create soft shadows based on opacity mapped objects. In Chapter 12, *Advanced Lighting*, we will explore the features of **Area Shadows** and **Advanced Ray Traced Shadows**.

EXERCISE 11.4: Opacity Map

In this tutorial, you'll put your name on the glass paperweight you worked with earlier.

1. Load the file **PaperweightGlass.max** from the disc that comes with this book.

2. Open the Material Editor. Select the first sample slot for the material **Paperweight Glass**.

3. Next, you'll make your name appear on the glass. Assign the file **MyName.tga** to the **Opacity** map of the material.

   Your name appears on the sphere in the sample slot. The material transparency is now calculated entirely from the opacity map, and not at all from the **Opacity** parameter on the **Blinn Basic Parameters** rollout.

4. In the **Coordinates** rollout, change the **Tiling** values for the **MyName.tga** opacity map. Assign a **U Tiling** value of **4**, and a **V Tiling** value of **8**.

5. Scroll down to the **Bitmap Parameters** rollout. In the **Mono Channel Output** section, select the **Alpha** option. This sends the alpha channel of the Targa file to the Opacity map, instead of the color channels.

6. Render the Camera view. Your name appears all over the paperweight.

Figure 11-27: Names on paperweight

7.   On the **Coordinates** rollout of the Material Editor, uncheck the **U** and **V Tile** check boxes.

8.   Render the Camera view again. Your name appears on the paperweight once.

The names are a little transparent. This is one of the quirks of Additive transparency. Even though the alpha channel defines the name area as completely opaque, Additive transparency always adds a little transparency to the opaque area. You can try using **Filter** transparency instead.

Figure 11-28: Name on paperweight once

9.   When you are happy with the result, render an image with the filename **PaperweightName.tga** in your folder.

10. Save the scene in the file **PaperweightName.max** in your folder.

FALLOFF EXTENDED PARAMETER

On the **Extended Parameters** rollout, you can set up **Falloff** for the transparency. Falloff causes the material's transparency to vary from the inside of the object to the outside. When **In** is selected, the object appears to be more transparent on the inside than the outside. The opposite is true for the **Out** option.

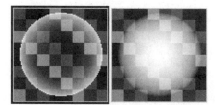

*Figure 11-29: The sample at left has a **Falloff** type set to **In**, and the sample at right is set to **Out**.*

The Falloff **Amt** parameter sets the amount of variation in transparency from the inside to the outside of the object. The Falloff effect will only work if the Amt value is more than **0**. To achieve the greatest variation in transparency, set the **Opacity** parameter on the Basic Parameters rollout to **100**, and set the Falloff **Amt** to **100**.

FALLOFF MAP TYPE

A **Falloff** map type is a procedural map that creates a gradient between two colors according to scene coordinates, such as the viewing direction or the direction in which the faces of the object are pointing.

The Falloff map can be used to create many interesting effects, such as an X-ray look. To vary the opacity of an object near its silhouette edges, apply a Falloff map to the Opacity attribute, and leave the Falloff map parameters at their default values.

REFLECTION MAPPING

A scene looks most alive when shiny objects have some sort of reflection on them. There are many ways to set up reflections in 3ds Max. Each one has a different use. You must consider the objects in your scene and decide which way will work best.

Before looking at the mapping options, let's take a look at how reflection works.

BASIC OPTICS

Whether an object shows up in a reflection is determined by the *angle of incidence* of both the camera and the object. The angle of incidence is the angle between the normal of the reflective surface and the object that is casting light. This applies to objects that generate their own illumination, such as a light bulb, and it also applies to objects that reflect light off of their surfaces.

The *angle of reflectance* is the angle between the normal of the reflective surface and the viewpoint, such as a camera lens or eye. The angle of reflectance is always equal to the angle of incidence.

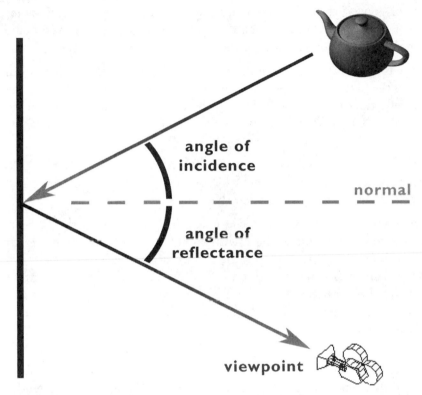

Figure 11-30: The angle of reflectance is always equal to the angle of incidence.

If the object and camera are both aligned at the same angle relative to the reflective surface, then a light ray traveling from the object will bounce off of the mirror and reach the camera lens. If the camera is not aligned properly, then it will not see an image of the object reflected in the mirror.

This principle applies regardless of the camera or object's distance from the mirror. The angle of incidence is the only factor that determines whether the object is visible in the mirror.

SHININESS AND REFLECTION

It goes without saying that a real-world object must have some shininess to reflect images. However, 3ds Max will be perfectly happy to let you set up a reflective material with a **Specular Level** or **Glossiness** value of **0**. It's up to you to make sure the material has a realistic combination of specularity and reflectance.

REFLECT/REFRACT MAP TYPE

When used as a reflection map, a **Reflect/Refract** map automatically reflects the scene around it. It is suitable for curved surfaces, whereas the **Flat Mirror** map type is designed for perfectly flat polygonal surfaces. Both of these map types expect that there is something in the scene to reflect onto the material. If you place an object with an automatic reflection map in an empty environment, the object will just reflect blackness.

The Reflect/Refract map creates reflections by rendering six images of the scene from the point of view of the reflective object, and then applying those images back onto the object as bitmap reflection maps. It's actually as if there was a little camera inside the reflective object, taking snapshots of the environment and mapping those snapshots onto the object.

The Reflect/Refract map is not as accurate as the **Raytrace** map. Also, it won't do multiple reflections, where a reflective object can be seen reflected in another object. However, Reflect/Refract renders a lot quicker than the Raytrace map.

If you're rendering an animation, you should decide whether you want the Reflect/Refract map to take fresh snapshots of the environment on every frame or not. If it's important to see moving objects reflected on the surface of another object, then turn on the **Every Nth Frame** option in the **Automatic** section. Otherwise, leave the setting at its default, which is **First Frame Only**. The Reflect/Refract map will then calculate the automatic reflection map just once, and save the snapshots into memory, to be used again on subsequent frames.

Since these snapshots are bitmap images, they have a resolution. If the reflection appears too blocky or chunky, increase resolution with the **Size** parameter on the **Reflect/Refract Parameters** rollout. Be careful not to increase the resolution too high, or your render times will be very long! The default Size of **100** is probably too low if you're shooting a closeup of a reflective object. In that case, try a value of **256**.

BLUR AND BLUR OFFSET

A typical problem with reflection maps is *aliasing*, or jagged edges in a bitmap. Even if you increase the resolution of the map, you can still get aliasing. One way to try to get around this is by blurring the map. In the Reflect/Refract map, you'll see two parameters: **Blur** and **Blur Offset**. The **Blur** parameter applies blurring to the map depending on how far away the object is from the camera. Farther distances result in more blurring.

Blur Offset applies a blur to the map without regard to camera distance. This parameter is extremely sensitive, and a value of **0.005** will soften up most aliasing. Higher values will result in a blurry reflection, like a dull, matte finish— which is a nice effect.

Blur and Blur Offset are available for any bitmap, not just automatic Reflect/Refract maps. You can use it if bitmap textures are "dancing" or aliasing during an animation.

EXERCISE 11.5: Automatic Reflections with Reflect/Refract Map

In this tutorial, you'll learn how to use the **Reflect/Refract** map in a scene.

1. Load the file **AtticReflections.max** from the disc that comes with this book.

   This file is a variation on the attic scene you worked with in earlier chapters. The focal point of the scene is the apple, which we will make into a reflective object.

2. Render the Camera view to give you a comparison point for further renderings.

3. Open the Material Editor. Select the first sample slot, which is the material **Apple**. The material applied to the apple is a shiny gray material.

4. On the **Maps** rollout, assign a **Reflect/Refract** map to the Reflection attribute. Leave all the settings at their default values.

   The sphere in the sample slot doesn't change when a **Reflect/Refract** map is assigned.

   If necessary, turn on the sample slot Background [image], and you'll see the checkered background reflected onto the sample sphere.

5. Re-render the Camera view. The renderer takes a few moments to calculate the automatic reflections. The **Rendering** progress dialog reads, "Rendering Reflect/Refract maps."

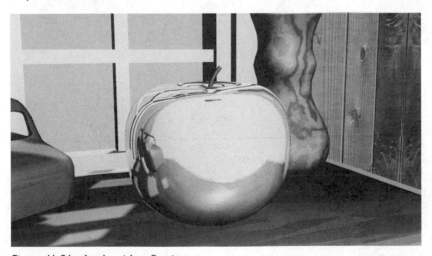

Figure 11-31: Apple with reflection

The largest part of the reflection consists of a large blue area. To see why this is happening, close the **Material Editor** and look at the scene in other viewports.

The only elements in the scene are those that appear in the **Camera01** view. In order to have reflections with the **Reflect/Refract** map, you must have other objects in the scene.

6. Go to the **Display** panel and click **Unhide All.** The remainder of the scene appears.

   The scene has more elements against the far wall such as pictures and a table with boxes. It also now has a floor, roof, and a far wall.

7. Re-render the Camera view at 640x480. Many more reflected objects appear on the apple. Notice that there are jagged, aliased pixels visible in the reflection.

8. Increase the **Blur Offset** parameter to **0.01** and re-render. The aliased pixels are blurred. Save the rendered image as **AtticAuto-reflect.tga** in your folder.

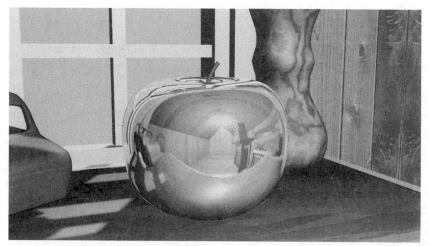

Figure 11-32: Apple with more reflection

9. Save the scene with the filename **AtticAuto-reflect.max** in your folder.

REFLECTION DIMMING

It is very common for a reflection map to look too bright, resulting in little or no areas of darkness on the surface of the object. This makes the object look flat and computer generated. If you find that your reflections are too bright, you can adjust them with **Reflection Dimming** in the **Extended Parameters** rollout. Turn on the Apply button and adjust the parameters.

The **Dim Level** parameter is the contrast, or black level. If it is at **0** (the default), the object will have maximum reflection dimming. Paradoxically, raising the Dim Level to **1.0** results in no reflection dimming at all. The **Refl. Level** parameter is the brightness, or white level. The default of **3.0** is usually too high. Reduce the Refl. Level to **1** or **2**, and you will get more realistic shading across the surface of the reflective object.

RAYTRACE REFLECTION MAP

The **Raytrace** reflection map traces light rays around the scene to figure out what will reflect what. This is the only type of map that will give you multiple reflections– reflective objects visible on the surfaces of other reflective objects. Like the Reflect/Refract map, the Raytrace map requires that there are objects surrounding the scene, or else the empty areas of the scene will render as black areas on the reflective objects.

You must use a Raytrace reflection map when you have two or more closely placed, very shiny objects as the focal point of the scene. Otherwise, you could simply use the Reflect/Refract map type and save yourself a lot of rendering time. However, the Raytrace map does look a little better than the Reflect/Refract map, and you may decide that the extra rendering time is worth it.

EXERCISE 11.6: Raytraced Reflections

In this exercise, you'll replace the **Reflect/Refract** map you set up earlier with a **Raytrace** map, and observe the results.

1. Load the file **AtticAuto-reflect.max** from the disc that comes with this book.

2. Render the Camera view so you have a comparison point for new renderings.

3. Open the Material Editor. Select the **Apple** material.

4. While at the **Reflect/Refract** map level, click the **Map Type** button on the Material Editor toolbar, which is currently labeled **Reflect/Refract**. Choose **Raytrace** as the map type. In the **Replace Map** pop-up dialog, choose **Discard Old Map**.

   The **Raytracer Parameters** rollout appears on the **Material Editor**. Leave all settings at their default values.

5. Render the Camera view.

   The renderer takes some time to calculate the raytrace reflections, going through each object in the scene. When the rendering appears, the reflection on the apple is a little more precise than it was with the Reflect/Refract map. However, there is still some visible aliasing. See the following illustration.

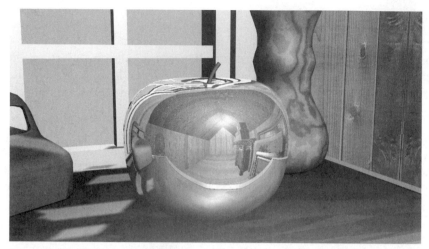

Figure 11-33: Raytraced reflections

6. To conquer the aliasing problem, go to the Main Menu and choose **Rendering > Raytracer Settings**. The **Render Scene** dialog opens with the **Raytracer** tab active. In the **Global Ray Antialiaser** section, activate the **On** check box. Leave all other settings at their defaults. Close the dialog.

7. Re-render the scene. The aliasing is mostly gone, and the image looks better than it did with the Reflect/Refract map. However, it took at least twice as long to render. Save the rendered image as **AtticRaytrace.tga** in your folder.

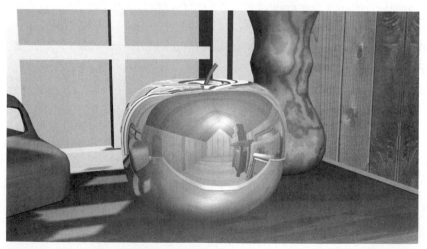

Figure 11-34: Antialiased raytraced reflections

8. Save the scene as **AtticRaytrace.max** in your folder.

TIP: It's up to you to decide if raytracing is worth the extra rendering time. For general purpose animation projects, you can achieve acceptable results with the Reflect/Refract map. For still images, or animations requiring close scrutiny of reflections, raytracing may be necessary. Just be prepared for extremely long render times, especially if you enable antialiasing within the Raytracer Settings dialog.

REFRACTION MAPPING

Refraction is the bending of light as it passes through transparent media such as glass or water. Refraction causes objects behind the refractive media to appear distorted.

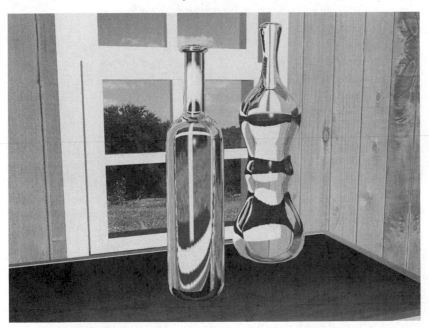

*Figure 11-35: Bottles with **Reflect/Refract** applied as a refraction map*

The use of refraction in renderings makes a great difference in the realism of transparent objects. Without refraction, glass looks like plastic, and water looks like air: invisible.

All refraction maps are assigned through the **Refraction** map on the **Maps** rollout. There are many combinations of maps that can be used for refraction. Some look better than others, some render faster than others. As with reflections, the general rule of thumb is this: the longer it takes to render, the better it will look. The methods covered here are listed roughly in order of the speed at which they render, with the quickest being first.

THIN WALL REFRACTION MAP

As the name indicates, the **Thin Wall Refraction** map type is good for thin, flat transparent surfaces such as window glass. Rather than calculating the bending of light, this map type simply uses an offset value to shift the image seen through the object. This takes very little time to render. Increase the **Thickness Offset** parameter to increase the refraction effect.

Thin Wall Refraction works very well with bump maps, creating a realistic effect with very little increase in rendering time. If the material has a bump map, adjust the **Bump Map Effect** parameter within the Thin Wall Refraction map.

Figure 11-36: Thin Wall Refraction with a bump map

REFLECT/REFRACT MAP

The **Reflect/Refract** map, when assigned to the **Refraction** map attribute, works in a similar manner as it does when assigned as a reflection map. It renders the surrounding scene to figure out what the refraction should look like. After the refraction calculation, the refraction image is pasted onto the object.

The pasted-on image appears to be a refraction, but it's really opaque, so you can't see anything behind it. This means that if there is liquid in a glass bottle, putting this type of refraction on the bottle will make it appear to be empty. Any labels on the back of the bottle will be obscured and won't render.

The advantage of using the Reflect/Refract map type is that it renders extremely fast in comparison to ray traced refractions. Reflect/Refract is physically inaccurate, but it can give a convincing illusion under the right circumstances. Use a Reflect/Refract map whenever you want to render a solid transparent object that does not have something inside it.

As with automatic reflections generated with Reflect/Refract, automatic refractions created with the **Reflect/Refract** map are very sensitive to other objects in the scene. A scene without much going on behind the camera will result in limited refraction.

INDEX OF REFRACTION

When the Reflect/Refract map is used for the Refraction map attribute, the refraction effect is calculated according to the *index of refraction*. The index of refraction (IOR) is a number between 0.1 and 3.0 used by physicists to indicate how much an object refracts light. Glass, for example, might have an IOR of 1.3 or 1.5. Air has an IOR of 1.0.

In 3ds Max, the index of refraction is set with the **Index of Refraction** value on the **Extended Parameters** rollout. This number corresponds in a general way to the numbers used by physicists.

The real-life IOR values for a few common substances can be found in the 3ds Max manual. However, this list has limited usefulness. When the Reflect/Refract map is used, the IOR settings in 3ds Max usually don't produce the same effects as real-life IORs. With the **Reflect/Refract** map, the only way to find the right IOR for your scene is through experimentation.

RAYTRACE MAP

The **Raytrace** map, when used as a refraction map, provides beautifully stunning realism, but takes a long time to render. Depending on the complexity of the scene, the Raytrace map can take up to 20 times longer to render than the Reflect/Refract map.

When the Raytrace map is used, the Index of Refraction value on the Extended Parameters rollout produces an effect that is very similar to the same IOR in physical materials. You can use IOR numbers from the real world as a starting point for your material design, but ultimately what matters is whether the scene looks good to you.

EXERCISE 11.7: Falloff and Raytraced Refractions

In this exercise, you'll put water in the paperweight you created earlier.

Falloff and Blur for Glass

1. Load the file **PaperweightGlass.max** from your folder or from the disc. Render the camera view.

2. The change made by the Falloff will be subtle. When you render the scene with the new material later on, it will be easier for you to see the effect if you can compare it with the original version. Click the **Clone Rendered Frame Window** button. Don't close either of the Rendered Frame Windows. Now you have two copies of the current rendering.

3. Open the Material Editor. Select the sample slot with the material **Paperweight Glass**. Drag the material to a new sample slot, and rename it **NewGlass**. Assign the material **NewGlass** to the object **PaperweightGlass**.

4. Increase the **Glossiness** parameter to **60**. Increase the **Opacity** parameter to **100**. On the **Extended Parameters** rollout, increase the **Falloff Amt** parameter to **100**.

5. Go into the **CHROMIC.JPG** relfection map for the **NewGlass** material. In the **Coordinates** rollout, increase the **Blur Offset** to **0.05**. This will soften up the bitmap, giving the paperweight glass a more realistic look.

6. Render the scene. The Rendered Frame Window is overwritten with the new rendering, but the clone is not affected. Compare the version with Falloff to the version without Falloff. The edges of the paperweight appear more solid, like glass. The reflection map is more subtle and looks less computer-generated.

Figure 11-37: Paperweight with more solid edges

Refractive Water

1. Right-click in the viewport and choose **Unhide All** from the Quad Menu.

   An object called **Paperweight Water** appears in the scene. This is a water-shaped object that sits inside the paperweight. It currently has a solid material assigned to it.

2. In the Material Editor, make a copy of the **NewGlass** material, and rename it **Paperweight Water**. Assign it to the water.

3. Reduce the **Falloff Amt** of the **Paperweight Water** material to **0**.

4. Assign a **Raytrace** map to the **Refraction** of the **Paperweight Water** material. On the **Raytracer Parameters** rollout, turn off the **Enable Self Reflect/Refract** check box.

5. Re-render the camera view.

Figure 11-38: Refractive water in paperweight

6. Experiment with the water material. Render and save an image called **PaperweightWater.tga**.

7. Save the scene with the filename **PaperweightWater.max** in your folder.

TIP: Try giving the water material a different IOR parameter, or animate the IOR parameter for different effects.

SPECIAL SHADERS AND MATERIALS

When first starting to use 3ds Max, you will use the **Standard** material and the **Blinn** shading algorithm. Standard materials are fine for the vast majority of your needs. Blinn is the most versatile of all shaders, and can produce almost any surface, from skin to metal. However, there are a few specialized materials and shading algorithms that may come in handy on occasion.

SHADING ALGORITHMS

Phong

One of the oldest shading algorithms around is **Phong**. It was developed in 1974 by a researcher named Bui-Toung Phong. This shader gives a plastic look that can be readily identified in many early computer animation pieces. Blinn is basically an improved version of Phong, and the only advantage of the Phong shader is quicker render times.

Oren-Nayar-Blinn

A more useful shader than Phong, **Oren-Nayar-Blinn** is helpful for rough surfaces such as unpolished stone, clay, and cloth. It provides two **Advanced Diffuse** controls, **Diffuse Level** and **Roughness**, that can be adjusted to achieve the look of a matte finish.

Figure 11-39: Blinn shader on the left; **Oren-Nayar-Blinn** shader on the right

Metal and Strauss

The **Metal** shader is designed to emulate the effect of metallic surfaces. It doesn't have a Specular Color swatch, because metal highlights are always the same color as the light that shines upon the object. In most cases, the Blinn shader will yield perfectly acceptable results for metal surfaces. The only case in which you'll really need the Metal shader is when you're trying to achieve the look of a very dull metal surface. In this case, use a low Glossiness value and a high Specular value. The highlights on the surface will be darker in the center and brighter near the edge.

Figure 11-40: Blinn shader on the left; **Metal** *shader on the right*

Strauss is a shader that is designed for highly reflective, polished metal surfaces. It has a very simple interface and far fewer parameters than other shaders. There's no compelling reason to use it instead of Blinn.

Translucent

The Translucent Shader is good for materials that allow light to partially transmit through their surfaces. Examples are small plastic or stone objects, or thin cloth. Increase the brightness of the Translucent Clr to see the effect.

If you are trying to reproduce translucent materials such as candle wax, you could place a map into the Translucent Clr attribute. That way, you can make the candle more translucent near the flame and more opaque away from the flame.

Anisotropic

The **Anisotropic** shader creates elliptical highlights. *Anisotropic* means "not the same shape in all directions." Instead of the perfectly circular highlights produced by Blinn or Phong shading, the Anisotropic shader results in streaked highlights. This is great when you need a glossy highlight that conforms to the surface of the object. Classic examples are hair and silk, which never look right if they are rendered with round highlights.

Figure 11-41: Blinn shader on the left; **Anisotropic** *shader on the right*

Multi-Layer

The Multi-Layer shader is designed to give you more control over specular highlights. You have two separate layers of highlights, and you can determine color and shape of each independently.

SPECIAL MATERIALS

Raytrace

Previously, you learned to use the Raytrace *map* in the Reflection or Refraction slot of a Standard material. In 3ds Max, raytracing may also be used in a scene via the Raytrace *material.*

The Raytrace material has many more controls than the Raytrace map. However, in most cases, you won't need them. The Raytrace material takes longer to render, unless you are careful to turn off the options you don't need. For that reason, you are usually better off using the Raytrace map in a Standard material.

Ink 'n Paint

One of the most exciting features in 3ds Max is the **Ink 'n Paint** material. This is commonly referred to in the computer graphics industry as a "toon shader"— a rendering technique which creates images resembling cartoons. It can be used to create a variety of rendered effects, including sketchy charcoal drawings and engineering-style blueprints. It only works in Camera or Perspective views.

Since Ink 'n Paint is a material, you can combine realistic and nonrealistic rendering in the same scene. It is a fun technique that is ripe for experimentation!

Figure 11-42: **Standard** and **Ink 'n Paint** materials with similar parameters

Ink 'n Paint has two main components: the Ink and the Paint. Paint consists of the **Lighted, Shaded,** and **Highlight** components. These correspond to the Diffuse, Ambient, and Specular components in a Standard material. Any of these may be mapped. For example, you can place a bitmap texture in the Lighted map slot and produce an interesting hybrid of "drawn" and photographed imagery.

If the Shaded component check box is on, then the color in the Lighted swatch will appear throughout the entire surface of the object. The number next to Shaded determines how much influence the lighting in the scene will have over the shading. Higher values will "flatten" the paint out, reducing contrast. A value of **100** results in a single flat wash of color in the Paint component. Lower values give more contrast.

If the Shaded check box is off, a color swatch appears. This is essentially the ambient color of the Paint. Now you have two colors (or maps) for the paint, and the Shaded swatch is the color of the object where it is not directly illuminated. Choose a color that is different from the Lighted swatch for interesting gradient effects.

The **Paint Levels** spinner is an important one. This determines how many bands of color will appear across the surface of the object. A value of **1** effectively disables the Shaded component, and you only see the Lighted component. Increase the Paint Levels parameter to give more detail to the shading. If you increase the Paint Levels high enough, to **20** or more, the Paint begins to resemble a Standard material.

The **Highlight** component is optional. If it is checked off, no hot spots appear in the rendering. If it's enabled, you can choose a color or map, and the **Glossiness** parameter determines the size of the highlight.

If both Ink and Paint are enabled, you'll get the classic cartoon look. If all Paint components are switched off, the material will render Ink only. This is very useful for renderings that look like line drawings.

Figure 11-43: **Ink Width** controlled by a **Noise** map

Ink has more controls than Paint. The most important parameter group is Ink Width. Here you can determine how thick the lines will render. If desired, you can use a variable width ink line. If you want 3ds Max to decide how to vary the line, simply enable the **Variable Width** check box and set a **Min** and **Max** width value. Or, you can place a map into the Ink Width map slot.

Ink lines can be generated in various ways. The remaining parameters on the **Ink Controls** rollout define how the 3D scene produces ink lines. You can enable or disable the following Ink lines: **Outline, Overlap, Underlap, SmGroup,** and **Mat ID**.

The most basic Ink line is Outline, which draws a line around the edges of objects. Overlap causes a line to be drawn wherever an object overlaps itself. Underlap does the same thing, except the line is drawn on the far surface instead of the near surface.

SmGroup allows you to force lines to be drawn based on boundaries between the smoothing groups on an object. Mat ID works the same way, forcing lines to be drawn at the boundaries of object material IDs. If you want a line to appear somewhere, assign the faces of the object different smoothing groups or material IDs.

Ink 'n Paint is a great addition to the 3ds Max toolset. However, be aware that it is quite slow to render, especially if you use Ink. The **Ink Quality** spinner will increase render times dramatically.

In addition, Ink lines must be *supersampled* to look good. This is controlled in the **Supersampling/Antialiasing** rollout. Usage of this tool is discussed later in this chapter.

COMPOUND MATERIALS

A **Compound** material is composed of two or more sub-materials. The Multi/Sub-Object material is an example of a compound material.

Top/Bottom

A **Top/Bottom** material uses an object's local axes or the world axes to assign one material to one side of the object, and another to the other side. You have the option of orienting the material to the object's **Local** axes, or to the **World** coordinate system.

Double Sided

A **Double Sided** material applies one material to the sides of faces with normals, and another to faces without normals. In other words, it works like a 2-Sided material, except that there is one material on one side of the face, and another on the other side.

Translucency controls the blending of the two materials. When Translucency is **0**, the materials are opaque. When Translucency is **50**, the materials are blended equally. When Translucency is **100**, the materials are effectively swapped. If your object is not transparent, leave Translucency at **0**. For objects that are partially or fully transparent, set Translucency to a value between **0** and **50** to blend the materials.

Shellac

If you need to simulate a thick coat of varnish or clear paint, the **Shellac** material is just what you need. Shellac lets you add two sub-materials to one another. The brightness values of the two sub-materials are added to produce the final result. This allows you to create highlights on the overlaying material that are very different from the diffuse and ambient portions of the underlying material.

For example, you can use bump mapping on the underlying material, but not on the overlay material. This results in a varnish or shellac effect, because the highlights on the overlay material are soft and round, while the diffuse shading of the underlying material is bumpy and rough.

*Figure 11-44: The teapot at right uses a **Shellac** material to overlay smooth highlights over a bumpy diffuse surface. Compare it to the teapot at left, which has bumpy highlights.*

Blend and Composite

The **Blend** and **Composite** materials are similar to the compositor maps discussed earlier. These types of material are very useful for situations in which it would be too difficult or time-consuming to use Multi/Sub-Object materials, or if you need a gradual transition from one material to another.

The **Blend** material works exactly the same as the Mix map, except that you can place a material in each slot instead of merely a map. You control the placement of each material over the surface of the object with a grayscale map. As discussed later in this chapter, you can also use a Blend material to animate a transition from one material to another.

The **Composite** material is very similar to the map of the same name, except that it relies on material transparency rather than alpha channels for compositing. The sub-materials within the Composite material must have transparency built into them, usually in the form of opacity maps.

Each sub-material in a Composite map can be added, subtracted, or mixed with the sub-material behind it. To choose which method to use for a sub-material, click a **Composite Type** button labeled **A**, **S**, or **M**. The default is **Add**, which overlays sub-materials on top of one another based on their opacity. Opaque areas of a material will obscure the materials behind it. If you choose **Mix**, the sub-materials will simply be superimposed, and you can choose a mix amount for each.

WARNING: Remember, the data flow in the Material Editor is from top to bottom, so the materials higher on the Composite panel will actually render behind the materials that are lower on the panel.

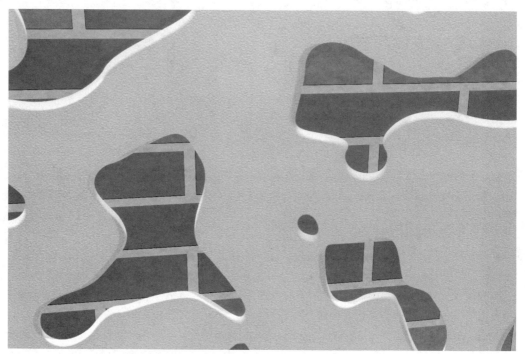

*Figure 11-45: Two materials combined with the **Composite** material*

ARCHITECTURAL MATERIALS

One material type, **Architectural**, is specifically designed to create physically accurate simulations of materials. If you have installed the Architectural materials, you can open them by loading their material library files, which are located in the **3dsMax8\matlibs** folder.

Architectural materials produce highly realistic results, but they only work with advanced lighting features such as **Photometric** lights, **Radiosity**, and the **mental ray** renderer. These features are discussed in Chapter 12, *Advanced Lighting*.

ANIMATED MATERIALS

ANIMATED PARAMETERS

The simplest way to animate a material in 3ds Max is to animate the colors on a material. To do this, simply move to a frame other than **0**, turn on the **Auto Key** button, and change a color. You can animate any spinner parameter, such as the **Amount** for a map. Any technique for animation in 3ds Max can be applied to material parameters, such as Function Curve editing, procedural Controllers, and Parameter Wiring.

In some procedural maps and atmospheres, you can animate the **Phase** value to make the pattern change form over time. An example would be a **Smoke** map. Animating the **Phase** value causes the smoke to change form over time.

MAPPING COORDINATES

Animating the Gizmo of a **UVW Map** modifier causes the map to "move through" the object. You can achieve a similar effect by animating the **Offset, Tiling**, and **Angle** parameters on the **Coordinates** rollout of the Material Editor.

When animating the Offset value, remember that each whole number is a single tile of a map. If you animate from **0.0** to **1.0**, the map returns to the exact same place. This only applies to maps with repeating patterns, such as Bitmap and Gradient. Nonrepeating maps, such as Noise and Planet, can be Offset infinitely without tiling.

ANIMATED BITMAPS

A movie file can be mapped onto a material just like a bitmap. Simply assign a **Bitmap** as the map, and browse to the movie file, such as an **.AVI** or QuickTime file. You can also use a sequentially numbered list of still images as a moving map.

Once you have assigned a movie or image sequence to a map, you can alter its parameters in the **Time** rollout at the bottom of the Material Editor's Bitmap panel. You can set a **Start Frame**, which is when the first frame of the movie will begin in the 3ds Max scene. The End Condition options let you choose how the movie will behave before the start frame and after the movie has played out.

You may also change the **Playback Rate** by entering a value in the spinner field. A value of **1** will play the movie at the scene's frame rate, which is set in the 3ds Max **Time Configuration** dialog. Higher values will play the movie faster, lower values will play it slower. A value of **2** plays the movie at double speed, and a value of **0.5** plays it at half speed.

MIX AND BLEND

The compositor maps and compound materials discussed earlier can be used to combine images over the surfaces of objects. They can also be used to change material parameters over time, such as fading from one map to another. The **Mix** map and the **Blend** material let you animate the **Mix Amount** spinner if there is nothing in the **Mix Amount** or **Mask** slot. Simply animate the Mix Amount from **0** to **100** to fade from one material or map to another.

MORPHER MATERIAL

The Mix map or Blend material are limited to just two maps or materials. The **Morpher** material works with the Morpher *modifier* to provide up to 100 channels of materials, which can be faded up and down with the Morpher modifier channels.

The Morpher material will only work if an object in the scene has a Morpher modifier applied to it. You must click the **Choose Morph Object** button on the Morpher material panel, then click an object in the scene that already has a Morpher modifier applied.

The channels displayed for the Morpher material have a one-to-one relationship with the channels in the Morpher modifier. Each modifier channel can be used for a geometric morph target or for a sub-material. You don't need to morph any geometry to use the Morpher material. It's perfectly legal to apply a Morpher modifier to an object that doesn't change shape, and only animate the materials.

The Morpher material will only superimpose sub-materials. It mixes them, but does not overlay them like the Composite material does.

ANIMATED MATERIAL PREVIEWS

You can create a preview of an animated material, rather than rendering the scene. This is much quicker than creating an **.AVI** of a rendered scene. Click **Make Preview** on the Material Editor toolbar, and enter your rendering preferences in the pop-up dialog.

When the preview is finished rendering, the preview will play back automatically. To view the preview again later, choose **Play Preview** from the flyout under the Make Preview button.

When you create a material preview, the preview is always saved in a filenamed **_MEDIT. AVI**. Only the most recently created preview is saved in this file. To prevent a preview from being overwritten the next time a material preview is made, click **Save Preview** from the Make Preview flyout and resave the preview with a new name.

ANIMATING MATERIAL IDS

You can also animate material IDs. This is accomplished with the **Volume Select** modifier. Use the Volume Select modifier to make a selection of faces, which is then passed up the stack. Then apply a **Material** modifier to set a material ID for the selected faces. When you animate the Volume Select Gizmo, the selection being passed up the stack changes. In this way, the material IDs for faces can change over time.

ADVANCED MAPPING

SUPERSAMPLING

In earlier chapters, you learned that bitmap images often exhibit jagged edges, a problem called *aliasing*. Techniques for smoothing out the jaggies are called *antialiasing* algorithms. When you render an image, you have many antialiasing options in the **Render Scene** dialog. These all apply some form of blurring to the image to smooth over the aliasing. However, sometimes you will see aliasing within a texture map, and you won't be able to fix it by changing the render settings.

To solve this problem, you can use the **Blur** and **Blur Offset** parameters in the Material Editor. However, this does not always work, especially for procedural textures. In fact, sometimes the Blur and Blur Offset spinners are visible on the **Coordinates** rollout, but they don't do anything at all!

Supersampling is often the only way to fix aliasing of procedural textures and raytraced reflections and refractions. Supersampling is a technique whereby an image is rendered at high resolution, and then the resolution is reduced at the final output stage. In this way, the jagged edges can be smoothed out by averaging the values of nearby pixels.

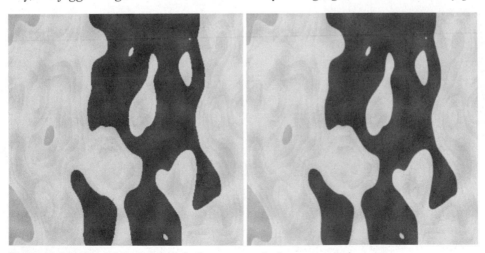

*Figure 11-46: Aliased texture at left, **Supersampled** texture at right*

When you use the **SuperSampling** rollout in the Material Editor, you have several algorithms to choose from. Some are "smarter" than others, but take longer to render. The default sampler, **Max 2.5 Star**, is good in most cases, but the adaptive samplers are better. The only way to figure out what works is by trying all of the options and seeing what's best in your particular situation.

Supersampling works at the material level. Unfortunately, you can't tell 3ds Max to supersample some mapped attributes and not others within the same material. It's either on or off.

MAP CHANNELS

Just as you can have more than one material on a single object, you can also have more than one set of UVW mapping coordinates per object. This is accomplished through the use of **Map Channels**.

Let's say you want to apply a diffuse map and an opacity map to an object, but you want them to have different mapping coordinates. Maybe you need the diffuse map to have spherical coordinates, but the opacity map should have planar coordinates. All you have to do is assign two **UVW Map** modifiers to the object, one for the diffuse map and one for the opacity map.

In each UVW Map modifier, assign a **Map Channel** ID number. Then, in the Material Editor, assign each map a Map Channel ID number in the **Coordinates** rollout. The Map Channel ID numbers determine which UVW Map modifier is used for which mapped material attribute.

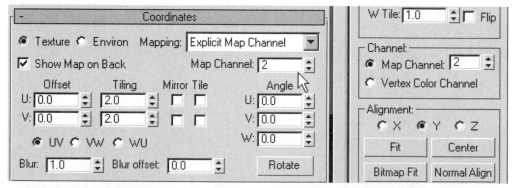

Figure 11-47: ***Map Channels*** *in the Material Editor and in the UVW Map modifier*

You can also use this technique to specify placement of maps within a compositor map. For example, you can use Composite or Mix as a diffuse map, and assign each sub-map a different Map Channel. Then you can use multiple UVW Map modifiers to adjust the placement of the submaps on the surface of the object.

UNWRAP UVW MODIFIER

Inevitably, you will arrive in a situation in which the standard primitive forms of mapping available through the UVW Map modifier will not meet your needs. Complex models with branching architecture cannot be satisfactorily mapped with default Planar, Cylindrical, Spherical, or Shrink Wrap coordinates. If you try to map a tree or a human with standard coordinates, you'll get pixels stretching across the surface of the object.

One way to deal with this is to use compositor maps and mapping channels, as described in the last section. You can place a Composite map into the Diffuse Color channel, and assign multiple Map Channels to UVW Map modifiers. If you use an alpha channel or other map transparency method, you can blend the edges of the maps where the UVW Map Gizmos overlap. However, this method is tedious, and has very real limitations.

What is needed in these cases is some way of directly editing the UVW mapping coordinates, similar to sub-object editing of a mesh object. The perfect tool for this is the **Unwrap UVW** modifier. The heart of the Unwrap UVW modifier is the **Edit UVWs** dialog. With it, you can "tweak" UVW coordinates to seamlessly place maps anywhere, on even the most complex of models.

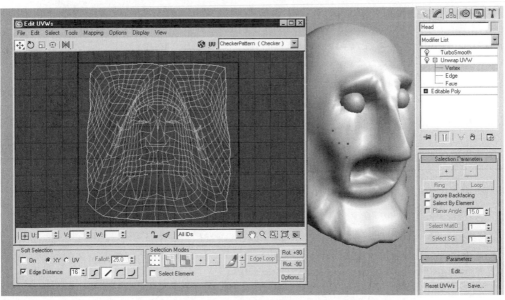

*Figure 11-48: Using **Unwrap UVW** to edit complex mapping coordinates*

The Unwrap UVW modifier is very powerful, and it takes some practice to get good results. The number one rule is to never have any overlapping UVW edges, as this will result in massive tearing and stretching of your map.

Custom Paint Job

A common use of the Unwrap UVW modifier is to prepare a model for a custom paint job. Build a low-polygon model, so you will have fewer UVW vertices to worry about. You can add a TurboSmooth or other subdivision surface modifier later. Assign a material that has a Checker map in the Diffuse Color channel. The checker pattern makes it easy to visualize UV coordinate space on the surface, and to correct any stretching or tearing.

You could begin by assigning basic mapping coordinates with the UVW Map modifier, but that's not really necessary, because Unwrap UVW has the tools to add planar, cylindrical, and spherical mapping. One of the most exciting features in the Unwrap UVW toolset is **Pelt** mapping, which is a quick and easy way to flatten UV coordinates by simulating the way a pelt or skin stretches.

Most of your time is spent in the Edit UVWs dialog. The goal is to use the many tools in this dialog to produce an evenly distributed checker pattern on the object, and get rid of any distorted checkered areas. This will make it a lot easier to paint a custom texture later.

When you are finished editing the UV coordinates, choose **Tools > Render UV Template** from the menu in the Edit UVW dialog. This tool renders the UV coordinates to a bitmap which you can use as a guide for creating custom maps in a 2D paint program. Then simply replace the checker map in your 3ds Max material with the bitmap you created in the paint program. Your custom paint job is finished!

EXERCISE 11.8: Painting a Face

In this exercise you'll use the Unwrap UVW modifier to apply and edit UV coordinates in preparation for creating a bitmap texture in a 2D paint program. UV editing is an art that takes a lot of skill and practice. Here, we'll take a look at the basics.

1. Load the file **unwrap_uvw_start.max** from the disc that comes with this book.

2. Select the **head** object and add an **Unwrap UVW** modifier. The modifier creates default UV mapping coordinates for the object. Seams in the mapping are displayed as thick green lines. See the following illustration.

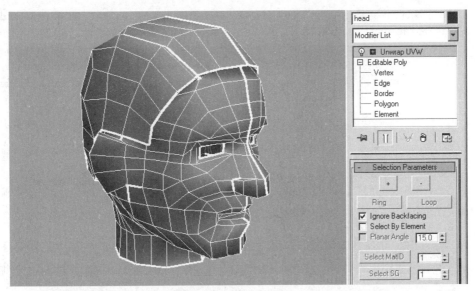

Figure 11-49: **Unwrap UVW** *applies default mapping coordinates with seams*

3. In the Unwrap UVW modifier, enter **Face** sub-object mode. Hold down **<CTRL>** and press **<A>** to select all of the polygons. Scroll the Modify panel to the **Map Parameters** rollout. Click the **Cylindrical** button to add cylindrical mapping coordinates to the selected faces. Click the **Reset** button to restore the mapping gizmo to its default transforms.

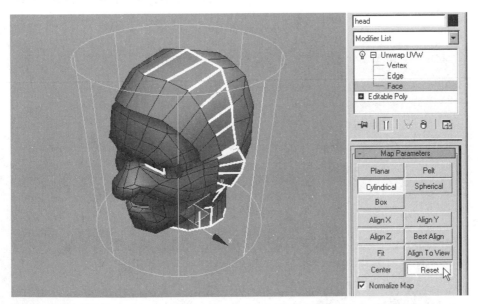

Figure 11-50: Add cylindrical mapping coordinates and click **Reset**

4.   Now there is a series of seams along the side of the head. Press the **<A>** key to activate Angle Snap. Use the Select and Rotate tool to turn the cylindrical mapping gizmo
so that its seam, indicated by a white vertical line on the gizmo, is precisely at the back
of the head. The seams on the object update when you release the mouse button.
When your screen looks like the following illustration, click the Cylindrical button again
to turn it off.

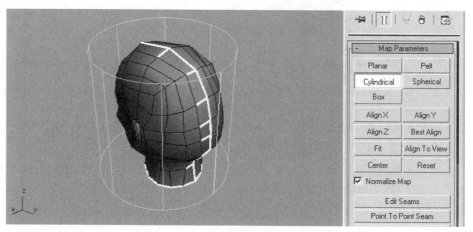

Figure 11-51: Cylindrical mapping gizmo rotated so that seams are at the back of the head

5.   Exit sub-object mode. In the **Parameters** rollout, click the **Edit...** button. The **Edit
UVWs** dialog appears. Turn off the **Show Map** button at the top right of the
dialog. Click an empty place in the UVW editing area to deselect all of the UV vertices.
The dialog should look like the following illustration.

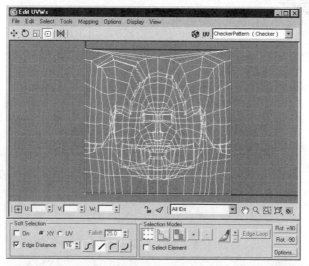

Figure 11-52: Edit UVWs dialog showing the results of cylindrical mapping

6. Choose the **Move** tool at the top of the Edit UVWs dialog. Enter **Face** sub-object mode by clicking the button in the **Selection Modes** section at the bottom of the dialog. Choose **Select > Select Inverted Faces** from the menu at the top of the dialog. All of the faces that are flipped relative to the cylindrical mapping gizmo are selected in the dialog and in the viewport.

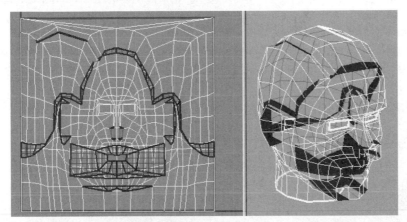

Figure 11-53: **Select Inverted Faces**

7. These are the problem faces on the object. In order to paint accurately in a 2D application, these UV coordinates must be completely flattened. In the Modify panel, disable **Ignore Backfacing**. Hold the **<ALT>** key and click and drag in the viewport to deselect the faces around the nose and mouth. In the Edit UVWs dialog, choose the menu item **Tools > Relax Dialog**. In the Relax Tool dialog, click the **Apply** button. The selected faces are relaxed.

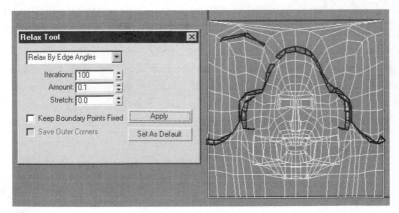

Figure 11-54: *The result of using the* **Relax Tool** *on selected faces*

8. Close the Relax Tool dialog, and choose Select > Select Inverted Faces again. A different set of faces is selected this time. **Choose Select > Select Overlapped Faces** to see the faces that overlap one another. These are also problem areas.

Enter **Vertex** sub-object mode and move the UV vertices in the dialog so that there are no inverted or overlapping faces. Unfortunately, there is no interactive update to indicate whether a face is inverted or overlapping, so you have to move a few UV vertices, go back into Face sub-object mode, and choose an item from the Select menu to check your work. This can be very tedious, but it's still an improvement from the days when the dialog had *no* indication of inverted or overlapping faces whatsoever. To speed the process, use the hotkeys **<1>**, **<2>**, and **<3>** to switch among sub-object modes.

Figure 11-55: Problem area around the ear corrected on the left side of the Edit UVWs dialog

To make matters worse, the Select Inverted Faces menu item is not terribly intuitive when working with non-triangular faces such as quadrilateral polygons in Editable Poly. The polygon may be facing toward the mapping gizmo, but the triangles within that polygon may not. Likewise, the Select Overlapped Faces command may not behave in a predictable or intuitive fashion.

To help you in your work, the Edit UVWs dialog has a special transform gizmo called

Freeform Mode . It is a combination tool that lets you move, rotate, and scale selected sub-objects by clicking various parts of the gizmo. However, sometimes it's better to simply use the standard Move, Rotate, and Scale tools found at the top of the dialog. To scale in both X and Y axes of the dialog, click and drag with the Scale tool active. To scale in just one axis, hold down the **<SHIFT>** key while dragging with the Scale tool. Movement can also be constrained to just one axis with the <SHIFT> key.

Edit UVWs also has a **Soft Selection** feature, which helps when transforming clusters of UV vertices. It is found at the bottom of the dialog, and it works very similarly to Soft Selection in the Modify panel. Adjust the **Falloff** parameter to extend or tighten the influence of Soft Selection.

Correcting all of the problems in the UV mapping coordinates may take a while, perhaps a half-hour or more. Use all of the tools discussed in this tutorial to clean up the overlapping and inverted faces. Do your best to keep the image symmetrical. Make sure that all of the vertices at the borders of the UV space line up precisely with the dark blue grid lines. This is where the edge of the bitmap will be. Keeping UV vertices at the edge of the bitmap makes it easier to paint. You may also need to **Weld** UV vertices among faces that are broken. **Weld Selected** and **Target Weld** are found in the Tools menu.

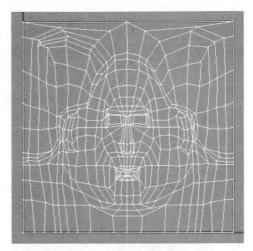

Figure 11-56: UV vertices moved to correct inverted and overlapping faces

9. Exit sub-object mode. Open the Material Editor. Assign the material labeled **Checker** to the model. Make sure that Show Map in Viewport is enabled in the Material Editor. View the model in shaded mode in the Perspective viewport. Make fine adjustments to the vertices in the Edit UVWs dialog to correct the size and shape of the checker pattern on the model in the viewport. A good approach is to even out the size of the checkers. If you need more detail in a certain area of your map, such as around the eyes, make the UV faces larger in the Edit UVWs. This makes the checkers smaller on the model in the viewport.

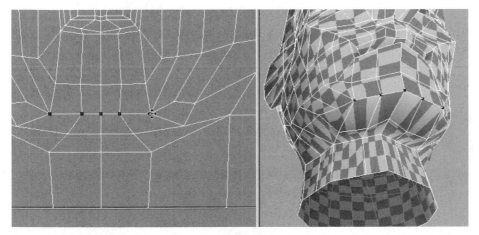

Figure 11-57: Edit UV vertices to produce an even checker pattern on the model

It's often very helpful to use **Constant Update in Viewports** when doing this. Enable Constant Update from the dialog menu item **Options > Advanced Options**, or from the Options toolbar, which is made visible by clicking the **Options** button at the bottom of the dialog. Be aware that this will slow down viewport interaction.

10. When the checker pattern is satisfactory, it's time to export an image to paint. In the Edit UVWs menu, choose **Tools > Render UV Template**. The **Render UVs** dialog appears.

    Change the resolution by entering a value of **512** for the **Width** and **Height**. For gaming and simulation applications, it's best to use bitmaps that are square, and whose resolutions are powers of two. For example, 256 x 256, 512 x 512, and 1024 x 1024 are common bitmap resolutions for games.

    Choose **Normal** as the **Mode** in the **Fill** section of the dialog.

11. Click the **Render UVW Template** button. The **Render Map** window appears. By default, overlapping faces are highlighted in red. This window is not as forgiving as the Select Overlapping Faces command. If necessary, adjust the UV vertices in the Edit UVWs window, and render another image. Click the **Save Bitmap** button at the top of the Render Map window, and save the file as `head_diffuse_template.tga`.

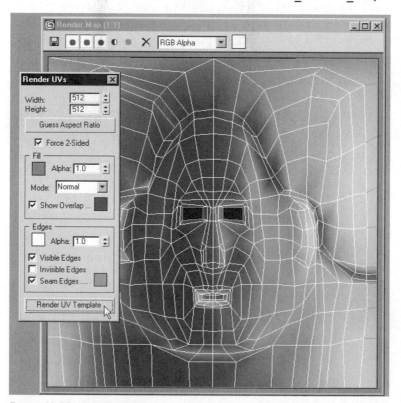

*Figure 11-58: **Render UVs***

12. Close all open dialogs. Right click and choose **Convert To > Convert to Editable Poly** from the Quad Menu. Save the scene as **unwrap_edited.max** in your folder.

13. Open **head_diffuse_template.tga** in an image editor such as Photoshop. Paint over the image on another layer. Save the image out in the image editor's native format, to perserve the layers. For example, in Photoshop, save the document as a **.PSD** file.

    Also save a flattened copy in Targa format to use for the bitmap in 3ds Max. Save the image as **head_diffuse.tga**.

14. Create a new material for the head, and assign **head_diffuse.tga** as the Diffuse Color map. You may add a TurboSmooth modifier. The mapping coordinates from the low-resolution model are propagated to the subdivided model.

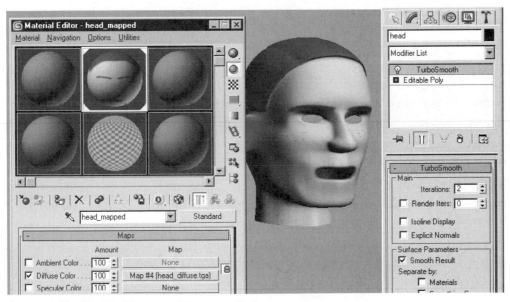

Figure 11-59: Model with custom bitmap applied

15. Save the scene as **unwrap_finished.max** in your folder.

SUMMARY

3ds Max can be used to create bitmap labels, including bitmaps with alpha channels.

Bitmaps can be composited together with the **Composite**, **Mix**, and **Mask** maps.

Opacity makes an object more or less transparent. Any map can be used for opacity. A **Falloff** map makes an excellent opacity map for glass objects.

Reflection is desirable in scenes to lend more realism. **Flat Mirror Reflection** causes reflection on flat surfaces, while the **Reflect/Refract** and **Raytrace** maps are good for reflection on rounded objects.

Refraction is the bending of light through a transparent medium such as glass or water. The **Refraction** map can be used to set up different kinds of refraction, such as **Thin Wall Refraction** and **Raytrace** refraction. The **IOR** parameter on the **Extended Parameters** rollout controls the degree of refraction.

Raytracing is implemented in 3ds Max as both a map and a material. Both take a long time to render, so automatic reflections and refractions should be created with the Reflect/Refract map whenever possible.

3ds Max offers several specialized materials and maps, such as the **Ink 'n Paint** material. This material produces a nonrealistic rendering suitable for cartoons or simulated drawings.

Materials can be animated in various ways. You can animate any color or spinner parameter in the Material Editor, or use moving images as maps. The **Morpher** material lets you fade among up to 100 materials.

Advanced mapping includes **Supersampling** to smooth over jagged pixels, and manual editing of mapping coordinates with the **Unwrap UVW** modifier.

REVIEW QUESTIONS

1. Which file format should you render to in order to save an alpha channel?

2. How do you use a bitmap's RGB intensity in place of an alpha channel?

3. What does the **Mix** map do?

4. What is **Falloff**?

5. What is the *angle of incidence*?

6. How is the reflection calculated in a **Reflect/Refract** map?

7. What is the advantage of using Reflect/Refract as a reflection map, instead of Raytrace?

8. What is the advantage of using **Raytrace** as a refraction map, instead of Reflect/Refract?

9. What property must the sub-materials in a **Composite** material have in order to reveal the sub-materials behind them?

10. What is the *index of refraction* in real life? How does this value correspond to the **IOR** parameter on the Extended Parameters rollout?

11. Does an object need to change shape in order to take advantage of the **Morpher** material?

12. What is **Supersampling**? How is it different from the **Antialiasing** options on the Render Scene dialog?

13. What tool should you use to manually edit mapping coordinates?

Chapter 12
Advanced Lighting

OBJECTIVES

In this chapter, you will learn about:

- Advanced Shadows
- Photometric Lights
- Exposure Control
- Light Tracer
- Skylight
- Radiosity
- Mental Ray

ABOUT ADVANCED LIGHTING

One of the most exciting features in 3ds Max is the suite of advanced lighting tools. With these, it is now possible to create nearly perfect photorealism with less effort. However, to effectively use these tools, it is necessary for the artist to have a strong familiarity with how lighting works in the real world, and how it is simulated in 3ds Max.

It is common for the advanced lighting features to cause very long render times. In some cases, such as with Advanced Shadows, this is unavoidable, because there is no other way to achieve the desired visual result in 3ds Max. In other cases, it may be possible to get good results using the conventional lighting tools. Just as with ray traced materials, it is up to you to decide if the extra rendering time is worth it. Don't use advanced lighting for its own sake; make sure that the end result justifies the rendering time involved.

ADVANCED SHADOWS

ADVANCED RAY TRACED SHADOWS

In chapter 11, *Advanced Materials*, you learned that opacity does not work with Shadow Maps. A light that uses Shadow Maps always produces solid, opaque shadows, despite the material's Opacity attribute. This is a problem if you need soft shadows to be cast by a semitransparent or opacity-mapped object. Standard Ray Traced Shadows do work with opacity maps, but can only produce hard, crisp shadow edges.

This problem is overcome through the use of **Advanced Ray Traced Shadows**. This is a shadow type available for any light in 3ds Max. To use this feature, select a light and go to the Modify panel. Choose **Adv. Ray Traced** from the pulldown list in the **Shadows** section of the **General Parameters** rollout. Then adjust the controls in the **Adv. Ray Traced Params** and **Optimizations** rollouts.

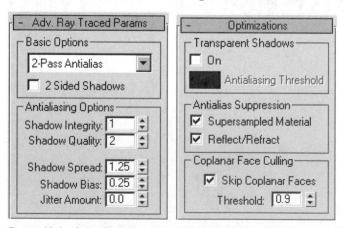

Figure 12-1: **Adv. Ray Traced Params** *and* **Optimizations** *rollouts*

Figure 12-2: **Advanced Ray Traced Shadows** *from an opacity-mapped object*

The controls in the Adv. Ray Traced Params rollout determine the quality of the rendering. The most important factor is the number of shadow rays to be cast. More rays generally results in a higher quality rendering, and the expense of much longer render times.

In the **Basic Options** section, you can choose the raytracing mode from the pulldown list. The **Simple** option disables most of the parameters in the rollout, and the shadow will be the same as a standard Ray Traced Shadow: always sharp and aliased. This is useful for test renders, so you don't have to wait for a lengthy calculation.

To render soft shadows, choose **1-Pass Antialias** or **2-Pass Antialias**. The **1-Pass Antialias** option produces basic soft shadows. You control the accuracy of the shadows by changing the number of rays. To do this, adjust the **Shadow Integrity** spinner in the **Antialiasing Options** section. If small objects or parts of objects are not correctly generating shadows, increasing the Shadow Integrity value will fix the problem.

WARNING: Use caution when increasing the Shadow Integrity and Shadow Quality parameters. The spinners have a range from **1** to **15**. Higher values will increase rendering time without necessarily improving render quality.

If you choose the **2-Pass Antialias** mode, you have more control over the look of the shadow at its edge. The **Shadow Quality** spinner is activated, and you can adjust its value to produce higher-quality shadow edges. The Shadow Quality value controls how many additional rays will be cast to clean up the shadow edge. It should always be equal to or greater than the Shadow Integrity value. Again, beware of high settings, because render times will be excessively long.

Shadow Spread is similar to the Sample Range parameter within a Shadow Map. It is the blur factor for the shadow. Higher values result in softer shadows. However, in an Advanced Ray Traced Shadow, the Shadow Spread value is measured in pixels. So, if you change the resolution of the rendered image, the look of your shadow edges will change. You'll need to remember to adjust the Shadow Spread parameter for all lights that use Advanced Ray Traced Shadows. For example, if you double the resolution of your rendering, you'll also need to double the Shadow Spread value to maintain the same look.

Shadow Bias works the same as it does for other shadow types: it moves the shadow toward or away from the shadow-casting object. **Jitter Amount** randomizes the ray tracing to produce a scattering effect on the shadow. This helps eliminate visible patterns in the shadow edge. Jitter is a very sensitive parameter, and usually you won't need a value above **1.0**. Higher values will introduce too much scattering to the shadow.

The most important parameter in the **Optimizations** rollout is the **Transparent Shadows** check box. Turn it on if you wish to project soft shadows from an opacity mapped object. The rest of the Optimization parameters should normally be left at their defaults.

One of the other main advantages of an Advanced Ray Traced Shadow is that you can render shadows that change focus depending on distance. In the real world, a shadow is sharp where the shadow-casting and shadow-receiving objects are close to one another, and blurry where the objects are far apart. The distance to the light source also affects the shadow focus; nearby lights tend to produce blurry shadows, and distant lights tend to produce sharp shadows.

To achieve these effects, you can use an Advanced Ray Traced Shadow with an Area Light, which we will discuss later in this chapter. However, you can use any light type to create distanced-based shadows if you choose Area Shadows.

AREA SHADOWS

Distance-based soft shadows are achieved with a shadow type called **Area Shadows**. This works with any light in 3ds Max. An Area Shadow is a special form of Advanced Ray Traced Shadow. An Advanced Ray Traced Shadow behaves as if the light is coming from a single point, but an Area Shadow simulates the effect of a shadow from a light source which takes up some amount of space, like a real light bulb.

The controls in the **Area Shadows** rollout area are very similar to those in the Adv. Ray Traced Params rollout. The difference is that Area Shadows have parameters to determine the shape and size of the simulated light source.

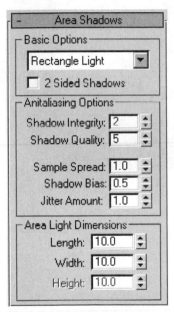

Figure 12-3: **Area Shadows** *rollout*

In the **Basic Options** section, you choose a shadow mode from the pulldown list. The **Simple** option essentially disables the Area Shadow and converts it to a standard Ray Traced Shadow. Other shadow modes in the pulldown list determine the shape of the light for shadow-casting purposes. You can choose among Rectangle, Disc, Box, and Sphere. In the **Area Light Dimensions** rollout, you can control the size of the light source for purposes of the shadow effect. Larger values create softer shadows.

NOTE: The settings in the Area Shadows rollout do not affect the illumination of the scene, they only affect the shadows. A Standard light such as an Omni or Spot light always projects illumination from a single point, and therefore it has no size. To get the effect of a large source of light projected onto diffuse surfaces, use a light type such as Area, Linear, or Skylight. These are discussed later in this chapter.

EXERCISE 12.1: Area Shadows

In this exercise, you will use Area Shadows to create realistic soft shadows.

1. Open the file **PoleShadowMap.max** from the disc that comes with this book.

2. Render the Camera view. The cylinder casts a standard Shadow Map shadow, which has the same amount of blurriness all around the shadow edge.

*Figure 12-4: Default **Shadow Map** shadow*

3. Select the Spot light and open the Modify panel. In the General Parameters rollout, choose **Area Shadows** from the pulldown list. Re-render the scene. The shadow is extremely diffuse.

*Figure 12-5: Default **Area Shadow***

4. Scroll down to the **Area Shadows** rollout. In the **Area Light Dimensions** section, reduce the **Length** and **Width** to 5. Re-render the scene.

    The shadow is now considerably sharper, especially where the cylinder meets the ground plane.

5. Change the shadow mode to **Disc Light** by selecting it from the pulldown list at the top of the Area Shadows rollout. Re-render the scene.

    The shadow changes shape, and its soft edge is more uniform.

6. Increase the **Shadow Integrity** to 5, and the **Shadow Quality** to 7. Re-render the scene. The Area Shadow is starting to shape up, but rendering takes longer.

Figure 12-6: ***Disc Light*** *with smaller* **Light Dimensions**, *higher* **Antialiasing** *settings*

7. Increase the Shadow Quality to **10**. Re-render the scene.

    The shadow edge looks a little better, but takes a lot longer to render.

8. Experiment with the various Area Shadows parameters until you get the look you like.

9. Save a rendered image as **PoleAreaShadows.tga** in your folder.

10. Save the scene as **PoleAreaShadows.max** in your folder.

PHOTOMETRIC LIGHTS

There is another category of lights available from the pulldown list in the Create > Lights panel: **Photometric** lights. These are designed to simulate real-world lighting instruments, such as incandescent bulbs or fluorescent tubes of a certain manufacture.

If you wanted to, you could specify exact real-world light measurements for a photometric light, either by entering information into the spinners, or by loading a file that defines the exact properties of a particular lighting instrument. This is very useful for architecture and visual simulation, so that lighting tests can be conducted on a virtual environment before it is built in the real world.

However, for most of us, this is overkill. There is little need for scientific precision in animation for film and video, nor for most 3D illustration. All that matters is that the image on the screen or on the page looks the way we want it to. The scientific basis of Photometric lights does not matter much to us, but the superior image quality they can achieve is very important.

Because Photometric lights are based on physics, attenuation is built in. To adjust the falloff in intensity of a Photometric light, use the **Distribution** controls, which are discussed later in this chapter.

 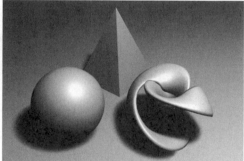

*Figure 12-7: The same scene rendered twice: a single Standard **Omni** light at left, and a single Photometric **Point** light at right. Note how flat and computer-generated the Standard lighting looks.*

UNIT SCALE

Photometric lights rely on real-world measurements of distance and area. To use them most effectively, you should choose some unit of measure for your scene. 3ds Max Generic units don't give you any sense of scale, so you might try to light a gigantic scene with a Photometric light, and only get a black screen. Before you work with Photometric lights, it's a good practice to change the **Display Unit Scale** in the **Customize > Units Setup** dialog. It doesn't matter whether you choose Meters or Feet, as long as you choose some unit of measure, to give you a reference for the size of the scene.

LIGHT TYPES

There are three types of general-purpose Photometric lights: **Point, Linear**, and **Area**. Each comes in a Target and a Free variety. Just as with Standard lights, you can convert among the different light types on the **General Parameters** rollout.

Photometric **Point** lights radiate light energy from a single point source, just like Standard lights. **Linear** lights are in the shape of a straight line, and **Area** lights are rectangular. Linear and Area lights give a more realistic effect, because lights in the real world always have some amount of area. For example, a light bulb emits light from its entire surface, not from an infinitely small point at the center of the bulb. Higher settings on the **Linear Light Parameters** or **Area Light Parameters** rollout will result in softer, more diffuse lighting.

There are a few other special types of lights in 3ds Max, but these only work with Advanced Lighting plugins, which we will discuss later in this chapter.

CONTROLS

Photometric Lights have controls that are very different from Standard lights. You set the parameters for a Photometric light in the **Intensity/Color/Distribution** rollout.

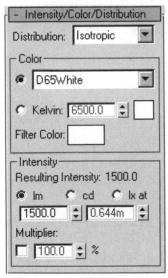

Figure 12-8: **Intensity/Color/Distribution** rollout

Intensity

When assigning intensity to Photometric lights, you are working with real-world units of measure, such as the *candela*. Usually it's not important for you to use precise measurements, but you should have a general idea of the scale of these units. A candela is roughly equivalent to the light emitted by a single wax candle.

The default Intensity for a 3ds Max Photometric light is **1500** candelas. This is roughly equivalent to the light energy output of ten 100-watt light bulbs. As you increase the **Intensity** spinner (under the **lm** button) you increase the number of candelas the light generates. (You may also specify intensity in *lumens*, or as a number of *lux* at a certain distance.) Note that viewport brightness is not accurate with Photometric lights.

Color

The color of a Photometric light can be chosen from a pulldown list, which reads **D65White** by default. This is the color of ordinary daylight, and corresponds to an RGB setting of **255, 255, 255** for a Standard light. The various presets on the pulldown indicate different types of lighting instruments, such as standard **Incandescent** light bulbs, or **Fluorescent** tubes. A sample of the color you have chosen appears in the swatch directly below the pulldown.

You may customize the color of a light by selecting the option labeled **Kelvin**, and entering a value in the spinner. When you do this, you are controlling the *color temperature* of the light source. Color temperature is a way of calibrating light color, and it is expressed in degrees Kelvin. Higher values are more blue, lower values are more orange. A preview of the light color appears in the adjacent swatch.

If you wish to use ordinary RGB or HSV values to define your light color, choose the D65White preset and adjust the **Filter Color** by clicking on the sample swatch.

Distribution

Distribution refers to how the intensity of a Photometric light changes depending on direction. Your choice on the Distribution pulldown determines whether a Photometric light shines equally in all directions or not.

Point lights can behave similarly to omnidirectional or spot lights, depending on Distribution mode. Choose **Isotropic** to send light out in all directions from the Point light. Note that the look of an Omni light is different from a Point light using Isotropic Distribution. The Point light is much more realistic.

Choose **Spotlight** Distribution to access the familiar **Spotlight Parameters** rollout. There, you can define angles for the **Hotspot/Beam** and the **Falloff/Field.** There is no Overshoot option for Photometric Spotlights, as there is for Standard Spotlights.

Linear and Area lights don't have the option to use Isotropic Distribution. Instead, they use **Diffuse** Distribution, which is similar to the way Standard lights affect the diffuse surfaces of materials. The lack of an Isotropic option isn't really much of a hindrance to good image quality, because Area lights have their own soft look.

Finally, for any Photometric light, you can choose **Web** distribution. This option lets you load a file that describes the directional properties of a light. These files are available from lighting manufacturers if you need to simulate a certain lighting instrument.

EXERCISE 12.2: Area Light

In this exercise, you'll compare a Standard Omni light to an Area light. You'll also use Advanced Ray Traced Shadows to produce a soft lighting effect.

Figure 12-9: ***Area*** *light with* ***Advanced Ray Traced*** *shadows*

1. Load the file **PrimitiveScene.max** from the disc that comes with this book.

2. Render the Camera view. Click the **Clone** button to make a copy of the Rendered Frame Window, so you can compare the various renderings. Don't close either of the Rendered Frame Windows; just minimize them to get them out of the way.

    The scene only has a single Omni light, so the rendering is not very interesting.

3. From the Main Menu, choose **Tools > Light Lister**. Turn off the **Omni** light and turn on the **Photometric** light. Re-render the camera view.

    Compare the two renderings. Note how the Photometric Point light has automatic attenuation, so the light fades off into the distance. The overall quality of the photometric rendering is superior.

4. Select the **Photometric** light and open the Modify panel. In the **General Parameters** rollout, change the **Light Type** to **Area**. Re-render the scene.

    The scene is slightly darker, and the highlights on the objects are not as focused.

5. Increase the **Intensity** to **3000 cd**. Re-render the scene.

6. Change the Shadow type to **Adv. Ray Traced**. Re-render the scene.

   Make certain that you don't render an image larger than about 400 pixels wide. The rendering takes much, much longer because 3ds Max is calculating soft light and soft shadows. Higher resolutions may take hours to render a single image.

7. Save the rendering as `AreaLight.tga` in your folder.

8. Save the scene as `PrimitiveSceneAreaLight.max` in your folder.

EXPOSURE CONTROL

Even if you set up a scene with Photometric lights to the correct scale and an appropriate real-world light intensity, you still might not get anything resembling what you'd see with the naked eye. In fact, you might get a black screen.

This isn't a bug or mistake within 3ds Max. The same thing would happen if you took a photograph with the wrong camera settings. In a real camera, you determine the brightness of the recorded image by adjusting the exposure settings, such as f-stop and shutter speed. In 3ds Max, you accomplish this by changing the settings in the **Exposure Control** section of the Environment and Effects dialog.

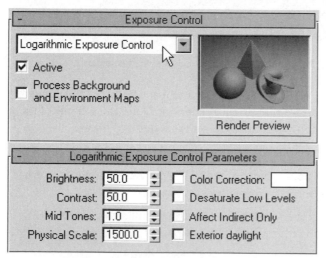

Figure 12-10: Exposure Control settings on the Environment and Effects dialog

To use Exposure Control, choose a type from the pulldown list. In most cases you should use **Logarithmic Exposure Control**; other types can cause flickering in animated scenes. Then adjust the **Brightness**, **Contrast**, and **Mid Tones** parameters to achieve the look you're after. Click the **Render Preview** button to see a thumbnail of what the rendered image will look like.

Exposure Control works with both Photometric and Standard lights, but it's most helpful when using Photometric lights. It's especially necessary with very dark or very bright scenes, or scenes with extreme contrast. If you use **Radiosity**, you'll probably need Exposure Control. Radiosity is discussed later in this chapter.

Figure 12-11: No Exposure Control at left, **Logarithmic Exposure Control** *at right*

ADVANCED LIGHTING PLUGINS

3ds Max has two **Advanced Lighting** plugins. These are methods for calculating *global illumination*, which accounts for the way light is transferred from one object to another. Standard 3D rendering, called *scanline rendering*, does not attempt to simulate how light interacts among multiple objects, but global illumination algorithms do.

Standard raytracing is a form of global illumination which is excellent at reproducing specular reflections, transparency, and refraction. However, it does not simulate the transfer of light energy among *diffuse* surfaces. As you remember from Chapter 4, *Materials*, the diffuse component of a material is the color of the object where it is directly illuminated. In the real world, a diffuse surface reflects light to its surroundings.

Advanced Lighting controls are found in their own dialog, which is launched from the **Rendering > Advanced Lighting** item on the Main Menu. In the Advanced Lighting dialog, you can choose the **Light Tracer** or **Radiosity** plugin. 3ds Max can now render highly realistic diffuse surfaces. Just remember that, as always, render times increase as you add more sophisticated lighting and rendering techniques to the scene.

TIP: You can combine Advanced Lighting plugins with the Raytrace map to achieve images of startling realism and beauty. Advanced Lighting handles the diffuse surfaces, and ray tracing handles the specular components, reflections, and refractions.

LIGHT TRACER

The 3ds Max **Light Tracer** is an Advanced Lighting plugin which is designed to produce the effect of light bouncing off of diffuse surfaces. It is not physically based, and can work with either Standard or Photometric lights. In general, it is better for outdoor scenes than for indoor scenes. If you want to use Advanced Lighting for indoor scenes, you're better off with the **Radiosity** plugin.

To use the Light Tracer, go to the Main Menu and choose **Rendering > Advanced Lighting > Light Tracer**. This opens the Render Scene dialog, selects the **Advanced Lighting** tab, and assigns the Light Tracer as the current plugin.

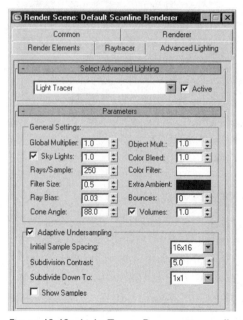

*Figure 12-12: Light Tracer **Parameters** rollout on the **Advanced Lighting** tab of the Render Scene dialog*

The Light Tracer **Parameters** rollout has a fair number of controls. Most of these are optimizations for render quality. The most important setting is the **Bounces** spinner. This controls how many times a light ray will bounce off of diffuse surfaces. Usually, a Bounce setting of **1** is sufficient for most scenes. At a Bounce setting of **0**, the Light Tracer will not allow light energy to propagate through the scene at all.

Skylight

Although you can use the Light Tracer with any lights in 3ds Max, it becomes most useful in conjunction with a **Skylight**. The Skylight is dome light that simulates the effect of ambient light coming from everywhere in the sky. It will only work if you enable one of the Advanced Lighting plugins.

Using a Skylight is a much more sophisticated method of creating simple ambient lighting than adjusting the Ambient Light color swatch in the Environment dialog. As you learned in Chapter 5, *Cameras and Lights*, the Ambient Light setting tends to wash out the scene. A Skylight, on the other hand, produces a very soft and natural look, even if the Bounces setting on the Light Tracer Parameters rollout is set to **0**.

A Skylight alone is not sufficient to produce a realistic outdoor scene, unless you are trying to reproduce a totally overcast day, and there are no shiny, specular surfaces in the scene. To reproduce a clear, sunny day, you'll need to add a Direct light also, preferably with Ray Traced shadows. The combination of a Direct light for specular highlights and shadows, and a Skylight for ambient bounced light, produces a very pleasing look to a sunny scene.

EXERCISE 12.3: Light Tracer

In this exercise, you'll use a Skylight and Light Tracer to add realism to an outdoor scene.

1. Open the file **FountainNew.max** from the disc that comes with this book.

    Currently, there is no lighting in the scene.

2. To keep render times reasonable, render the following exercise at 320x240.

    Render the camera view. It appears flat and uninteresting.

3. Create a **Target Direct** light in the Top viewport. Position the light so it is shining down on the scene from the left of the camera, as shown in the following.

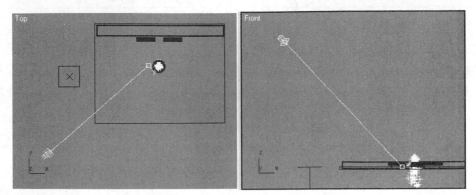

*Figure 12-13: Create and position a **Direct** light*

4. Enable **Ray Traced Shadows** for the Direct light. Turn on the **Overshoot** option. Increase the **Falloff** parameter to enclose the entire scene, so the shadows will appear everywhere.

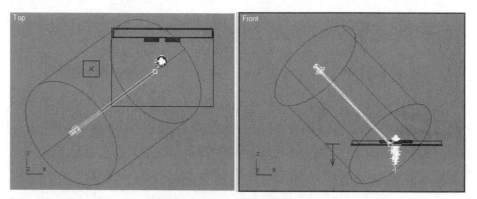

*Figure 12-14: Increase the **Falloff** parameter to enclose the scene*

5 Re-render the scene. The Direct light alone gives a poor approximation of sunlight.

*Figure 12-15: Scene rendered with a single **Direct** light only*

6. Turn off the Direct light. Create a **Skylight** off to the side of the scene geometry. The location of the Skylight does not affect the rendering. Check the Multiplier value of the Skylight; it should be set to **1.0**.

7. In the Main Menu, select **Rendering > Advanced Lighting > Light Tracer**. The **Render Scene** dialog appears with the **Advanced Lighting** tab visible. Leave all of the settings at their defaults.

8. Re-render the scene. Enabling Advanced Lighting results in much longer rendering time.

*Figure 12-16: Scene rendered with a single **Skylight** only*

It looks like a completely overcast day, because there are no hard shadows. Also, there are no specular highlights on the water materials for the fountain, and this does not look right.

Now you will combine the Direct light with the Skylight to achieve the effect of a sunny day with realistic ambience.

9.   Reduce the **Multiplier** value of the Skylight to **0.5**.

The combined illuminance of a Direct light and the Skylight at full intensities would wash out the scene. You could use Exposure Control to correct this, but there is no need to do so in this case, and it would only increase render times unnecessarily.

10.   Turn the Direct light back on, and re-render the scene.

*Figure 12-17: Scene rendered with **Direct** light and **Skylight***

The scene looks much more convincing, because the Direct light renders shadows and specular highlights, and the Skylight renders realistic diffuse surfaces.

Finally, you will finish the job by enabling diffuse reflections.

11.   In the Advanced Lighting dialog, increase the **Bounces** value to **1**. Re-render the scene.

*Figure 12-18: Scene rendered with a **Bounce** value of **1***

Notice how all of the diffuse surfaces are slightly brighter. Also, shadow areas are filled in more pleasantly. The Light Tracer is simulating the bounced light from nearby diffuse surfaces.

12.   Save the rendered image as **FountainLightTracer.tga** in your folder.

13.   Save the scene as **FountainLightTracer.max** in your folder.

IES Sky and IES Sun

3ds Max also offers a Photometric version of the Skylight, which is called **IES Sky**. This is designed to be used with an **IES Sun** light. If you use these Photometric lights, make sure you enable Logarithmic Exposure Control, because they have very high intensity, just like real sunlight.

You can also employ the **Daylight** system from the Systems panel, which automatically positions an IES Sky and IES Sun light in the scene. The location of the lights depends on parameters such as date, time, and geographic location. After you have created the system, you can access its parameters from the Motion panel.

The Daylight system is designed for architects who need to visualize what a building will look like at certain times of the day and of the year. For general-purpose animation and graphic design applications, the combination of a Direct light and a Standard Skylight is more effective and easier to control than the Daylight system.

NOTE: IES stands for the Illuminating Engineering Society, which publishes recommendations for the lighting industry.

RADIOSITY

Another Advanced Rendering plugin for global illumination is called **Radiosity**. This method is appropriate for indoor scenes. It is more complicated to set up than the Light Tracer, but this setup time pays off in the form of much shorter render times.

To use radiosity, you must calculate a radiosity solution, which may take some time– at least a few minutes, and possibly a half hour or more. Select the **Radiosity** plugin from the pulldown on the **Advanced Lighting** tab of the Render Scene dialog. Then click the **Start** button on the **Radiosity Processing Parameters** rollout. The **Initial Quality** parameter determines how accurate the solution will be, and the default value of **85%** is usually a good setting. Higher Initial Quality values usually won't help the image quality much, but will take a lot longer to calculate.

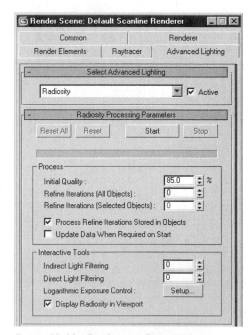

Figure 12-19: **Radiosity Processing Parameters** *on the* **Advanced Lighting** *tab in the Render Scene dialog*

When the solution is found, diffuse lighting information is stored in the vertices of the scene. If the objects in the scene don't move, you can render multiple images using the same radiosity solution. This is perfect for moving-camera "walkthoughs" of architectural models.

If you move any object or light, or change any parameters in the scene, the radiosity solution is no longer valid, and you will need to click the **Reset All** button and recalculate the solution.

If some areas of the scene render with dark spots or splotches, you can use the **Refine Iterations** parameters. These controls will cause the Radiosity plugin to perform additional calculations on the entire scene, or on the selected objects. If you adjust the Refine Iterations parameters, you'll have to reset the radiosity solution and recalculate it.

To further improve the quality of a radiosity rendering, use **Filtering**. This is a post-processing effect applied after the solution is found. It removes splotches and dark spots from radiosity-rendered objects. There are filter settings for both **Direct Light Filtering** and **Indirect Light Filtering**. Direct illumination comes from lights in the scene, and indirect illumination bounces off of surfaces. Values of **1** to **5** should be fine for most scenes. You don't need to re-calculate the radiosity solution if you change the Filtering settings.

Photometric Lights

Although the 3ds Max Radiosity plugin will work with Standard lights, it achieves more realistic results if you use Photometric lights and Exposure Control. Whenever you use Photometric lights, you should model your scene to the proper scale. If you decide to use Photometric lights and radiosity on an existing scene, you should scale it to the proper real-world size before placing lights and calculating radiosity.

Radiosity and Modeling

Radiosity is a mesh-based approach. In general, more complex models will yield better results. The diffuse surfaces of the models are considered on a per-face basis. This means that a model with a high level of detail will produce a more accurate radiosity solution.

If you don't have the time or inclination to remodel your scene, or subdivide the faces of individual models, you can make the Radiosity plugin do that for you. Open the **Radiosity Meshing Parameters** rollout on the Advanced Lighting tab of the Render Scene dialog, click the **Enabled** option, and enter values in the **Maximum Mesh Size** and **Minimum Mesh Size** spinners. Smaller Meshing Size values will produce more accurate radiosity solutions, at the expense of longer calculation times.

By default, meshing is adaptive, meaning that radiosity triangles are smaller in areas of higher contrast. As seen in the following illustration, the radiosity mesh is more dense in shadow areas than in areas of flat illumination. This results in a more accurate simulation of how light is transferred among surfaces.

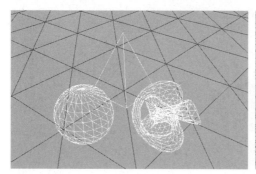 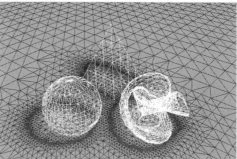

Figure 12-20: A scene before and after radiosity meshing

Also, since radiosity depends on the location of faces and vertices, the actual structure of your scene does matter. In some situations, you might see "light leaks." This term describes a rendering problem in which radiosity faces are too bright, as if light had leaked through one object to illuminate another object.

To avoid this situation, try to model your scene with radiosity in mind. Make sure that ceilings, walls, and floors meet perfectly, with no overlaps or gaps in geometry. It's best to weld

Figure 12-21: A "light leak"

the vertices, and use Multi/Sub-Object materials for the different surfaces. If you run into light leaks that can't be fixed by remodeling, you can use the **Light Painting** feature in the Advanced Lighting tab of the Render Scene dialog. This allows you to add or subtract light from the selected object using a paint tool.

Radiosity and Materials

Radiosity is very sensitive to materials. If you create a material with dark color components, objects with that material will not contribute much bounced light to the radiosity solution. Some artists prefer to use very dark colors in the **Ambient** color swatch of a material. However, this doesn't work well with radiosity. To ensure that materials are allowing light to bounce, use a bright Ambient color. Locking the Ambient and Diffuse color swatches helps.

This is where the **Architectural** material type mentioned in Chapter 11 comes in. These materials have been designed to physically simulate the way light interacts with surfaces. To use Architectural materials, go to the Material Editor and click **Get Material** . In the Material/Map Browser, click **Browse From: > Mtl Library**. Click **File > Open** in the Material/Map Browser and choose one of the Architectural libraries. You'll find preset materials that work great with Photometric lights and Radiosity. Use of Architectural materials with Standard lights won't work well and is not recommended by Autodesk.

Also, 3ds Max offers a material type called **Advanced Lighting Override**. This is an alternative to Architectural materials, so you can you use existing materials. Advanced Lighting Override adds new parameters to Standard or other material types. These let you specify additional material properties related to radiosity, such as Reflectance Scale, which helps control much diffuse light will be reflected. The Luminance Scale parameter also allows a self-illuminated object to cast light into the scene, as if it were a light object. This is great for neon signs and other oddly shaped objects that emit light.

Radiosity and Animation

It is possible to apply radiosity to animated objects such as characters. To do so, you can enable the option labeled **Compute Advanced Lighting when Required** on the **Render Scene** dialog. However, this is usually not practical, because the solution must be calculated for each frame. The time required for radiosity calculation on each frame can easily put a project past its deadline, so this technique is not used often in production.

Radiosity can be used in conjunction with standard rendering, to avoid the extra rendering time. To do so, exclude characters from being processed by the Radiosity plugin. Right-click an object and choose **Properties** from the Quad Menu. Then select the **Adv. Lighting** tab in the **Object Properties** dialog, and check the option labeled **Exclude from Adv. Lighting Calculations**. Then you can calculate a radiosity solution for the architecture and props just once, and render your scene with animated objects using the standard scanline renderer. Just be aware that it may be difficult to match the look of the animated characters to the look of the radiosity-rendered objects.

The Adv. Lighting tab in the Object Properties dialog has other controls, such as refinement. You can force an object to be refined a certain number of times during the calculation of the radiosity solution. This overrides the settings on the Advanced Lighting dialog, and is very useful for animations, to ensure that the same number of refinements is made for each frame.

EXERCISE 12.4: Radiosity

In this exercise, you'll use a Photometric IES Sun light to illuminate a scene with the Radiosity plugin.

1. Load the scene `AtticEmpty.max` from the disc that comes with this book.

    This is a modified version of the attic scene you worked on earlier. It has just one light source, an IES Sun light with default parameters, shining through the window. Ray Traced Shadows are enabled for this light already.

2. Render the scene. With no Advanced Lighting enabled, the scene is black except for the window and a spot of white light on the floor.

3. On the Main Menu, choose **Rendering > Advanced Lighting > Radiosity**. A dialog appears asking if you wish to use Camera Exposure Control. Click **No**.

4. Click the **Start** button in the **Radiosity Processing Parameters** rollout to begin calculating the solution. When it is finished, render the scene again.

Figure 12-22: **Radiosity** *rendering with default parameters*

The result is not very good. The problem is that the scene is very simple, and there isn't enough detail to the geometry.

5. Open the **Radiosity Meshing Parameters** rollout. Click the **Enabled** option, and change the **Maximum Mesh Size** value to **0.2m**.

6. In the **Radiosity Processing Parameters** rollout, click the **Reset All** button, then click the **Start** button to recalculate the solution. A confirmation dialog appears; click **Yes** to reset the radiosity solution.

   The solution takes longer this time, because the scene geometry is being automatically subdivided, and the diffuse brightness of each face is being considered in the calculation.

*Figure 12-23: Radiosity with **Meshing** enabled*

   You can see geometric patterns of dark and light on the objects in the scene. This is the result of radiosity meshing. Now you will refine the solution to improve visual quality.

7. On the Radiosity Processing Parameters, increase the value of **Refine Iterations (All Objects)** to **3**. Then press the **Continue** button. The radiosity solution is refined for the entire scene.

   There is no need to reset the radiosity solution.

8. Re-render the camera view. The variations in brightness are still visible, but they are not as extreme as before.

9. Increase the **Indirect Light Filtering** parameter to **5**, and re-render the scene.

The scene is almost finished. However, the overall brightness needs to be adjusted. Notice how the spot of light on the floor is solid white, and the rest of the scene is too dark. To fix this, we will use Exposure Control.

10. In the **Interactive Tools** section of the Radiosity Processing Parameters rollout, there is a button labeled **Setup**. Click this button, and the **Environment** dialog is opened automatically.

    In the Exposure Control section of the Environment dialog, select **Logarithmic Exposure Control** from the pulldown list. Set the **Brightness** parameter to **50**.

11. Re-render the scene.

*Figure 12-24: Rendering with **Refine Iterations** and **Logarithmic Exposure Control***

12. Save the rendered image as `AtticRadiosity.tga` in your folder.

13. Save the scene as `AtticRadiosity.max` in your folder.

MENTAL RAY

This book has focused on the 3ds Max **Default Scanline Renderer**. However, it is easy to use other rendering technologies with 3ds Max. Other rendering solutions are available on the market as plugins, such as Brazil, finalRender, and VRay. 3ds Max also ships with an alternate advanced renderer, **mental ray**, included for free.

To use mental ray, open the **Render Scene** dialog. You can use the **<F10>** hotkey. On the **Common** tab of the dialog, open the rollout labeled **Assign Renderer**. Click the button with three dots next to the currently listed Production Renderer, which is the Default Scanline Renderer. Choose **mental ray Renderer** from the pop-up dialog. Now whenever you render, the mental ray software is invoked.

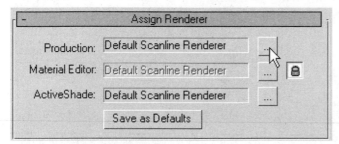

Figure 12-25: **Assign Renderer** *rollout on the Common tab of the Render Scene dialog*

Mental ray is the product of a company called mental images. This software works with competing 3D graphic applications as well. It excels at creating realistic reflections, refractions, and has been used in big productions for many years. Mental ray is a highly advanced renderer, including a software developers' kit, so people can write their own shading algorithms.

The quality of images rendered in mental ray differs from that of the Default Scanline Renderer. In general, reflections and refractions look better in mental ray. Unfortunately, mental ray scenes take longer to set up and render. The render times for mental ray scenes can be very long, but the results are always good and sometimes amazing.

Certain advanced features are not available in the Default Scanline Renderer, or if they are, they're not practical to use. A clear example is the phenomenon of *caustics*, which are focused reflections and refractions. Mental ray simulates the way light is bent when it reflects from a curved surface, and also simulates light passing through a transparent substance. In either case, the result is focused patterns of light projected on nearby surfaces. The Default Scanline Renderer wasn't designed to account for this effect. Although you can get caustics with the Light Tracer Advanced Lighting plugin, it's difficult and takes too long to render.

In general, the integration of mental ray into 3ds Max is good, but it's not 100% perfect. Some features won't work together. For example, you can't use Light Tracer or Radiosity with mental ray. A few parameters within a map or material might not work with mental ray. For example, mental ray does not support the Morpher material.

The interface and parameters in the **Renderer** tab of the Render Scene dialog are completely different when mental ray is active. The **Advanced Lighting** and **Raytracer** tabs are replaced by the **Indirect Illumination** and **Processing** tabs, and their parameters are also completely different from those of the Default Scanline Renderer. You've probably noticed the **mr Area Omni** and **mr Area Spot** light types. These only work with the mental ray renderer. You can also choose **mental ray Shadow Map** as a type of shadow for any light.

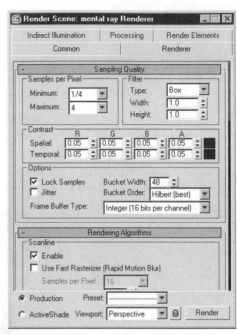

Figure 12-26: Renderer tab with mental ray

JOSEPH WEHLAND

*Figure C-1: Detail of **Rock Castle** by Joseph Wehland*

MATT HIGHISON

Figure C-2: **Abandoned Passage** *by Matt Highison. For more examples of Matt's work, visit www.moonlitbasement.com.*

Figure C-3: Character model by Matt Highison

Figure C-4: Detail of **Abandoned Passage** *by Matt Highison*

Figure C-5: Brooklyn, New York proposal for Gene Kaufman, architect. Image by Planet 9 Studios.
www.planet9.com

Figure C-6: San Jose, California proposal for Skidmore Owings and Merrill, architects. Image by Planet 9 Studios.

Figure C-7: Result of Exercise 12.3: Light Tracer

Figure C-8: Result of Exercise 12.4: Radiosity

AARON ROSS

Figure C-9: **Bronwen I, Rex Mustelidae** *by Aaron Ross*

SUMMARY

3ds Max offers many advanced lighting features. All of them will add to rendering time, so it is important to use these features sparingly and conservatively. If acceptable results can be achieved using Standard lights and shadows, then you should use them.

Opacity mapping does not work with Shadow Maps, and standard Ray Traced shadows always have hard edges. **Advanced Ray Traced** shadows can have soft edges and still render transparency correctly.

Area shadows render the effect of shadows becoming softer over distance. However, when used with a Standard Omni, Spot, or Direct light, only the shadows look soft. To produce soft highlights representative of a large, diffuse light source, you need to use an **Area light**. An Area light with Advanced Ray Traced shadows produces a very pleasant diffuse lighting effect, but it takes a long time to render.

Area lights are found in a separate category, called **Photometric** lights. Photometric lights are designed to work with scenes which are modeled to scale. They are physically based, and you can use them for architectural lighting simulations. Attenuation is calculated automatically when you set a Photometric light's **Intensity** and **Distribution**.

When using Photometric lights, it's often necessary to adjust the brightness and contrast of the rendering with **Exposure Control**. This feature is found in the Environment and Effects dialog. It lets you map the wide range of Photometric light energy into a renderable image.

Global illumination is a technique for simulating the transfer of light among many objects in a scene. Ray tracing is one form of global illumination that excels at rendering specular highlights, reflections, and refractions. For accurate simulation of diffuse reflections, 3ds Max offers two **Advanced Lighting** plugins: the **Light Tracer** and **Radiosity**.

The **Light Tracer** is great for outdoor scenes. Although it works with any kind of light in 3ds Max, it is very useful in conjunction with a **Skylight**. A Skylight simulates true ambient light coming from the entire hemisphere of the sky. For an outdoor scene, use a single Direct light, a Skylight, and the Light Tracer.

The **Bounces** spinner in the Light Tracer determines how many times diffuse reflections will be allowed to bounce off of objects. A setting of **0** or one is usually adequate.

The other type of diffuse global illumination in 3ds Max is called **Radiosity**. This is most useful when employed with Photometric lights in indoor scenes. It requires a lot of setup, but the results can be very realistic. Modeling, materials, and animation all have special considerations when using radiosity. Generally speaking, it is best for objects that don't move, because animated radiosity solutions take a very long time to calculate.

For advanced reflections, refractions, and caustics, the **mental ray** renderer can be more effective and look better than the Default Scanline Renderer.

REVIEW QUESTIONS

1. True or false: You should use advanced lighting techniques whenever possible.

2. What is the advantage of **Advanced Ray Traced** shadows?

3. What is an **Area** light?

4. What is the main difference between **Photometric** lights and Standard lights?

5. Why is **Exposure Control** necessary?

6. What is *global illumination?*

7. What are the two methods for calculating diffuse reflections in 3ds Max?

8. Which **Advanced Lighting** plugin is best for outdoor scenes? Which is best for indoor scenes?

9. What is a **Skylight?**

10. What is the name of the advanced rendering software that ships with 3ds Max?

CUSTOMIZING 3DS MAX

3ds Max abounds in methods for customizing its user interface and functionality. Some of the most useful and commonly used methods are described in this appendix.

USER INTERFACE SCHEMES

When 3ds Max is loaded for the first time, the default UI (user interface) is displayed. If you wish, you can change this. 3ds Max ships with several other UI *schemes*, which use different colors and icons. All of the interface elements are still present and in their default locations.

To load a scheme, go to the Main Menu and choose **Customize > Load Custom UI Scheme**. Then choose a **.UI** file from the **3dsMax8/UI** folder. The next time you start 3ds Max, the scheme you chose will appear again.

The **.UI** file is a collection of all of the different customizations to the 3ds Max user interface. There are several different files for each custom setting. For example, a **.KBD** file stores keyboard shortcuts, and a **.CLR** file stores changes to the UI colors.

If you make changes to the 3ds Max interface, you should save the changes to your own UI scheme. Choose **Customize > Save Custom UI Scheme** from the Main Menu. Make sure you choose a new filename, and don't save over the default 3ds Max factory UI files! If you do, and need to get back to factory settings, you will have to delete all of the files in the **UI** folder. Then use the **Repair** feature on the 3ds Max installation disc to reinstall the UI files.

When you click **OK** in the Save Custom UI Scheme dialog, a dialog labeled **Custom Scheme** pops up. Here you are able to select which of the customization files you wish to save. Click **Save** in the Custom Scheme dialog, and 3ds Max creates all of the necessary files. All of these files have the filename you specified earlier, but they have different extensions. Make sure that you create backups of all of these files, so if you have to reinstall 3ds Max, you can restore the files to the **UI** folder.

TOOLBARS

A *toolbar* is a collection of icons that appear in a small panel. The **Main Toolbar** and **reactor** are examples of toolbars. Others include **Layers**, **Snaps**, and **Brush Presets**. You can move existing toolbars around the 3ds Max interface to suit your preferences.

At the end of each toolbar is a vertical or horizontal handle. Click and drag the handle to move the toolbar. If you release the mouse button while the cursor is located near the center of the screen, the toolbar becomes detached. This is called a *floating* toolbar.

Each toolbar has a pop-up menu, which is accessed by right-clicking an empty area of the toolbar or its handle. To float a toolbar, you may choose **Float** from the right-click menu, instead of dragging the handle.

Another right-click menu item is called **Dock**. A docked toolbar is stuck to the top, bottom, left, or right of the viewport area. The Main Toolbar and Command Panels are docked to the top and right of the screen by default. The Dock right-click menu item parks the toolbar at one side of the viewport area. You can also dock a toolbar by dragging it to the edge of the 3ds Max interface. The mouse cursor changes when the toolbar is in position to dock.

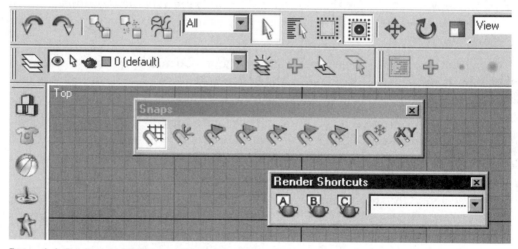

Figure A-1: Docked and floating toolbars

To prevent accidental changes to your existing layout, enable the Main Menu item **Customize > Lock UI Layout.**

CREATING A CUSTOM TOOLBAR

To create a custom toolbar, go to the Main Menu and choose **Customize > Customize User Interface.** You can also right-click an empty area of any toolbar and choose **Customize** from the pop-up menu. The **Customize User Interface** dialog appears.

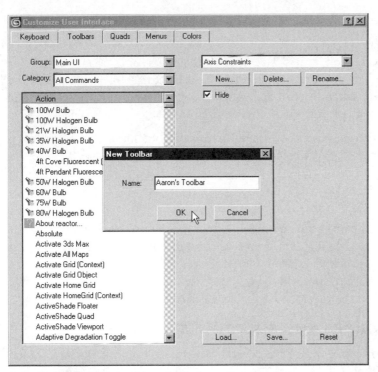

Figure A-2: **Customize User Interface** *dialog*

This dialog lets you customize many parts of the 3ds Max interface. To create a new toolbar, choose the **Toolbars** tab, and click the button labeled **New**. Enter a name for your new toolbar in the pop-up dialog, and click **OK**. A small empty toolbar appears to the left of the dialog, floating over the viewports.

To make some room in the toolbar to place buttons, move the cursor over the rightmost end of the new toolbar until the double-headed arrow appears, and click and drag to the right to extend the toolbar size.

Adding Commands

A custom toolbar button can be created to issue nearly any 3ds Max command. One way to add commands is to simply copy icons from other toolbars onto your new toolbar. If the Customize User Interface dialog is open, you can easily copy icons from one toolbar to another by dragging and dropping. Hold down the <**CTRL**> key, and click and drag to copy an icon. To move an icon, hold down the <**ALT**> key, click and drag.

For commands that aren't already found on toolbars, use the **Action** pane of the Customize User Interface dialog. Select a **Group** of commands, and a **Category** within that group. Search through the list of Actions until you find the one you want, and drag and drop it onto your custom toolbar. Be warned that the naming conventions of the commands are not consistent.

Editing Buttons

If you add a command from the Actions pane of the Customize User Interface dialog, it will appear with a text button at first. To edit the button text, or to assign an icon for the new button, right-click the button and choose **Edit Button Appearance** from the pop-up menu. The **Edit Macro Button** dialog appears. Here you can choose alternate text for the button by typing it into the **Label** field.

If you prefer an icon button, select the **Image Button** option. Then choose a **Group** of icons from the pulldown list. When you find a suitable icon, select it, and then click **OK**.

Using the Toolbar

When you have finished setting up the new toolbar, make sure to **Save** it from the Customize User Interface dialog. All custom toolbars present in the current 3ds Max session are saved in a single interface layout file, which has the extension **.CUI**. Choose a new filename, and don't save over the 3ds Max factory default file. Close the Customize User Interface dialog, and the toolbar is available for use.

CUSTOMIZING COLORS AND SHORTCUTS

The Customize User Interface dialog has several other tabs. Each works in a similar manner to the Toolbars tab.

The **Keyboard** tab lets you assign custom keyboard shortcuts. In the **Action** pane, you will find the same commands that you saw in the Toolbars tab. Next to some of the commands are numbers or letters. These are the currently assigned shortcuts. To assign a new shortcut, highlight an Action and type a key into the **Hotkey** field. If that key is already assigned (nearly all of them are), the current assignment is listed. Click the **Assign** button. Remember to **Save** your changes to a new filename; don't overwrite the 3ds Max defaults!

Likewise, the **Colors** tab allows customizations to the 3ds Max colors. The bottom of the Colors tab assigns colors to generic interface elements such as windows and tooltips. The top part of the dialog assigns colors to viewports and panels. Select an interface category from the **Elements** pulldown, such as **Gizmos,** then choose an interface element, such as **Camera Cone**. Click the color swatch to the right, and assign a color.

Click the **Apply Colors Now** button to update your changes in the 3ds Max interface. As always, **Save** your changes to a new filename.

CONCLUSION

The 3ds Max user interface is highly customizable, but the system is not perfect. There are different files stored for each customization, such as **.CUI** files for toolbars, and **.MNU** files for menus. If you use the **Customize > Save Custom UI Scheme** Main Menu item, you can save all of these files automatically.

aliasing • Undesirable jagged lines, edges, or maps in computer images.

algorithm • A procedure or program that performs a task, usually invisible to the computer user.

alpha channel • A channel of data in bitmap images that stores the transparency of the image, generally used for compositing.

ambient color • The color of an object where it is not directly illuminated. It is influenced by the brightness and color of **ambient light**.

ambient light • General lighting that illuminates the entire scene, without source or direction. It attempts to mimic the effect of light bouncing off of all objects to illuminate all other objects, but is not a true simulation such as **radiosity**.

animation hierarchy • Links among objects, used for controlling animations. An object may have many children, but can have only one parent. In standard animation hierarchies (also called *forward kinematics*), a child object inherits the transforms of its parent. See **inverse kinematics**.

antialiasing • An algorithm that attempts to correct **aliasing**. Blurring the image is one such method. A more advanced method is called **supersampling**.

aspect ratio • The proportion of image width to image height. Aspect ratio describes the shape of an image, not its size. The resolution of an image is not related to the image aspect ratio. See **pixel ratio**.

Bezier • A type of **spline** or **patch** that uses adjustable tangent handles to control the curvature near each control vertex.

bitmap • A computer graphic image produced by a fixed grid of pixels. Examples of bitmap file formats include **.JPEG** (Joint Photographic Experts Group), **.TIF** (Tagged Image File Format), and **.TGA** (Truevision Targa).

Boolean • A compound object that is created by performing a symbolic logical operation on two objects, called *operands*. There are three operations: *union*, *intersection*, and *subtraction*.

Cartesian coordinate system • A method of locating points in 2D or 3D space by measuring distances parallel to intersecting straight-line axes. The coordinate axes are perpendicular to one another, defining a 2D grid or 3D cubic lattice.

diffuse color • The color of an object where it receives direct illumination.

edge • A straight line in a mesh object. An edge connects two vertices and is bounded by a face on either side.

element • In 3ds Max, an element is a group of connected faces within a mesh object.

extrude • A simple process of converting 2D shapes into 3D. A copy of the 2D shape is moved perpendicular to the original, then connected to create a closed surface.

face • A renderable triangle. This term is often used interchangeably with **polygon**, but in 3ds Max, there is a difference. A face is a triangle, the simplest possible 2D shape. All **mesh** objects are ultimately composed of faces.

fields • Interlaced subframes in a video frame. There are two fields in each video frame. Rendering animations to fields improves the smoothness of fast-moving objects when the animation is viewed on interlaced video formats. See **interlaced scan**.

flat shading • A rendering method that does not smooth any edges among faces, resulting in a faceted look. Also known as *constant* shading.

function curve • A graph used for editing **keyframe interpolation**. The vertical dimension of the graph is the value of the animation track, the horizontal dimension is time.

gamma • The contrast curve of a recorded image. Increasing the gamma of an image makes it appear brighter (more washed out); decreasing the gamma makes the image darker (muddier).

global illumination • A rendering algorithm that accounts for the ways in which light is transferred among surfaces in the scene. Examples: **raytracing**, **radiosity**.

interlaced scan • A video display mode that uses alternating **fields** to create the image. NTSC video is interlaced; most computer monitors are not. See **progressive scan**.

interpolation • Automatic generation of values between known values. Examples include **spline** curves and animation **function curves**. The computer user explicitly defines certain values, such as spline control vertex locations or animation keyframes. The interpolation algorithm automatically creates all of the in-between values.

inverse kinematics (IK) • An intuitive alternative to a standard **animation hierarchy**. In IK, the parent inherits the transforms of the child. The final transform state of the linked objects is called the *IK solution*.

keyframe • An animation value that is explicitly defined at a certain time. Computer animation relies on **interpolation** between keyframes.

Klaatu barada nikto • A line from the 1951 science fiction film, *The Day the Earth Stood Still*.

material • The surface properties of an object, such as color, reflectivity, and opacity.

map • A bitmap or algorithm used to create surface detail on a 3D object. **Bitmap** maps require mapping coordinates, **procedural maps** usually do not.

mesh • A 3D object composed of faces or polygons. A mesh object has no true curvature. The appearance of curvature is achieved by increasing the number of polygons or faces (called *mesh density*), and by surface **smoothing**.

normal • A nonrendering line that points out perpendicular to the surface of a face. Ordinary faces have only one normal, which determines which side of the face is renderable. If a normal is pointed away from the camera, that face will not render. Also called face normals, polygon normals, or surface normals.

NURBS • Non-Uniform Rational Basis Spline. A special type of spline, which has weighted control points. The more weight a control vertex (CV) has, the more the curve is attracted to it, and the sharper the curve bends.

orthographic projection • A view of a scene projected onto a screen, in which the lines of projection are perpendicular to the screen. In 3D graphics, orthographic views are always aligned with the axes of the global Cartesian coordinate system.

patch • A deformable surface, useful for creating curved objects. **Bezier** patches are based on Bezier spline math, so the curvature is controlled by the position of control vertices and adjustable tangent handles. **NURBS** surfaces are also built from patches.

perspective projection • Representation of a scene in which parallel lines are depicted as converging, in order to give the illusion of depth and distance.

pixel • Abbreviation of *picture element*: the smallest spatial division of a bitmap image. A bitmap image is defined by a fixed number of pixels arranged in a 2D grid or mosaic. Many small pixels blend together to give the illusion of a continuous image.

pixel ratio • The proportion of pixel width to pixel height. In most cases, pixels are square, giving a pixel ratio of 1:1. However, some display formats use nonsquare pixels, such as the NTSC serial digital video standard, which has a resolution of 720 x 486, with a screen aspect ratio of 1.33:1. This results in a pixel ratio of 0.9:1, meaning the pixels are slightly taller than they are wide. This creates distortion when displaying NTSC serial digital images on standard computer monitors, which are only capable of displaying square pixels.

polygon • A closed 2D plane bounded by straight lines. In some computer graphics applications, the terms **face** and polygon are used interchangeably, but in 3ds Max a polygon is a group of connected triangular faces that all lie in the same plane.

procedural map • An algorithm that generates maps directly within a 3D graphics application. Procedural textures can be 2D or 3D. 3D procedural maps usually do not require mapping coordinates.

progressive scan • A video display mode that draws each scanline sequentially, and does not use alternating **fields**. This is the opposite of an **interlaced scan**.

radiosity • Rendering algorithm that simulates diffuse reflections. Radiosity renders the effects of light energy bouncing off of diffuse surfaces to illuminate other objects. See **global illumination**.

raytracing • Rendering algorithm that simulates rays of light moving through the scene. Standard raytracing accurately renders specular reflections, refractions, and transparency. See **global illumination**.

reflection map • A simpler method of generating the effect of reflective surfaces than the computationally expensive raytracing technique. An image, usually a bitmap, is mapped onto a reflective surface. As the object moves, the reflection map maintains its orientation to the global Cartesian coordinate system, giving the illusion of a scene reflected in the object.

refraction • Bending of light rays passing from one substance (such as air) into another (such as glass). Can be rendered using a *refraction map* (similar to a reflection map), or through raytracing.

rendering • The process of producing finished 2D computer graphic images. *Realtime rendering* occurs when viewing objects in the 3ds Max viewports, or when interacting with a game or simulation. *Production rendering* draws a much higher quality image, at the expense of speed. A production render may take anywhere from a few seconds per frame to several hours per frame.

resolution • The number of pixels in a bitmap image. Resolution is not directly related to image quality; it is easy to create a poor quality image with a high resolution.

smoothing • A rendering algorithm that creates smooth-looking surfaces from mesh objects. Without smoothing, all mesh objects would have a faceted appearance.

specular color • The color of a highlight on a shiny object. The color of a specular highlight is often determined more by the color of the light than by the object's color. See **diffuse color** and **ambient color**.

spline • A line whose shape and curvature is determined by control vertices.

supersampling • An antialiasing algorithm that internally renders at high resolution, then averages pixel values for a lower resolution output. Significantly improves some CG images at the expense of longer render times.

transform • The reassignment of points to new locations. The three types of transforms are position, rotation, and scale.

vector graphics • Computer graphic images that are defined by points and lines, instead of pixels. Examples include type fonts, PostScript files, and Flash movies. All 3D models exist in 3D vector space and are projected into pixel space for display.

vertex • A fancy word for "point." Vertices are merely markers in space; they have no dimension. They serve many purposes in computer graphics, from defining the contours of objects to controlling animation though **function curves**.

INDEX